INSTABILITIES OF FLOWS AND TRANSITION TO TURBULENCE

TAPAN K. SENGUPTA

CRC Press
Taylor & Francis Group
Boca Raton   London   New York

CRC Press is an imprint of the
Taylor & Francis Group, an **informa** business

CRC Press
Taylor & Francis Group
6000 Broken Sound Parkway NW, Suite 300
Boca Raton, FL 33487-2742

First issued in paperback 2018

ISBN 13: 978-1-138-07621-1 (pbk)
ISBN 13: 978-1-4398-7944-3 (hbk)

Library of Congress Cataloging-in-Publication Data

Sengupta, Tapan Kumar, 1955-
  Instabilities of flows and transition to turbulence / Tapan K. Sengupta.
    p. cm.
  Includes bibliographical references and index.
  ISBN 978-1-4398-7944-3 (hardback)
  1. Turbulence. 2. Transition flow. 3. Stability. I. Title.

TA357.5.T87S46 2012
532'.0527--dc23   2012005268

Visit the Taylor & Francis Web site at
http://www.taylorandfrancis.com

and the CRC Press Web site at
http://www.crcpress.com

Contents

Symbol Description

F	Non-dimensional physical frequency	m	Falkner–Skan pressure gradient parameter
$F_I(\alpha)$	Unilateral Laplace transform	u_1	Streamwise velocity profile
$F_{II}(\alpha)$	Bilateral Laplace transform	w_1	Cross flow profile
$F(\omega_0)$	Fourier transform	Γ	Vortex strength/circulation
H	Shape factor	Ω	Angular velocity
Re	Reynolds number	ξ, η	Transformed plane variables
T	Transfer function	ρ	Density
T_u	Turbulence intensity	ψ	Stream function
U	Streamwise mean flow	ω_0	Circular frequency
U_1	Unit step function	$\omega, \omega_d, \omega_m$	Vorticity
U_e	Boundary layer edge velocity		
W	Mean crossflow		
c	Complex phase speed		
$f(t)$	Fourier spectrum		

Chapter 5

f, ϕ, h, π	Complex amplitude function
y_1 to y_6	Compound matrix variables

A, A_m	Blowing-suction amplitude
E_d	Disturbance mechanical energy

Full list:

F	Non-dimensional physical frequency
$F_I(\alpha)$	Unilateral Laplace transform
$F_{II}(\alpha)$	Bilateral Laplace transform
$F(\omega_0)$	Fourier transform
H	Shape factor
Re	Reynolds number
T	Transfer function
T_u	Turbulence intensity
U	Streamwise mean flow
U_1	Unit step function
U_e	Boundary layer edge velocity
W	Mean crossflow
c	Complex phase speed
$f(t)$	Fourier spectrum
f, ϕ, h, π	Complex amplitude function
y_1 to y_6	Compound matrix variables
y_j	Compound matrix variable
Γ	Vortex strength/circulation
Φ_I	Wall mode
Φ_I	Free stream mode
α	Wavenumber
β_h	Hartree parameter
δ	Boundary layer thickness
δ^*	Displacement thickness
γ_j	Bromwich contour
π	Disturbance pressure
ϕ	Disturbance amplitude of v'
ψ	Stream function
ω_0	Circular frequency
ω_i	Amplification rate in the temporal theory
ω, ω_b	Vorticity

Chapter 4

E	Total mechanical energy
H	Shape factor
H_1	Fixed height
Q_∞	Oncoming free stream flow
U_∞	Free stream flow in streamwise direction
U_s	Surface speed on rotating cylinder
V, V_d, V_m	Velocity
W_∞	Free stream cross flow
c	Convection speed of vortex
d	Diameter of rotating and translating cylinder
h_1, h_2	Scale factors of transformation

Chapter 5

A, A_m	Blowing-suction amplitude
E_d	Disturbance mechanical energy
E_v	Disturbance kinetic energy
F	Non-dimensional wall excitation frequency
N	Exponential amplification factor
Re	Reynolds number
$Re_{\delta*}$	Reynolds number based on displacement thickness
U	Mean flow profile
U_i	Mean flow component
$U_1(t)$	Heaviside function
V_e	Energy propagation speed
V_g	Group velocity
V_s	Signal speed
f	Non-dimensional stream function for Blasius solution
k	Complex wavenumber
u_d	Streamwise disturbance velocity
u_i	Disturbance flow velocity
x_{ex}	Wall exciter location
α	Wavenumber
α_1	Wall excitation amplitude factor
β	Circular frequency
β_0	Fixed frequency
β_1, β_2	Bandwidth of the excitation
ζ	Similarity variable
ϕ	Amplitude of stream function
ϕ_1, ϕ_3	Inviscid and viscous components of ϕ
ϕ_{10}, ϕ_{30}	Inviscid and viscous components of ϕ at wall
ψ	Stream function
$\omega, \bar{\omega}, \omega_d$	Vorticity

Chapter 6

A_e	Equilibrium amplitude
$A_j(t)$	Instability amplitude function
A_p	Acceleration parameter
D	Diameter of cylinder
En	Enstrophy
N_j	Nonlinear interaction term
P_t	Total pressure
R_{ij}	Two-point correlation function
Re	Reynolds number
Re_{cr}	Critical Reynolds number
Tu	Turbulence intensity
U_∞	Free stream velocity
$\bar{U}_\infty$	Final velocity
Y	Compound matrix variable
a_j	POD amplitude function
e_t	Normal distribution
f_0	Shedding frequency of vortices
h_1, h_2	Scale factors of transformation
l	Landau coefficient $(l_r + il_i)$
u, v	Contravariant components of velocity
u', v'	Disturbance velocity components
u_c	Convective speed of disturbance at outflow
u_r	Radial component of velocity
x_t	FST model for small scale fluctuation
α_f	Moving average model constant
α, α_j	LSE equation coefficients, linear terms
β, β_j	LSE equation coefficients, nonlinear terms
η	Radial directions
λ_j	POD eigenvalues
μ_j	Moments of fluctuating velocity
ν	Kinematic viscosity
σ_r	Linear temporal amplification rate
ϕ_m	Eigenvectors of correlation matrix
ψ	Stream function
ω'	Disturbance vorticity
ω_1	Circular frequency parameter
$\omega, \bar{\omega}, \omega_d$	Vorticity

Chapter 7

A, A_m	Amplitude of wall excitation
C_p	Specific heat at constant pressure
D	Dispersion relation $(D_r + iD_i)$

F	Non-dimensional wall excitation frequency
Gr	Grashof number
K	Buoyancy parameter
K_{cr}	Critical buoyancy parameter
P	Non-dimensional pressure
Pr	Prandtl number
Re	Reynolds number
Re_{cr}	Critical Reynolds number
Ri	Richardson number or Archimedes number
T, T_w, T_∞	Temperature
U_c	Convection velocity at outflow
U_∞	Free stream velocity
U, V	Velocity components in boundary layer
f	Non-dimensional stream function
h	Amplitude of temperature fluctuation
k	Wavenumber
l	Length scale
p	Pressure
q	Heat generation term
r	Compound matrix vector
u, v	Contravariant velocity components
u_d	Streamwise disturbance velocity
$\mathbf{y_j}$	Compound matrix variables
Θ	Non-dimensional temperature
Θ'	Wall-normal derivative of temperature
Φ_v	Viscous diffusion
Ψ	Stream function
α	Thermal diffusivity
β_0, β_L	Circular frequency
β_t	Volumetric thermal expansion coefficient
γ	Amplitude of streamwise velocity fluctuation
δ^*	Displacement thickness
η	Boundary layer similarity coordinate
ξ, ζ	Transformed orthogonal coordinate
π	Amplitude of pressure fluctuation
ϕ	Amplitude of wall-normal velocity
ψ	Stream function
ω	Vorticity

Chapter 8

F, G	Non-dimensional stream function, Falkner–Skan–Cooke profile
H	Boundary layer shape factor
Q_∞	Free stream velocity
Re	Reynolds number
Re_θ	Reynolds number based on momentum thickness
U_1	Boundary layer streamwise velocity profile
U_ψ	Projection of velocity
V_g	Group velocity
W_1	Boundary layer cross flow velocity profile
X, Y, Z	Wing fixed coordinates
f	Non-dimensional vector potential component
f^*	Wave frequency in Hz
g, g_1	Non-dimensional vector potential component
m	Falkner–Skan pressure gradient parameter
m_j	Coefficients of boundary layer equations
p	Static pressure
r_e	Radius of curvature of external streamline
u_e	Boundary layer streamwise edge velocity
u, v, w	Boundary layer velocity profiles
v_w	Transpiration velocity at wall
w_e	Boundary layer spanwise edge velocity
x, y, z	External streamline fixed coordinates
Γ	Equivalent streamwise growth
Λ	Sweepback angle
α	Wavenumber in streamwise direction
α_1	Angle of attack
α_ψ	Wavenumber projection in ψ direction
β	Wavenumber in z direction
β_h	Hartree parameter
δ^*	Boundary layer displacement thickness
ϵ	Cross flow angle
ϵ_I	Critical cross flow angle
η	Boundary layer similarity coordinate
θ	Boundary layer momentum thickness
θ_0	Slope of external streamline
λ^*	Streak wavelength of stationary waves
μ	Coefficient of viscosity
ν	Kinematic viscosity
ρ	Density
ϕ	Component of vector potential
ϕ_1	Wave propagation angle
φ	Sweepback angle
$\hat{\psi}$	Component of vector potential
ψ	Angle of wavenumber vector w.r.t. x axis

Chapter 9

C_p	Coefficient of pressure
F	Non-dimensional excitation frequency
H	Shape factor
L/D	Lift to drag ratio
M	Mach number
Re	Reynolds number
U_∞	Free stream speed
U_e	Boundary layer edge velocity
c	Chord of airfoil
c_d	Drag coefficient
c_l	Lift coefficient
$c_{l,max}$	Maximum lift coefficient
c_m	Pitching moment coefficient
m	Falkner–Skan pressure gradient parameter
u	Streamwise edge velocity
α	Angle of attack
δ_f	Flap deflection angle
θ	Momentum thickness
τ_0	Wall shear stress

Epilogue

$E(k)$	Energy spectrum
F	Non-dimensional excitation frequency
Re_{δ^*}	Reynolds number based on displacement thickness
$U_d(k)$	Bilateral Laplace transform of u_d

$U(k)$	Bilateral Laplace transform of azimuthal velocity component	α_1	Non-dimensional excitation amplitude
f	Applied body force for 2D DNS	α_e	Damping frequency due to Ekmann friction
k	Wavenumber		
u_d	Streamwise disturbance velocity amplitude	β_0	Circular frequency
u_{dm}	Streamwise maximum disturbance velocity amplitude	ν	Kinematic viscosity
		ξ, η	Transformed coordinates
v_d	Wall-normal excitation velocity	ψ	Stream function
x_{ex}	Exciter location	ω	Vorticity

List of Figures

List of Tables

Preface

Everything should be made as simple as possible, but not simpler – **Albert Einstein**

There are books and monographs available on this subject, including one where the present author collaborated in producing lecture notes on wider topics. Despite the availability of these, this book is written with a completely different outlook, incorporating recent results, which changes the perspective of the subject completely. This has been made possible with the work of the author's group, proposing new theoretical and computational results which establishes a deterministic route to turbulence.

The Navier–Stokes equation governs fluid flows and has been known since the first half of the nineteenth century, with a few exact solutions obtained using restrictive conditions. However, such solutions are not always observed in nature. For example, experimental data for pipe flow did not show any agreement with one such analytic solution. The reason for the mismatch was investigated by Reynolds, who explained that the *exact* solution of the Navier–Stokes equation, even when it exists, is unable to maintain its stability with respect to omnipresent small disturbances in the flow. An equilibrium solution ensures satisfaction of conservation principles of mass, momentum and energy balance, yet the flow in a pipe disintegrates into sinuous motion, eventually leading to turbulent flow. This has been attributed to the instability of flows.

Viewing fluid flows as dynamical systems, it is difficult to ignore the role of receptivity, as compared to instability. The former was emphasized by Schubauer and Skramstad [316] in reporting their classic experimental results which verified linear viscous instability of zero pressure gradient flow past a flat plate. They reported waves as an early marker of flow instability, which was theoretically predicted by Tollmien and Schlichting by stability calculations. While a vortical excitation validated instability theory, the latter does not require any specific excitation, except in prescribing the qualitative nature of the boundary conditions. Schubauer and Skramstad could not produce wavy solutions by acoustic excitation. One of the main features of this book is to relate receptivity with flow instability and show how different routes of excitation lead to different types of disturbance evolution. This has been achieved here theoretically, as well as computationally.

The subject of instability has seen torturous development, following early pioneering work by Helmholtz, Kelvin and Rayleigh. They set out conditions of inviscid instability, which are seen to work remarkably well, whenever the conditions are right. Their observations on inviscid mechanisms are based on a heuristic explanation that viscous action can only lead to attenuation of disturbances. Thus, when viscous stability results appeared with a wavy solution, these were not immediately embraced with enthusiasm. Viscous instability results also suffered credibility due to the use of a parallel flow assumption for the equilibrium flow, along with linearization. This criticism was muted by experimental verification provided in [316], much later. However, there are many other flows for which linear theory results are at variance with experimental observations. The authors in [411] have questioned the success of instability theories for shear driven flows, which *include to a lesser degree, Blasius boundary layer flow*. This is also the flow which many cite as the success story of linear stability theory! Pipe and Couette flows are seen to be always stable by linear theories. Reynolds, in reporting his famous pipe flow experiments, attributed such a failure to nonlinearity.

Despite many attempts made over more than a century and a half, exact routes to turbulence are not so clear for many flows. This prompted Morkovin [235] to state: *"One hundred eight years after O. Reynolds demonstrated turbulence in a circular pipe, we still do not understand the nature of the irregular fluctuations at the wall nor the formation of larger coherent eddies convected downstream further from the wall. Neither can we describe the mechanisms of the instabilities that lead to the onset of turbulence in any given pipe nor the Reynolds number (between about 2000 and 100,000) at which it will take place. It is sobering to recall that Reynolds demonstrated this peculiar non-laminar behaviour of fluids before other physicists started on the road to relativity theory, quantum theory, nuclear energy, quarks etc."* While additional researches have clarified key concepts, the situation on flow transition in pipes remains the same.

There is another major drawback of linear instability theory which was not followed seriously until very recently. As a matter of expediency, flows have been studied for either temporal or spatial instability, with an assumption that these approaches are interchangeable. Initially, most studies related to temporal growth of disturbances, as it is performed relatively easily. For 3D flows, disturbance flow direction may not be aligned with equilibrium flow and this adds another degree of freedom, complicating spatial instability studies further. Researchers have also devoted their attention to looking at spatio-temporal growth of disturbances and some efforts have been reported for 3D flows. However, in recent times, the author's group has been successful in reporting spatio-temporal growth for 2D flows — which is more generic than what has been reported for 3D flows. In a linear framework with parallel flow assumption, it has been shown that spatio-temporal modes are present which grow continually, by the Bromwich contour integral method, developed by the author's group. This is another important highlight of the present book.

Progress in solving the Navier–Stokes equation for direct simulation of flow instability allows one to remove the limiting assumptions implicit with linear instability theory. However, this requires developing very high accuracy methods which are dispersion relation preserving, i.e., the numerical and physical dispersion relations are almost identical. This has been done by developing methods whose numerical group velocity matches the physical group velocity of the chosen model equation. Having made this crucial improvement in computing, tremendous impact has been made in advancing our understanding of instability and transition. The present book is replete with results showing effects of nonlinearity and growth of boundary layer for the canonical flow past a flat plate, with and without heat transfer. Also, flow past a natural laminar flow airfoil is studied to understand bypass transition. This book has multiple chapters dealing with the important area of bypass transition. Bypass transition has been systematically explored here, using experimental, theoretical and computational results. An energy-based receptivity theory has also been described and used to explain the classical and bypass route of transition. One of the major highlights of this new theory is that this is obtained directly from the Navier–Stokes equation without making any limiting assumptions.

While transition to turbulence is a consequence of multiple instabilities, not all flow instabilities of equilibrium flow lead to turbulence. For example, flow past bluff bodies at a low Reynolds number experiences primary instability, following which the flow settles down to another equilibrium state where vortex streets are noted in the wake. This is a consequence of the action of nonlinearity during the evolution of disturbances. This nonlinear effect in moderating linear temporal instability is often modeled by the Landau equation. However, in recent times, this flow has been investigated for other modes, which has been aided by proper orthogonal decomposition of the accurate solution of the Navier–Stokes equation.

This has assisted in developing a dynamical system theory for flow instability past bluff bodies, as presented here for the first time.

This book has evolved into an account of the personal research interests of the author over the years. A conscious effort has been made to keep the treatment at an elementary level requiring rudimentary knowledge of calculus, Laplace–Fourier transform and complex analysis, which should be equally amenable to undergraduate students, as well as serious researchers in the field. In Chap. 2, readers are familiarized with the analysis of any numerical method to decide if it is capable of computing physically unstable and transitional flows. It is to be noted that this is tougher than DNS of fully developed turbulent flows. In Chap. 3, basic rudiments of receptivity and instability have been covered by a unified analysis. This is neglected in most books covering this subject. Any beginner would benefit most in understanding the classical aspects of the subject through these. At the same time, the solution of the Navier–Stokes equation would convince readers about many prevalent misconception in the subject. In this and subsequent chapters, readers will appreciate that the role of Tollmien–Schlichting waves is tremendously overstated. What has appeared in the literature may have been considered good, but based on the accounts provided here, it is not good enough for a proper understanding of transition. This book reveals to readers that without good computing, this subject will be poorer in linking spatio-temporal growth, instability at low frequencies and the actual physical basis of transition. In providing computational results from receptivity to a fully developed turbulent stage of 2D flows, a case has been made that will answer many questions, including whether turbulence is stochastic or deterministic. It will spur newer understanding in nonlinear dynamics. In closing this discussion, we note that the subject has matured very rapidly in recent times with the advent of very high accuracy methods of computing, which will lead to an explosive growth of activities in the subject field.

The author is deeply indebted to all his co-authors of various journal papers, who thought differently and took a difficult journey with the author. Among the international peer group, I would like to thank Profs. T. T. Lim, W. Schneider, H. Steinrueck, T. Bridges, M. Deville, Bernd Noack, K. R. Sreenivasan, P. J. Strykowski, Y. T. Chew and Dr. J. M. Kendall for providing data and having stimulating discussions. It will also be most appropriate to acknowledge Prof. M. Gaster, who encouraged me to develop a proper theory of receptivity, based on transform techniques. He, along with Prof. D. G. Crighton, provided a challenging environment in which to work during the initial phase of the work started at Cambridge. The lessons learnt there made later researches that much easier. Without those, the contents of the book related to the author's group would not have been possible.

Special mention must be made of students who worked with me at IIT Kanpur and NUS Singapore on various aspects: Sandeep Nijhawan, Manish Ballav, Vivek Rana, Manoj Nair, A. K. Rao, K. Venkatasubbaiah, Manojit Chattopadhyay, Z. Y. Wang, V. K. Suman, P. Mohanamuraly, V. Lakshmanan, Vikram Singh, Sarvagy Shukl, S. B. Talla, Anurag Dipankar, N. A. Sreejith, Neelu Singh, Shakti Saurabh and A. P. Sinha. More recent and important contributions were made by Swagata Bhaumik, Yogesh Bhumkar, V. V. S. N. Vijay, Manoj Rajpoot and S. Unnikrishnan. It is my pleasant duty to specially thank V. V. S. N. Vijay, who read various chapters and provided vital comments and corrections; Yogesh Bhumkar and Swagata Bhaumuk provided figures tirelessly, whenever requested, and has always been ready for discussions on various topics. I also feel very fortunate to have many other students of HPCL, the Aerospace Engineering Department at IIT Kanpur, all of whom have contributed indirectly to creating some of the contents of this book. Finally, I would like to thank my secretarial assistants, Baby Gaur and Shashi Shukla, who have really worked tirelessly in preparing this manuscript. Especially, Ms. Gaur who has spent endless hours beyond the call of duty in preparing texts and figures. It is a pleasure to acknowledge her dedication to work. It is also my pleasant duty to acknowledge the team of experts from CRC Press and Dr. Gagandeep Singh, especially, who have been instrumental in seeing this edition of the book in print. I am also very fortunate to have a family who all shared my moments of elation and somber moods during this long sojourn.

1

Introduction to Instability and Transition

1.1 Introduction

This book covers material ranging from (i) classical linear hydrodynamic instability; (ii) receptivity and instability of different types; (iii) vortex induced instability and bypass transition; (iv) transient growth and spatio-temporal instability; (v) bifurcation and dynamical system theory of nonlinear instability for different flows; (vi) instability of mixed convection flow by restricted heat transfer; (vii) instability of three dimensional flows and (viii) applications of instability and transition for flow past airfoils. These topics are admixtures of linear and nonlinear aspects of flow instabilities studied via analytical and computational routes. For this latter reason, we have detailed discussions on computing related to unstable, transitional and turbulent flows. Thus, theoretical results are supplemented by computational results obtained by specifically developed high accuracy direct numerical simulation (DNS) techniques. For this purpose, special attention is devoted to understanding waves and the role of computing errors. However, over-riding all the special aspects, one must understand concepts of fundamental principles of receptivity and instability.

Fluid dynamics represents the tougher part of continuum mechanics; the equations of motion were first written down for inviscid flow by Euler in 1752. In 1840, the equation of fluid motion in the presence of friction - the Navier–Stokes equation was published. The correctness of this equation is now well accepted in the continuum regime of fluid flow. Apart from the fact that correct governing equations are needed, one has also to ensure that correct boundary and initial conditions are used in the solution process. One notes that the no-slip boundary condition is a modeling approximation and has never been proven rigorously. Batchelor [22] has noted for Newtonian fluid flow that *the absence of slip at a rigid wall is now amply confirmed by direct observation and by the correctness of its many consequences under normal conditions.* Despite this, we do not question the correctness of the no-slip condition for continuum flows.

After the Navier–Stokes equation was written down in the first half of the nineteenth century, few exact solutions were obtained for few fluid flows. In one such case, Stokes compared theoretical prediction with available experimental data for pipe flow and found no agreement whatsoever. Now we know that the theoretical solution of Stokes corresponded to undisturbed laminar flow, while the experimental data given to him corresponded to a turbulent flow. This was investigated by Osborne Reynolds, who explained the reason for such mismatches by his famous pipe flow experiments [297]. It was shown that the basic flow obtained, as a solution of the Navier–Stokes equation, is unable to maintain its stability with respect to omnipresent small disturbances in the flow. The mere mathematical existence of a solution does not guarantee its physical realization and observation. The existence of a mathematical solution shows a possibility of a solution (that we will refer as the equilibrium solution), as it embodies satisfaction of conservation laws satisfying mass, momentum and energy balance. However, additionally one needs to study the stability of each and every such solution to ascertain their observability. Reynolds demonstrated experimentally that the equilibrium parabolic profile disintegrates into sinuous motion of water in the pipe, which eventually leads to turbulent or chaotic flow.

Landau & Lifshitz [192] aptly noted that *the flow that occurs in nature must not only follow the equations of fluid dynamics, but also be stable.* If solutions are not *observable*, then the corresponding equilibrium flows are not *stable.* Here, implication of flow *instability* is in the context of continuous deviation of the instantaneous solution from the equilibrium solution caused by growth of infinitesimally small perturbations present in the surroundings of the system. This sensitive dependence on the disturbance environment makes the subject of instability very interesting, as well as challenging. The smallness of background disturbances allows one to study the problem of growth of these from a small perturbation approach. This greatly helps, if the governing nonlinear equations can be solved for the equilibrium solution with ease and then its stability can be studied by linearizing the governing equation for the perturbation field.

The dye experiments performed in [297] are perhaps the first recorded thorough experimental observations of flow instability. Reynolds took pipes of different diameters fitted with a trumpet shaped mouth-piece or bell-mouth, which accelerated the flow at entry. Such acceleration creates a favorable pressure gradient, attenuating background disturbances, as will be explained in Chap. 3. To reduce disturbances, experiments were performed late at night to avoid noise from daytime vehicular traffic. Reynolds observed that the rapid diffusion of dye with surrounding fluid depends on the non-dimensional parameter Va/ν, with V as the center-line velocity in the pipe of diameter a, and ν is the kinematic viscosity. This ubiquitous non-dimensional parameter is now called the Reynolds number (Re) to underscore the singular importance of this famous pipe flow experiment. Reynolds found that the flow can be kept orderly or laminar up to $Re = 12,830$. He noted that this value is very sensitive to the disturbances in the flow before it enters the tube and he noted prophetically that *this at once suggested the idea that the condition might be one of instability for disturbance of a certain magnitude and stable for smaller disturbances.* The relationship of instability with disturbance amplitude is a typical attribute of non-linear instability. It is now well established that pipe and plane Couette flows are linearly stable for all Reynolds numbers, when the viscous linear stability analysis is performed. This suggests that either nonlinear and/or different unknown linear mechanism(s) of instabilities are at play for these flows.

Critical Re was further raised later for pipe flows, establishing that there is perhaps no upper limit above which transition to turbulence cannot be prevented. This example also suggests the importance of receptivity of flow to different types of input disturbances to the system. In the absence of input on a fluid dynamical system of a particular kind that triggers instability, response will be orderly, even if the system is unstable to that kind of input. Thus, if the basic flow is receptive to a particular disturbance, then the equilibrium flow will not be observed in the presence of such disturbances.

Reynolds' experiment pointed out the instability as the main reason for non-observability of equilibrium flows, but it did not yield the steps following which one gets to the turbulent stage. Like any other field, associated scientific ideas, laws, and their discovery as narrated in textbooks on flow instabilities are a mere distillation of complex, multi-faceted, subtle and convoluted historical narrative. Initial impetus in the field is due to a combination of missed starts, intense debates, dead ends, etc., and followed by mistakes and sophistries including self-fulfilling prophesies and deceptions which make the journey a maze with errors appearing understandable, as an after-thought. The present book will take the reader through all these steps.

There were significant contributions initially in [286, 287] using inviscid analysis, all of which are still very relevant. In their quest to justify inviscid analysis, it was stated that viscous action due to its very dissipative nature can only be stabilizing and need not be included in instability studies, even though in describing the equilibrium flow one may be required to include viscous actions. Such was the impact of this that when Heisenberg [143] submitted his dissertation solving perturbation equations including viscous terms for a boundary layer, the examination committee could find nothing wrong in the analysis, yet found it difficult to accept that viscous action can add to instability. It makes perfect sense now to realize that viscous action can provide a correct phase shift for positive feedback in destabilizing the flow. The same fate awaited the researchers from the Göttingen school led by Prandtl [282] in developing viscous linear stability theory which explained many aspects of the early stages of the transition process involving formation and growth of waves, attributed to Tollmien and Schlichting. This theory is based on the key equation developed independently by Orr [260] and Sommefeld [380] and is named after them. It took the pioneering experimental effort of Dryden and his associates in establishing viscous linear stability theory by detecting the so-called Tollmien–Schlichting waves through the famous vibrating ribbon experiments of [316]. Like the experiments of Reynolds, this also did not explain the crucial later stages of transition. In this book, we will note these stages with the help of accurately computed solution of the Navier–Stokes equation.

Linear stability theory results match well with controlled laboratory experiments for thermal and centrifugal instabilities. But, instabilities dictated by shear force do not match so well, e.g., linear stability theory applied to plane Poiseuille flow gives a critical Reynolds number of 5772, while experimentally, such flows have been observed to become turbulent even at $Re = 1000$, as shown in [80]. Couette and pipe flows are also found to be linearly stable for all Reynolds numbers; the former was found to suffer transition in a computational exercise at $Re = 350$ [208] and the latter was found to be unstable in experiments for $Re \geq 1950$.

Flow instability of attached boundary layers has been predicted with some success and the corresponding empirical transition prediction methodologies have matured to such an extent that they are now routinely used in the aircraft industry. A flat plate placed in a stream with a moderately low ambient disturbance level (turbulence intensity below 0.5%), flow transition is noted to take place at a distance x from the leading edge given by

$$Re_{tr} = \frac{U_\infty x}{\nu} = 3.5 \times 10^5 \quad \text{to} \quad 10^6$$

The onset of instability is predicted theoretically at $Re_{cr} = 519$ (based on displacement thickness as the length scale). It is important to realize that instability and transition are not synonymous. The actual process of transition begins with the onset of instability, but the completion may depend upon multiple factors; those form the basis for adjunct topics like secondary, tertiary and nonlinear instabilities. Hence, it is difficult to predict Re_{tr} as compared to Re_{cr}.

It is also important to note that the linear stability theory studies a particular class of problems, where the disturbances decay as one moves away in the wall-normal direction from one of the boundaries. Thus, the developed theory is mainly for disturbances that originate at the wall. The problem of destabilizing a shear layer by disturbances outside the shear layer has not received sufficient attention or been adequately tackled in the past. This is also one of the major focuses of this book, discussed in Chaps. 3 and 4.

A search for complete understanding on the origin and nature of turbulence continues with the hope that the numerical solution of the full Navier–Stokes equation without any modeling, as in DNS, may provide insight to it. A multitude of published DNS results in the literature suffers from a major drawback though, with most of them not requiring any explicit forcing of the flow via a definitive and realistic input disturbance field. Most of these depend upon computational noises and/or "random noise."We wish to point out that, unlike in mathematical closed-form solutions, a numerical solution is always visited upon by *numerical noise* of the method and the hardware. Thus, the implicit assumption in DNS that these *numerical noise* sources will produce a *turbulent flow* which is same as the physical turbulent flow is a statement of hope and yet to be established rigorously. In fact, the contrary seems to be the case, as is shown in Chaps. 4, 5, 6 and 7 via DNS of bypass transition, where physical processes are traced following explicit excitation.

1.2   What Is Instability?

To study a physical problem analytically, one first obtains governing equations which model the phenomenon adequately. If the auxiliary equations pertaining to initial and boundary conditions are well posed, then obtaining the solution is straightforward. Mathematically, one is concerned with the *existence* and *uniqueness* of the solution. Yet not every solution of equations of motion, even if it is exact, is observable in nature. This is at the core of many physical phenomena where *observability* of solution is of fundamental importance. If the solution is not *observable*, then the corresponding basic flow is not *stable*. Here, the implication of *stability* is in the context of the solution with respect to infinitesimally small perturbations.

In studying the stability of a problem with respect to ambient disturbances, it is hardly ever possible that one can incorporate all the contributing factors in a given physical scenario for posing a physical problem. Occasionally these neglected *causes* can be incorporated by *process noise* and results are made to correlate with the physical situation. This is possible when the *causes* are statistically independent and then it follows upon using the Central Limit Theorem.

1.3   Temporal and Spatial Instability

Instability of an autonomous system is strictly for time-dependent systems that display growth of disturbances with time. This may also mean that either we are studying the stability of a flow at a fixed spatial location or the full system displays identical variation in time for each and every spatial location. Fluid flow instabilities are treated as if the disturbance growth is either in space or in time. This approach is merely for expediency and not based on any assessment of reality. It is ideally suited to treat the growth in a spatio-temporal framework, but no theoretical framework exists, except the use of the Bromwich contour integral method. This method is used in a linear framework and is capable of tracking growth in space and time simultaneously. This method will be described and results shown in Chaps. 3 and 5. A disturbance originating from a fixed location in space can grow as it convects downstream. Such a disturbance is termed unstable if it grows as it moves downstream. This is the convective instability. This type of instability is considered for wall-bounded shear layers, exemplified by the classic vibrating ribbon experiment of [316]. This experiment was performed in a very quiet facility to create Tollmien–Schlichting (TS) waves by vibrating a ribbon inside a flat plate boundary layer, a detailed description of which is provided in Chap. 3. This experiment was the first to show the existence of a viscous unstable wave which was predicted earlier theoretically in [143, 408, 310], but was not supported immediately in an experiment. Hence, the existence of TS waves was doubted before the experimental results in [316] were published. Additionally, this experiment was also the first to display receptivity of wall-bounded shear layers to vortical disturbances created inside a shear layer, while showing the inadequacy of acoustic excitation in creating TS waves. One of the major aims of the book is to show that the classical picture of convective instability is not correct, and the transition is shown to be created by spatio-temporal wave fronts.

In many flows, it can happen that the disturbance can grow first in time at a fixed location, before it is convected downstream. Such growth of disturbances both in space and time are seen in many free shear layers and bluff-body flows. If we subject the equilibrium solution of such an unstable fluid dynamical system to a localized impulse, then the response field spreads both upstream and downstream of the location with respect to the local flow where it originated while growing in amplitude. Such instabilities have been termed *absolute instabilities*. Here, an additional distinction needs to be made between convective and absolute instabilities. On application of an impulse, both situations display disturbances in upstream and downstream directions. However, in a convectively unstable system, the growth of the disturbance is predominantly in one direction, while for an absolutely unstable system the growth will be omni-directional.

1.4   Some Instability Mechanisms

Here two simple cases of instabilities are considered to emphasize the concepts described above. We begin by distinguishing the difference between static and dynamic instability by considering the stability of the atmosphere as an example.

When a parcel of air in the atmosphere is moved rapidly from an equilibrium condition and its tendency to come back to its undisturbed position is noted, then we term the atmosphere statically stable. The movement of the packet is considered as impulsive, to preclude any heat transfer from the parcel to the ambience. This tendency of static stability, when if exists, is due to the buoyancy force caused by the density differential due to temperature variation with height and such a body force acts upon the displaced air parcel. In static stability studies, we do not look for detailed time-dependent motion of the parcel following the displacement (as the associated accelerations are considered negligible).

1.4.1   Dynamic Stability of Still Atmosphere

For the dynamic stability study, we consider once again a parcel of air to be at equilibrium, at a height z and displaced to a height $(z + \xi)$ to follow its detailed time history of motion. We note that the motion is caused again by the buoyancy force, caused by the temperature variation of the ambient air given by $T = T(z)$. Here, the vertical displacement of the air parcel is ξ and the dynamics follow the force balance equation

$$\rho' \, \ddot{\xi} = g \, (\rho - \rho')_{z+\xi} \tag{1.1}$$

where $\ddot{\xi}$ is the instantaneous acceleration experienced by the parcel of air. The density of the displaced parcel is considered to be given by ρ', while the ambient fluid has the density ρ, so that the right hand side of the above equation represents the buoyancy force. Equilibrium thermodynamics tells us that for a simple compressible substance with only one mode of work, any state property can be represented by any two other properties and let us consider them to be the pressure (p) and the entropy (s). Thus, using a Taylor series we can relate the density of the ambient air at the two heights as

$$\rho(z+\xi) = \rho(z) + \left(\frac{\partial\rho}{\partial p}\right)_s [p(z+\xi) - p(z)] + \left(\frac{\partial\rho}{\partial s}\right)_p [s(z+\xi) - s(z)] + \dots \quad (1.2)$$

Once again, we will assume that the displacement process of the air parcel is isentropic (there are no viscous or heat losses associated with the rapid movement of the parcel) and thus for the air parcel

$$\rho'(z+\xi) = \rho'(z) + \left(\frac{\partial\rho}{\partial p}\right)_s [p(z+\xi) - p(z)] \quad (1.3)$$

In Eqs. (1.2) and (1.3), mechanical equilibrium ensures the same δp and $\rho(z) = \rho'(z)$. Thus, the density differential causing the buoyancy is given by

$$(\rho - \rho')_{z+\xi} = \left(\frac{\partial\rho}{\partial s}\right)_p [s(z+\xi) - s(z)] = \left(\frac{\partial\rho}{\partial s}\right)_p \frac{ds}{dz}\xi \quad (1.4)$$

We can also relate the density of the air parcel at the two heights as

$$\rho'(z+\xi) = \rho(z) + \left(\frac{\partial\rho}{\partial p}\right)_s \frac{dp}{dz}\xi = \rho(z) + \frac{1}{c^2}\frac{dp}{dz}\xi$$

where c is the speed of sound. Equation (1.1) can also be written in terms of the specific volume ($v = \frac{1}{\rho}$), using Eq. (1.4), as

$$\ddot{\xi} = \frac{g}{\rho'}(\rho - \rho')_{z+\xi} = -\left[\frac{g}{v}\left(\frac{\partial v}{\partial s}\right)_p \frac{ds}{dz}\xi\right] / \left[1 + \frac{v\xi}{c^2}\frac{dp}{dz}\right]$$

From mechanical equilibrium, $\frac{dp}{dz} = -\rho g$ and the above can be simplified to

$$\ddot{\xi} = -\left[\frac{g}{v}\left(\frac{\partial v}{\partial s}\right)_p \frac{ds}{dz}\xi\right] / \left[1 - \frac{g\xi}{c^2}\right]$$

We can further simplify by using the thermodynamic relations: $\left(\frac{\partial v}{\partial s}\right)_p = \left(\frac{\partial T}{\partial s}\right)_p \left(\frac{\partial v}{\partial T}\right)_p$ and $\frac{ds}{dz} = \left(\frac{\partial s}{\partial T}\right)_p \frac{dT}{dz} + \left(\frac{\partial s}{\partial p}\right)_T \frac{dp}{dz}$, noting that $\left(\frac{\partial s}{\partial T}\right)_p = \left(\frac{\partial s}{\partial h}\right)_p \left(\frac{\partial h}{\partial T}\right)_p = \frac{C_p}{T}$. From Maxwell's relation, $\left(\frac{\partial s}{\partial p}\right)_T = -\left(\frac{\partial v}{\partial T}\right)_p$ and we obtain $\frac{ds}{dz} = \frac{C_p}{T}\frac{dT}{dz} + \rho g\left(\frac{\partial v}{\partial T}\right)_p$.
All these simplifications lead to

$$\ddot{\xi} = -\frac{g\xi}{v}\left(\frac{\partial v}{\partial T}\right)_p \frac{T}{C_p}\left[\frac{C_p}{T}\frac{dT}{dz} + \frac{g}{v}\left(\frac{\partial v}{\partial T}\right)_p\right] / \left[1 - \frac{g\xi}{c^2}\right]$$

If we consider air as a perfect gas with $p = \rho RT$, then $\left(\frac{\partial v}{\partial T}\right)_p = v/T$ and the above further simplifies to

$$\ddot{\xi} = -\frac{g}{T}\left(\frac{dT}{dz} + \frac{g}{C_p}\right)\xi / \left[1 - \frac{g\xi}{c^2}\right]$$

If we further consider the speed of sound (c) to be very large, then the above equation can be further approximated to

$$\ddot{\xi} + N^2\xi = 0 \tag{1.5}$$

where

$$N^2 = \frac{g}{v}\left(\frac{\partial v}{\partial s}\right)_p \frac{ds}{dz}$$

We can consider the following possibilities:

Case 1: If $N^2 > 0$ then the dynamics of the displacement will be purely oscillatory, implying neutral stability of the static atmosphere.

Case 2: If $N^2 < 0$, then the vertical displacement will vary as

$$\xi(t) = Ae^{|N|t} + Be^{-|N|t}$$

where the first component clearly indicates instability. N is called the *Brunt-väisälä* or buoyancy frequency. As given above, following [406], we can obtain this frequency with air treated as an ideal gas by

$$N^2 = \frac{g}{T}\left[\frac{dT}{dz} + \frac{g}{C_p}\right] \tag{1.6}$$

For dry air, $\frac{g}{C_p} = -0.01$ K/meter and hence for stability of dry atmosphere the temperature distribution has to be such that $\frac{dT}{dz} > -0.01$ K/meter. Thus $\frac{dT}{dz} = 0.01$ represents the border line of instability and the numerical value on the right hand side within the square bracket is termed the dry adiabatic lapse rate, because this ensures $\frac{ds}{dz} = 0$.

1.4.2 Kelvin–Helmholtz Instability

This arises when two layers of fluids (may not be of the same species or density) are in relative motion. Thus, this is an interfacial instability and the resultant flow features due to imposed disturbance will be much more complicated due to relative motion. The physical relevance of this problem was seized upon by Helmholtz [144], who observed that the interface as a surface of separation tears the flow *asunder*. Sometime later Kelvin [171] posed this problem as one of instability and solved it. We follow this latter approach here. The basic equilibrium flow is assumed to be inviscid and incompressible - as two parallel streams having distinct density and velocity - flowing one over the another, as depicted in Fig. 1.1.

Before any perturbation is applied, the interface is located at $z = 0$ and subsequent displacement of this interface is expressed parametrically as

$$z_s = \hat{\eta}(x, y, t) = \epsilon\eta(x, y, t) \tag{1.7}$$

where ϵ is a small parameter, defined to perform a linearized perturbation analysis. One can view the interface itself as a shear layer of vanishing thickness. For the considered inviscid irrotational flow, velocity potentials in the two domains are given by

$$\tilde{\phi}_j(x, y, z, t) = U_j x + \epsilon\phi_j(x, y, z, t) \tag{1.8}$$

The governing equations in either of the flow domains are given by

$$\nabla^2\tilde{\phi}_j = 0 \tag{1.9}$$

The potentials must satisfy the following far-stream boundary conditions

$$\phi_j's \quad \text{are} \quad \text{bounded} \quad \text{as} \quad z \to \pm\infty \tag{1.10}$$

Another set of boundary conditions is applied at the interface, which is the no-fluid through the interface condition, i.e.,

$$\frac{\partial \hat{\eta}}{\partial t} - \frac{\partial \tilde{\phi}_j}{\partial z} = -\frac{\partial \hat{\eta}}{\partial x}\frac{\partial \tilde{\phi}_j}{\partial x} - \frac{\partial \hat{\eta}}{\partial y}\frac{\partial \tilde{\phi}_j}{\partial y} \tag{1.11}$$

In addition, in the absence of surface tension pressure must be continuous across the interface. Upon linearization, the interface boundary condition of Eq. (1.11) simplifies to

$$\frac{\partial \eta}{\partial t} + U_j\frac{\partial \eta}{\partial x} - \frac{\partial \phi_j}{\partial z} = 0 \quad \text{for} \quad j = 1, 2 \tag{1.12}$$

where $\tilde{\phi}_j$ and ϕ_j are as related in Eq. (1.8). Defining the pressure on either flow domain by the unsteady Bernoulli's equation, one can write

$$p_j = C_j - \rho_j\left\{\frac{\partial \tilde{\phi}_j}{\partial t} + \frac{1}{2}(\nabla\tilde{\phi}_j)^2 + g\hat{\eta}\right\} \tag{1.13}$$

Simplifying and retaining up to $0(\epsilon)$ terms, we get the following conditions

$$0(1) \quad \text{condition}: \quad C_1 - \frac{1}{2}\rho_1 U_1^2 = C_2 - \frac{1}{2}\rho_2 U_2^2 \tag{1.14a}$$

$$0(\epsilon) \quad \text{condition}: \quad \rho_1\left\{\frac{\partial\phi_1}{\partial t} + U_1\frac{\partial\phi_1}{\partial x} + g\eta\right\} = \rho_2\left\{\frac{\partial\phi_2}{\partial t} + U_2\frac{\partial\phi_2}{\partial x} + g\eta\right\} \tag{1.14b}$$

One can consider a very general interface displacement given in terms of a bilateral Laplace transform as

$$\eta(x,y,t) = \int\int F(\alpha,\beta,t)\, e^{i(\alpha x + \beta y)} d\alpha\, d\beta \tag{1.15}$$

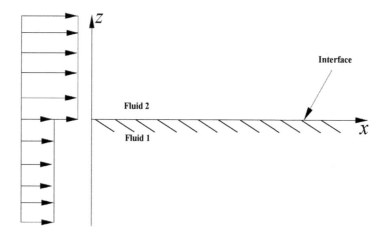

FIGURE 1.1
Kelvin–Helmholtz instability at the interface of two flowing fluids.

Correspondingly, the perturbation velocity potential is expressed as

$$\phi_j(x, y, z, t) = \int \int Z_j(\alpha, \beta, z, t) \, e^{i(\alpha x + \beta y)} \, d\alpha \, d\beta \tag{1.16}$$

Writing $k^2 = \alpha^2 + \beta^2$ and using Eq. (1.16) in Eq. (1.9), one gets the solution that satisfies the far-stream boundary conditions of Eq. (1.10) as

$$Z_j = f_j(\alpha, \beta, t) \, e^{\pm kz} \quad \text{for} \quad j = 1 \text{ and } 2 \tag{1.17}$$

Using Eq. (1.15) in the interface boundary condition of Eq. (1.12) one gets

$$\dot{F} + i\alpha U_1 F - k f_1 = \dot{F} + i\alpha U_2 F + k f_2 = 0 \tag{1.18}$$

where the dots denote differentiation with respect to time. If we denote the density ratio $\rho = \rho_2 / \rho_1$, then the linearized pressure continuity condition of Eq. (1.14b) gives

$$\frac{\partial \phi_1}{\partial t} - \rho \frac{\partial \phi_2}{\partial t} + U_1 \frac{\partial \phi_1}{\partial x} - \rho U_2 \frac{\partial \phi_2}{\partial x} + (1 - \rho) g \eta = 0 \tag{1.19}$$

Using Eqs. (1.15) and (1.16) in the above one gets

$$\dot{f}_1 - \rho \dot{f}_2 + i\alpha U_1 f_1 - i\alpha \rho U_2 f_2 + (1 - \rho) g F = 0 \tag{1.20}$$

Eliminating f_1 and f_2 from Eq. (1.20) using Eq. (1.18), one gets, after simplification,

$$(1 + \rho) \ddot{F} + 2i\alpha(U_1 + \rho U_2) \dot{F} - \{\alpha^2(U_1^2 + \rho U_2^2) - (1 - \rho) g k\} F = 0 \tag{1.21}$$

This ordinary differential equation for the time variation of the interface displacement F can be understood better in terms of its Fourier transform defined by

$$F(., t) = \int \hat{F}(., \bar{\omega}) \, e^{i\bar{\omega} t} \, d\bar{\omega} \tag{1.22}$$

One obtains the following dispersion relation by substitution of Eq. (1.22) in Eq. (1.21) as

$$-\bar{\omega}^2(1 + \rho) - 2\alpha\bar{\omega}(U_1 + \rho U_2) + (1 - \rho) g k - \alpha^2(U_1^2 + \rho U_2^2) = 0 \tag{1.23}$$

This provides the characteristic exponents in Eq. (1.22) as

$$\bar{\omega}_{1,2} = -\frac{\alpha(U_1 + \rho U_2)}{(1 + \rho)} \mp \frac{\sqrt{gk(1 - \rho^2) - \alpha^2 \rho(U_1 - U_2)^2}}{(1 + \rho)} \tag{1.24}$$

Based on this dispersion relation the following sub-cases are considered:

CASE 1: If the interface is disturbed in the spanwise direction only, i.e., $\alpha = 0$, then

$$\bar{\omega}_{1,2} = \mp \sqrt{g\beta \frac{(1 - \rho)}{(1 + \rho)}} \tag{1.25}$$

Thus, the streaming velocities U_1 and U_2 do not affect the response of the system. If in addition, $\rho > 1$, i.e., a heavier liquid is over a lighter liquid, then the buoyancy force causes temporal instability (if β is considered real) as is the case for Rayleigh-Taylor instability (see Chandrasekhar [59]).

CASE 2: For a general interface perturbation if $gk(1 - \rho^2) - \alpha^2 \rho(U_1 - U_2)^2 < 0$, then the interface displacement will grow in time. This condition can be alternately stated as a condition for instability as $(U_1 - U_2)^2 > \frac{gk}{\alpha^2}\left(\frac{1 - \rho^2}{\rho}\right)$.

Thus, for a given shear at the interface given by $(U_1 - U_2)$ and for a given oblique disturbance propagation direction at the interface indicated by the wavenumber vector k, instability would occur for all wavenumbers k^*, given by

$$k^* > \left(\frac{k^*}{\alpha}\right)^2 \frac{g}{(U_1 - U_2)^2} \left(\frac{\rho_1}{\rho_2} - \frac{\rho_2}{\rho_1}\right)$$

Note that the wavenumber vector makes an angle γ with the x-axis, such that $\cos \gamma = \frac{\alpha}{k^*}$ and the above condition can be conveniently written as

$$k^* > \frac{g}{(U_1 - U_2)^2 \cos^2 \gamma} \left(\frac{\rho_1}{\rho_2} - \frac{\rho_2}{\rho_1}\right) \tag{1.26}$$

The lowest value of wavenumber $(k^* = k_{min})$ would occur for two-dimensional disturbances, i.e., when $\cos \gamma = 1$ and this minimum is given by

$$k^*_{min} = \frac{g}{(U_1 - U_2)^2} \left(\frac{\rho_1}{\rho_2} - \frac{\rho_2}{\rho_1}\right) \tag{1.27}$$

CASE 3: Consider the case of shear only of the same fluid in both domains, i.e., $\rho = 1$. The characteristic exponents then simplify to

$$\bar\omega_{1,2} = -\alpha \frac{U_1 + U_2}{2} \mp \frac{i\alpha}{2}(U_1 - U_2) \tag{1.28}$$

The presence of an imaginary part with a negative sign implies temporal instability for all wave lengths. Also, note that since the group velocity and phase speed in the y direction are identically zero, therefore the Kelvin–Helmholtz instability for pure shear always will lead to two-dimensional instability.

2

Computing Transitional and Turbulent Flows

2.1 Fluid Dynamical Equations

The basic equations of fluid mechanics are nothing but the statements of conservation principles for mass, momentum and energy, which lead to space-time dependent quasi-linear partial differential equations. The following steps are implemented in obtaining the governing equations:

(i) One chooses the physical principles to be satisfied, e.g., one might just be focusing upon incompressible flow without heat transfer. Then one would just be concerned with mass and momentum conservation, without the need to worry about energy conservation in the domain of interest.

(ii) Based on the stated physical principles, one selects an appropriate model for the flow.

(iii) Finally, from combined application of (i) and (ii), one obtains the mathematical equations. Use of computational tools comes next. Transitional and turbulent flow computations require special approaches distinctly different from the CFD methodologies one usually comes across for engineering analysis.

One concentrates on (ii) first, i.e., chooses a suitable model that incorporates all the important physical principles. For transitional flows, even this is done in two stages. One obtains the equilibrium flow, which can be obtained by considering the full Navier–Stokes equation or some simplified form of it. Historically, even inviscid irrotational flows have been adopted for obtaining equilibrium flow. While these approaches have provided invaluable insights, recent research efforts have used the Navier–Stokes equation to obtain the equilibrium flows, especially in the context of studying nonlinear receptivity analysis of basic time dependent equilibrium flows. In the second stage, stability of the equilibrium flow is investigated with respect to applied disturbances. In the context of small perturbations, the governing Navier–Stokes equation can be used, after linearizing it. This may lead to a typical eigenvalue problem and has been the focus of most of the earlier studies in the field and will also be considered in this book. In this approach, one is not required to characterize the disturbance environment. However, if one is interested in studying the response of a dynamical system to a specific class of disturbances, then one adopts what has been identified as receptivity analysis. Receptivity analysis can also be performed in the linearized framework and has been advanced in [322, 349, 350] via the Bromwich contour integral and is taken up for detailed discussion in Chap. 5, in the context of the parallel flow assumption. Irrespective of the amplitude of perturbations, one can also use the full Navier–Stokes equation for receptivity study. This is undertaken here in Chaps. 3 to 7. Thus, it is imperative that readers appreciate the nuances of solution techniques of various space-time dependent equations.

Unlike in solid mechanics problems, note that fluid mechanics problems are characterized by a relatively larger number of degrees of freedom. This is because fluid mechanics is characterized by very low physical dissipation. This dissipation cannot be completely ignored from the whole flow domain and we would note here that in the spectral plane the maximum of dissipation occurs at higher wavenumbers. Thus, for simulating transitional flows one would have to be very careful in resolving very high wavenumbers. The importance of this is often lost on CFD practitioners who insist on performing large eddy simulations of transitional flows, by considering only the energy spectrum.

We enunciate the conservation principles for mass, momentum and energy for fluid flows. We begin by stating the conservation of mass.

2.1.1 Equation of Continuity

This physical principle provides a balance between mass entering and leaving the *control volume* through the *control surfaces* with the mass that may be created inside the *control volume*. For unsteady compressible flows, this leads to

$$\frac{D\rho}{Dt} + \rho \, \nabla \cdot \vec{V} = \frac{\partial \rho}{\partial t} + \nabla \cdot (\rho \, \vec{V}) = 0 \tag{2.1}$$

For incompressible flows, $\rho = $ constant and Eq. (2.1) simplifies to

$$\nabla \cdot (\vec{V}) = 0 \tag{2.2}$$

Any vector field that satisfies the equivalent condition given by Eq. (2.2) is called the *divergence-free* or *solenoidal* field.

2.1.2 Momentum Conservation Equation

Here, the physical principle for conservation of translational momentum for flows is considered, which is nothing but Newton's second law of motion applied to a control volume system. As one follows the *control volume* in motion, then one is considering *a control-mass system* for which the mass does not change with time. Hence, the rate of change of momentum is nothing but the mass multiplied by the acceleration experienced by the fluid in the *control volume*. One can write only the x-component of this relation as

$$F_x = ma_x \tag{2.3}$$

with the notation carrying their conventional meanings. Looking closely at the left-hand side of Eq. (2.3), one notes that the constituent forces are body forces acting on the control volume from a distance and do not depend on the geometry of the body and the surface forces acting directly on the *control surfaces*. These surface forces can be further sub-divided into *normal* and *shear forces* caused by corresponding stresses.

For an elementary volume element $(dx \, dy \, dz)$, if the local density is ρ, then the body force acting in the x-direction can be simply written as

$$= \rho f_x (dx \, dy \, dz) \tag{2.4}$$

where f_x is the associated acceleration in the x-direction.

The surface forces can be accounted for, with τ_{ij} denoting the stress tensor acting in the j-direction on a plane whose normal is in the i-direction. One can account for all the contributory stresses giving rise to surface force in the x-direction as given by

$$\left(-\frac{\partial p}{\partial x} + \frac{\partial \tau_{xx}}{\partial x} + \frac{\partial \tau_{yx}}{\partial y} + \frac{\partial \tau_{zx}}{\partial z} \right) dx \, dy \, dz \tag{2.5}$$

Thus, the force term on the left-hand side of Eq. (2.3) is the vector sum of Eqs. (2.4) and

(2.5). Considering that the mass of the fluid element is $(\rho\, dx\, dy\, dz)$ and the acceleration of the moving element is given by the substantive derivative, one can rewrite Eq. (2.3) as

$$\rho\frac{Du}{Dt} = \left(-\frac{\partial p}{\partial x} + \frac{\partial \tau_{xx}}{\partial x} + \frac{\partial \tau_{yx}}{\partial y} + \frac{\partial \tau_{zx}}{\partial z}\right) + \rho f_x \tag{2.6}$$

One can similarly obtain the y- and z-components of the momentum equation as

$$\rho\frac{Dv}{Dt} = \left(-\frac{\partial p}{\partial y} + \frac{\partial \tau_{xy}}{\partial x} + \frac{\partial \tau_{yy}}{\partial y} + \frac{\partial \tau_{zy}}{\partial z}\right) + \rho f_y \tag{2.7}$$

$$\rho\frac{Dw}{Dt} = \left(-\frac{\partial p}{\partial z} + \frac{\partial \tau_{xz}}{\partial x} + \frac{\partial \tau_{yz}}{\partial y} + \frac{\partial \tau_{zz}}{\partial z}\right) + \rho f_z \tag{2.8}$$

Equations (2.6) to (2.8) are known as the Cauchy equations. These can also be written in vector form as

$$\rho\frac{D\vec{V}}{Dt} = -\nabla p + \nabla \cdot \tau_{ij}\,\delta_{ij} + \rho\vec{F} \tag{2.9}$$

where δ_{ij} is the Kronecker delta function.

It is possible to express the x-component of the Cauchy equations in *conservation form*, by using the vector form of the continuity equation given by

$$\frac{\partial(\rho u)}{\partial t} + \nabla \cdot (\rho\, u\vec{V}) = \left(-\frac{\partial p}{\partial x} + \frac{\partial \tau_{xx}}{\partial x} + \frac{\partial \tau_{yx}}{\partial y} + \frac{\partial \tau_{zx}}{\partial z}\right) + \rho f_x \tag{2.10}$$

Similarly, one can write the other two components of the momentum equation in *conservative form*. These equations are not usable without information about the general stress system in the flow field. In fluids, strain rates are often related to velocity gradients by a linear relationship for Newtonian flows. For Newtonian fluids, Stokes [390] has shown that

$$\tau_{ij} = \lambda\delta_{ij}\frac{\partial v_k}{\partial x_k} + \mu\left(\frac{\partial v_i}{\partial x_j} + \frac{\partial v_j}{\partial x_i}\right) - p\delta_{ij} \tag{2.11}$$

where μ is the molecular viscosity, λ is the bulk viscosity coefficient and p is the thermodynamic pressure or hydrostatic stress. If p_m is the mechanical pressure, then by definition

$$p_m = \frac{\tau_{ii}}{3}$$

Stokes again hypothesized a relationship between μ and λ by relating thermodynamic and mechanical pressure by

$$3\lambda + 2\mu = 0 \tag{2.12}$$

The ratio of molecular viscosity (μ) with the density of the fluid (ρ) defines the kinematic viscosity (ν) and is used in the Navier–Stokes equation. Using the constitutive relation given in Eq. (2.11), one can write the final form of the Navier–Stokes equation for compressible flow as

$$\frac{\partial(\rho u)}{\partial t} + \frac{\partial(\rho u^2)}{\partial x} + \frac{\partial(\rho uv)}{\partial y} + \frac{\partial(\rho uw)}{\partial z} = -\frac{\partial p}{\partial x} +$$

$$\frac{\partial}{\partial x}\left(\lambda\nabla\cdot\vec{V} + 2\mu\frac{\partial u}{\partial x}\right) + \frac{\partial}{\partial y}\left(\mu\left[\frac{\partial v}{\partial x} + \frac{\partial u}{\partial y}\right]\right) + \frac{\partial}{\partial z}\left(\mu\left[\frac{\partial u}{\partial z} + \frac{\partial w}{\partial x}\right]\right) + \rho f_x \tag{2.13}$$

$$\frac{\partial(\rho v)}{\partial t} + \frac{\partial(\rho u v)}{\partial x} + \frac{\partial(\rho v^2)}{\partial y} + \frac{\partial(\rho v w)}{\partial z} = -\frac{\partial p}{\partial y} +$$

$$\frac{\partial}{\partial y}\left(\lambda \nabla \cdot \vec{V} + 2\mu \frac{\partial v}{\partial y}\right) + \frac{\partial}{\partial x}\left(\mu\left[\frac{\partial v}{\partial x} + \frac{\partial u}{\partial y}\right]\right) + \frac{\partial}{\partial z}\left(\mu\left[\frac{\partial v}{\partial z} + \frac{\partial w}{\partial y}\right]\right) + \rho f_y \qquad (2.14)$$

$$\frac{\partial(\rho w)}{\partial t} + \frac{\partial(\rho u w)}{\partial x} + \frac{\partial(\rho w v)}{\partial y} + \frac{\partial(\rho w^2)}{\partial z} = -\frac{\partial p}{\partial z} +$$

$$\frac{\partial}{\partial z}\left(\lambda \nabla \cdot \vec{V} + 2\mu \frac{\partial w}{\partial z}\right) + \frac{\partial}{\partial x}\left(\mu\left[\frac{\partial w}{\partial x} + \frac{\partial u}{\partial u}\right]\right) + \frac{\partial}{\partial y}\left(\mu\left[\frac{\partial v}{\partial z} + \frac{\partial w}{\partial y}\right]\right) + \rho f_z \qquad (2.15)$$

One can simplify these equations further for incompressible flows. In Eqs. (2.13)-(2.15), one sees that the terms associated with the bulk viscosity coefficient drop out. If one also considers incompressible flows without heat interaction, then the coefficient of molecular viscosity can be treated as a constant. One finally obtains the vector form of the Navier–Stokes equation for incompressible flow as

$$\frac{\partial \vec{V}}{\partial t} + (\vec{V} \cdot \nabla)\vec{V} = -\frac{\nabla p}{\rho} + \nu \, \nabla^2 \vec{V} + \vec{F} \qquad (2.16)$$

These are also called the equations in *primitive variables*.

2.1.3 Energy Conservation Equation

Here, this is the first law of thermodynamics stated for a control volume system which states that: The rate of change of energy inside the *control volume* must be due to the heat interaction across the *control surface* plus the work done reversibly due to boundary displacement by body and surface forces. Detailed derivations can be obtained in many text books on fluid mechanics and computations; see, e.g., [356]. The constituent terms are given next.

The work done terms:

The rate of work done by body forces for the *control volume* of mass ρ *(dx dy dz)* is given in a mixed form as

$$\left[-\nabla \cdot (p\vec{V}) + \frac{\partial}{\partial x_j}(u_i \tau_{ji})\right] dx \, dy \, dz + \rho \, \vec{F} \cdot \vec{V} \, dx \, dy \, dz \qquad (2.17)$$

The heat transfer terms:

The net flux of heat is due to volumetric heating such as absorption or emission of radiation and heat transfer across the control surface due to thermal conduction. If one defines the rate of volumetric heat addition per unit mass as $\dot{q}$, then the volumetric heating of the element is

$$= \rho \, \dot{q} \, dx \, dy \, dz \qquad (2.18)$$

Using Newton's law, one can obtain directional conductive heat transfer with temperature gradient using Fourier's law by

$$\dot{q}_j = -\kappa \frac{\partial T}{\partial x_j}$$

where κ is the thermal conductivity.

Hence the total heat interaction term is obtained by using Newton's law to provide

$$\left[\rho \, \dot{q} + \frac{\partial}{\partial x_j} \left(\kappa \frac{\partial T}{\partial x_j} \right) \right] dx \, dy \, dz \tag{2.19}$$

The rate of change of energy terms:

The total energy of a moving fluid per unit mass is the sum of its internal energy per unit mass (e) and its kinetic energy per unit mass $(\frac{V^2}{2})$. Since we are following a moving fluid element, the time rate of change of energy per unit mass is given by the substantive derivative

$$\rho \frac{D}{Dt} \left(e + \frac{V^2}{2} \right) dx \, dy \, dz \tag{2.20}$$

The final form:

The final form of the energy equation is obtained by collating the terms given above to obtain the *non-conservation* or *convective* form of the energy equation as

$$\rho \frac{D}{Dt} \left(e + \frac{V^2}{2} \right) = \left[\rho \, \dot{q} + \frac{\partial}{\partial x_j} \left(\kappa \frac{\partial T}{\partial x_j} \right) \right] - \nabla \cdot (p\vec{V}) + \frac{\partial}{\partial x_j} (u_i \tau_{ji}) + \rho \, \vec{F} \cdot \vec{V} \tag{2.21}$$

2.1.4 Alternate Forms of the Energy Equation

Occasionally, the energy equation is written in terms of internal energy only. This is obtained as

$$\rho \frac{De}{Dt} = \rho \, \dot{q} + \frac{\partial}{\partial x_j} \left(\kappa \frac{\partial T}{\partial x_j} \right) - p\nabla \cdot \vec{V} + \lambda (\nabla \cdot \vec{V})^2 + 2\mu \left[\left(\frac{\partial u}{\partial x} \right)^2 + \left(\frac{\partial v}{\partial y} \right)^2 + \left(\frac{\partial w}{\partial z} \right)^2 \right]$$

$$+ \mu \left[\left(\frac{\partial u}{\partial y} + \frac{\partial v}{\partial x} \right)^2 + \left(\frac{\partial u}{\partial z} + \frac{\partial w}{\partial x} \right)^2 + \left(\frac{\partial v}{\partial z} + \frac{\partial w}{\partial y} \right)^2 \right] \tag{2.22}$$

The energy equation in conservation form is obtained from the above by noting

$$\rho \frac{De}{Dt} = \frac{\partial}{\partial t} (\rho e) + \nabla \cdot (\rho e \vec{V})$$

Terms involving λ and μ constitute the dissipation term Φ_0.

2.1.5 Equations of Motion in Terms of Derived Variables

One of the problems of primitive variable formulation is the absence of unique and definitive boundary conditions for pressure and this is often attempted by using an alternative formulation that does not have pressure dependent terms explicitly. Despite the fact that the pressure is a tensor of rank two, one can eliminate it by taking a curl of Eq. (2.16). Thus, one obtains the vorticity transport equation as

$$\frac{\partial \vec{\omega}_I}{\partial t} + (\vec{V} \cdot \nabla)\vec{\omega}_I = (\vec{\omega}_I \cdot \nabla)\vec{V} + \nu \nabla^2 \vec{\omega}_I \tag{2.23}$$

This equation can also be written in the partial conservative or Eulerian form as

$$\frac{\partial \vec{\omega}_I}{\partial t} + \nabla \times (\vec{\omega}_I \times \vec{V}) = \nu \nabla^2 \vec{\omega}_I \tag{2.24}$$

We have used $(\vec{V}\cdot\nabla)\,\vec{V} = \nabla(\frac{V^2}{2}) - \vec{V}\times(\nabla\times\vec{V}) = \nabla(\frac{V^2}{2}) - \vec{V}\times\vec{\omega}$ in writing this partial con-servative form. This is also called the Laplacian form of velocity-vorticity formulation. This equation helps one follow the evolution of the divergence of the vorticity field. Solenoidality or the divergence free condition of the vorticity field is traced by taking divergence of Eq. (2.23) to yield

$$\frac{\partial D_\omega}{\partial t} = \nu\nabla^2 D_\omega \qquad (2.25)$$

where $D_\omega = \nabla\cdot\vec{\omega}_I$ and the above equation represents unsteady diffusion of the solenoidality condition. This equation tells one that if the vorticity field is divergence free at $t = 0$, then the right-hand side of Eq. (2.25) is identically zero and that would imply that D_ω will be time invariant and will continue to be divergence free. However, this has to be treated with caution as vorticity is continually generated at the no-slip wall. It is not apparent how the generated wall vorticity can be made divergence free and this will be a continual source of error if left unchecked; the evolution of this vorticity is given by the diffusion Eq. (2.25).

In velocity-vorticity formulation, one augments the transport equations by deriving an auxiliary relation from the kinematic definition of vorticity. For example, taking a curl of vorticity expressed in the inertial frame, one obtains

$$\nabla^2\vec{V} = -\nabla\times\vec{\omega}_I \qquad (2.26)$$

Equations (2.23), (2.24) and (2.26) constitute the governing velocity-vorticity equa-tions in the derived variable formulation. In using the velocity-vorticity formulation, some researchers have not used the vector Poisson equation. For example, authors in [115, 116, 117, 261] have directly used the continuity equation and the kinematic defini-tion of vorticity, along with the vorticity transport equation. One of the reasons for using the stream function-vorticity formulation for two-dimensional flows over the primitive vari-able formulation is due to its ability to satisfy mass conservation exactly. Such benefits are not directly available for flows in three dimensions. This is the motivation for developing a highly accurate form of governing equation using derived variables.

2.2   Some Equilibrium Solutions of the Basic Equation

In this section we will mostly consider those low speed flows for which the transport co-efficients and density are treated as constants unless stated otherwise. For the sake of convenience, we restate the basic conservation equations in dimensional form below

$$\nabla \bullet \vec{V} = 0 \qquad (2.27)$$

$$\rho\frac{D\vec{V}}{Dt} = -\nabla\hat{p} + \mu\nabla^2\vec{V} \qquad (2.28)$$

$$\rho c_p\frac{DT}{Dt} = k\nabla^2 T + \Phi_0 \qquad (2.29)$$

where $\hat{p} = p + \rho gz$ and Φ_0 represent the viscous, dissipation function. In the following, we first look for exact solutions of the above equation set, under some assumptions or restrictions imposed on the flow.

The basic equations, as given in Eqs. (2.27) to (2.29), are represented in a coordinate frame invariant form. These equations take a simple form when represented in a Cartesian frame. However, we may be required to write these equations in other orthogonal coordinate systems. For this reason we note expressions for representing some operators which appear in basic equations, for cylindrical and spherical coordinate systems.

For the cylindrical coordinate system, we use the following vector representations. The gradient operator, operating on a scalar ϕ, is given by the r-, Φ- and z-components shown here separated by commas as

$$\nabla \phi = \left(\frac{\partial \phi}{\partial r}, \frac{1}{r} \frac{\partial \phi}{\partial \Phi}, \frac{\partial \phi}{\partial z} \right) \tag{2.30}$$

Similarly, the divergence operator acting on a vector $\vec{V}$ with components $(v_r,\ v_\Phi,\ v_z)$ is given by

$$\nabla \bullet \vec{V} = \frac{1}{r} \frac{\partial}{\partial r}(r v_r) + \frac{1}{r} \frac{\partial v_\Phi}{\partial \Phi} + \frac{\partial v_z}{\partial z} \tag{2.31}$$

Other useful identities are

$$\vec{V} \bullet \nabla = v_r \frac{\partial}{\partial r} + \frac{v_\Phi}{r} \frac{\partial}{\partial \Phi} + v_z \frac{\partial}{\partial z} \tag{2.32}$$

$$\nabla^2 \phi = \frac{1}{r} \frac{\partial}{\partial r}\left(r \frac{\partial \phi}{\partial r} \right) + \frac{1}{r^2} \frac{\partial^2 \phi}{\partial \Phi^2} + \frac{\partial^2 \phi}{\partial z^2} \tag{2.33}$$

For the spherical coordinate system, similar expressions are obtained for the scalar and the vector with components $(v_r,\ v_\theta,\ v_\Phi)$, as given below

$$\nabla \phi = \left(\frac{\partial \phi}{\partial r}, \frac{1}{r} \frac{\partial \phi}{\partial \theta}, \frac{1}{r \sin \theta} \frac{\partial \phi}{\partial \Phi} \right) \tag{2.34}$$

$$\nabla \bullet \vec{V} = \frac{1}{r^2} \frac{\partial}{\partial r}(r^2 v_r) + \frac{1}{r \sin \theta} \frac{\partial}{\partial \theta}(v_\theta \sin \theta) + \frac{1}{r \sin \theta} \frac{\partial v_\Phi}{\partial \Phi} \tag{2.35}$$

$$\vec{V} \bullet \nabla = v_r \frac{\partial}{\partial r} + \frac{v_\theta}{r} \frac{\partial}{\partial \theta} + \frac{v_\Phi}{r \sin \theta} \frac{\partial}{\partial \Phi} \tag{2.36}$$

$$\nabla^2 \phi = \frac{1}{r^2} \frac{\partial}{\partial r}\left(r^2 \frac{\partial \phi}{\partial r} \right) + \frac{1}{r^2 \sin \theta} \frac{\partial}{\partial \theta}\left(\sin \theta \frac{\partial \phi}{\partial \theta} \right) + \frac{1}{r^2 \sin^2 \theta} \frac{\partial^2 \phi}{\partial \Phi^2} \tag{2.37}$$

First, we discuss the class of flow problems which are driven by the motion of one of the bounding walls. Such flows were originally reported by Couette and are called the Couette flows.

We note the Navier–Stokes equation in the polar coordinate system, which is useful for a few applications. If v_r and v_θ are the radial and azimuthal components of the velocity vector and X_r and X_θ are the components of the body force in the radial and azimuthal directions, then the radial component of the Navier–Stokes equation is given by

$$\frac{\partial v_r}{\partial t} + v_r \frac{\partial v_r}{\partial r} + \frac{v_\theta}{r} \frac{\partial v_r}{\partial \theta} - \frac{v_\theta^2}{r} = X_r - \frac{1}{\rho} \frac{\partial p}{\partial r}$$

$$+ v \left(\frac{\partial^2 v_r}{\partial r^2} + \frac{1}{r} \frac{\partial v_r}{\partial r} - \frac{v_r}{r^2} + \frac{1}{r^2} \frac{\partial^2 v_r}{\partial \theta^2} - \frac{2}{r^2} \frac{\partial v_\theta}{\partial \theta} \right) \tag{2.38a}$$

And the azimuthal component of the Navier–Stokes equation is

$$\frac{\partial v_\theta}{\partial t} + v_r \frac{\partial v_\theta}{\partial r} + \frac{v_\theta}{r} \frac{\partial v_\theta}{\partial \theta} + \frac{v_r v_\theta}{r} = X_\theta - \frac{1}{\rho r} \frac{\partial p}{\partial \theta}$$

$$+ v \left(\frac{\partial^2 v_\theta}{\partial r^2} + \frac{1}{r} \frac{\partial v_\theta}{\partial r} - \frac{v_\theta}{r^2} + \frac{1}{r^2} \frac{\partial^2 v_\theta}{\partial \theta^2} + \frac{2}{r^2} \frac{\partial v_r}{\partial \theta} \right) \tag{2.38b}$$

2.2.1   Couette Flow between Parallel Plates

Here, the two-dimensional flow is established by the motion of the top plate, while the lower plate is static. The temperature condition is as indicated in Fig. 2.1, with the adopted coordinate system as indicated.

For the given geometry, a possible equilibrium solution can be obtained by noting that the y-component of velocity is identically zero as one possibility. Then from the continuity equation, one gets $\frac{\partial u}{\partial x} = 0$, which in turn implies a parallel flow whose general form is $u(y)$. Such flows which do not depend upon the streamwise coordinate are also called *fully developed flows*. For such a flow field, the y-momentum equation is also identically satisfied. The x-momentum equation reduces to

$$\mu \frac{d^2 u}{dy^2} = 0$$

which, satisfying the boundary conditions, yields the solution

$$u = \frac{U}{2}\left(1 + \frac{y}{h}\right) \tag{2.39}$$

Note that the shear stress is given here as $\tau = \mu \frac{du}{dy} = \frac{\mu U}{2h} = \text{constant}$

For the parallel flow, the dissipation function (Φ_0) takes the simplified form and the energy equation is given by

$$k \frac{d^2 T}{dy^2} + \mu \left(\frac{du}{dy}\right)^2 = 0$$

which, satisfying the boundary conditions, yields the solution

$$T = \frac{T_1 + T_0}{2} + \frac{T_1 + T_0}{2}\frac{y}{h} + \frac{\mu U^2}{8\kappa}\left(1 - \frac{y^2}{h^2}\right) \tag{2.40}$$

If T is non-dimensionalized by $(T_1 - T_0)$, then a non-dimensional dissipation parameter is introduced, named the Brinkman number, as

$$Br = \frac{\mu U^2}{\kappa(T_1 - T_0)}$$

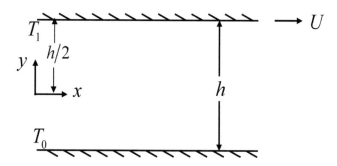

FIGURE 2.1
Couette flow between two parallel plates at different temperatures. The top wall moves at a constant speed to the right.

The temperature distribution can be written in terms of Br as

$$T^* = \frac{1}{2}\left(\frac{T_1 + T_0}{T_1 - T_0}\right) + \frac{1}{2}y^* + \frac{Br}{8}(1 - y^{*2}) \tag{2.41}$$

where h has been used as the length scale in defining the non-dimensional quantities with an asterisk. The last term is due to viscous dissipation and is significant for highly viscous fluids only.

The rate of heat transfer at the walls can be calculated as

$$q_w = \left[\frac{dT}{dy}\right]_{\pm h} = \frac{\kappa}{2h}(T_1 - T_0) \pm \frac{\mu U^2}{4h} \tag{2.42}$$

If we define the heat transfer coefficient as

$$\zeta = \frac{q_w}{(T_1 - T_0)}$$

Then one can define a non-dimensional heat transfer coefficient, the Nusselt number, as

$$Nu_{2h} = \frac{2\zeta\,h}{\kappa} = 1 \pm \frac{Br}{2} \tag{2.43}$$

2.2.2 Flow between Concentric Rotating Cylinders

Consider the flow between two concentric cylinders, with an inner cylinder of radius a_1 rotating at an rpm of Ω_1, inside the outer cylinder of radius a_2 rotating at an rpm of Ω_2. Inner and outer cylinder temperatures are held fixed at T_1 and T_2, respectively. For this problem, it is convenient to work with the basic equations defined in the polar coordinate system. This is obtained from Eqs. (2.27), (2.28) and (2.29) using the identities given by Eqs. (2.30)-(2.33) with the proviso that the flow is only in the (r, θ)-plane.

Once again in the search for an exact solution, we consider the special flow for which there is no flow in the axial direction, so that the flow can be considered as two dimensional. Furthermore, consider the flow to be perfectly azimuthal, i.e., the flow streamlines are considered to be concentric streamlines for the steady flow considered here.

From the mass conservation equation, it is easy to see that the azimuthal velocity is a function of a radial coordinate only, i.e., $v_\Phi = v_\Phi(r)$. Such a flow can be established by the centrifugal force being balanced by a radial pressure gradient that one can easily verify from the radial momentum equation as given below

$$\frac{\partial p}{\partial r} = \frac{\rho v_\Phi^2}{r} \tag{2.44}$$

The azimuthal momentum equation in this takes the simplified form

$$\frac{d^2 v_\Phi}{dr^2} + \frac{d}{dr}\left(\frac{v_\Phi}{r}\right) = \frac{1}{\mu r}\frac{\partial p}{\partial \theta} \tag{2.45}$$

As v_Φ is a function of r only, Eq. (2.45) can be rewritten as

$$\frac{\partial p}{\partial \theta} = \rho r f(r)$$

which upon integration would give

$$p = \rho r f(r)\,\theta + g(r)$$

However, from the periodicity of the solution, one has $p(\theta) = p(\theta + 2\pi)$ and this condition therefore requires $f(r) \equiv 0$ and thus we have only a radially varying pressure field.

Similarly, the energy equation takes the form

$$\frac{\kappa}{r} \frac{d}{dr}\left(r \frac{dT}{dr}\right) + \mu \left(\frac{dv_\Phi}{dr} - \frac{v_\Phi}{r}\right)^2 = 0 \tag{2.46}$$

Note that the last term in Eq. (2.46) is the dissipation function for this simplified flow. The solution of Eq. (2.45) is

$$v_\Phi = Ar + \frac{B}{r} \tag{2.47a}$$

which upon satisfying the boundary conditions

$$v_\Phi = \Omega_1 a_1 \quad \text{at} \quad r = a_1$$

and

$$v_\Phi = \Omega_2 a_2 \quad \text{at} \quad r = a_2$$

yields

$$A = \frac{(\Omega_2\, a_2^2 - \Omega_1\, a_1^2)}{(a_2^2 - a_1^2)} \quad \text{and} \quad B = \frac{(\Omega_1 - \Omega_2)a_1^2\, a_2^2}{(a_2^2 - a_1^2)} \tag{2.47b}$$

From Eq. (2.44), with the pressure established as a function of r only, one can obtain

$$p = \rho \int_{a_1}^{r} \frac{v_\Phi^2}{r} dr + C$$

with $C = \frac{p(a_1)}{\rho}$, either a known quantity or it can be measured. Using Eqs. (2.47a) and (2.47b), one can obtain the pressure distribution as

$$\begin{aligned}
p = p\,(a_1) + \frac{\rho}{(a_2^2 - a_1^2)^2} &\left[(\Omega_2\, a_2^2 - \Omega_1\, a_1^2)^2 \left(\frac{r^2 - a_1^2}{2}\right) \right. \\
&- 2a_1^2\, a_2^2\, (\Omega_1 - \Omega_2)\, (\Omega_2\, a_2^2 - \Omega_1\, a_1^2)\, Ln\frac{r}{a_1} \\
&\left. - \frac{1}{2} a_1^4\, a_2^4\, (\Omega_2 - \Omega_1)^2 \left(\frac{1}{r^2} - \frac{1}{a_1^2}\right) \right]
\end{aligned} \tag{2.48}$$

Using Eq. (2.47a), one can evaluate the dissipation function term of Eq. (2.46) as $-4\mu \frac{B^2}{r^4}$ and thus Eq. (2.46) takes the form

$$\frac{\kappa}{r} \frac{d}{dr}\left(r \frac{dT}{dr}\right) = -4\mu \frac{B^2}{r^4}$$

which can be integrated twice with respect to r to get the temperature distribution as

$$T = -\frac{\mu B^2}{\kappa r^2} + C_1\, Ln\, r + C_2$$

The constants C_1 and C_2 can be obtained from the fixed wall temperatures, T_1 and T_2. The final temperature distribution in the annulus can be obtained in the simplified non-dimensional form as

$$\frac{T - T_2}{T_1 - T_2} = PrEc\frac{a_1^4(1 - \Omega_1/\Omega_2)^2}{a_1^4 - a_2^4}\left(1 - \frac{a_2^2}{a_1^2}\right)\left[1 - \frac{Ln(r/a_2)}{Ln(r/a_1)}\right] + \frac{Ln(r/a_2)}{Ln(r/a_1)} \tag{2.49}$$

Once again the first term in Eq. (2.49) is due to viscous dissipation and the second term is due to conduction. Note that $Br = Pr\, Ec$, where $Pr = \frac{\mu c_p}{\kappa}$ and $Ec = \frac{U^2}{c_p(T_1 - T_0)}$ are the Prandtl and Eckart numbers.

Note: 1) If one considers flow past a rotating circular cylinder, then it is easy to set $a_2 \to \infty$ and $\Omega_2 \to 0$ to have the steady flow due to a rotating cylinder of radius a_1 in tranquil fluid as

$$v_\Phi = \frac{a_1^2\, \Omega_1}{r} \tag{2.50a}$$

and

$$p = p\,(a_1) + \frac{\rho}{2}a_1^2\, \Omega_1^2\left(1 - \frac{a_1^2}{r^2}\right) \tag{2.50b}$$

This is a special case where the flow is *irrotational*. However, this irrotational flow satisfies the *no-slip* condition!

2) The solution given by Eqs. (2.47a) and (2.48) shows it to be independent of viscosity and thus the viscous solution is independent of Re! In other words, no boundary layer (as will be discussed later) forms in this case. However, the torque exerted on the cylinders still depends upon the coefficient of viscosity. For example, the viscous stress acting on the inner cylinder is given by

$$\tau = \mu\left[r\frac{d}{dr}\left(\frac{v_\Phi}{r}\right)\right]_{r=a_1}$$

As the inner surface area is $2\pi a_1$ and the moment arm is a_1, one calculates the moment exerted on the inner cylinder as

$$M_{\text{inner}} = 4\pi\mu\frac{a_1^2\, a_2^2}{a_2^2 - a_1^2}(\Omega_2 - \Omega_1) \tag{2.51}$$

One can also estimate the moment exerted on the outer cylinder. However, without estimating the same, one can reason out that $M_{\text{outer}} = -M_{\text{inner}}$.

2.2.3 Couette Flow between Parallel Plates, Driven by Pressure

Again we are discussing the problem that was discussed in Sec. 2.2, with the difference that the flow is driven by a pressure gradient in the streamwise direction. We derive the solution as rigorously as possible, with governing equations given by

$$\frac{\partial u}{\partial x} + \frac{\partial v}{\partial y} = 0 \tag{2.52}$$

$$\frac{\partial u}{\partial t} + u\frac{\partial u}{\partial x} + v\frac{\partial u}{\partial y} = X_x - \frac{1}{\rho}\frac{\partial p}{\partial x} + \nu\left(\frac{\partial^2 u}{\partial x^2} + \frac{\partial^2 u}{\partial y^2}\right) \tag{2.53}$$

$$\frac{\partial v}{\partial t} + u\frac{\partial v}{\partial x} + v\frac{\partial v}{\partial y} = X_y - \frac{1}{\rho}\frac{\partial p}{\partial y} + \nu\left(\frac{\partial^2 v}{\partial x^2} + \frac{\partial^2 v}{\partial y^2}\right) \tag{2.54}$$

Once again, we look for a solution of the form

$$v(x, y, t) = 0$$

With this trial solution, Eqs. (2.52)–(2.54) simplify to

$$\frac{\partial u}{\partial x} = 0 \tag{2.55}$$

$$\frac{\partial u}{\partial t} + u\frac{\partial u}{\partial x} = X_x - \frac{1}{\rho}\frac{\partial p}{\partial x} + \nu\left(\frac{\partial^2 u}{\partial x^2} + \frac{\partial^2 u}{\partial y^2}\right) \tag{2.56a}$$

$$0 = X_y - \frac{1}{\rho}\frac{\partial p}{\partial y} \tag{2.56b}$$

From Eq. (2.55) it is obvious that $u = u(y,t)$ only. Therefore, the second term on the left-hand side of Eq. (2.56a) also must be omitted and once again we have a *fully developed flow*. If we talk about constant body force, then from Eq. (2.56b)

$$p = \rho X_y\, y + C(x,t)$$

from which one can obtain $\frac{\partial p}{\partial x} = \frac{\partial C}{\partial x}$, which, upon substitution in Eq. (2.56a), provides

$$\frac{\partial C}{\partial x} = \rho\, X_x + \rho\left(\nu\frac{\partial^2 u}{\partial y^2} - \frac{\partial u}{\partial t}\right) \tag{2.57}$$

In this equation, the second term on the right-hand side is only a function of y and t, while by assumption the first term is a constant. However, by our assumption C can only be a function of x and t only. Thus

$$\frac{\partial C}{\partial x} = \frac{\partial p}{\partial x} = F(t)$$

Substitution of this in Eq. (2.57) gives us

$$\frac{\partial u}{\partial t} = \left[X_x - \frac{F(t)}{\rho}\right] + \nu\frac{\partial^2 u}{\partial y^2} \tag{2.58}$$

Now we want to solve this equation for $u(y,t)$ subject to the following boundary conditions at $y = 0$:

$$u(0,t) = U_1(t) \tag{2.59a}$$

and at $y = H$:

$$u(H,t) = U_2(t) \tag{2.59b}$$

The initial condition is given by

$$u(y,0) = U_0(y) \tag{2.60}$$

In a general framework, one would look for a solution of Eq. (2.58) subject to boundary and initial conditions given by Eqs. (2.59a), (2.59b) and (2.60). Let us restrict our attention to specific simple cases.

Case 1: Steady Channel Flow
For steady flow, the left-hand side of Eq. (2.58) is identically zero and the governing equation is

$$\mu\frac{d^2 u}{dy^2} = (F - \rho X_x) \tag{2.61}$$

where F is not a function of time - instead we just have a constant pressure gradient that drives the flow. Furthermore, if we take the simpler case of $U_1(t) = 0$, then the solution of Eq. (2.61) is given as

$$\frac{u}{U_2} = \frac{y}{H} - \frac{H^2}{2\mu U_2}(F - \rho X_x)\left(\frac{y}{H}\right)\left(1 - \frac{y}{H}\right) \tag{2.62}$$

One can construct a non-dimensional number indicating the pressure gradient

$$P_0 = \frac{H^2}{2\mu U_2}\left(\rho X_x - F\right) \tag{2.63}$$

Then the solution in Eq. (2.62) can be viewed as a combination of flow due to a shear (as given by the first term) and that due to an imposed pressure gradient. The superposition principle worked because Eq. (2.61) turns out to be linear. In Fig. 2.2, we have sketched some typical velocity profiles for different values of the non-dimensional parameter P_0. Based on the figure, the following observations can be drawn.

(a) For $P_0 = 0$: The velocity profile is a straight line, as was the case for Couette flow discussed in Sec. 2.2.1.

(b) For $P_0 > 0$: The velocity profile shows a faster flow as compared to the case of $P_0 = 0$. This is because of the fact that the shear and the pressure gradient operate in the same direction.

(c) For $P_0 < 0$: The velocity profile slows down as compared to the zero pressure gradient case. As a result the wall shear stress keeps decreasing. In this case, the applied pressure gradient opposes the shear imposed by the bottom plate movement. Additionally, the following sub-cases can occur.

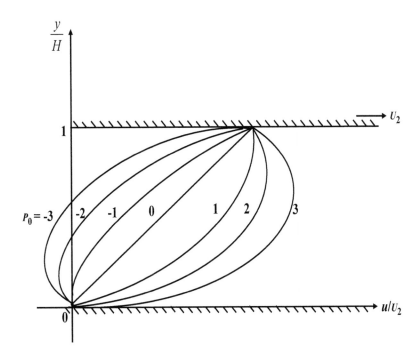

FIGURE 2.2
Combined Couette–Poiseuille flow driven by shear and non-dimensional pressure gradient, $P_0 = \frac{H^2}{2\mu U_2}(\rho X_x - F)$, with U_2 as the speed of the top wall; $F(t)$ is the time-dependent streamwise pressure gradient; X_x is the body force and H is the distance between the plates.

(i) For $-1 < P_0 < 0$: The velocity profile slows down without any back-flow, at any height.

(ii) For $P_0 < -1$: There is a height band close to the static wall where the flow reverses direction.

An additional sub-case can be considered for which $U_2 \equiv 0$, known as plane Poiseuille flow, given by

$$u = \frac{H^2}{2\mu}(\rho\,X_x - F)\left(\frac{y}{H}\right)\left(1 - \frac{y}{H}\right) \tag{2.64}$$

The nomenclature of this flow follows from the fact that the early investigation of flow driven by a pressure gradient was performed by Poiseuille in circular pipes.

Case 2: Steady Channel Flow with Blowing and Suction

The flow is between two permeable plates located at $y = 0$ and $y = H$. Once again, we consider a simple case where a constant $v = v_0$ velocity exists everywhere in the flow domain, i.e., the flow emanates from the lower plate and is absorbed by the upper plate. Additionally, let the upper plate move with constant velocity, U_2, to the right, which will drive the flow in the presence of surface transpiration. Also, let there be a constant pressure gradient, $F = \frac{\partial p}{\partial x}$, applied in the streamwise direction.

Under the action of these flow drivers and the trial solution $v = v_0$, the governing equation simplifies to

$$v_0\frac{du}{dy} = X_x - \frac{1}{\rho}\frac{\partial p}{\partial x} + \nu\frac{d^2u}{dy^2} \tag{2.65}$$

$$\frac{\partial p}{\partial y} = \rho X_y \tag{2.66}$$

Because of the constant body force assumption, one can establish in the same way as it was established in the previous section that

$$\frac{\partial p}{\partial x} = F = (\text{constant})$$

This allows us to solve Eq. (2.65) subject to the boundary conditions $u = 0$ at $y = 0$ and $u = U_2$ at $y = H$, which is given by

$$\frac{u}{U_2} = \left[1 + \frac{C}{v_0/U_2}\right]\frac{1 - e^{\frac{Re\,v_0\,y}{U_2H}}}{1 - e^{\frac{Re\,v_0}{U_2}}} - \frac{C}{v_0/U_2}\left(\frac{y}{H}\right) \tag{2.67}$$

where

$$C = \frac{F - \rho X_x}{\rho U_2^2/H}$$

and $Re = \frac{\rho U_2 H}{\mu}$ are the two non-dimensional parameters relevant to this problem.

2.2.4   Steady Stagnation Point Flow

Here, we investigate the steady flow impinging on a flat surface, as shown in Fig. 2.3 with the adopted coordinate system. The governing equations are given in Eqs. (2.52)–(2.54). However, we neglect the body forces and rewrite these as

$$\frac{\partial u}{\partial x} + \frac{\partial v}{\partial y} = 0 \tag{2.52}$$

$$\frac{\partial u}{\partial t} + u\frac{\partial u}{\partial x} + v\frac{\partial u}{\partial y} = -\frac{1}{\rho}\frac{\partial p}{\partial x} + \nu\left(\frac{\partial^2 u}{\partial x^2} + \frac{\partial^2 u}{\partial y^2}\right) \tag{2.68}$$

$$\frac{\partial v}{\partial t} + u\frac{\partial v}{\partial x} + v\frac{\partial v}{\partial y} = -\frac{1}{\rho}\frac{\partial p}{\partial y} + \nu\left(\frac{\partial^2 v}{\partial x^2} + \frac{\partial^2 v}{\partial y^2}\right) \tag{2.69}$$

On the flat plate, one needs to satisfy the no-slip condition. Very far from the plate, i.e., at $y \to \infty$, conditions are stated later.

As at $y = 0$:

$$v = 0 \text{ for all } x$$

One can try the following ansatz

$$v = -f(y) \tag{2.70}$$

From Eq. (2.52), using this ansatz one gets

$$u = x\left(\frac{df}{dy}\right) + F(y)$$

However, the solution must be symmetric about the y-axis and thus one must have $F(y) = 0$. Therefore

$$u = x\left(\frac{df}{dy}\right) \tag{2.71}$$

Using the expressions for u and v in Eq. (2.69) one obtains

$$\frac{1}{\rho}\frac{\partial p}{\partial y} = -f\frac{df}{dy} - \nu\frac{d^2 f}{dy^2}$$

Integrating this, one obtains an expression for pressure and subsequently differentiating it with respect to x would reveal that $\frac{\partial p}{\partial x}$ is strictly a function of x only.

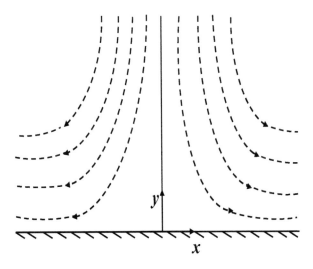

FIGURE 2.3
Streamlines sketched for the stagnation point flow, with the stagnation point at the origin.

Therefore, using Eqs. (2.70) and (2.71) in Eq. (2.68), one obtains

$$\left(\frac{df}{dy}\right)^2 - f\frac{d^2f}{dy^2} - \nu\frac{d^3f}{dy^3} = C \text{ (const.)} \tag{2.72}$$

It is possible to show

$$\frac{p}{\rho} = -\frac{Cx^2}{2} + D \tag{2.73}$$

To solve Eq. (2.72), we need three boundary conditions for f. Two of these are obtained at the wall ($y = 0$) as the no-slip condition given by

$$f, \ \frac{df}{dy} = 0 \tag{2.74}$$

And for $y \to \infty$, the boundary condition is obtained by imposing the boundedness condition on u and v. For u to remain bounded (which is a first derivative of f), one must have the second derivative equal to zero. The same must be true for all other higher order derivatives. Therefore, from Eq. (2.72) the boundary condition is given as

$$\lim_{y\to\infty}\left(\frac{df}{dy}\right)^2 = b^2 = C_1$$

Then the requisite boundary condition is

$$\lim_{y\to\infty}\frac{df}{dy} = b \tag{2.75}$$

From the definition of u, as given by Eq. (2.71), the consequence of the above will be

$$\lim_{y\to\infty} u = bx \tag{2.76}$$

Also, utilizing Eq. (2.75) and the definition of v, one gets the asymptotic behavior of this velocity component as

$$\lim_{y\to\infty} v = -by - c \tag{2.77}$$

Thus, one needs to solve Eq. (2.72) using the boundary conditions given by Eqs. (2.74) and (2.75). This is solved numerically, as the equation is a non-linear ordinary differential equation

$$\Phi_1''' + \Phi_1\Phi_1'' - \Phi_1'^2 + 1 = 0$$

This is obtained by introducing the similarity transformation for the independent variable $\eta = \alpha y$ and a dependent variable as $f = A\Phi_1$. Primes indicate differentiation with respect to η. A typical numerical solution is shown in Fig. 2.4 for unknown (f) and its derivatives.

2.2.5 Flow Past a Rotating Disc

This flow can be viewed as an extension of the stagnation point flow that was described in the previous subsection. This refers to axisymmetric flow, while the stagnation point flow is planar. Here, we have a large infinite disc in the $z = 0$ plane, as shown in Fig. 2.5, which rotates about the z-axis with speed Ω. The resultant steady flow field created above the plate will be analyzed for the case with no variation in the azimuthal direction, i.e., all flow variables would display $\frac{\partial}{\partial\theta} = 0$.

The governing equations are then given by

$$\frac{\partial v_r}{\partial r} + \frac{v_r}{r} + \frac{\partial v_z}{\partial z} = 0 \tag{2.78}$$

$$v_r\frac{\partial v_r}{\partial r} + v_z\frac{\partial v_r}{\partial z} - \frac{v_\Phi^2}{r} = -\frac{1}{\rho}\frac{\partial p}{\partial r} + \nu\left(\frac{\partial^2 v_r}{\partial r^2} + \frac{1}{r}\frac{\partial v_r}{\partial r} - \frac{v_r}{r^2} + \frac{\partial^2 v_z}{\partial z^2}\right) \tag{2.79}$$

$$v_r\frac{\partial v_\Phi}{\partial r} + v_z\frac{\partial v_\Phi}{\partial z} + \frac{v_r v_\Phi}{r} = \nu\left(\frac{\partial^2 v_\Phi}{\partial r^2} + \frac{1}{r}\frac{\partial v_\Phi}{\partial r} - \frac{v_\Phi}{r^2} + \frac{\partial^2 v_\Phi}{\partial z^2}\right) \tag{2.80}$$

$$v_r\frac{\partial v_z}{\partial r} + v_z\frac{\partial v_z}{\partial z} = -\frac{1}{\rho}\frac{\partial p}{\partial z} + \nu\left(\frac{\partial^2 v_z}{\partial r^2} + \frac{1}{r}\frac{\partial v_z}{\partial r} + \frac{\partial^2 v_z}{\partial z^2}\right) \tag{2.81}$$

Boundary conditions applicable at the disc ($z = 0$) are

$$v_r = 0; \quad v_\Phi = r\Omega \quad \text{and} \quad v_z = 0 \tag{2.82a}$$

For away from the disc ($z \to \infty$)

$$v_r, v_\Phi \quad \text{and} \quad v_z \quad \text{remains bounded.} \tag{2.82b}$$

The following ansatz was suggested in [20], but originally was posed alongside three other associated ansatzs for this problem and solved in [169] without any explanation

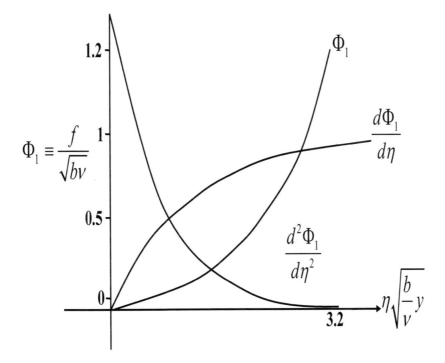

FIGURE 2.4
Solution of stagnation point flow, shown in the transformed η coordinate. Various quantities are as described in the text.

$$v_z = f(z) \tag{2.83}$$

Using Eq. (2.83) in Eq. (2.78) one gets

$$\frac{1}{r}\frac{\partial}{\partial r}(rv_r) = -\frac{df}{dz}$$

which can be integrated, noting that v_r is finite at $r = 0$, as

$$v_r = -\frac{r}{2}\frac{df}{dz} = rg(z) \tag{2.84}$$

Also, using the ansatz Eq. (2.83) in the z-momentum Eq. (2.81), one gets

$$\frac{1}{\rho}\frac{\partial p}{\partial z} = \nu\frac{d^2 f}{dz^2} - f\frac{df}{dz}$$

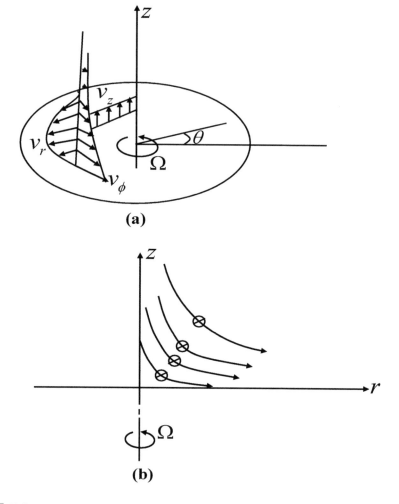

FIGURE 2.5
(a) Velocity profiles and (b) streamlines for the axisymmetric flow configuration.

This equation can be integrated once with respect to z to get

$$\frac{p}{\rho} = v\frac{df}{dz} - \frac{f^2}{2} + h(r) \tag{2.85}$$

Substituting Eqs. (2.84) and (2.85) in Eq. (2.79), one gets

$$\frac{1}{r}\frac{dh}{dr} - \frac{v_\Phi^2}{r^2} = -g^2 - f\frac{dg}{dz} + v\frac{d^2g}{dz^2} = F(z) \tag{2.86}$$

Substituting the boundary condition of Eq. (2.82a) into the above, one obtains

$$\frac{dh}{dr} - r\Omega^2 = rF(0) = rc_1$$

Integrating this equation with respect to r, one gets

$$h(r) = (\Omega^2 + c_1)r^2 + c_2 \tag{2.87}$$

Using Eq. (2.87) in Eq. (2.86), one notices that

$$F(z) = \frac{1}{r}\frac{dh}{dr} - \frac{v_\Phi^2}{r^2} = (\Omega^2 + c_1) - \frac{v_\Phi^2}{r^2} \tag{2.88}$$

Thus, v_Φ/r is a function of z only, i.e.,

$$v_\Phi = r\,G(z) \tag{2.89}$$

Using this notation in Eq. (2.86), one obtains

$$\Omega^2 + c_1 - G^2 = -g^2 - f\frac{dg}{dz} + v\frac{d^2g}{dz^2} \tag{2.90}$$

As v_r and v_z are related to g and f, respectively, one can obtain the required behavior as prescribed in Eqs. (2.82b) and (2.84) as

$$\lim_{y\to\infty}\frac{df}{dy},\ g,\ \frac{dg}{dz},\ \frac{d^2g}{dz^2} = 0 \tag{2.91}$$

Using the conditions given in Eq. (2.91) in Eq. (2.90), one notices

$$\Omega^2 + c_1 = G^2 \quad for\ (z \to \infty)$$

The above right-hand side is a measure of v_Φ for $z \to \infty$. Thus, v_Φ is zero far away from the plate and then $\Omega^2 + c_1 = 0$.

From Eq. (2.90), one obtains

$$-G^2 = v\frac{d^2g}{dz^2} - f\frac{dg}{dz} - g^2 \tag{2.92}$$

Also, from Eqs. (2.85) and (2.87)

$$\frac{p}{\rho} = v\frac{df}{dz} - \frac{f^2}{f} + c_2 \tag{2.93}$$

The right-hand side of the above is a function of z only. It is possible to non-dimensionalize the introduced dependent variables as given in the following

$$f^* = \frac{f}{\sqrt{\nu\Omega}}; \ g^* = \frac{g}{\Omega}; \ G^* = \frac{G}{\Omega}; \ p^* = \frac{p}{\rho\nu\Omega}$$

For the non-dimensional independent variable, $z^* = z/\sqrt{\nu/\Omega}$, collating all the governing equations, we get

$$\frac{df^*}{dz^*} + 2f^* = 0 \tag{2.94}$$

$$\frac{d^2g^*}{dz^{*2}} - f^*\frac{dg^*}{dz^*} - f^{*2} = -G^* \tag{2.95}$$

$$\frac{d^2G^*}{dz^{*2}} - f^*\frac{dG^*}{dz^*} - 2f^*G^* = 0 \tag{2.96}$$

and

$$p^* = \frac{df^*}{dz^*} - \frac{f^{*2}}{2} + \frac{c_2}{\nu\Omega} \tag{2.97}$$

These equations were solved in [68], subject to the following boundary conditions:

$$z^* = 0: \ f^* = 0, \ g^* = 0, \ G^* = 1 \tag{2.98a}$$

and

$$z^* \to \infty: \ g^*, \ \frac{dg^*}{dz^*}, \ \frac{d^2g^*}{dz^{*2}}, \ G^*, \ \frac{dG^*}{dz^*} = 0 \tag{2.98b}$$

Note: (1) As only the first derivative of f^* is involved in the governing equations, only one boundary condition is adequate. A typical solution is shown in Fig. 2.6.

(2) The solution for f^* at $z^* \to \infty$ comes out numerically and indicates the axial flow very far from the rotating disk. The fact that such a velocity in Fig. 2.6 comes out as a negative quantity reveals the physical fact that flow is directed toward the disc from a large distance and is ejected out radially by the rotating disc.

2.3   Boundary Layer Theory

In the previous sections, we have discussed cases where simplifications were made to the governing equations using geometric features or similarity considerations for the flow. This allowed us to obtain a closed form solution. There exists another possibility of simplifying the governing equation based on flow parameters.

For example, at extremely low speed or highly viscous flow problems, it is possible to neglect the non-linear inertia terms in the Navier–Stokes equation, rendering the problem linear. Such flows, representing creeping motion, are called Stokes flow and is equivalent to assuming vanishingly small Reynolds numbers. We will not discuss them here, because such flows do not suffer from instabilities. For such flows, there is a perfect balance between the pressure and viscous terms of the Navier–Stokes equation. Solutions have been found for such cases and these tend to agree with experimental observations, validating the assumptions in simplifying the governing equation.

In contrast, for large Reynolds number flows, a similar reasoning will show the predominance of inertia terms over viscous terms. In the absence of a body force term, Eq. (2.16) then simplifies to

$$\frac{\partial \vec{V}}{\partial t} + \vec{V} \bullet \nabla \vec{V} = -\nabla p / \rho \tag{2.99}$$

This is Euler's equation of inviscid motion. It would be tempting to think that the above equation will reproduce the actual flow behavior for high Reynolds numbers ($Re \to \infty$). However, there are two problems with such simplifications. First, it is really not much of a simplification, as we are still left with an incomplete parabolic differential equation system, which is also non-linear and therefore not amenable to close-form analytic solution. Second, neglecting the viscous terms causes the highest order derivative terms to drop off from the momentum conservation equation, reducing the order of the partial differential equation. Thus, the solution procedure would require satisfaction of fewer boundary conditions as compared to that required for the full Navier–Stokes equation. As one is discarding the viscous term at high Reynolds numbers, it is natural to remove the no-slip boundary condition and retain the impermeability condition to be satisfied. However, this raises a few conceptual problems. While the governing equation depends continuously upon Reynolds number, the boundary conditions do not have any such dependence. Hence, it is paradoxical to talk about removal of boundary conditions depending on Reynolds number, as they continue to be relevant for all Reynolds numbers. Consequently one must retain the viscous terms always, so as to be able to satisfy all boundary conditions. It may happen that the region where viscous effects are important will keep changing depending upon the Reynolds number. But one would never come across a situation where the viscous region will completely

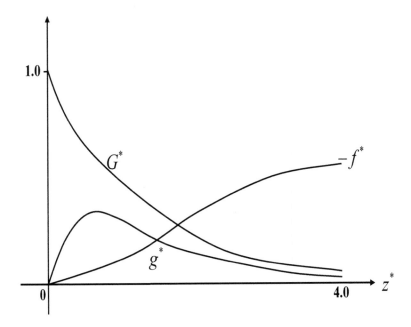

FIGURE 2.6
Solution components (as defined in the text) for the solution of Eqs. (2.94) to (2.97), with G^* as the tangential, $-f^*$ as the axial and g^* as the radial components of induced equilibrium flow.

disappear. For flow problems involving a solid boundary, it is possible that a small region would exist very close to the boundary where viscous effects will be dominant even at high Reynolds numbers. In this section, we will probe the existence of such region(s). For many flow conditions such a region is seen to exist as a thin layer next to the boundary, and it is called the *boundary layer*. Outside this layer the governing equation simplifies to that given in Eq. (2.99).

This major contribution was made by Prandtl [281] at the Third International Math. Congress in Heidelberg and is one of the major cornerstones of the modern edifice of fluid mechanics. He showed the rational way of taking the high Reynolds number limit of the Navier–Stokes equation and simplifying the governing partial differential equation. He propounded the boundary layer theory to show the essential influence of viscous forces in fluid flows at high Reynolds numbers for a certain class of geometries and opened the way for obtaining approximate solutions to the Navier–Stokes equation.

For the Navier–Stokes equation at high Reynolds numbers, as the viscous terms are multiplied by $(1/Re)$ in the non-dimensional form, a thin shear layer forms where these terms remain relevant. In boundary layer theory this is called the inner layer. Also Euler's equation given by Eq. (2.99) plays the role of the *outer solution*. Next we would like to obtain the *inner solution*. The existence of a boundary layer, or more generally a thin shear layer, depends on the smallness of a parameter like $(1/Re)$. In defining Re, we have artificially introduced a single length scale. It is important to note that in many problems multiple length scales exist. It is natural for one to ask which length scale is to be chosen. Once a length scale is chosen, should it adequately represent the flow in all the regions with different length scales?

One should realize that adopting the formalism of mathematical techniques only models an observed physical phenomenon. The existence of a shear layer is in fact related to the structure of the solution which is modeled by the singular perturbation technique. It is much more revealing to note that whenever a thin shear layer exists, it introduces a new length scale in the problem. This length scale is much smaller than the integral dimension of the flow field. One can estimate it from the Navier–Stokes equation by an order of magnitude analysis. Let us say that the integral length L is the scale of the geometry which is kept in an external flow to create the flow field. Thus, it could be the diameter of a cylinder or a chord of an airfoil and this length scale defines the inertial terms in the momentum equation as

$$\vec{V} \bullet \nabla \vec{V} \sim U^2/L \qquad (2.100)$$

The reasoning that the viscous forces are negligible for high Reynolds numbers breaks down in a layer close to the physical body, because the flow develops an internal length scale which is much smaller than L. This is the viscous length scale, δ, such that the viscous forces have the following order of magnitude

$$\nu \nabla^2 \vec{V} \sim \nu U/\delta^2 \qquad (2.101)$$

Inertial and viscous forces will thus be comparable in the boundary layer if the estimates obtained in Eqs. (2.100) and (2.101) are of the same order, i.e.,

$$U^2/L \sim \nu U/\delta^2$$

which simplifies to

$$\frac{\delta}{L} \sim (\frac{UL}{\nu})^{-1/2} = Re^{-1/2} \qquad (2.102)$$

This gives a relative estimate of the two length scales when there exists a shear layer, which progressively thins as the Reynolds number increases. Interestingly enough, this does not tell us about the structure of the shear layer that information has to come from phenomenology. For example, for an infinitesimally thin flat plate placed in an uniform flow U_∞ in the x-direction, Euler's Eq. (2.99) shows that the velocity everywhere above the plate remains unaltered, as if the plate is not there. In an actual physical situation this is what would be perceived too, except very near the flat plate where one would notice a dominant shear stress ($\mu\frac{\partial u}{\partial y}$), as the flow field must adjust itself to satisfy the no-slip boundary condition on the surface ($y = 0$). Thus, on either side of the plate one would see the formation of the shear layer, which is very thin. At the edge of the shear layer, the shear stress must become negligible, thereby showing that inside the shear layer $\frac{\partial^2 u}{\partial y^2}$ is going to be very high. This makes viscous forces dominant inside the shear layer. The flow asymptotically merges to its free stream value, as one approaches the edge of the shear layer. It is necessary to provide a mathematical definition of the shear layer thickness. A common procedure is to prescribe this thickness as that value of y where the streamwise velocity reaches 99% of its free stream value. For a semi-infinite flat plate, there is no natural choice of the integral length scale other than taking it as the distance of any point from the leading edge of the plate. Thus, according to Eq. (2.102), one should have

$$\frac{\delta}{x} \sim \left(\frac{U_\infty x}{\nu}\right)^{-1/2} \tag{2.103}$$

Equation (2.102) or (2.103) indicates the presence of a shear layer at high Reynolds numbers. There is another way of stating the same thing by noting that outside the shear layer $\frac{\partial u}{\partial y}$ is of the same order as $\frac{\partial u}{\partial x}$. But within the shear layer

$$\frac{\partial u}{\partial y} >> \frac{\partial u}{\partial x} \tag{2.104}$$

This is often stated as the *thin shear layer* (TSL) assumption. This makes ample sense when we consider flows without the presence of a solid body and still there is a narrow region inside the flow where the above inequality given by Eq. (2.104), holds, as in mixing layers or jets. The thin shear layer approximation is also often stated as $\frac{d\delta}{dx} << 1$, for a flow in predominantly the x-direction. This discussion is qualitative in nature, without reference to whether the flow is laminar or turbulent. Although the shear layer thickness is larger in turbulent flows as compared to laminar flows, the above definition of the shear layer approximation holds for all flows.

We develop the TSL equation for a two-dimensional steady incompressible flow. Suppose the shear layer that is forming is of negligible curvature so that the Cartesian coordinates can be used to describe the governing equations with x in the flow direction and y normal to the wall. Consider a flow velocity U_e outside the shear layer is prescribed as a function of x. Let the corresponding static pressure be given by P_e. The shear layer has length scales L and δ in the x- and y-directions, respectively. Let the corresponding velocity scales be U and V, by which we will perform an order of magnitude analysis first. Due to the TSL approximation of the kind defined above by Eq. (2.104), it is expected that the pressure gradient in the streamwise and wall-normal directions will be different and one would therefore use the order of magnitude of pressure drop in the x- and y-direction by $\prod$ and Γ, respectively.

Now, we will consider each of the conservation equations, one by one, and label each and every term in these equations with their order of magnitude estimates given below them. For the continuity equation

$$\frac{\partial u}{\partial y} + \frac{\partial u}{\partial x} = 0 \tag{2.105}$$

$$\frac{U}{L} \qquad \frac{V}{\delta}$$

As both the terms are of equal importance, it suggests $V \sim U(\delta/L)$. Thus, the normal velocity component is small compared to the streamwise component. For high Reynolds numbers, one can set $\epsilon = Re^{-1/2}$ and then $V \sim O(\epsilon)$, if $U \sim O(1)$.

The x-momentum equation is given along with its order of magnitude analysis

$$u\frac{\partial u}{\partial x} + v\frac{\partial u}{\partial y} = -\frac{1}{\rho}\frac{\partial p}{\partial x} + \nu\frac{\partial^2 u}{\partial x^2} + \nu\frac{\partial^2 u}{\partial y^2} \tag{2.106}$$

$$\frac{U^2}{L} \qquad \left(\frac{VU}{\delta} \sim \frac{U^2}{L}\right) \qquad \frac{\Pi}{\rho L} \qquad \frac{\nu U}{L^2} \qquad \frac{\nu U}{\delta^2}$$

It is seen that the wall-normal diffusion term is much larger compared to the streamwise momentum diffusion term. Other terms are of same order of magnitude, i.e.,

$$\frac{\Pi}{\rho L} \sim \frac{U^2}{L} \sim \frac{\nu U}{\delta^2} \tag{2.107}$$

The y-momentum equation and the order of magnitude of each term are given below as

$$u\frac{\partial v}{\partial x} + v\frac{\partial v}{\partial y} = -\frac{1}{\rho}\frac{\partial p}{\partial x} + \nu\frac{\partial^2 v}{\partial x^2} + \nu\frac{\partial^2 v}{\partial y^2} \tag{2.108}$$

$$\frac{UV}{L} \qquad \frac{V^2}{\delta} \qquad \frac{\Gamma}{\rho\delta} \qquad \frac{\nu V}{L^2} \qquad \frac{\nu V}{\delta^2}$$

Eliminating V using the order of magnitude analysis in Eq. (2.105), the above order of magnitudes are rewritten as

$$\frac{U^2\delta}{L^2} \qquad \frac{U^2\delta}{L^2} \qquad \frac{\Gamma}{\rho\delta} \qquad \frac{\nu U^2\delta}{L^3} \qquad \frac{\nu U}{L\delta}$$

Thus, in this equation all the terms are of the same importance. At the same time, all the terms are sub-dominant, i.e., smaller compared to the retained x-momentum equation terms. It therefore implies that this equation is trivially satisfied at the high Reynolds number limit. Also, from the order estimates of Eqs. (2.106) and (2.108), it is easy to see that

$$\frac{\Gamma}{\Pi} \sim \left(\frac{\delta}{L}\right)^2$$

that is, the pressure difference across the shear layer is much smaller than the pressure difference in the streamwise direction. Thus, the *pressure in a direction normal to the shear layer is practically constant*. It is said that the pressure is *impressed* upon the shear layer by the outer flow.

Thus, the approximation of the Navier–Stokes equation for high Reynolds numbers reduces it to

$$\frac{\partial u}{\partial x} + \frac{\partial v}{\partial y} = 0 \tag{2.109}$$

$$u\frac{\partial u}{\partial x} + v\frac{\partial v}{\partial y} = -\frac{1}{\rho}\frac{\partial p}{\partial x} + \nu\frac{\partial^2 u}{\partial y^2} \tag{2.110}$$

$$\frac{\partial p}{\partial y} = 0 \tag{2.111}$$

It is customary to drop Eq. (2.111) and convert the partial derivative of the pressure in Eq. (2.110) by an ordinary derivative. Equations (2.109)−(2.111) are approximate equations for the high Reynolds number limit which is valid in the boundary layer. The outer solution is obtained from the equivalent Euler's equation

$$\frac{\partial U_e}{\partial t} + U_e \frac{\partial U_e}{\partial x} = -\frac{1}{\rho} \frac{dp}{dx} \tag{2.112}$$

Thus, in a formal boundary layer solution procedure, one would solve the decoupled equation sets of the outer and the inner solution. The outer solution is provided by Euler's equation, which in the case of a flow past a flat surface is given by Eq. (2.112). Having obtained this solution, it is easy to calculate the pressure distribution at the edge of the shear layer and from this pressure distribution, one can calculate, $\frac{dp}{dx}$ for solving Eqs. (2.109) and (2.110) together for the inner solution. In summarizing this section, we note that the fluid particles acquire vorticity only by viscous diffusion for two-dimensional flows. The no-slip condition is responsible for the creation of vorticity at the wall that is convected and diffused in the shear layer. At high Reynolds numbers the convection predominates over diffusion and hence the vorticity remains confined in a narrow layer next to the wall.

2.4   Control Volume Analysis of Boundary Layers

In the previous section while defining the boundary layer thickness, we took it at $y = \delta$, where $u(y) = 0.99 U_e$. This implies that joining the edge of the shear layers for different x will not yield a streamline, because however far one may move out of the shear layer, the normal component of velocity does not become zero - as a consequence of mass conservation and the application of the no-slip boundary condition. Because of the non-uniform velocity profile, it is possible to define estimates of length scales which would explain defects introduced in the transport of mass, momentum and energy due to the presence of the shear layer. To illustrate it, we consider flow past a flat plate so that the edge velocity at all streamwise locations is given by U_∞. A sketch of the chosen control volume is illustrated in Fig. 2.7. For convenience we consider a coordinate system fixed at the indicated leading edge of the plate. At the inflow, located at the leading edge of the plate, there is a uniform flow with velocity U_∞, entering the control volume through the inflow. The top of the control volume is located far away from the plate so that the velocity at the edge of the control volume is everywhere equal to U_∞. Also, if the top lid is a streamline, then there is no mass flow across this lid. The effect of the plate is to produce a shear at the wall which would be experienced as a frictional drag by the plate. Thus, one of the effects of the shear layer would be to cause a momentum defect of the oncoming flow that gives rise to drag. Also, the velocity will rise monotonically from zero at the wall to its edge value at the top of the control volume. This will cause the streamlines which are parallel to each other before the plate to curve upwards to satisfy mass conservation.

The presence of the shear layer cause mass defect due to shearing of a velocity profile. To account for mass and momentum defects, we will introduce two thicknesses called displacement and momentum thickness, respectively.

2.4.1   Displacement Thickness

If the conservation of mass for the control volume shown in Fig. 2.7 is applied, one gets

$$\iint_{CS} \rho \vec{v} \bullet \hat{n} \, dS = \int_0^{Y_0} \rho u \, dy - \int_0^{Y_i} \rho U_\infty \, dy = 0 \tag{2.113}$$

Simplifying and reorganizing, this can be written as

$$U_\infty (Y_0 - Y_i) = \int_0^{Y_0} (U_\infty - u) \, dy \tag{2.114}$$

The displacement thickness (δ^*) signifies the amount by which the boundary is *displaced* outwards due to the presence of the shear layer. Thus, $Y_0 - Y_i = \delta^*$ and therefore the mathematical definition of boundary layer displacement thickness is given by

$$\delta^* = \int_0^\infty \left(1 - \frac{u}{U_\infty}\right) dy \tag{2.115}$$

Note the upper limit in Eq. (2.115) has been changed, as the integrand is zero outside the shear layer. Second, the above analysis is for the case of a flat plate experiencing no pressure gradient. In the presence of a pressure gradient also, we will use the same expression with U_∞ replaced by U_e, the boundary layer edge velocity. Finally, note that the boundary layer displacement thickness is a function of x only.

2.4.2   Momentum Thickness

For the control volume shown in Fig. 2.7, if one applies conservation of momentum for a steady two-dimensional flow, one gets

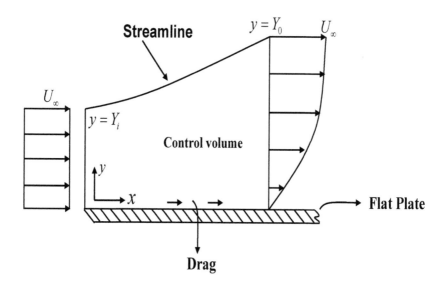

FIGURE 2.7
Control volume for flow past a plate to show the displacement effect. Note that the top segment is a streamline.

$$\iint_{\delta\Omega} \hat{n}_x \, \vec{\vec{\sigma}} \, dS = -D = \int_{\delta\Omega} (\rho\vec{v}) \, u \, dy = \int_0^{Y_0} u \, \rho \, u \, dy - \int_0^{Y_i} U_\infty \, \rho \, U_\infty \, dy \qquad (2.116)$$

where D is the drag experienced by the flat plate. The above can be simplified using Eq. (2.116), by noting $Y_i = \int_0^{Y_0} \frac{u}{U_\infty} \, dy$, to

$$D = \int_0^{Y_0} \rho \, u \, (U_\infty - u) \, dy \qquad (2.117)$$

Dividing both sides by ρU_∞^2, one notices that the resultant quantity has a dimension of length. We define this quantity as the momentum thickness (θ) and is thus expressed using $Y_0 \to \infty$ as

$$\theta = \frac{D}{\rho \, U_\infty^2} = \int_0^{Y_0} \frac{u}{U_\infty} \left(1 - \frac{u}{U_\infty} \right) dy \qquad (2.118)$$

Like the displacement thickness, the momentum thickness definition is also extended to flows in the presence of a pressure gradient by replacing U_∞ by the shear layer edge velocity (U_e). Comparing the expressions for displacement and momentum thicknesses as given by Eqs. (2.115) and (2.118), it is obvious that the latter is always smaller than the former, as the integrand of Eq. (2.115) is multiplied by $\frac{u}{U_\infty}$ to get the integrand for Eq. (2.118), a quantity which is always less than equal to one for all heights. The ratio of these two thicknesses is called the *Shape Factor* (H) and defined as

$$H = \frac{\delta^*}{\theta} > 1 \qquad (2.119)$$

As θ is a measure of the skin friction drag experienced by a shear layer, it is apparent that H for a turbulent boundary layer would be significantly lower than its value for laminar flows.

2.4.3 Separation of a Steady Boundary Layer

It is possible to comment about the properties of the boundary layer equation obtained. One observation stems from the fact that the boundary layer equations, as given by Eqs. (2.109)−(2.111), bring in qualitative change to the mathematical property of the solution. The original problem of solving the Navier–Stokes equation is one of solving an initial-boundary value problem. This transforms to a marching problem, where one solves the boundary layer equation. Thus, instead of solving a problem involving all the points of the domain in the (x, y)-plane, one marches in the streamwise direction. It was a virtual revolution when this idea was put forth and it allowed many problems to become amenable to solution despite nonlinearity, which otherwise were intractable. However, such a marching procedure is doomed to fail if the flow develops a tendency to go in the reverse direction. This is understandable and one would note the equations becoming singular at those points. This is termed boundary layer separation. There are other complementary views of the phenomenon which we will not go through. Instead, we give a qualitative picture of the phenomenon, based partly on the physical implication and partly on the mathematical description of such a flow field. Physically, boundary layer separation occurs if the inertia of fluid particles is retarded by the joint action of viscous traction and the present pressure gradient. Usually, such a scenario occurs when the pressure is increasing with distance downstream. Following Eq. (2.112) for steady outer flow, this happens when

$$\frac{dU_e}{dx} < 0 \qquad (2.120)$$

A pressure of this type creates an adverse pressure gradient as opposed to a favorable pressure gradient which ameliorates any tendency of the flow to separate. Thus, flow separation is natural where the stream tube cross section area keeps increasing, as on the lee side of a bluff body or as in a diffuser.

Qualitatively it is explained that the presence of a pressure gradient causes the particles to lose their kinetic energy as they travel downstream. This adverse effect has its detrimental behavior prominently displayed close to the solid surface where the particles already have less kinetic energy because of viscous action. The pressure gradient can thus bring such particles to a halt and further downstream the applied pressure can push the fluid close to the wall in the upstream direction. This causes the wall streamline to break away from the physical surface. The point where the flow shows an incipient tendency to break away from the surface is called the point of separation. A heuristic argument of this kind led Prandtl to set the following criterion for incipient separation of steady flow as

$$\tau_{\text{wall}} = 0 \tag{2.121}$$

From the boundary layer equation, Eq. (2.110) applied at the wall gives

$$\left(\frac{\partial^2 u}{\partial y^2} \right)_{\text{wall}} = \frac{1}{\mu} \left(\frac{dp}{dx} \right) \tag{2.122}$$

The curvature of the velocity profile at the wall is determined by the impressed pressure gradient upon the shear layer. It is also known that as one approaches the edge of the shear layer, the shear (τ) monotonically decreases to zero. Thus, at the edge of the shear layer ($y \to \infty$)

$$\left(\frac{\partial \tau}{\partial y} \right)_{\infty} < 0, \quad \text{i.e.,} \quad \left(\frac{\partial^2 u}{\partial y^2} \right)_{\infty} < 0 \tag{2.123}$$

Equations (2.122) and (2.123) provide the velocity profile curvature at the wall and free stream, respectively. With this information one can look at different flow types.

2.4.3.1   Accelerated Flows ($dp/dx < 0$)

This type of flow has been termed favorable pressure gradient flow. Here, the velocity profile and its first two derivatives are sketched as shown in Fig. 2.8. Note that the velocity profile shows a monotonic growth with distance from the wall and the shear stress decays monotonically from its wall value to a vanishingly small value at the free stream. For the pressure gradient associated with the accelerated flows the curvature at the wall is negative as indicated in Fig. 2.8. The same curvature is negative and negligibly small at the free stream. This variation of $\frac{\partial^2 u}{\partial y^2}$ is shown where it is negative across the whole of the shear layer. Thus, from the middle panel, it is clear that such flows do not suffer separation as τ_w is nowhere zero.

2.4.3.2   Retarded Flow ($dp/dx > 0$)

This is also called the adverse pressure gradient flow. The velocity profile and its first two derivatives are as shown in Fig. 2.9. Here, the applied adverse pressure gradient tends to slow the flow within the shear layer. Here, $\frac{\partial^2 u}{\partial y^2}$ is positive at the wall, while it is negative in the free stream. Thus, there must be a location within the shear layer where $\frac{\partial^2 u}{\partial y^2} = 0$. Such a point is called the *point of inflection*. The existence of such a point decides importantly on the instability of such velocity profiles as will be seen in Chap. 3. The location of the *inflection point* is associated with the maximum of the shear stress noted in the flow, as shown in the middle sketch. Here, the velocity profile is sketched for a case when the flow has not separated. It should, however, be pointed out that the existence of a point where $\frac{\partial^2 u}{\partial y^2} = 0$ is not the only criterion for flow instability. Because for zero pressure gradient flows, the curvature of the profile is zero at the wall, still it is not unstable for all Reynold numbers.

As the pressure gradient becomes more adverse, the wall shear (τ_w) approaches zero and the velocity profile becomes steeper near the wall. Any pressure gradient which is more than necessary to cause separation causes the velocity profile to take an S-shape near the wall. A persistent adverse pressure gradient makes the velocity profile more and more S-shaped as one travels downstream. In Fig. 2.10, a set of representative velocity profiles is shown during boundary layer separation along with the associated streamline line pattern. The above description of flow separation is representative of flow separation for steady laminar flows. The criterion given by Eq. (2.121) holds to decide whether a flow has separated or

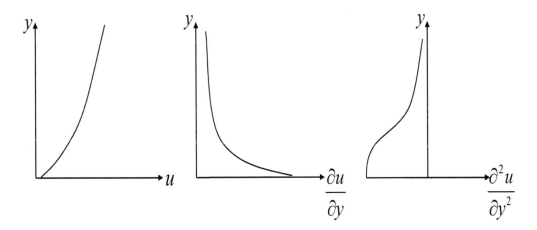

FIGURE 2.8
Velocity, shear and second derivative profile for an accelerated flow.

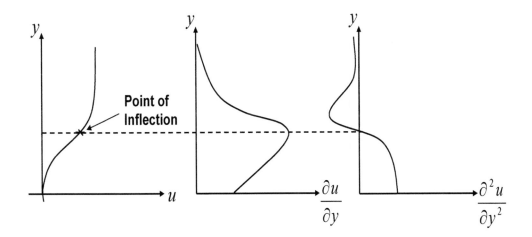

FIGURE 2.9
Velocity, shear and second derivative profile for a retarded flow with the point of inflection marked.

not. This is an empirical model for predicting separation. Later on we will redefine this criterion in terms of a stability property and the existence of upstream propagating modes. Also, it must be remembered that the empirical criterion of Eq. (2.121) does not hold for unsteady and turbulent flows.

2.5   Numerical Solution of the Thin Shear Layer (TSL) Equation

We have already noted in the last section that thin shear layer approximation makes the incompressible flow governing equation *parabolic* from *elliptic* partial differential equation. In physical terms, this implies that the solution, at a given location, can be marched in the direction of shear layer growth. This, in fact, allows converting the partial differential equation into an ordinary differential equation for unseparated flows. This process makes use of the similarity of velocity profiles in an appropriate coordinate system. This was originally performed by Blasius, as described in [312].

It is to be noted that even when the velocity profiles are not similar, *similarity transformations* can be used for the numerical solution of non-similar boundary layers, as using the Falkner–Skan transformation [97]. However, we briefly introduce the basic ideas of similarity transformation first.

Consider a two-dimensional laminar flow in which the boundary layer grows. We can write the solution of the boundary layer equations, given by Eqs. (2.109)–(2.111), in a general form by

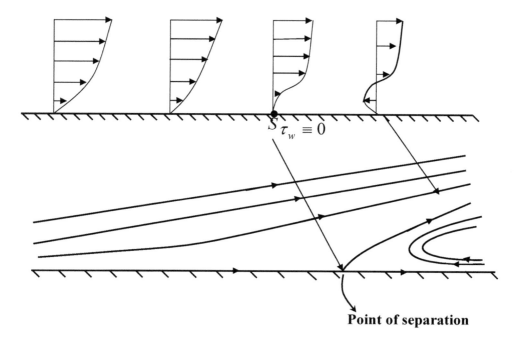

Point of separation

FIGURE 2.10
Sketch of a velocity profile under the influence of an adverse pressure gradient, with S as the point of separation, for steady flow ($\tau_w = 0$). On the bottom are shown the streamlines.

$$\frac{u}{U_e} = \Phi\left(x, y\right) \tag{2.124}$$

There are special cases of flows for which Eq. (2.124) can be rewritten as a function of a single variable, i.e.,

$$\frac{u}{U_e} = \Phi\left(\eta\right) \tag{2.125}$$

where η is the similarity variable and is a special function of x and y. Such velocity profiles, when they exist, are called *similarity or affine profiles*. In Eq. (2.124) or (2.125), use of U_e does not restrict similarity profiles only for wall bounded flows. For example, for jets, one can conveniently use the jet center-line velocity (U_c) as U_e.

Similarly for wakes, we can rewrite Eq. (2.125) as

$$\frac{U_c - u}{\Delta U} = \Phi\left(\eta\right) \tag{2.126}$$

where $\Delta U = U_c - U_e$.

Without going through the detailed steps (which can be found in [55]), we introduce a new independent variable

$$\eta = \left(\frac{U_e}{\nu x}\right)^{1/2} y \tag{2.127}$$

and a non-dimensional stream function $f(\eta)$ through the following

$$\psi = (U_e \nu x)^{1/2} f(\eta) \tag{2.128}$$

These transformations were used by Blasius in his analysis of boundary layers for the first time and the details are in [312]. These transformations, as given by Eqs. (2.127) and (2.128), are an illustration of a similarity transform which exists when the shear layer grows in a constant pressure gradient or in a zero pressure gradient environment.

For non-similar flows, the shear layer has explicit dependence on streamwise location, in addition to the similarity variable given by Eq. (2.127). For such flows, it is customary to use the following transformation for the stream function

$$\psi = (U_e \nu x)^{1/2} f(x, \eta) \tag{2.129}$$

In solving the TSL equation, we will thus be working in the non-orthogonal (x, η)-plane instead of the Cartesian frame. The derivatives in these two planes are related by

$$\left(\frac{\partial}{\partial x}\right)_y = \left(\frac{\partial}{\partial x}\right)_\eta + \frac{\partial \eta}{\partial x}\left(\frac{\partial}{\partial \eta}\right)_x \tag{2.130a}$$

and

$$\left(\frac{\partial}{\partial y}\right)_x = \frac{\partial \eta}{\partial y}\left(\frac{\partial}{\partial \eta}\right)_x \tag{2.130b}$$

The Cartesian components of velocity are then given by

$$u = \left(\frac{\partial \psi}{\partial y}\right)_x = (U_e \nu x)^{1/2}\left(\frac{U_e}{\nu x}\right)^{1/2} f' = U_e f' \tag{2.131a}$$

and

$$-v = \left(\frac{\partial \psi}{\partial x}\right)_y = \left[f \frac{d}{dx}(U_e \nu x)^{1/2} + (U_e \nu x)^{1/2} \frac{\partial f}{\partial x}\right] + (U_e \nu x)^{1/2} f' \frac{\partial \eta}{\partial x} \tag{2.131b}$$

For brevity of notation, we have indicated a derivative with respect to η by a prime. Similarly

$$\frac{\partial^2 \psi}{\partial y^2} = U_e \left(\frac{U_e}{\nu x}\right)^{1/2} f'' \tag{2.132a}$$

$$\frac{\partial^3 \psi}{\partial y^3} = U_e \left(\frac{U_e}{\nu x}\right) f''' \tag{2.132b}$$

Since $\frac{\partial \eta}{\partial x} = -\frac{\eta}{2x}$, the following quantities are estimated in the transformed plane to be substituted in the TSL equation of Eq. (2.110)

$$v = -\frac{\partial}{\partial x}\left[(U_e \nu x)^{1/2} f\right] + \frac{\eta}{2}\left(\frac{U_e \nu}{x}\right)^{1/2} f'$$

and

$$\frac{\partial u}{\partial x} = U_e \frac{\partial f'}{\partial x} + f' \frac{dU_e}{dx} - \frac{\eta U_e}{2x} f''$$

Therefore

$$u \frac{\partial u}{\partial x} = U_e^2 f' \frac{\partial f'}{\partial x} + U_e f'^2 \frac{dU_e}{dx} - \frac{\eta U_e^2}{2x} f' f'' \tag{2.132c}$$

$$v \frac{\partial u}{\partial y} = \frac{U_e \eta}{2x} f' f'' - U_e^2 f'' \frac{\partial f}{\partial x} - \frac{U_e^2}{2x} f f'' - \frac{U_e}{2} \frac{dU_e}{dx} f f'' \tag{2.132d}$$

Thus, the convection terms of Eq. (2.110) are added together as

$$u \frac{\partial u}{\partial x} + v \frac{\partial u}{\partial x} = U_e^2 f' \frac{\partial f'}{\partial x} + U_e \frac{dU_e}{dx} f'^2 - U_e^2 f'' \frac{\partial f}{\partial x} - \frac{U_e^2}{2x} f f'' - \frac{U_e}{2} \frac{dU_e}{dx} f f'' \tag{2.133}$$

The diffusion and pressure gradient terms are similarly simplified, so that the right-hand side of Eq. (2.110) is given by

$$= U_e \frac{dU_e}{dx} + \frac{U_e^2}{x} f''' \tag{2.134}$$

Thus, the TSL equation is

$$\cdot \quad U_e^2 f' \frac{\partial f'}{\partial x} + U_e \frac{dU_e}{dx} f'^2 - U_e^2 f'' \frac{\partial f}{\partial x} - \left(\frac{U_e^2}{2x} + \frac{U_e}{2} \frac{dU_e}{dx}\right) f f''$$

$$= U_e \frac{dU_e}{dx} + \frac{U_e^2}{x} f''' \tag{2.135}$$

Multiplying both sides of Eq. (2.135) by x/U_e^2, we get

$$x f' \frac{\partial f'}{\partial x} + \frac{x}{U_e} \frac{dU_e}{dx} f'^2 - x f'' \frac{\partial f}{\partial x} - \left(\frac{1}{2} + \frac{x}{2U_e} \frac{dU_e}{dx}\right) f'' = \frac{x}{U_e} \frac{dU_e}{dx} + f''' \tag{2.136}$$

It has been shown in [97] that a similarity solution exists if the outer inviscid flow can be represented by

$$U_e = Cx^m \tag{2.137}$$

Such an outer flow is achieved by many flows, including the wedge flow (see Cebeci [57] for a general treatment of such outer flows), where the wedge angle is given by $\frac{\beta\pi}{2}$ and $\beta = \frac{2m}{m+1}$. For outer flows given by Eq. (2.137), one can set

$$\frac{x}{U_e}\frac{dU_e}{dx} = m$$

Using this in Eq. (2.136) one gets the TSL equation as

$$f''' + m(1 - f'^2) + \frac{m+1}{2}ff'' = x\left(f'\frac{\partial f'}{\partial x} - f''\frac{\partial f}{\partial x}\right) \tag{2.138}$$

It is noted that Eq. (2.138) is not restricted to wedge flow only. Any outer flow edge velocity distribution has an equivalent m, which can be used in Eq. (2.138) to analyze the associated TSL.

To solve Eq. (2.138), one needs to satisfy the following boundary conditions.

At $y = 0$:

$$u = 0; \quad \text{and} \quad v = v_w(x) \tag{2.139a}$$

where $v_w(x)$ is the prescribed wall transpiration velocity. As the edge of the shear layer is approached, it should satisfy the following

$$u = U_e \quad \text{as} \quad y \to Y_e \tag{2.139b}$$

These boundary conditions translate into the following for the integration of Eq. (2.138) in the (x, η)-plane for $\eta = 0$:

$$f' = 0 \quad \text{and} \quad f_w = -\frac{1}{(U_e\nu x)^{1/2}}\int_0^x v_w\, dx \tag{2.140a}$$

And for $\eta \to \eta_e$:

$$f' = 1 \tag{2.140b}$$

A few words are necessary for the solution of Eq. (2.138) subject to the boundary values of Eqs. (2.140a) and (140b). For external flows, one starts off from the stagnation point with a solution which corresponds to that obtained in Sec. 2.2.4 for steady stagnation point flow. Then the TSL can be solved along the upper and lower surfaces, starting from the stagnation point in solving Eq. (2.138). The right-hand side is discretized by one-sided formulae, due to the parabolic nature of Eq. (2.138).

2.5.1 Falkner–Skan Similarity Profile

If the flow discussed in the previous section is truly similar, then f is not an explicit function of x and the right-hand side of Eq. (2.138) is identically equal to zero. The governing differential equation for such a similar flow is given by

$$f''' + m(1 - f'^2) + \frac{m+1}{2}ff'' = 0 \tag{2.141}$$

Equation (2.141) was originally developed in [97] for similar flow past wedges. It is noted that for similar flows, the boundary conditions as given by Eq. (2.140a), cannot be dependent on x and thus f_w must be a constant.

2.5.2   Separation Criterion for Wedge Flow

From the definition of m, it can be seen that this criterion is related to the local pressure gradient. For decelerating flows, m takes negative values until $f''_w = 0$, i.e., at that value the local wall shear becomes zero, signifying flow separation and thus ending the validity of Eqs. (2.138) and (2.141). It has been found that $m = -0.0904$ represents the value for which the TSL separates.

2.5.3   Blasius Profile

One can obtain the zero pressure gradient case equation that was obtained by Blasius for the first time. This is obtained by setting $m = 0$ in Eq. (2.141). Thus, the governing equation for the zero pressure gradient similarity profile is given as the solution of

$$f''' + \frac{1}{2} f\, f'' = 0 \qquad\qquad (2.142)$$

This represents flow past a flat plate at zero angle of attack. For similarity to hold, the solution should be sought at larger x, away from the leading edge of the plate, where the solution is independent of x.

2.5.4   Hiemenz or Stagnation Point Flow

This case was discussed in Sec. 2.2.4, where it was shown that the external flow turns by 90^0 at the stagnation point. Thus, this case can be considered as a special case of flow described by Eq. (2.141). Here, for the specified flow turning, one must have $\beta = 1$ and thus $m = 1$. From Eq. (2.141), one gets the governing differential equation for stagnation point flow as

$$f''' + f\, f'' - f'^2 + 1 = 0 \qquad\qquad (2.143)$$

Compare this equation with Eq. (2.72) and its simplified form for Φ_1 developed for stagnation point flow from the first principle. In this case, the external or outer flow has $U_e = C\,y$ and $V_e = -Cx$ (note that x and y interchange for stagnation point flow).

2.6   Laminar Mixing Layer

The mixing layer is a typical example of free shear flows which exhibit the presence of shear in the absence of solid walls. The shear is a consequence of velocity gradients created by an upstream mechanism, like the mixing of two dissimilar flows, as discussed for analyzing Kelvin–Helmholtz instability in Chap. 1, as an example of inviscid instability. In the presence of viscous action, discontinuity prior to the process of mixing will be smoothed out.

Let us consider mixing of flow of different streams of the same fluid with identical transport properties. In Fig. 2.11, a typical sketch of a mixing layer arrangement and coordinate system is shown. Two parallel streams meet at $x = 0$ and as the flow convects, profile on the right is sketched for a downstream station.

Here also we look for a similarity solution for the shear layer whose governing equation is given by

$$\frac{\partial u}{\partial x} + \frac{\partial v}{\partial y} = 0 \qquad\qquad (2.144)$$

and

$$u\frac{\partial u}{\partial x} + v\frac{\partial v}{\partial y} = \nu\frac{\partial^2 u}{\partial y^2} \tag{2.145}$$

subject to the boundary conditions

$$y \to \infty : \quad u \to U_1 \tag{2.146a}$$

$$y \to -\infty : \quad u \to U_2 \tag{2.146b}$$

To solve Eqs. (2.144)-(2.145) subject to the boundary conditions (2.146a,b), we introduce a stream function which will automatically satisfy Eq. (2.144)

$$\psi(x, y) = U_1 \delta(x) f(\eta) \tag{2.147}$$

The non-dimensional stream function f depends upon the similarity variable

$$\eta = \frac{y}{\delta(x)} \tag{2.148}$$

where $\delta(x)$ is some hitherto unspecified length scale. For convenience the velocity scale has been chosen as U_1. From Eqs. (2.147) and (2.148), noting that $\frac{\partial \eta}{\partial x} = -\frac{\eta}{\delta}\frac{d\delta}{dx}$, we obtain

$$u = U_1 f'(\eta) \tag{2.149a}$$

and

$$v = U_1 \left[f'\eta - f \right] \frac{d\delta}{dx} \tag{2.149b}$$

It is noted that the choice of x-axis does not have to be the line of symmetry at all stations. Using the expressions for the velocity components in Eq. (2.149), we obtain the following

$$\frac{\partial u}{\partial x} = -\frac{U_1}{\delta}\eta\frac{d\delta}{dx}f''$$

$$\frac{\partial u}{\partial y} = \frac{U_1}{\delta}f''$$

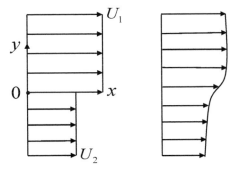

FIGURE 2.11
Formation of a laminar mixing layer starting from two uniform flows with velocity U_1 and U_2, shown on the left.

$$\frac{\partial^2 u}{\partial y^2} = \frac{U_1}{\delta^2} f'''$$

Using the above in the shear layer equation, Eq. (2.145), and simplifying, we obtain

$$f''' + \frac{U_1 \delta}{\nu} \frac{d\delta}{dx} f \, f'' = 0 \qquad (2.150)$$

Thus, to obtain the similarity solution we must have

$$\frac{U_1 \delta}{\nu} \frac{d\delta}{dx} = C \text{ (const.)}$$

Upon integration one obtains

$$\frac{\delta^2}{2} = C\left(\frac{\nu x}{U_1}\right)$$

If we arbitrarily choose $C = 1/2$, then

$$\delta = \left(\frac{\nu x}{U_1}\right)^{1/2} \qquad (2.151)$$

Thus, to obtain the similarity solution we must solve the following equation

$$f''' + \frac{1}{2} f \, f'' = 0 \qquad (2.152)$$

This is a third order ordinary differential equation and we would require three boundary conditions. Out of the three, the following conditions apply at the edges of the shear layer on either side as

$$\eta = \eta_e : \quad f' = 1 \qquad (2.153a)$$

and

$$\eta = -\eta_e : \quad f' = \frac{U_2}{U_1} = \lambda_1 \text{ (a parameter)} \qquad (2.153b)$$

The third boundary condition can be chosen as a zero shear value at the top edge of the computing domain, i.e.,

$$\eta = \eta_e : \quad f'' = 0 \qquad (2.153c)$$

2.7   Plane Laminar Jet

We consider the case of a plane two-dimensional jet which emerges from a long (in the spanwise direction) narrow slit and mixes with the surrounding fluid. As the jet emerges into still ambient fluid at constant pressure, the jet grows in a zero pressure gradient ambiance. The governing equation for the shear layer is the same as in Eqs. (2.144) and (2.145). The jet must also have constant momentum flux across any streamwise location, i.e.,

$$J = \rho \int_{-\infty}^{\infty} u^2 dy = \text{constant} \qquad (2.154)$$

Equation (2.154) is obtained by a control volume analysis performed in Sec. 2.4.2 considering no drag for the zero pressure gradient condition. The emerging jet entrains surrounding fluid which is at rest. The entrainment is due to the action of shear stress in a real fluid. If we place the x-axis along the jet center-line, then the necessary boundary conditions to solve Eqs. (2.144)-(2.145) are at $y = 0$:

$$v = 0 \text{ and } \frac{\partial u}{\partial y} = 0 \tag{2.155a}$$

and as $y \to \pm\infty$:

$$u = 0 \tag{2.155b}$$

Here also we look for the similarity solution via a non-dimensional stream function

$$\psi = U_c(x)\,\delta(x)\,f(\eta) \tag{2.156}$$

where U_c is the jet center-line velocity at each station and δ is a length scale, which could be conveniently considered as the jet width. Here η is defined in the same way as it was in Eq. (2.148). Here $u = U_c f'$ and hence

$$J = 2\rho M \int_0^\infty f'^2 \, d\eta \tag{2.157}$$

where $M = U_c^2(x)\,\delta(x)$. As the momentum flux (J) is constant, so will be M. Using Eq. (2.156) in Eq. (2.145) and simplifying, one gets

$$\frac{U_c\delta}{2\nu}\frac{d\delta}{dx}\left[f'^2 + f\,f''\right] = -f''' \tag{2.158}$$

Again, to obtain the similarity solution we require

$$\frac{U_c\delta}{2\nu}\frac{d\delta}{dx} = C \text{ (const.)} \tag{2.159}$$

Putting this constant equal to 1, we get the governing equation for the similarity problem as

$$f'^2 + f\,f'' + f''' = 0 \tag{2.160}$$

subject to the boundary conditions

$$\eta = 0: \quad f = f'' = 0 \tag{2.161a}$$

$$\eta = \eta_e: \quad f' = 0 \tag{2.161b}$$

From Eq. (2.159)

$$\frac{U_c\delta}{2\nu}\frac{d\delta}{dx} = 1 \tag{2.162}$$

Since $M = U_c^2\delta$, Eq. (2.162) can be rewritten as

$$\frac{\sqrt{\delta M}}{2\nu}\frac{d\delta}{dx} = 1$$

which upon integration provides the following estimate of jet width as

$$\delta = \left(\frac{9\nu^2 x^2}{M}\right)^{1/3}$$

Thus the jet width varies as $x^{2/3}$ and the jet center-line velocity is estimated as

$$U_c = \sqrt{\frac{M}{\delta}} = \frac{M^{2/3}}{(3\nu x)^{1/3}}$$

Thus the jet center-line velocity varies as $x^{-1/3}$. Equation (2.160), although non-linear, can be integrated directly. Integrating it once, we get

$$f'' + f f' = C_2 \text{ (constant)} \tag{2.163}$$

Using the boundary conditions of Eq. (2.161a) in Eq. (2.163) yields $C_2 = 0$. Integrating Eq. (2.163) and using the above we get

$$f = \sqrt{2} \tanh\left(\frac{\eta}{\sqrt{2}}\right) \tag{2.164}$$

Using this solution in Eq. (2.157), we get $M = \frac{3}{4\sqrt{2}}\left(\frac{J}{\rho}\right)$. Finally, we get the jet width and jet center-line velocity (in terms of the jet momentum flux) as

$$\delta = \left(\frac{12\sqrt{2}\,\nu^2 x^2}{J/\rho}\right)^{1/3} \tag{2.165}$$

$$U_c = \left[\frac{3}{32}\left(\frac{J}{\rho}\right)^2 \frac{1}{\nu x}\right]^{1/3} \tag{2.166}$$

This solution can also be used to investigate non-similar jets by using

$$\eta = \frac{y}{\delta} = \frac{y}{x^{2/3}}\left(\frac{J/\rho}{12\sqrt{2}\,\nu^2}\right)^{1/3} \tag{2.167}$$

$$\psi = \left(\frac{9\sqrt{2}}{8}\frac{J\nu x}{\rho}\right)^{1/3} f(x, \eta) \tag{2.168}$$

It is left as an exercise to show the governing equation for the non-similar case as

$$f'^2 + f\,f'' + f''' = 3\xi\left(f'\frac{\partial f'}{\partial \xi} - f''\frac{\partial f}{\partial \xi}\right) \tag{2.169}$$

where $\xi \equiv x$.

2.8   Issues of Computing Space–Time Dependent Flows

It is clear that the vast multitude of flow phenomena observed in nature are governed by conservation laws which could be stated by five mathematical equations. Different flow behaviors which we perceive for different flows are thus not due to different governing principles — differences arise due to boundary and initial conditions. In general, one mostly comes across either Dirichlet or Neumann conditions. In convective heat transfer problems, one encounters Robin or mixed conditions. One notes that derivative boundary conditions are a major source of numerical errors, as computationally representing a derivative is always associated with approximation, which leads to loss of information.

In computing governing equations accurately for transitional and turbulent flows, one has to consider together space-time dependence of the problem together. This aspect of computing is very often overlooked, where spatial and temporal discretization is often decoupled, which can lead to serious dispersion errors. This has been demonstrated in [335, 349].

2.8.1   Waves — Building Blocks of a Disturbance Field

Before discussing dispersion, we highlight aspects of disturbance evolution which takes a flow from the laminar to the turbulent state. Irrespective of the mechanism being linear or non-linear, it is always possible to explain disturbance as being composed of Fourier–Laplace transforms. For space-time dependent systems, such disturbances can often be viewed as plane waves. Waves can be dispersive or non-dispersive, governed by hyperbolic or non-hyperbolic partial differential equations.

The prototype of hyperbolic waves is often taken as the multiple dimension convection equation

$$u_{tt} = c^2 \, \nabla^2 u \tag{2.170}$$

where u can represent any disturbance quantity, e.g., the displacement of an interface perturbation of pressure or density. If the disturbance propagates only in the x-direction, Eq. (2.170) can be described by

$$u_{tt} = c^2 \, u_{xx} \tag{2.171}$$

Consider the propagation of disturbances subject to initial conditions

$$u(x,0) = f(x) \quad \text{and} \quad u_t(x,0) = gg(x) \quad \text{for} \quad -\infty < x < \infty \tag{2.172}$$

The unbounded spatial domain makes this a Cauchy problem. In Eq. (2.172), $f(x)$ and $gg(x)$ are considered continuous functions. Introducing two new independent variables

$$\xi = x + ct \quad \text{and} \quad \eta = x - ct \tag{2.173}$$

in Eq. (2.171) and directly solving, one gets

$$u = F(\xi) + G(\eta)$$

where F and G are arbitrary, twice continuously differentiable functions. This leads to the well-known D'Alembert solution of the wave equation

$$u(x,t) = F(x + ct) + G(x - ct) \tag{2.174}$$

Using this general solution in the initial conditions of Eq. (2.172), one obtains

$$u(x,t) = \frac{1}{2} \left[f(x+ct) + f(x-ct) + \frac{1}{c} \int_{x-ct}^{x+ct} gg(y)\, dy \right] \tag{2.175}$$

The straight lines $\xi = x + ct = $ constant and $\eta = x - ct = $ constant are the characteristics of the waves for Eq. (2.171), representing two branches of solution, one going upstream and another going downstream, with respect to the origin. Thus, these are also called left- and right-running waves, respectively.

The case of zero initial velocity $[gg(x) \equiv 0]$ is of special interest. It corresponds to an initial displacement of the surface $f(x)$, which is then left to propagate, and the general solution reduces to

$$u(x,t) = \frac{1}{2}\left[f(x+ct) + f(x-ct)\right] \tag{2.176}$$

The nature of this solution indicates that half of the initial disturbance propagates to the right and the other half propagates to the left. The characteristic amplitude is half of the initial amplitude and the shape of both these solutions is exactly identical to the initial condition. Note the phase variation of these two branches of solution indicate propagating plane waves.

The building blocks of any arbitrary aggregation of plane waves can be understood by defining certain wave parameters for a single-periodic function

$$u(x,t) = a\sin\left[\frac{2\pi}{\bar{\lambda}}(x-ct)\right] \tag{2.177}$$

One can identify this as a specific solution for Eq. (2.176) which has a non-zero right-running wave solution. In the above, the quantity in the square brackets represents the phase of the wave and a represents the amplitude of the wave. The quantity $\bar{\lambda}$ is the wavelength, since u does not change when x is changed by $\bar{\lambda}$, with t held fixed. One defines wavenumber $k(=\frac{2\pi}{\bar{\lambda}})$, which provides the number of full waves in a length 2π. Thus, the representation of Eq. (2.177) can be alternately written as

$$u(x,t) = a\sin\left[k(x-ct)\right] \tag{2.178}$$

Keeping one's gaze fixed at a single point, the least time after which $u(x,t)$ retains the same value determines the time period T, and this is also the time required for the wave to travel one wavelength: $T = \frac{\bar{\lambda}}{c}$. The number of oscillations at a point per unit time is the frequency given by $f_0 = \frac{1}{T}$. One can define a circular frequency $\bar{\omega}$ by noting

$$\bar{\omega} = kc \tag{2.179}$$

Thus, $c\,(=\bar{\omega}/k)$ has a dimension of speed and is appropriately called the phase speed, the rate at which the phase of the wave propagates. Such movement is not always physical and most often illusory. Equation (2.179) is known as the physical dispersion relation for obvious reasons.

2.8.2   Plane Waves

We have defined the wave parameters for a one-dimensional plane wave above. A three-dimensional plane wave can be written in the general form as

$$u(\vec{r},t) = a\sin(kx + ly + mz - \bar{\omega}t)$$

$$= a\sin(\vec{\mathbf{k}}\cdot\vec{r} - \bar{\omega}t) \tag{2.180}$$

where the wavenumber vector $\vec{\mathbf{k}}$ has components k, l and m, such that $|\vec{\mathbf{k}}| = (k^2 + l^2 + m^2)^{\frac{1}{2}}$ and the wavelength for the planar wave of Eq. (2.177) is given by $\bar{\lambda} = 2\pi/|\vec{\mathbf{k}}|$.

Note that the phase speed for the plane wave is given by

$$c = \frac{\bar{\omega}}{|\vec{\mathbf{k}}|}\hat{e}_k \tag{2.181}$$

where the unit vector in the wave propagation direction is given by $\hat{e}_k = \vec{\mathbf{k}}/|\vec{\mathbf{k}}|$. In general, the phase speed components are given as

$$c_x = \frac{\bar{\omega}}{k}, \quad c_y = \frac{\bar{\omega}}{l} \quad \text{and} \quad c_z = \frac{\bar{\omega}}{m}$$

We note the curious feature that $|c_x|, |c_y|, |c_z| > |c|$, i.e., the components are larger than the resultant and vector rules of addition and subtraction do not apply to phase speed.

2.9 Wave Interaction: Group Velocity and Energy Flux

In developing linear theories, it is often assumed that each harmonic can be studied individually and effects can be superposed. This has been termed normal mode analysis in linear stability theory. This has a serious lacuna and is discussed next in the context of two neighboring modes interacting with each other.

If we look at the disturbance energy spectrum $(E(k))$ of many physical systems, as sketched in Fig. 2.12, then it is apparent that a continuum of wavenumbers is present in the system. For such a space-time dependent system, it is important to know how the disturbance energy is transmitted or if the propagation of disturbances and the propagation of associated energy are different. These queries can be satisfactorily answered if we can

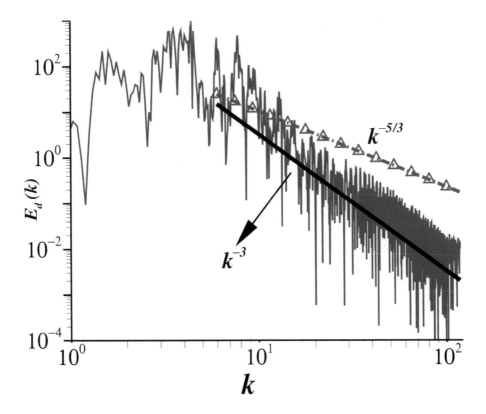

FIGURE 2.12
A typical energy spectrum of a two-dimensional flow over a flat plate. Note that the spectrum varies as k^{-3} for an intermediate wavenumber range [323].

work out the details of how neighboring elements in the spectrum relate or interact with each other. Let us say that the spectrum and the dispersion relation $\bar{\omega} = \bar{\omega}(k)$ are continuous functions of their argument.

Let us now track two such closely spaced neighboring wavenumbers k_1 and $k_2(= k_1 + dk)$. Corresponding circular frequencies are also closely spaced with values $\bar{\omega}_1$ and $\bar{\omega}_2(= \bar{\omega}_1 + d\bar{\omega})$. Also, the wavenumbers are so close that the amplitudes of the individual harmonic components are also taken to be the same and their superposition gives rise to the wave-form

$$Y = a\cos(k_1 x - \bar{\omega}_1 t) + a\cos(k_2 x - \bar{\omega}_2 t)$$

$$= \left[2a\cos\left(\frac{dk}{2}x - \frac{d\bar{\omega}}{2}t\right)\right]\cos\left[\left(k_1 + \frac{dk}{2}\right)x - \left(\bar{\omega}_1 + \frac{d\bar{\omega}}{2}\right)t\right] \tag{2.182}$$

While the phase of the second factor resembles the phase of the original harmonic elements, it is the first factor that is of significant interest, which represents an amplitude varying slowly in space (with wavelength $4\pi/dk$) and time (with time period $4\pi/d\bar{\omega}$).

This slow modulation of the resultant amplitude occurs via its phase variation and the corresponding $x/t = $ constant line moves with the speed

$$V_g = \frac{d\bar{\omega}}{dk} \tag{2.183}$$

which is defined as the group velocity. This is nothing but the slope of the dispersion relation. A typical sketch for the disturbance variation of this group, composed of k_1 and k_2 is as shown in Fig. 2.13.

It can be shown that the energy of the system propagates at this speed, V_g. Heuristically, note the location of the nodes where the resultant amplitude is zero. As this amplitude propagates with V_g and the energy of a system is given by the square of the amplitude, hence the energy also propagates with this speed.

2.9.1   Physical and Computational Implications of Group Velocity

We have noted the physical implications of V_g as the speed with which energy travels in a system displaying a wide-band spectrum. One notes that the group velocity is a consequence of the dispersion property, i.e., the circular frequency is a function of wavenumber. This has been recognized by researchers across many disciplines of science and engineering and studied in [17, 43, 162, 198, 440]. Rayleigh [288] laid the foundation for group velocity as opposed to phase speed − although it was discussed earlier in [140]. The carrier waves use the phase speed for phase variation, while the group velocity is associated with the propagation of amplitude. According to Brillouin [43], group velocity is the velocity of energy propagation and this was identified as signal speed by Rayleigh [289].

Consider the propagation of waves governed by the 1D convection equation

$$\frac{\partial u}{\partial t} + c\frac{\partial u}{\partial x} = 0; \qquad c > 0 \tag{2.184}$$

for which one can substitute the trial solution $u = ae^{i(kx - \bar{\omega}t)}$ to obtain the dispersion relation as given by Eq. (2.179).

One readily notes that

$$\frac{d\bar{\omega}}{dk} = V_g = c \tag{2.185}$$

This signifies that the phase speed and group velocity are indistinguishable for non-dispersive systems like the one given in Eq. (2.184).

For a general dispersive system

$$V_g = c + k \frac{\partial c}{\partial k} \tag{2.186}$$

2.9.2 Wave Packets and Their Propagation

In many real-life applications, we do not have unending streams of packets – as indicated in Fig. 2.13. Instead we have a packet with the same carrier wavelets, but with very compact support. A typical example is given by

$$\eta_w(x,t) = e^{-\alpha(\xi - \xi_0)^2} \cos k\xi \tag{2.187}$$

where $\xi = x - ct$ and ξ_0 is the value at $t = 0$ and is the location where the packet has maximum amplitude. Here, compact support is provided by the Gaussian modulation given by the first factor on right-hand side of Eq. (2.187). We use such packets to analyze numerical schemes as they mimic vortical structures or fronts in fluid flows.

For surface waves, if the initial surface displacement is given as

$$\eta_w = a(x) \cos kx \tag{2.188}$$

then in [273], it has been shown that for small times the wave evolves as

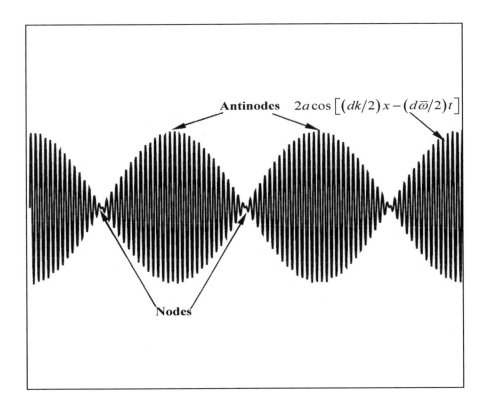

FIGURE 2.13
Phenomenon of modulation in a group of waves.

$$\eta_w(x,t) = a(x - V_g t) \cos(kx - \bar{\omega}t) \qquad (2.189)$$

again demonstrating that the amplitude of the wave packet travels with V_g.

2.10   Issues of Space–Time Scale Resolution of Flows

Most practical flows are essentially at high Reynolds numbers and laminar flows become unstable at Reynolds numbers above critical values. So, flows are either transitional or turbulent. Laminar flows can be either steady or unsteady, but transitional and turbulent flows are inherently unsteady. These flows exhibit space and time scales with broad spectra. Resolving these spectra is the challenge in computing these flows. Thus, the computing tools employed in solving these problems must be *spectrally accurate* to resolve these scales. Not only must the spatial and temporal scales be resolved together, their relationship must be obeyed, as the length and time scales are dependent via the dispersion relation. For example, in solving Eq. (2.184), one must preserve the dispersion relation given by Eq. (2.179). While in linear stability theory, one obtains such dispersion relation, it is not available for general nonlinear flows.

However, let us explore the spatial and temporal scales independently below for turbulent flows. We note that the existence of broad band spectra is a consequence of nonlinearity in the governing equations. For example, the first term on the right-hand side of the vorticity transport Eq. (2.23) for general three-dimensional flows represents the stretching term that transfers energy from large to small scales. Such features are intrinsic to the Navier–Stokes equation and can be expected to be noted for later stages of transitional flows, and of course, for turbulent flows. Similarly for two-dimensional flows, one expects an enstrophy cascade (see Davidson [79] for details) which produces a reverse migration of energy from small to large scale — known as the inverse cascade process. Therefore, the following description for turbulent flows is instructive in noting computational requirements for transitional and turbulent flows.

2.10.1   Spatial Scales in Turbulent Flows

Scales of turbulent flows are related to eddy sizes in the evolving flow field. The largest scale is associated with the integral dimension of the fluid dynamical system, denoted by l, at which the flow is fed with energy.

In general for turbulent flows, one represents kinetic energy density (kinetic energy per unit volume) in the spectral plane. The spectral plane is preferred due to easier interpretation and because a developed theory is available for homogeneous turbulence in wavenumber space. As noted for internal and external flows, the dissipation peak is located at a higher wavenumber *as compared to the peak of energy spectrum*. This is due to the fact that the dissipation is given by $\nu||\nabla u||_2^2$. This can be shown from the energy budget of the disturbance field. In general, the energy spectrum depends on the wavenumber (k), dissipation (ε) and kinematic viscosity (ν). If we define u as the velocity (which represents kinetic energy per unit mass) in the large scale, then we can define a Reynolds number given by

$$Re = \frac{ul}{\nu}$$

It has been shown by Kolmogorov (see Tennekes & Lumley [403]) that the smallest excited length scale, known as the Kolmogorov scale, is given by

$$\eta_K = (\nu^3/\varepsilon)^{\frac{1}{4}} \tag{2.190}$$

Thus the largest and the smallest length scales are related by

$$\frac{l}{\eta_K} = (Re)^{\frac{3}{4}} \tag{2.191}$$

In turbulent boundary layers, there is a region very close to the wall where the velocity varies linearly with the distance from the wall. This is called the *viscous sublayer* and its thickness ξ is related to l by

$$\frac{l}{\xi} = Re \tag{2.192}$$

For flow computations at high Reynolds numbers via DNS, these scales are resolved. If the cut-off wavenumber is represented by k_c (related to η_K), then Eq. (2.191) can also be written as

$$k_c l \approx Re^{\frac{3}{4}} \tag{2.193}$$

This equation is used to state grid requirements for DNS. For three-dimensional flows, this shows that the resolution requirement scales as $(Re^{3/4})^3$ or roughly about Re^2.

In deriving Kolmogorov's scaling theory, it is said that there exist length scales shorter than those are directly excited (l), but larger than the Kolmogorov scale (η_K), for which the energy spectrum is independent of the viscous dissipation mechanism. At these intermediate scales – the *inertial subrange* – the structure of the energy spectrum ($E(k)$) is determined solely by nonlinear energy transfer (via the stretching term given by the first term on the right-hand side of Eq. (2.23)) by a cascade process and the overall energy flux is shown to depend as

$$E(k, \varepsilon) = C_k k^{-\frac{5}{3}} \varepsilon^{\frac{2}{3}} \tag{2.194}$$

The existence of an *inertial subrange* suggests some form of universality of flow structure, below this length scale. This is exploited in LES where the flow is computed by resolving all the way up to the *inertial subrange* and anything smaller than this is modeled via sub-grid scale (SGS) stress models.

2.10.2   Two- and Three-Dimensional DNS

In the vorticity transport equation, the first term on the right-hand side of Eq. (2.23), $(\omega_I \cdot \nabla)\vec{V}$, is due to vortex stretching and is present only for three-dimensional flows. For two-dimensional flows, this is identically zero. Due to stretching, the cross sectional area of an eddy decreases and thereby in trying to conserve angular momentum for inviscid flow, increases the eddy's vorticity in that direction. Thus, the enstrophy of the flow increases due to vortex stretching (See [89], pages 49–53). It is also shown in [89] that the enstrophy is proportional to $\int k^2 E(k)\, dk$, while the kinetic energy density ($\int E(k)\, dk$) remains conserved for inviscid flow. Thus, in the inviscid limit, these requirements are simultaneously met if $E(k)$ migrates to higher and higher values of k. This is the well-known forward scatter of energy due to vortex stretching in three-dimensional flows, while the actual energy spectrum is as given in Eq. (2.194) and a sketch of it is shown in the top frame of Fig. 2.14. For such an energy spectrum, the dissipation function $D(k) = \nu ||\nabla V||^2$ can be shown to be proportional to $k^{1/3}$ – as shown in the bottom frame of Fig. 2.14.

It has been shown in [79, 89] that there are important two-dimensional flows which display turbulence structures which are totally different from three-dimensional turbulent flows. For such inviscid flows, the forward scatter due to vortex stretching is absent and thus both the energy and enstrophy remains conserved. Thus, if both $\int E(k)\,dk$ and $\int k^2 E(k)\,dk$ are constants of motion, then any forward scatter must be compensated by an inverse scatter or backscatter. However, for viscous flows, enstrophy increases due to continual vorticity generation at the wall (which may not be compensated by dissipation). Enstrophy can be created or destroyed when large eddies break by Kelvin–Helmholtz instabilities and/or vortex interaction or coalescence. It has been reasoned and demonstrated in [184] that enstrophy plays the same role in two-dimensional flows as the energy plays in three-dimensional flows. In an analogous manner, one can formulate a high wavenumber dissipation scaling for two-dimensional turbulent flow. The details are given in [184] and [89], which show the high wavenumber energy spectrum given by

$$E(k,\chi) = \chi^{2/3} k^{-3} \qquad\qquad (2.195)$$

where χ is the enstrophy dissipation rate, given by $\|\nabla\omega\|_2^2$ (see [89] for details).

This has prompted [89] to sketch a composite energy spectrum, as shown in Fig. 2.15, in which the $k^{-5/3}$ spectrum is followed by k^{-3} dependence at higher wavenumbers. Are these dependencies actually seen in real flows? Atmospheric data collected by [244] shown in Fig. 2.16, depicts the opposite trend, showing two-dimensionality preceding three-dimensional

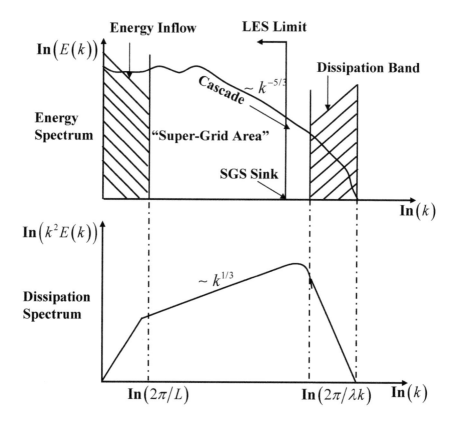

FIGURE 2.14
Schematic of the energy spectrum in Kolmogorov's theory is shown at the top. The dissipation spectrum as used in Kolmogorov's theory is shown at the bottom.

variation in the spectrum and is perhaps more important to the latter. It is important to note that the flow in atmosphere is essentially two-dimensional, following the geostrophic assumption and the Taylor–Proudman theorem (see [413]).

2.11   Temporal Scales in Turbulent Flows

We have already noted that temporal scales are dependent upon spatial scales through the dispersion relation. However, if one considers temporal scales as an independent variable, then it is related to the frequency of externally induced or self induced unsteadiness. In flow induced vibration at high Reynolds numbers, these two unsteadinesses are often coupled. In general, unsteadiness can be *stochastic, highly organized or mixed (interacting)*. Computationally, the highest frequency determines the time step. For unsteady flows with organized unsteadiness, if the dominant frequency is f_0, then one defines a Strouhal number by

$$St = \frac{f_0\, l}{U}$$

For forced flows, such as in rotor wing aerodynamics or in turbomachines, the dominant driven response is sinusoidal with higher harmonics excited, as the blades/airfoils pass through trailing vortices of the preceding blades/airfoils. These flows are characterized by

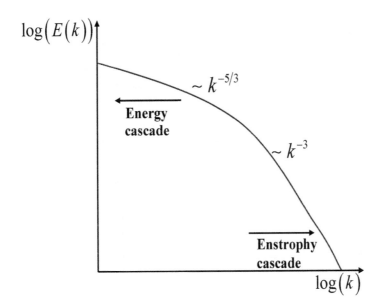

FIGURE 2.15
Sketch of the turbulent flow energy spectrum, exhibiting both two-dimensional and three-dimensional flow natures, as drawn in some monographs, showing energy and enstrophy cascades. (Reproduced with permission from Doering and Gibbon, *Applied Analysis of the Navier–Stokes Equations*, Copyright (1995) Cambridge University Press.)

a non-dimensional frequency parameter defined as

$$\bar{\omega}_f = \frac{\bar{\omega}_0 c}{2\pi}$$

where $\bar{\omega}_0$ is the forcing circular frequency and c is the chord of the airfoil section. Let us enumerate some typical St for unsteady flows:

1. For a body oscillating in its own plane, a shear layer forms whose thickness is of the order of $(O\sqrt{\nu/f_0})$, where ν is the kinematic viscosity.

2. For a rigid stationary circular cylinder in uniform flow with a Reynolds number in the range of $10^3 < Re < 5 \times 10^5$, unsteadiness is approximately given by $St \approx 0.2$ and for the range $5 \times 10^5 < Re < 10^7$, the unsteadiness varies in the range $St \approx 0.2-0.4$. Thus, the time scales correspond to low frequency events associated primarily with Benard–Karman vortex shedding — as discussed in [444]. At even higher Reynolds numbers, vortices are not predominantly discernible, but low frequencies are noted in the time series of physical variables in the near wake.

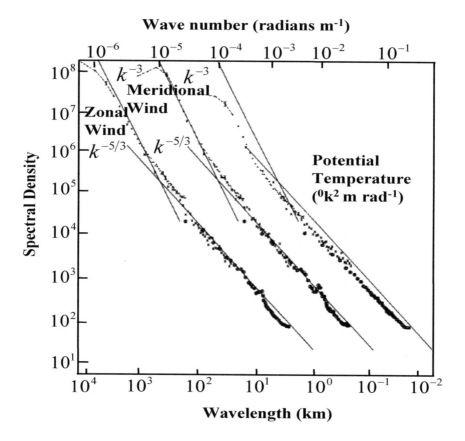

FIGURE 2.16
Power spectra of wind and potential temperature in the atmosphere measured at tropopause by an aircraft (data taken from [244]). (Reproduced with permission from Galperin and Orszag, *Large Eddy Simulation of Complex Engineering and Geophysical Flows*, Copyright (1993) Cambridge University Press.)

3. For a sudden change in a boundary layer, a Stokes layer is formed at the wall and its thickness is always less than the boundary layer thickness. The diffusive time scale is $O(\delta^2/\nu)$ and pertains to diffusion of momentum and vorticity from the body surface where it is usually created.

4. For an impulsive start of a physical body, once again a Stokes layer is formed whose thickness is $O(\sqrt{\nu t})$ for small times.

A detailed description of these classes of flows is provided in [402, 312].

2.12   Computing Time-Averaged and Unsteady Flows

Consider incompressible flow at very high Reynolds numbers, so that the flow is fully turbulent. The governing equations are rewritten here as

$$\nabla \cdot \vec{V} = 0 \tag{2.196}$$

$$\frac{\partial \vec{V}}{\partial t} + (\vec{V} \cdot \nabla)\vec{V} = -\nabla p/\rho + \nu \nabla^2 \vec{V} \tag{2.197}$$

If the flow variables exhibit unsteadiness at high frequencies, as in high Reynolds number flows, then one can double decompose the velocity field by

$$\vec{V}(\vec{X}, t) = U(\vec{X}) + v(\vec{X}, t) \tag{2.198}$$

On the right-hand side of Eq. (2.198), variables are vector quantities and upper case variables are time independent or time average of the velocity field. This means that lower case components are truly random fluctuating quantities, i.e., these have zero time averages. This type of splitting variables for turbulent flows was suggested in [298] and the process is called Reynolds averaging. Here, the time averaging is defined as

$$\langle V(\vec{X}, t) \rangle = U(\vec{X}) = \lim_{T \to \infty} \frac{1}{T} \int_0^T \vec{V}(\vec{X}, t)\, dt \tag{2.199}$$

The angular brackets denote the time averaging operation. If we split the velocity field as in Eq. (2.198), then the mean field U satisfies time independent boundary conditions which are prescribed for $\vec{V}(\vec{X}, t) = \vec{V}_b(\vec{X})$ and $v(\vec{X}, t)$ satisfies homogeneous boundary conditions, irrespective of whether the boundary conditions are Dirichlet or Neumann type. If $\vec{V}$ is prescribed as periodic in some direction, then v is also periodic in that direction.

Now, if time averages of time derivatives vanish and time averaging commutes with spatial derivative operations, then we can average Eqs. (2.196) and (2.197) to obtain

$$U \cdot \nabla U + \langle v \cdot \nabla v \rangle = -\nabla p/\rho + \nu \nabla^2 U \tag{2.200}$$

$$\nabla \cdot U = 0 \tag{2.201}$$

Note the presence of additional momentum flux terms $\langle \vec{v} \cdot \nabla \vec{v} \rangle$, on the left-hand side of Eq. (2.200), due to the presence of fluctuation terms. These are the Reynolds stresses, constituting a symmetric stress tensor, like the general stress system for fluids in motion. However, unlike viscous stresses, Reynolds stresses are properties of the flow and are not isotropic.

The time averaged Navier–Stokes equation, with additional Reynolds stress terms due to turbulent fluctuations, is called the **Reynolds Averaged Navier–Stokes (RANS) equation**. Many CFD procedures are based on this equation. There are two principal drawbacks with this approach. First, one doesn't know how to provide information about the Reynolds' stresses − since no additional principles are known about these flow-dependent *stress tensors*. In practice, these are modeled or parametrized via turbulence models. The second problem comes about defining time averages of turbulent signals. In using the definition given by Eq. (2.198), it is assumed that the time dependence of the flow field is via the fluctuating quantities $v(\vec{X}, t)$ and is at higher frequencies. This procedure is right for flows which have a truly time-independent mean − as may be the case of disturbed steady laminar flows. For unsteady equilibrium flows with high frequency contents, unsteadiness of the disturbance flow cannot be differentiated from the unsteadiness of the equilibrium flow. Thus, this procedure will fail when the averaging time interval (T) is less than the period of equilibrium unsteady motion. For turbulent flows, this is less of a problem, as the turbulent fluctuations are at significantly higher frequencies, as compared to the frequency of *mean motion*. This observation has been used where turbulence structures are unaffected by externally induced unsteadiness and there is a critical frequency below which steady flow turbulence models can be used for predicting unsteady (in the mean) turbulent flows. This is the area of unsteady RANS or URANS formulation.

It has been seen that for attached turbulent boundary layers, there are coherent structures in the near-wall region. It is noted that the organized structures in turbulent flows carry about 20% of total turbulent kinetic energy (TKE) and hence their role in determining turbulent flow dynamics cannot be underestimated. These organized structures show up as peaks and valleys in the near-wall region and in terms of wall units, they have lengths between 100 and 2000 units in the streamwise direction and have a spacing of about 50 units in the spanwise direction. These high energetic events occur at a height of about 20 to 50 units (i.e., in and around the buffer layer) − all these are statistical estimates. Additionally, these near-wall events are interspersed by bursting of these structures. After bursts, new intermediate scale motions ensue in the buffer layer; those are also streamwise and/or hair-pin vortices. Thus, the unsteadiness of the turbulent boundary layer is characterized by bursting frequency, even for an attached shear layer. For example, in the zero pressure gradient boundary layer, this critical frequency is roughly between 20 and 100% of the turbulence burst frequency (F_b), where

$$F_b = \frac{U_\infty}{5\delta}$$

with δ as the shear layer thickness. For adverse pressure gradient flows, this critical frequency is between 6 and 28% of F_b. Therefore it is not surprising that for flows past bluff bodies, the performance of RANS with a turbulence model is less than satisfactory. Also, in light of the existence of organized unsteady motion, it is not clear whether a double decomposition as resorted to in Reynolds averaging is a correct procedure. Perhaps a triple or multiple decomposition of the type given in [402] would be more appropriate for flows with dominant coherent eddies. For high Reynolds number flows, it is preferable to use LES, which do not use time averaging and instead work with spatial averages with subgrid scales parametrized.

In Fig. 2.17, various temporal scales excited in typical engineering flows are displayed for different speed regimes. For high Reynolds number flows, only the mean frequencies of large eddies are shown. It is quite clear that ranges of non-dimensional frequencies span over three orders of magnitude and numerical methods must resolve these scales for high Reynolds number flows.

2.13 Computing Methods for Unsteady Flows: Dispersion Relation Preserving (DRP) Methods

So far, we have discussed the concept of dispersion relation as applied to space-time dependent problems. In the context of transitional and turbulent flows, we have also noted the range of length and time scales in terms of the spectra of such flows. It is noted that DNS would require preservation of the physical dispersion relation in the numerical sense. We have noted the role of group velocity in transporting energy of a dynamical system. One of the main aims of the present section is to relate group velocity with the dispersion property of the system numerically. However, apart from matching physical and numerical group velocity, one would also like to ensure the overall accuracy of the numerical solution and relate the accuracy with the properties of adopted numerical methods.

In this context, it has been noted that Eq. (2.184) serves well the purpose of a model equation to check the accuracy of any numerical method involving space and time discretization. We have already established that this is a non-dissipative and non-dispersive equation

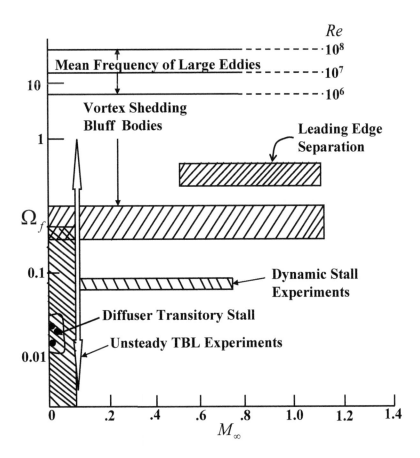

FIGURE 2.17
Ranges of temporal scales excited in flows of engineering interest for different speeds and the corresponding frequency ranges (Ω_f). Also, indicated is the band for mean frequencies of large eddies in high Reynolds number flows.

which convects the initial solution unattenuated to the right at the speed c representing both phase speed and group velocity.

As stated, this equation is non-dispersive and provides a simple, yet a tough test for any combined space-time discretization method's dispersion property. We reemphasize that the study of spatial and temporal discretizations separately is improper. Such studies can be found variously in [185, 397, 410]. Vichnevetsky & Bowles [425] have reported the dispersion property of a single finite difference and a finite element method by heuristic approaches. A proper estimate of the numerical dispersion property has been advanced in [356, 349, 335]. The explanation provided here is based on spectral analysis of numerical schemes developed over the full computational domain in [331].

In this analysis method, a dependent variable is represented by a hybrid expansion

$$u(x_m, t^n) = \int U(kh, t^n)\, e^{ikx_m}\, dk$$

2.13.1   Spectral or Numerical Amplification Factor

The spectral amplification factor or the gain is defined by

$$G(\Delta t, kh) = \frac{U(kh, t^n + \Delta t)}{U(kh, t^n)} \qquad (2.202)$$

Note that G is a complex quantity, as the Fourier–Laplace amplitude $U(kh, t)$ is complex. In the continuum limit of $\Delta t \to 0$, one must have

$$|G| \equiv 1 \qquad (2.203)$$

However, real and imaginary parts of G depend on the differential equation solved and the discretization methods adopted for spatial and temporal derivatives. Thus, this simple observation is heuristically correct, yet cannot be proven rigorously. It is at the core of any study involving numerical stability and clearly establishes that any numerical method must be neutrally stable.

Analyses of numerical methods and associated error propagation have been performed following different routes by many researchers. From the discretized governing equation, one can work backwards to obtain an equivalent differential equation that has been effectively tackled by the discrete equation. This approach is used in [367]. According to this, if the truncation error terms are retained for space-time dependent problems, then the corresponding differential form is called the Γ-form. If the retained truncation error terms are for space derivatives alone, then the corresponding differential form is called the Π-form. While most analyses in the literature have used the Π-form, only in a few the Γ-form has been utilized. The present approach follows the same route, where both space and time discretization are considered simultaneously.

In the classical approach attributed to von Neumann, the evolving error of the discretized differential equation with linear constant coefficients is assumed to follow an identical discrete equation. In this approach, the difference between an exact and a computed solution arises due to round-off error and error in the initial data. For periodic problems, error is furthermore decomposed into Fourier series and the individual normal modes are investigated. This approach is also extended for nonlinear systems and linear systems with variable coefficients. For nonlinear systems, the principle of superposition does not hold and additionally problem of aliasing arises due to nonphysical transfer of energy to wrong wavenumbers and frequencies in the computational plane, whenever products are numerically evaluated with finite grid resolution.

There have been many efforts in analyzing error dynamics, using the method attributed to von Neumann, as described in [60] and [72], which is readily applied to linear equations and to quasi-linearized forms of nonlinear equations. The main assumption for linear problems that the error and the signal follow the same dynamics appears intuitively correct. It has been unambiguously shown in [335] that this assumption is flawed for any discrete computing and the difference is due to dispersion and phase error and when the numerical method is not strictly neutrally stable. It is just adequate to show a violation of the assumption in the von Neumann method by any equation, irrespective of the method employed.

We demonstrate the violation with the help of Eq. (2.184) in analyzing space-time discretization schemes. This equation helps in testing numerical methods for solution accuracy, error propagation and, most importantly, the dispersion error − as has been variously attempted in [335, 349, 356, 425, 448].

We represent the unknown, by its Laplace transform at the jth node of a uniformly spaced discrete grid of spacing h as $u(x_j, t) = \int U(kh, t)\, e^{ikx_j}\, dk$. The exact spatial derivative at the same node is given by

$$\left[\frac{\partial u}{\partial x}\right]_{exact} = \int ikU\, e^{ikx_j}\, dk$$

While solving Eq. (2.184) by discrete methods, the spatial derivative u'_j (denoted by a prime) has been shown in [331, 356] to be equivalent to

$$\left[u'_j\right]_{numerical} = \int ik_{eq}U\, e^{ikx_j} dk \tag{2.204}$$

Numerically the same derivative is also estimated from $\{u'\} = \frac{1}{h}[C]\{u\}$. One obtains an appropriate $[C]$ matrix for finite-domain non-periodic discretized problems, with the dimension of $[C]$ corresponding to the number of nodes. This implies that the derivative at the jth node is evaluated as

$$u'_j = \frac{1}{h}\sum_{l=1}^{N} C_{jl}\, u_l$$

where $u_l = u(x_l, t) = \int U(kh, t)\, e^{ikx_l}\, dk$ is the function value at the lth node and N is the number of nodes. Using spectral representation, one can alternately write the numerical derivative as

$$u'_j = \int \frac{1}{h}\sum C_{jl}\, U(kh, t)\, e^{ik(x_l - x_j)}\, e^{ikx_j}\, dk \tag{2.205}$$

Comparing Eqs. (2.204) and (2.205), we note that

$$[ik_{eq}]_j = \frac{1}{h}\sum_{l=1}^{N} C_{jl}\, e^{ik(x_l - x_j)} \tag{2.206}$$

Although in physical plane computations C_{jl}'s are real, $[k_{eq}]_j$ is in general complex, with real and imaginary parts representing numerical phase and added numerical dissipation, respectively. It is determined by the numerical method fixing the entries of $[C]$.

Other important numerical properties are obtained via the spectral representation in Eq. (2.184), which gives

$$\int \left[\frac{dU}{dt} + \frac{c}{h}\sum UC_{jl}\, e^{ik(x_l - x_j)}\right] e^{ikx_j} dk = 0 \tag{2.207}$$

Since the above equation is true for all wavenumbers, the integrand must be zero for any k. The implicit condition of Eq. (2.207) can be reinterpreted as

$$\frac{dU}{U} = -\left[\frac{cdt}{h}\right] \sum_{l=1}^{N} C_{jl}\, e^{ik(x_l - x_j)} \tag{2.208}$$

The first factor on the right-hand side of Eq. (2.208) is nothing but the CFL number (N_c). As the right-hand side of Eq. (2.208) is node dependent, we express the left-hand side for the nodal numerical amplification factor (G_j) given by

$$G_j = G|_{(x=x_j)} = 1 - N_c \sum_{l=1}^{N} C_{jl}\, e^{ik(x_l - x_j)} \tag{2.209}$$

for the Euler time discretization scheme. From Eq. (2.206), one can replace the summed up quantity on the right-hand side as

$$G_j = G|_{(x=x_j)} = 1 - i N_c\, [k_{eq}h]_j \tag{2.210}$$

Thus, the numerical property can be expressed in terms of spatial discretization alone at each node for Euler time discretization. In general, spatial discretization alone cannot describe all numerical properties of the combined method for space-time discretization.

Euler time integration is not numerically stable, as $|G_j| > 1$. Two time level, multistage higher order methods are needed ideally for accuracy and stability - as explained with respect to the four stage, fourth order Runge–Kutta (RK_4) method. If one denotes the right-hand side of Eq. (2.184), by $L(u) = -c\frac{\partial u}{\partial x}$, then the steps used in (RK_4) are given by

$$\text{Step1}: \quad u^{(1)} = u^{(n)} + \frac{\Delta t}{2} L[u^{(n)}]$$

$$\text{Step2}: \quad u^{(2)} = u^{(n)} + \frac{\Delta t}{2} L[u^{(1)}]$$

$$\text{Step3}: \quad u^{(3)} = u^{(n)} + \Delta t L[u^{(2)}]$$

$$\text{Step4}: \quad u^{(n+1)} = u^{(n)} + \frac{\Delta t}{6} \{L[u^{(n)}] + 2L[u^{(1)}] + 2L[u^{(2)}] + L[u^{(3)}]\}$$

For the RK_4 time integration scheme, G_j is obtained as (see [335])

$$G_j = 1 - A_j + \frac{A_j^2}{2} - \frac{A_j^3}{6} + \frac{A_j^4}{24} \tag{2.211}$$

where

$$A_j = N_c \sum_{l=1}^{N} C_{jl}\, e^{ik(x_l - x_j)}$$

While G_j is a source of error, additional error arises due to dispersion, whose effects are subtle and often misunderstood.

2.13.2 Quantification of Dispersion Error

Using a Fourier transform for the analysis of a numerical method, one obtains a numerical dispersion relation by expressing the equivalent differential equation in the wavenumber-circular frequency plane. This dispersion relation is different from the physical dispersion relation and the difference between the two gives rise to dispersion error, as explained with the one-dimensional convection equation as a model of space-time dependent systems.

If we represent the initial condition for Eq. (2.184) as

$$u(x_j, t = 0) = u_j^0 = \int A_0(k) \, e^{ikx_j} \, dk \tag{2.212}$$

then the general solution at any arbitrary time can be obtained as

$$u_j^n = \int A_0(k) \, [|G_j|]^n \, e^{i(kx_j - n\beta_j)} \, dk \tag{2.213}$$

where $|G_j| = (G_{rj}^2 + G_{ij}^2)^{1/2}$ and $\tan(\beta_j) = -\frac{G_{ij}}{G_{rj}}$, with G_{rj} and G_{ij} as the real and imaginary parts of G_j, respectively. Thus, the phase of the solution is determined by $n\beta_j = kc_N t$, where c_N is the numerical phase speed. Although the physical phase speed is a constant for all k for the non-dispersive system, this analysis shows that c_N is, in general, k-dependent, i.e., the numerical solution is dispersive, in contrast to the non-dispersive nature of Eq. (2.184). The implications of this simple difference are profound, as demonstrated below.

The general numerical solution of Eq. (2.184) is denoted as

$$\bar{u}_N = \int A_0 \, [|G|]^{t/\Delta t} \, e^{ik(x - c_N t)} \, dk \tag{2.214}$$

The numerical dispersion relation is now given as $\omega_N = c_N k$, instead of $\bar{\omega} = ck$. Non-dimensional phase speed and group velocity (from the general definition given in Eqs. (2.185) and (2.186)) at the jth node are expressed as

$$\left[\frac{c_N}{c} \right]_j = \frac{\beta_j}{\omega \Delta t} \tag{2.215}$$

$$\left[\frac{V_{gN}}{c} \right]_j = \frac{1}{hN_c} \frac{d\beta_j}{dk} \tag{2.216}$$

If the computational error is defined as $e(x, t) = u(x, t) - \bar{u}_N$, then one obtains the governing equation for its dynamics in the following manner. Using Eq. (2.214) one obtains

$$\frac{\partial \bar{u}_N}{\partial x} = \int ik A_0 \, [|G|]^{t/\Delta t} \, e^{ik(x - c_N t)} \, dk \tag{2.217}$$

$$\frac{\partial \bar{u}_N}{\partial t} = - \int ik \, c_N \, A_0 \, [|G|]^{t/\Delta t} \, e^{ik(x - c_N t)} \, dk$$

$$+ \int \frac{Ln \, |G|}{\Delta t} A_0 \, [|G|]^{t/\Delta t} \, e^{ik(x - c_N t)} \, dk \tag{2.218}$$

Thus, the error propagation equation for Eq. (2.184) is given by [335]

$$\frac{\partial e}{\partial t} + c \frac{\partial e}{\partial x} = -c[1 - \frac{c_N}{c}] \frac{\partial \bar{u}_N}{\partial x} - \int \frac{dc_N}{dk} \left[\int ik' A_0 [|G|]^{t/\Delta t} e^{ik'(x - c_N t)} dk' \right] dk$$

$$- \int \frac{Ln \, |G|}{\Delta t} A_0 \, [|G|]^{t/\Delta t} \, e^{ik(x - c_N t)} \, dk \tag{2.219}$$

This is the correct error propagation equation, as opposed to that is obtained using the assumption made in von Neumann analysis, where the right-hand side is assumed to be identically equal to zero. This is on the premise that $c_N \cong c$, i.e., there is no dispersion error and the numerical method is perfectly neutral, so that the last term on the right-hand side of Eq. (2.219) is identically zero. Error can grow even faster when the numerical solution displays sharp spatial variation, due to the first term on the right-hand side of Eq. (2.219).

To understand the ramifications of Eq. (2.219) for the model equation, one needs to look at the numerical properties of a specific combination of spatial and temporal discretization methods. For this purpose, we show the properties of the compact scheme OUCS3 introduced in [331]. This scheme is for a non-periodic problem and here we show some typical results for a central and another node close to the inflow boundary, when RK_4 time integration strategy is used for solving Eq. (2.184). In Fig. 2.18, $|G|$, V_{gN}/c and $(1 - c_N/c)$ are plotted as contours in the indicated ranges of kh and N_c for a node adjacent to the boundary ($j = 2$) and one in the center of the domain ($j = 51$).

Results show the numerical properties of $j = 2$ are significantly different from those for $j = 51$. It is noted from the top frame for numerical amplification contours that the scheme is stable for $N_c \leq 1.301$, i.e., $|G| \leq 1$ for interior nodes. The scheme is neutrally stable for very small values of N_c and a limited range of kh − a property absolutely essential for DNS. One notes that the last term on the right-hand side of Eq. (2.219) vanishes for the neutrally stable case. In the middle frame of Fig. 2.18, V_{gN}/c contours display significant dispersion effects for high k's, which would invalidate long time integration results, even when $G_j = 1$ is ensured by computing with vanishingly small N_c. In fact, above $kh \geq 2.4$ the numerical solution will travel in the wrong direction, as $V_{gN} \leq 0$ for $N_c \cong 0$. Such spurious numerical waves are termed q-waves, which propagate upstream, in contrast to the physical or p-waves traveling downstream for Eq. (2.184). In Eq. (2.219), we note that the first term on the right hand side affects error evolution via the numerical property $(1 - \frac{c_N}{c})$. In the bottom frame of Fig. 2.18, contours of this quantity are shown. The bottom two frames of Fig. 2.18 indicate effects of dispersion error, which cannot be simply eliminated or reduced by grid refinement alone.

One can show the connection between q-waves as the extreme form of dispersion error in different numerical methods used for CFD. It is noted that q-waves travel in the opposite direction of p-waves. Even for a large range of k, corresponding p-waves travel at incorrect group velocity, constituting dispersion error. The q-wave is entirely due to dispersion in many discrete computations.

Flow transition is usually thought often to occur for external flows by Tollmien-Schlichting wave packets created as instability to small perturbations inside a shear layer. However, many flows bypass this classical route of transition and are collectively stated to have suffered bypass transition. In [336], one particular form of bypass transition is explained theoretically and experimentally, when the flow is excited by a disturbance source remaining always outside the shear layer. In this scenario, the resultant disturbance occurring inside the shear layer is seen to propagate upstream, with respect to the source of disturbance convecting outside the shear layer. Thus, there are flow transition scenarios where physical disturbances travel upstream and it is necessary to compute these, without the interference of q-waves created numerically for many numerical methods.

For attached flows, q-waves have been reported in [274] as very high k events. Thus, for DNS of bypass transitional flows, one would like to capture p-waves associated with bypass transition, while avoiding q-waves which arise solely due to dispersion error at high wavenumbers. For the zero pressure gradient boundary layer, upstream propagating damped waves were detected theoretically in [344]. We emphasize that in many flows, disturbances appear to move downstream with respect to an inertial frame − but an observer moving with the source at the convection speed will see the response inside the shear layer move upstream. Next, we quantify parameter combinations for which q-waves are generated in different numerical methods.

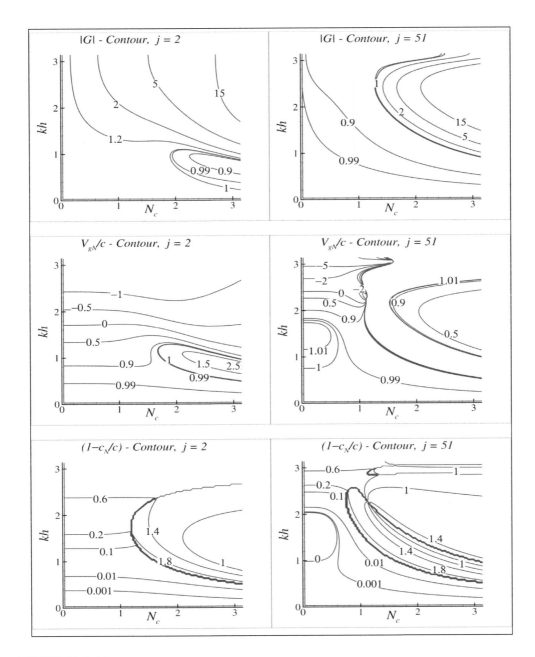

FIGURE 2.18
Comparison of numerical amplification factor ($|G|$), normalized numerical group velocity (V_{gN}/c) and phase error ($1-c_N/c$) contours for the near-boundary node $j = 2$ (left column) and central node $j = 51$ (right column) for the 1D convection equation solved by the OUCS3-RK_4 method.

2.14   DRP Schemes: Parameter Ranges for Creating q-Waves

Significant progress has been made in developing numerical methods for solving space-time dependent problems with DRP schemes. Early attempts suffered due to wrong identification of the numerical dispersion relation. Fourier spectral methods have the best DRP property up to the Nyquist limit, while for other discrete methods, closeness between the physical and numerical dispersion relation can be realized only for very limited ranges of space and time steps. Dispersion relation refers to the space-time dependence of the problem. In [331], a spectral analysis was developed to characterize numerical schemes in the full domain, with numerical group velocity used as a measure for quantifying dispersion error, as in [410]. Space-time discretization was treated independently in many earlier works, which led to incorrect numerical group velocity. This was latter corrected in [349, 335] and with the derivation of Eq. (2.219), a neutrally stable scheme was identified as essential for DNS and acoustics problems, apart from matching physical and numerical group velocity. These results were used in [285, 347] to optimize time discretization schemes for high accuracy spatial discretization schemes, by minimizing numerical error in the spectral plane.

Here, we have chosen representative methods from finite difference, finite volume and finite element approaches to explain q-waves and quantifying DRP properties. For the finite difference method, a compact scheme from [331] for spatial discretization, called the OUCS3 scheme, is chosen. In solving Eq. (2.184), the first spatial derivative, indicated by a prime, is obtained from the general representation

$$[A] \{u'\} = \frac{1}{h}[B] \{u\} \tag{2.220}$$

For the explicit discretization method, $[A]$ is the identity matrix and the $[C]$ matrix in Eqs. (2.205) and (2.206) is identical to the $[B]$ matrix. For implicit schemes, one identifies $[C]$ with the $[A]^{-1}[B]$ matrix. Having obtained $[C]$ for spatial discretization, one evaluates the complex amplification factor G_j for any time integration schemes. In particular, for the RK_4 method one can obtain G_j from Eq. (2.211) with

$$A_j = N_c \sum_{l=1}^{N} C_{jl} \, e^{ik(x_l - x_j)}$$

This approach is used to obtain G_j for explicit CD_2 and the optimized upwind compact scheme (OUCS3) for spatial discretization with RK_4 time integration scheme.

The first order spatial derivatives are represented in the OUCS3 scheme for the interior nodes by

$$p_{-1} \, f'_{j-1} + f'_j + p_1 \, f'_{j+1} = \frac{1}{h} \sum_{k=-2}^{2} q_k \, f_{j+k}, \quad 3 \le j \le N-2 \tag{2.221}$$

where $p_{\pm 1} = D \pm \frac{\eta}{60}$ $q_{\pm 2} = \pm \frac{F}{4} + \frac{\eta}{300}$ $q_{\pm 1} = \pm \frac{E}{2} + \frac{\eta}{30}$ and $q_0 = -\frac{11\eta}{150}$ with $D = 0.3793894912$, $E = 1.57557379$ $F = 0.18205192$ and $\eta = 0.0$. Here, η is the upwind parameter and a value of zero implies the corresponding central scheme. For boundary closure schemes, readers are referred to [331].

For a representative of a finite volume scheme, the QUICK scheme due to [196] is considered, which uses flux vector splitting of the convection term as

$$\int \left[\frac{\partial u}{\partial t}\right] dx + c \, (u^+_{j+1/2} - u^+_{j-1/2}) = 0 \tag{2.222}$$

In the above, RK_4 time integration is used for the jth cell and the superscript for the second set of terms indicates right moving quantities showing the balance of incoming and outgoing fluxes through the cell interfaces. One of the representative flux quantities is given by (see [148] for details)

$$u^+_{j-1/2} = u_{j-1} + \frac{1}{4}[(1 - \kappa)(u_{j-1} - u_{j-2}) + (1 + \kappa)(u_j - u_{j-1})] \qquad (2.223)$$

where $\kappa = 1/2$ for the QUICK scheme. Other flux terms $u^+_{j+1/2}$ can be similarly obtained. These expressions help one obtain the $[C]$ matrix in Eq. (2.205), which in turn yields G_j. Readers are referred to [334] for a similar analysis for a compact scheme based flux vector splitting finite volume method.

For the finite element method, the streamwise upwind Petrov-Galerkin (SUPG) method described in [134] is analyzed. Among finite element formulations, Galerkin methods belong to the class of solutions for partial differential equations in which solution residue is minimized, giving rise to the well-known weak formulation of problems. In this approach, dependent variable $u(x, t)$ for a one-dimensional space-time dependent problem is expressed in the form

$$u(x, t) = \sum_{j=1}^{N} \phi_j(x)\, u_j(t) \qquad (2.224)$$

where $\phi_1, ..., \phi_N$ are the chosen low order polynomials as the basis functions, localized about elements. Note that the weak formulation considers space-time dependent terms together, giving rise to better DRP properties for Galerkin formulations. However, it has been identified in [338] that the Galerkin methods display instability near the *inflow* when linear basis functions are used (G1FEM) for Eq. (2.184). Similarly, observations on the amplification factor holds for solving the 1D convection equation by quadratic basis functions by G2FEM.

One method of removing the problems of Galerkin FEM is to make the discrete equation dissipative. Authors in [83, 431, 432] have proposed dissipative Galerkin procedures. This procedure was furthermore adopted by [290], who advocated an optimal procedure following an approximate form of phase error proposed in [397] for one-dimensional convection equation. Brooks & Hughes [45] have adopted the same methodology in their Petrov–Galerkin formulation and called it the "Streamline Upwind/Petrov–Galerkin (SUPG)" formulation. Raymond & Garder [290] have also referred to "ghost waves" following [76]. A complete analysis of the "ghost" or q-waves is presented here for the SUPG formulation.

The discretized form of Eq. (2.184) for G1FEM is obtained using linear basis functions on a uniform grid as

$$\frac{h}{6}\left(\frac{du_{j+1}}{dt} + 4\frac{du_j}{dt} + \frac{du_{j-1}}{dt}\right) + \frac{c}{2}(u_{j+1} - u_{j-1}) = 0 \qquad (2.225)$$

for interior (j^{th}) nodes. For boundary nodes ($j = 1, N$) one obtains

$$j = 1: \qquad \frac{h}{3}\left(2\frac{du_1}{dt} + \frac{du_2}{dt}\right) + c(u_2 - u_1) = 0 \qquad (2.226)$$

$$j = N: \qquad \frac{h}{3}\left(2\frac{du_N}{dt} + \frac{du_{N-1}}{dt}\right) + c(u_N - u_{N-1}) = 0 \qquad (2.227)$$

Three aspects are evident from the above discrete equations for G1FEM: (i) the non-dissipative nature of the discrete equation for the interior nodes; (ii) instability at $j = 1$ due to the one-sided nature of the stencil, with the information propagating to the boundaries from the interior, which is contrary to the physical description given by Eq. (2.184) and (iii) the overly dissipative nature of the discrete equation at $j = N$.

Using the hybrid Fourier–Laplace representation of u in Eqs. (2.225)–(2.227), one obtains the effectiveness of the derivative discretization $k_{eq}^{(1)}$ for the G1FEM as

$$j = 1 : \qquad k_{eq}^{(1)} = \frac{3}{ih}\left(\frac{e^{ikh} - 1}{e^{ikh} + 2}\right) \tag{2.228}$$

$$2 \le j \le N - 1 : \qquad k_{eq}^{(1)} = \frac{3\sin kh}{[h(2 + \cos kh)]} \tag{2.229}$$

$$j = N : \qquad k_{eq}^{(1)} = \frac{3}{ih}\left(\frac{1 - e^{-ikh}}{2 + e^{-ikh}}\right) \tag{2.230}$$

If one adopts Euler time stepping for discretized Eqs. (2.225)–(2.227), the numerical amplification factor for G1FEM is given, as in [338]

$$j = 1 : \qquad G^{(1)} = 1 - 3N_c\left[\frac{e^{ikh} - 1}{e^{ikh} + 2}\right] \tag{2.231}$$

$$2 \le j \le N - 1 : \qquad G^{(1)} = 1 - 3iN_c\left[\frac{\sin kh}{2 + \cos kh}\right] \tag{2.232}$$

$$j = N : \qquad G^{(1)} = 1 - 3N_c\left[\frac{1 - e^{-ikh}}{2 + e^{-ikh}}\right] \tag{2.233}$$

If instead one uses quadratic basis functions (G2FEM) for a uniformly spaced grid, then the discrete equations for the Galerkin approximation of Eq. (2.184) are as given in [338]. Here, we report only the equation for the interior nodes as

$$3 \le j \le N - 2 : \quad \frac{h}{15}\left(-\frac{du_{l+2}}{dt} + 4\frac{du_{l+1}}{dt} + 24\frac{du_l}{dt} + 4\frac{du_{l-1}}{dt} - \frac{du_{l-2}}{dt}\right)$$

$$+\frac{c}{6}\left(u_{l-2} - 8u_{l-1} + 8u_{l+1} - u_{l+2}\right) = 0 \tag{2.234}$$

Using a hybrid Fourier–Laplace transform of the unknown u in the above, one obtains the effectiveness of the derivative discretization $k_{eq}^{(2)}$ for G2FEM for an interior node as

$$3 \le j \le N - 2 : \qquad k_{eq}^{(2)} = \frac{5\sin kh(4 - \cos kh)}{h(12 + 4\cos kh - \cos 2kh)} \tag{2.235}$$

If one adopts Euler time stepping for the discretized equations for G2FEM, one obtains the numerical amplification factor for G2FEM as

$$3 \le j \le N - 2 : \qquad G^{(2)} = 1 - 5iN_c\frac{\sin kh(4 - \cos kh)}{(12 + 4\cos kh - \cos 2kh)} \tag{2.236}$$

It is seen that G1FEM suffers from numerical instability at $j = 1$ and G2FEM suffers the same for $j = 1$ and 2. Also, it is shown in [338] that there is overstability at $j = N$ for G1FEM and at $j = (N - 1)$ and N for G2FEM.

To apply Galerkin methods for wave propagation problems, numerical instabilities occurring at and near boundaries must be cured – as mentioned in [83, 290, 431, 432]. Brooks & Hughes [45] adopted the prescription in [290] and renamed it as SUPG method. In Grescho & Sani [134], the discrete equation obtained by the SUPG method at an interior node ($2 \le j \le (N - 1)$) for Eq. (2.184) is given as

$$\frac{h}{6}\left[\left(1+\frac{\beta}{2}\right)\frac{du_{j-1}}{dt}+4\frac{du_j}{dt}+\left(1-\frac{\beta}{2}\right)\frac{du_{j+1}}{dt}\right]+\frac{c}{2}\left(u_{j+1}-u_{j-1}\right)$$

$$= \beta c\left(u_{j+1}-2u_j+u_{j-1}\right) \tag{2.237}$$

where β is the stream-wise diffusion parameter. An optimal value of β is given as $1/\sqrt{15}$ in [45] following the work in [290]. In [290], the analytical solution of Eq. (2.237) was taken from [185, 397] in arriving at the optimal value. However, the analytical solution was derived on the basis of a semi-discrete analysis, considering no error being committed in time discretization. Hence, this value is not universal and it varies from one time discretization method to another, a fact often ignored among practitioners. The right-hand side of Eq. (2.237) represents the lowest order dissipation term. The k_{eq}^{SUPG} of the SUPG method for discrete Eq. (2.237) can be obtained as

$$k_{eq}^{SUPG}=\left(\frac{6}{h}\right)\left(\frac{\sin kh-2i\beta(1-\cos kh)}{4+2\cos kh-i\beta\sin kh}\right) \tag{2.238}$$

Adopting Euler time stepping for Eq. (2.237), the numerical amplification factor for the SUPG method can be obtained as

$$G^{SUPG}=1-6\beta N_c\left[\frac{4(1-\cos kh)(2+\cos kh)-\sin^2 kh}{4(2+\cos kh)^2+\beta^2\sin^2 kh}\right]$$

$$-6iN_c\left[\frac{(4+2\cos kh)\sin kh+2\beta^2(1-\cos kh)\sin kh}{4(2+\cos kh)^2+\beta^2\sin^2 kh}\right] \tag{2.239}$$

In Figs. 2.19 and 2.20, we specifically show $G_j(N_c,kh)$ and V_{gN}/c, respectively, for the chosen four methods: (a) RK_4-OUCS3, (b) RK_4-CD_2, (c) RK_4-QUICK and (d) Euler-SUPG obtained using Eqs. (2.209), (2.211) and (2.216). In these figures, properties of these methods for only the interior nodes are shown. From the contour plots in Fig. 2.19, one notes a narrow range of N_c available for which the OUCS3 and CD_2 schemes have neutral stability for a full range of available kh. The region in the (N_c,kh)-plane where this is feasible is marked by hatched lines in Fig. 2.19. In comparison, neither the QUICK nor the SUPG scheme has a neutrally stable region and would not qualify for DNS.

The main interest here is to compare the dispersion property of these methods in terms of numerical group velocity and also find the reason for which q-waves are created numerically, in solving Eq. (2.184). The V_{gN}/c contours shown in Fig. 2.20 indicate that all these methods produce dispersion error, for the computational parameters, when the kh and N_c combination takes values away from the origin. The V_{gN}/c contours show for these four methods, presence of continuous line(s) along which V_{gN} is zero. For the CD_2 and OUCS3 schemes, this is a $kh=constant$ line, parallel to the N_c-axis, while for the other two methods, these lines are curved. From Eq. (2.216), the condition of zero group velocity corresponds to

$$\frac{d}{dk}\left[\tan^{-1}\left(-\frac{G_{ij}}{G_{rj}}\right)\right]=0$$

This can be further simplified to

$$G_{rj}\frac{\partial G_{ij}}{\partial k}=G_{ij}\frac{\partial G_{rj}}{\partial k}$$

For the RK_4 time integration and central spatial schemes, this condition further simplifies to

$$\frac{\partial k_{eq}}{\partial k}=0$$

as $A_j = -ik_{eq}N_c$. This explains the independence of the zero group velocity condition on N_c, while for the QUICK and SUPG methods, this line is a strong function of N_c. Above this line, the numerical group velocity of the methods for the solution of Eq. (2.184) is negative, i.e., the numerical waves would propagate upstream, despite the physical requirement of downstream movement. The region in the (N_c, kh)-plane where q-waves are created is marked by hatched lines in Fig. 2.20. Thus, in solving Eq. (2.184) by the RK_4-CD_2 method, q-waves are created for $kh > \pi/2$, and for the RK_4-OUCS3 method, q-waves are created for

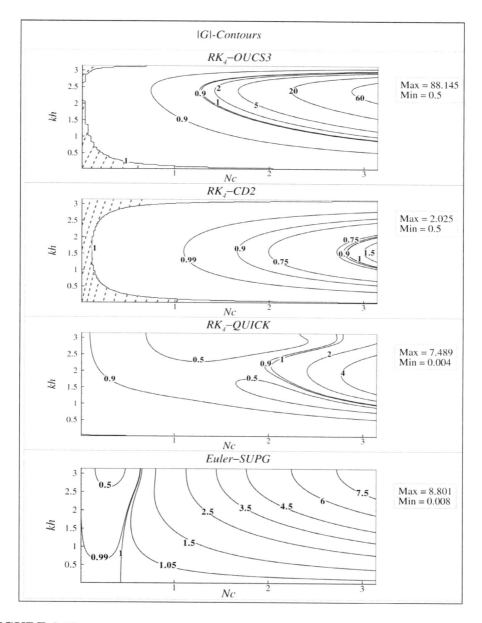

FIGURE 2.19

Comparison of the numerical amplification factor ($|G|$) for interior nodes by the indicated combinations of space–time discretization schemes.

$kh > 2.391$, for any choice of time steps. This critical value, $(kh)_{cr}$, for the OUCS3 method is significantly higher in comparison to the other three methods under consideration. In Fig. 2.20 for the CD_2 method, V_{gN}/c contours are symmetric about $(kh)_{cr}$ and also due to the symmetry of the stencil, minimum and maximum values of V_{gN}/c are same. Maximum and minimum values of numerical group velocities are lowest for the CD_2 method followed by the OUCS3 method. In comparison, the QUICK and SUPG methods have a significantly higher magnitude of maximum and minimum group velocity – essentially due to the addi-

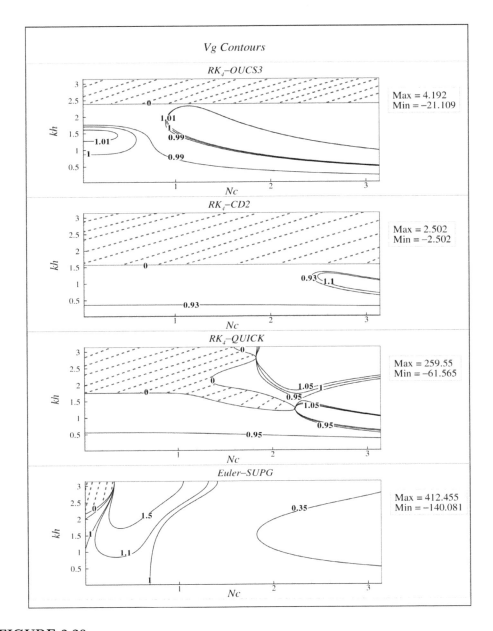

FIGURE 2.20
Comparison of normalized numerical group velocity (V_{gN}/c) for interior nodes for the indicated combinations of space–time discretizations.

tion of excessive numerical dissipation. It is often wrongly presumed that the addition of dissipation does not alter the dispersion property of numerical methods. The presence of q-waves indicates effects similar to what is noted as unsteady flow separation, which is often the harbinger of bypass transition, as shown in [336].

The presence or absence of q-waves depends not only upon the existence of a region with $V_{gN} < 0$, but also on the bandwidth of the implicit filter (as given by the real part of k_{eq}/k) and added dissipation (as given by the imaginary part of k_{eq}/k) of the basic numerical method. It is easy to reason that excessive filtering and damping removes q-waves. The appearance of q-waves is more likely for the OUCS3 method as compared to other methods, due to its lower filtering and less added dissipation. For the CD_2 method, q-waves will appear for $kh > \pi/2$, where the low-pass property of the method will filter the signal that is 38% at each time step, and in contrast, for the OUCS3 method at $(kh)_{cr} = 2.391$, corresponding filtering is equal to only 7%, and there is an additional small attenuation due to low numerical dissipation. It is seen that at $kh = 2.4$, the CD_2 method filters the signal (while spatially discretizing) by more than 70%. In comparison to these two methods, the QUICK and SUPG methods add excessive numerical dissipation which prevents the appearance of q-waves at the cost of numerical accuracy. For the SUPG method, q-waves are created for very small values of N_c and very high values of kh. However, at these high values of kh, the SUPG method introduces a very large amount of numerical dissipation, due to which q-waves are not seen, once again.

The above discussion shows the relative merits of different methods with respect to dispersion properties, including creation of spurious upstream propagating waves, as an example of an extreme form of dispersion error. It is shown that high accuracy compact schemes used with the RK_4 scheme provide the best option in terms of the DRP property. There are many other subtle issues of DRP schemes which can be noted in recent monographs and research articles. We end this chapter by noting that much progress is anticipated in the areas of transition and turbulence by the use of high accuracy techniques with good DRP properties, as will be shown in Chaps. 4 to 8.

3

Instability and Transition in Flows

3.1 Introduction

The basic aim of the material in this chapter is to acquaint readers with the state of the art in the study of receptivity and relate the same with instability and transition of fluid flow. This subject area remains the pacing item in understanding many natural phenomena, as well as in the analysis and design of many engineering systems. For example, this is pursued in civil aviation to design newer lifting surfaces with drag reduced by passive means. In this context, keeping the flow laminar (stable) over a wing, to as large an extent as possible, is the primary goal. With reduced drag, the aircraft speed and range can be increased for the same power consumed or have a less powerful engine for the same endurance and range of the flight envelope.

It is well known that for a zero pressure gradient flat plate boundary layer, the skin friction for laminar flow is given by $C_f = \frac{1.328}{\sqrt{Re}}$ that at a Reynolds number of 10^7 works out as 0.00043 [312] and the same profile drag increases to 0.0035 for the equivalent turbulent flow [417]. This is the rationale for trying to keep a flow laminar so that one can obtain an order of magnitude drag reduction in the relevant portion of the aerodynamic surface. Such drag reductions are also realizable for an airfoil − the quintessential lifting surface element of all aircraft wings. According to [428], flow past an airfoil at a moderate Reynolds number that is fully turbulent without any separation displays a profile drag coefficient of nearly 0.0085 and that can be reduced to 0.0010 if the flow over the airfoil is maintained fully laminar. Even a modest viscous drag reduction via transition delay can provide large benefits, if large numbers of aircrafts are involved, as is the case for the civil air transport industry.

Thus, transition delay for flow over aircraft wings takes on added importance when it is realized that by resorting to transition delay techniques on wings alone about 10 to 12% drag reduction is feasible on a modern transport aircraft. It is also now well established that transition to turbulence relates to the understanding of vorticity distribution in the shear layer and its reorganization in response to forcing by environmental disturbances. While the flow either in its spatio-temporal orderly form (in laminar flows) or in its chaotic form, is governed by the same generalized Navier–Stokes equation [124]. The late stages of transition processes or fully turbulent flows are not amenable to easy understanding due to the intractable nonlinearity of the Navier–Stokes equation [235]. However, the onset of the transition process (also known as the receptivity stage [231, 233, 234]) is bedeviled by our inability to catalog and quantify the background omnipresent disturbances. At this point in time, significant gains in understanding of the receptivity and the linear stage of transition have been made, which allow attempts made to design aircrafts with reduced drag via transition delay. In Fig. 3.1, we reproduce the system portrait given in [235] for the route to wall-bounded turbulent flows. As of today, the primary approaches in transition delay relate to suppressing disturbance growth during the receptivity and primary instability stages. Late stages of transition and turbulent states of flows can be controlled by active means.

This aspect is still an active area of research and is currently not used in any operational transport aircrafts. Flow control has gained in importance in recent times.

Primarily laminar boundary layers are sustained by either a small amount of suction [269] or by favorable pressure gradients. Stabilizing by suction though very efficient (the critical Reynolds number increases by 90 times to $Re_{cr} = 46,000$ for asymptotic suction, as compared to the no-suction case for the Blasius boundary layer for which $Re_{cr} = 520$ [154, 438]), is not practiced due to operational difficulties. For example, one must have the provision of a porous surface together with the necessary suction system – the complexity and maintenance of such a system makes this technology at present unattractive.

In contrast, stabilization of the boundary layer by contouring the airfoil surface to achieve a favorable pressure gradient as a passive way is found to be practical and attractive. The resultant section is known as the Natural Laminar Flow (NLF) airfoil and this is an area which has been under renewed investigation over the last three decades. Early efforts in designing airfoils to delay transition are as given in [1, 158, 399]. Typical examples of such airfoil design are given by the six digit NACA series airfoils. These airfoils were successful at low Reynolds numbers – as evidenced by their continued usage in gliders. This is due to the fact that early NLF airfoils exhibited low drag only for a narrow range of C_l's (designed considering cruise condition only). However, such a section does not perform optimally in other sectors of the flight envelope. This highlights that a practical NLF airfoil must not only have low cruise drag, but also must provide high lift characteristics – very essential during landing, take off and climb. Unfortunately, to maximize the desired performance in terms of low drag degrades high lift performance and vice versa.

The above discussion pertaining to drag reduction of external flow over an aircraft wing is equally relevant to the power requirements for the flow of water or oil in a pipeline. If the flow can be retained laminar and flow instability prevented and/or delayed, then there is a direct benefit in terms of energy efficiency. This is a justified motivation to discuss about flow instability, without the understanding of which it is not possible to design any system whose performance is dictated by fluid flow. Here we provide a general introduction to the topics of receptivity and flow instability, relevant to many engineering systems.

It is not straightforward to perform pure theoretical stability analyses for any given flow. To do this, one needs additional simplifying assumptions to make the problem tractable, with the major ones described next.

3.2   Parallel Flow Approximation and Inviscid Instability Theorems

The above discussion on the rudiments of instability indicates the need to obtain the equilibrium flow, followed by the study of its stability or instability. It so happens that obtaining the equilibrium flow itself could be arduous and that is compounded by the lack of a general procedure to study its instability, without making many simplifying assumptions. For example, instability studies have been facilitated by considering only those flows for which the shear layer grows very slowly, so that the streamlines within the shear layer can be approximated to be parallel to each other, the so-called parallel flow approximation. It should be noted that boundary layer flows under a mild pressure gradient remain unseparated and can be approximated as quasi-parallel flow for the purpose of studying their instability, even when the equilibrium flow has been obtained without such restrictions. A major part of linear (and weakly nonlinear) instability theories has been developed for quasi-parallel flows, starting with the pioneering work of [144, 171, 286].

To study the stability of a two-dimensional parallel flow that only supports a two-dimensional disturbance field, one considers the total flow field to be given by

$$u(x, y, t) = U(y) + \epsilon u'(x, y, t) \tag{3.1}$$

$$v(x, y, t) = \qquad \epsilon v'(x, y, t) \tag{3.2}$$

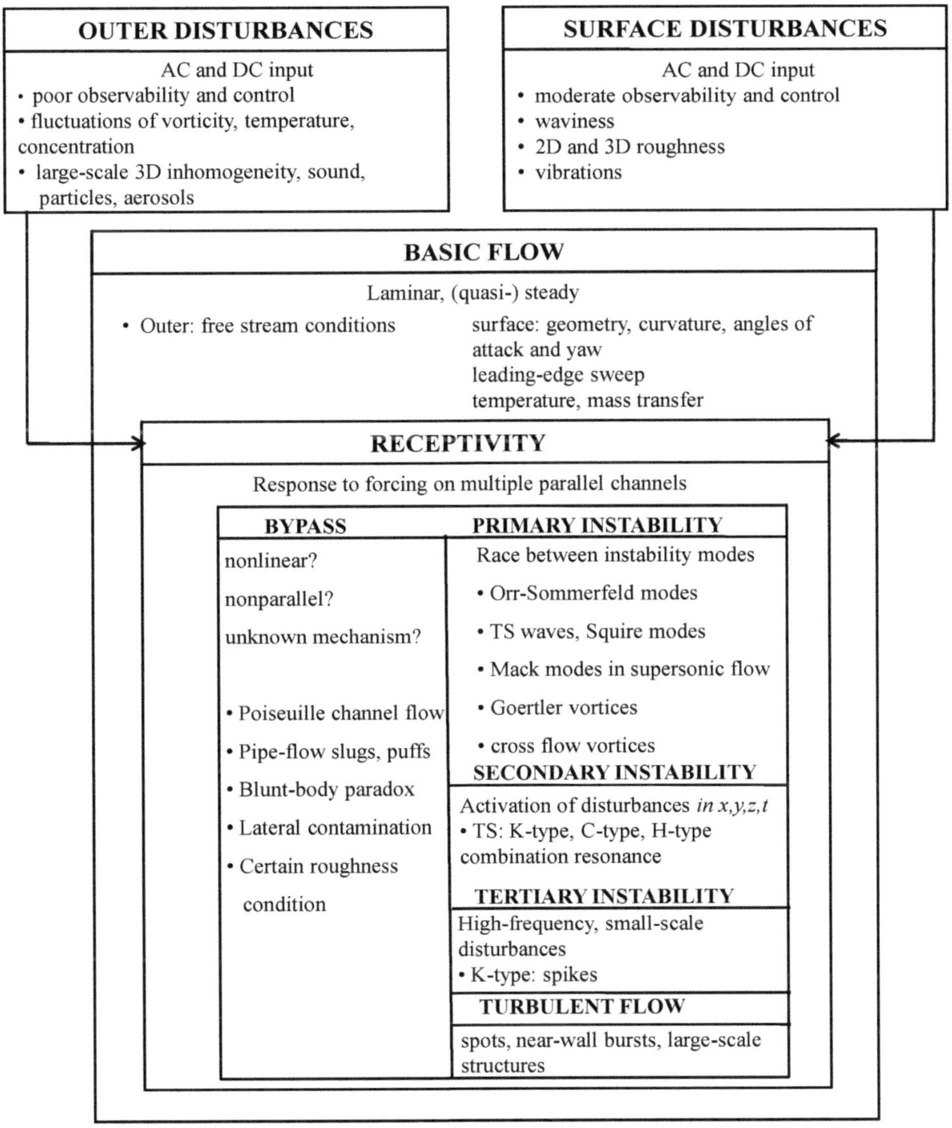

OUTER DISTURBANCES	SURFACE DISTURBANCES
AC and DC input	AC and DC input
• poor observability and control	• moderate observability and control
• fluctuations of vorticity, temperature, concentration	• waviness
• large-scale 3D inhomogeneity, sound, particles, aerosols	• 2D and 3D roughness
	• vibrations

BASIC FLOW

Laminar, (quasi-) steady

• Outer: free stream conditions   surface: geometry, curvature, angles of attack and yaw
leading-edge sweep
temperature, mass transfer

RECEPTIVITY

Response to forcing on multiple parallel channels

BYPASS	PRIMARY INSTABILITY
nonlinear?	Race between instability modes
nonparallel?	• Orr-Sommerfeld modes
unknown mechanism?	• TS waves, Squire modes
	• Mack modes in supersonic flow
• Poiseuille channel flow	• Goertler vortices
• Pipe-flow slugs, puffs	• cross flow vortices
• Blunt-body paradox	**SECONDARY INSTABILITY**
• Lateral contamination	Activation of disturbances *in x,y,z,t*
• Certain roughness	• TS: K-type, C-type, H-type combination resonance
condition	**TERTIARY INSTABILITY**
	High-frequency, small-scale disturbances
	• K-type: spikes
	TURBULENT FLOW
	spots, near-wall bursts, large-scale structures

FIGURE 3.1
System portrait of the road to turbulence [235].

$$p(x, y, t) = P(x, y) + \epsilon p'(x, y, t) \tag{3.3}$$

Note the space-time dependence for the perturbation field is without any restrictions at this stage. In the absence of body force for constant density flows, governing equations for the disturbance field in small perturbation analysis is obtained from the linearized perturbation equations given by

$$\frac{\partial u'}{\partial x} + \frac{\partial v'}{\partial y} = 0 \tag{3.4}$$

$$\frac{\partial u'}{\partial t} + U\frac{\partial u'}{\partial x} + v'\frac{dU}{dy} = -\frac{1}{\rho}\left(\frac{\partial p'}{\partial x}\right) \tag{3.5}$$

$$\frac{\partial v'}{\partial t} + U\frac{\partial v'}{\partial x} = -\frac{1}{\rho}\left(\frac{\partial p'}{\partial y}\right) \tag{3.6}$$

For the purpose of linear analysis, we represent the perturbation quantities by their Laplace–Fourier transform via

$$u'(x, y, t) = \int \bar{u}(y; \alpha, \omega_0)\, e^{i(\alpha x - \omega_0 t)} d\alpha\, d\omega_0 \tag{3.7}$$

$$v'(x, y, t) = \int \bar{v}(y; \alpha, \omega_0)\, e^{i(\alpha x - \omega_0 t)} d\alpha\, d\omega_0 \tag{3.8}$$

$$\frac{p'(x, y, t)}{\rho} = \int \bar{p}(y; \alpha, \omega_0)\, e^{i(\alpha x - \omega_0 t)} d\alpha\, d\omega_0 \tag{3.9}$$

In these representations, α and ω_0 are the wave number and circular frequency, respectively. One can use Eqs. (3.7)–(3.9) in Eqs. (3.4)–(3.6) and eliminate $\bar{u}$ and $\bar{p}$ from these equations to get a single differential equation for $\bar{v}$ as

$$\left(U - \frac{\omega_0}{\alpha}\right)\left(\frac{d^2\bar{v}}{dy^2} - \alpha^2\bar{v}\right) - \frac{d^2 U}{dy^2}\bar{v} = 0 \tag{3.10}$$

This is the celebrated Rayleigh's stability equation. To study this inviscid stability of a fluid dynamical system, one has to solve Eq. (3.10) subject to the homogeneous boundary conditions for $\bar{v}$. In general, for a fluid dynamical system admitting spatio-temporal growth of small perturbations, both α and ω_0 are complex. However, for ease of analysis, we study temporal growth, i.e., we consider α as real and ω_0 as complex. If we write $c = \omega_0/\alpha$, then the complex phase speed $(= c_r + ic_i)$ will determine the stability obtained as an eigenvalue of the equation given by

$$(U - c)\left(\frac{d^2\bar{v}}{dy^2} - \alpha^2\bar{v}\right) - \frac{d^2 U}{dy^2}\bar{v} = 0 \tag{3.11}$$

The criterion for instability then becomes: *There exists a solution with $c_i > 0$ for some positive α.*

3.2.1   Inviscid Instability Mechanism

Let $\bar{v}^*$ be a complex conjugate of $\bar{v}$, so that $\bar{v}\,\bar{v}^* = |\bar{v}|^2$. Multiplying Eq. (3.11) by $\bar{v}^*$ and integrating over a possible limit (say, $-\infty$ to $+\infty$), we get

$$\int_{-\infty}^{+\infty} \bar{v}^*\left[\frac{d^2\bar{v}}{dy^2} - \alpha^2\bar{v} - \frac{U''}{U - c}\bar{v}\right] dy = 0 \tag{3.12}$$

The above equation is non-singular, as we are looking for solutions with $c_i \neq 0$ such that the quantity in the denominator of the third term does not vanish. Integrating by parts the first term and simplifying Eq. (3.12), one gets

$$\int \left[\left| \frac{d\bar{v}}{dy} \right|^2 + \alpha^2 |\bar{v}|^2 \right] dy + \int \frac{U''|\bar{v}|^2}{|U-c|^2} (U-c)^* \, dy = 0 \tag{3.13}$$

In Eq. (3.13), the first term is real and positive. The imaginary part of this equation is

$$c_i \int \frac{U''}{|U-c|^2} |\bar{v}|^2 \, dy = 0 \tag{3.14}$$

This integral of Eq. (3.14) will vanish, if and only if, the integrand changes sign in the interval of integration. That is possible only when U'' changes sign. Thus, there must be a location where $U'' = 0$ for some $y = y_s$, within the limits of integration. This point is known as the inflection point and this leads to the following:

Rayleigh's Inflection Point Theorem: A *necessary condition* for instability is that the basic velocity profile should have an inflection point.

In this theorem, the inflection point refers to the existence of a point within a shear layer (at $y = y_s$ where the local velocity is given by U_s) and where the second derivative of the equilibrium flow profile vanishes, i.e., $\frac{d^2 U}{dy^2} = U'' = 0$, where y is the wall-normal coordinate inside the shear layer. A stronger version of Rayleigh's theorem was given later in [99].

Fjørtoft's Theorem: A *necessary condition* for instability is that $U''(U - U_s)$ is less than zero somewhere in the flow field.

Thus, if the velocity profile is a monotonically growing function of its argument with a single inflection point, then the above necessary condition for instability can be written for this velocity profile as $U''(U - U_s) \leq 0$ for the range of integration, with equality only at $y = y_s$, where $U(y_s) = U_s$. Both Rayleigh's and Fjørtoft's theorems are necessary conditions and they do not provide a sufficient condition for instability.

The above inviscid mechanism of instability is encountered in a free shear layers and jets. A fundamental difference between flows having an inflection point (such as in free shear layer, jets and wakes; cross flow component of some three-dimensional boundary layers) and flows without inflection points (as in wall-bounded flows in channel or in boundary layers) exists. Flows with inflection points are susceptible to temporal instabilities at very low Reynolds numbers. One can find detailed accounts of inviscid instability theories in [30, 91]. We will note in later chapters that temporal instability is also present in many flows with and without heat transfer. For flow past bluff bodies, this eventually leads to periodic vortex shedding.

3.3 Viscous Instability of Parallel Flows

In the beginning, when flow stability was being investigated, it was thought that the action of viscous forces is to dissipate energy and thus its effect is essentially stabilizing. This prompted early stability studies using Rayleigh's equation. However, Heisenberg and various researchers from Prandtl's school established (under the parallel flow assumption) that the action of viscosity can be destabilizing. This provides one mechanism for the instability of zero and favorable pressure gradient boundary layers which are otherwise stable with respect to the inviscid mechanism of Sec. 3.2.1. In this section, the stability of viscous flows (those are essentially parallel or those could be well-approximated as parallel flow) is discussed. Thus, one considers mean flow profiles given by

$$U = U(y)\,\hat{i} \tag{3.15}$$

Flow is considered to be in the x-direction and the velocity magnitude solely depends on the distance from the datum, $y = 0$.

Tollmien [408] calculated a critical Reynolds number (the lowest Reynolds number at which the flow past a flat plate first becomes unstable), details of which can be found in [312]. The value obtained for the critical Reynolds number by Tollmien was $(Re_x)_{crit} = \frac{U_\infty x}{v} = 60,000$ — a value much lower than the wind tunnel results that varied between $Re_{tr} = 3.5 \times 10^5$ (in noisy tunnels) and 10^6 (comparatively cleaner tunnels of that period). We have indicated Reynolds numbers by different subscripts for the purpose of explaining the discrepancy. The reason for it is that instability and transition are not synonymous. While a flow can become unstable early enough, it would take a while for these unstable disturbances to grow to sufficiently large amplitude to complete the process of transition. Schlichting [312] later calculated the amplitude ratio A/A_0 of the *most amplified frequency* as a function of Reynolds number for a flat plate boundary layer, where A_0 is the amplitude of disturbance at the onset of instability. He found out that the ratio A/A_0 varied between e^5 and e^9 at the observed Re_{tr}. These wave solutions are now called the TS waves, irrespective of whether they are amplified or attenuated. It is now generally conjectured that flow parameters such as pressure gradient, suction and heat transfer qualitatively affect transition in the same way that was predicted by viscous stability theory. In this theory, this is cast as an eigenvalue problem.

3.3.1   Eigenvalue Formulation for Instability of Parallel Flows

In this theory, equilibrium flow is obtained using the thin shear layer (TSL) approximation of the governing Navier–Stokes equation. However, to investigate the stability of the fluid dynamical system the disturbance equations are obtained from the full time dependent Navier–Stokes equations, with the equilibrium condition defined by the steady laminar flow. We obtain these in the Cartesian coordinate system given by

$$\frac{\partial u}{\partial t} + u\frac{\partial u}{\partial x} + v\frac{\partial u}{\partial y} + w\frac{\partial u}{\partial z} = -\frac{\partial p}{\partial x} + \frac{1}{Re}\nabla^2 u \tag{3.16}$$

$$\frac{\partial v}{\partial t} + u\frac{\partial v}{\partial x} + v\frac{\partial v}{\partial y} + w\frac{\partial v}{\partial z} = -\frac{\partial p}{\partial y} + \frac{1}{Re}\nabla^2 v \tag{3.17}$$

$$\frac{\partial w}{\partial t} + u\frac{\partial w}{\partial x} + v\frac{\partial w}{\partial y} + w\frac{\partial w}{\partial z} = -\frac{\partial p}{\partial z} + \frac{1}{Re}\nabla^2 w \tag{3.18}$$

$$\frac{\partial u}{\partial x} + \frac{\partial v}{\partial y} + \frac{\partial w}{\partial z} = 0 \tag{3.19}$$

Here the equilibrium flow is considered to be in the (x, z)-plane, with y indicating the wall-normal direction. The above equations are written in non-dimensional form with the shear layer edge velocity, U_e (or an appropriate velocity for other flows), as the velocity scale and L as the length scale so that $Re = \frac{U_e L}{\nu}$ is the Reynolds number.

To perform a linearized stability analysis of the fluid dynamical system, we express all flow quantities q into a steady mean (Q) and an unsteady disturbance term $(\epsilon q')$ which is considered one order of magnitude smaller than the mean quantities, so that

$$q(x, y, z, t) = Q(x, y, z) + \epsilon \, q'(x, y, z, t) \tag{3.20}$$

The smallness of the perturbation quantities is indicated by the small parameter, ϵ. Furthermore, the mean velocity field is assumed parallel/quasi-parallel, so that

$$U = U(y); \quad V \approx 0; \quad W = W(y) \tag{3.21}$$

If the splitting of variables, as indicated by Eq. (3.20), is substituted in Eqs. (3.16)–(3.19) and the $O(\epsilon)$ terms are collated, one gets the following disturbance equations

$$\frac{\partial u'}{\partial t} + U \frac{\partial u'}{\partial x} + W \frac{\partial u'}{\partial z} + v' \frac{dU}{dy} = -\frac{\partial p'}{\partial x} + \frac{1}{Re} \nabla^2 u' \tag{3.22}$$

$$\frac{\partial v'}{\partial t} + U \frac{\partial v'}{\partial x} + W \frac{\partial w'}{\partial z} = -\frac{\partial p'}{\partial y} + \frac{1}{Re} \nabla^2 v' \tag{3.23}$$

$$\frac{\partial w'}{\partial t} + U \frac{\partial w'}{\partial x} + W \frac{\partial w'}{\partial z} + v' \frac{dW}{dy} = -\frac{\partial p'}{\partial z} + \frac{1}{Re} \nabla^2 w' \tag{3.24}$$

$$\frac{\partial u'}{\partial x} + \frac{\partial v'}{\partial y} + \frac{\partial w'}{\partial z} = 0 \tag{3.25}$$

Next, one performs the normal mode analysis, i.e., the flow instabilities are governed by discrete eigenmodes that do not interact with each other and are studied separately. Equations (3.22)–(3.25) are variable coefficient linear partial differential equations, but they do not admit analytic solutions. However, with the help of normal mode analysis, this can be further simplified. As the coefficients of these equations are functions of the wall-normal coordinate, it is natural to expand the disturbance quantities in the following manner

$$\{u', v', w', p'\}^T = \{f(y), \varphi(y), h(y), \pi(y)\}^T e^{i(\alpha x + \beta z - \omega_0 t)} \tag{3.26}$$

Here, the disturbance amplitudes f, ϕ, h and π are the complex amplitude functions and ω_0 is the dimensionless circular frequency, $(= \omega_0^* L / U_e)$. When Eq. (3.26) is substituted in Eqs. (3.22)–(3.25), the following ordinary differential equations result

$$i[\alpha U + \beta W - \omega_0] f + U' \phi = -i\alpha \pi + \frac{1}{Re}[f'' - (\alpha^2 + \beta^2) f] \tag{3.27}$$

$$i[\alpha U + \beta W - \omega_0] \phi = -\pi' + \frac{1}{Re}[\phi'' - (\alpha^2 + \beta^2) \phi] \tag{3.28}$$

$$i[\alpha U + \beta W - \omega_0] h + W' \phi = -i\beta \pi + \frac{1}{Re}[h'' - (\alpha^2 + \beta^2) h] \tag{3.29}$$

$$i(\alpha f + \beta h) + \phi' = 0 \tag{3.30}$$

In these equations, primes indicate differentiation with respect to y. One can rewrite these equations as a set of six first order equations and thus one would require six boundary conditions to solve these simultaneously. For the stability analysis, the above equations are solved subject to homogeneous boundary conditions, which are compatible with the governing equations. For example, at the wall one uses the no-slip boundary conditions

$$f(0) = \phi(0) = h(0) = 0 \tag{3.31}$$

At the free stream (i.e., as $y \to \infty$), one requires the disturbance velocity components to decay to zero, i.e.,

$$f(y), \phi(y), h(y) \to 0 \quad \text{as} \quad y \to \infty \tag{3.32}$$

Non-trivial solutions of Eqs. (3.27)–(3.30), subject to the homogeneous boundary conditions of Eqs. (3.31) and (3.32), exist only for particular combinations of the parameters α, β, ω_0 and Re. This, then, produces the dispersion or eigenvalue relation as

$$g(\alpha, \beta, \omega, Re) = 0 \tag{3.33}$$

It is also possible to reduce Eqs. (3.27)–(3.30) to a single ordinary differential equation for the unknown ϕ. One can combine the first and the third equations in the set to form an equation for $(\alpha f + \beta h)$. This variable can be replaced by using Eq. (3.30), and after differentiation with respect to y and eliminating π' by using Eq. (3.28), one gets the following equation for ϕ as

$$\phi^{iv} - 2(\alpha^2 + \beta^2)\phi'' + (\alpha^2 + \beta^2)^2\phi =$$
$$i\,Re\left[(\alpha U + \beta W - \omega_0)[\phi'' - (\alpha^2 + \beta^2)\phi] - (\alpha U'' + \beta W'')\phi\right] \tag{3.34}$$

This is the well-known Orr–Sommerfeld equation, which forms the major tool for the investigation of flow instability. If one considers a two-dimensional disturbance field in a two-dimensional mean flow, then the above equation transforms to the simpler form

$$\phi^{iv} - 2\alpha^2\phi'' + \alpha^4\phi = i\,Re[(\alpha U - \omega_0)[\phi'' - \alpha^2\phi] - \alpha U''\phi] \tag{3.35}$$

Thus, the Orr–Sommerfeld equation is a fourth order ordinary differential equation and has the same form whether the mean flow is three dimensional or two dimensional. The same observation can be made with respect to the disturbance field, as well, for special cases. This fact can be exploited to relate more general cases to Eq. (3.35). It is illustrated below for two-dimensional mean flow with a three-dimensional disturbance field. Setting $W \equiv 0$ and using the following transformations in Eq. (3.34)

$$\alpha^2 + \beta^2 = \tilde{\alpha}^2; \quad \omega_0\,\tilde{\alpha} = \tilde{\omega}_0\,\alpha \quad \text{and} \quad Re\,\alpha = \tilde{Re}\,\tilde{\alpha} \tag{3.36}$$

one gets the following governing equation

$$\phi^{iv} - 2\tilde{\alpha}^2\phi'' + \tilde{\alpha}^4\phi = i\,\tilde{Re}\,[(\tilde{\alpha}\,U - \tilde{\omega}_0)[\phi'' - \tilde{\alpha}^2\phi] - \tilde{\alpha}U''\,\phi] \tag{3.37}$$

In essence, Eqs. (3.34) and (3.37) are identical — expressed for different parameters, where these are related via Eq. (3.36). The mean flow U is real and unchanged and if α and β are real, then a three-dimensional stability problem at a Reynolds number Re has been reduced to a two-dimensional problem at the lower Reynolds number $\tilde{Re}$. This is known as **Squire's theorem**, which is formally stated as:

> **Squire's Theorem**: In a two-dimensional boundary layer with real wave numbers, instability appears first for two-dimensional disturbances.

We must, however, note that the utility of Squire's theorem is lost if the mean flow is three dimensional, or if α and β are complex, as in spatial stability problems.

3.3.2 Temporal and Spatial Amplification of Disturbances

If α, β and ω_0 are all real, then from Eq. (3.26) one can see that the disturbances propagate through the shear layer with constant amplitude at all times. However, if α and β are real and ω_0 is complex, then according to Eq. (3.26) the disturbances will grow or decay with time. In contrast, if ω_0 is real and α, β are complex, then the disturbance amplitude will not change with time, but these will change with x and z. The former case, where the disturbance amplitude changes with time, is a subject dealt with in *temporal amplification theory* and the latter is the subject of *spatial amplification theory*. If all three quantities α, β and ω_0 are complex, then the disturbances grow in both space and time and are the subject matter of *spatio-temporal growth of disturbances*.

3.3.2.1 Temporal Amplification Theory

With $\omega_0 = \omega_r + i\omega_i$ and (α, β) real, the disturbance field can be written as

$$q'(x, y, z, t) = [\hat{q}(y) \, e^{\omega_i t}] \, e^{i(\alpha x + \beta z - \omega_r t)} \tag{3.38}$$

The magnitude of the wavenumber vector is $\bar{\alpha} = (\alpha^2 + \beta^2)^{1/2}$ and the angle between the wavenumber vector and the x-axis is given by $\psi = \tan^{-1}(\beta/\alpha)$. The phase speed of the disturbance field is given by $c_{ph} = \omega_r/\bar{\alpha}$. If A represents the magnitude of the disturbance at a particular height y, then it follows from Eq. (3.38)

$$\frac{1}{A}\frac{dA}{dt} = \omega_i \tag{3.39}$$

Thus, ω_i is the *amplification rate* in the temporal theory, as

$$A = A_0 \, e^{\omega_i t}$$

If

$$\omega_i < 0 \tag{3.40a}$$

then the disturbance is *damped*.
If, on the other hand,

$$\omega_i = 0 \tag{3.40b}$$

then the disturbance is *neutral*.
Finally, if

$$\omega_i > 0 \tag{3.40c}$$

then the disturbance *amplifies* with time.

3.3.2.2 Spatial Amplification Theory

In this theory, ω_0 is treated as real and the wavenumber components are complex, i.e.,

$$\alpha = \alpha_r + i\alpha_i \quad \text{and} \quad \beta = \beta_r + i\beta_i \tag{3.41}$$

Thus, one can write the disturbance field as

$$q'(x, y, z, t) = [\hat{q}(y) \, e^{-(\alpha_i x + \beta_i z)}] \, e^{i(\alpha_r x + \beta_r z - \omega_0 t)} \tag{3.42}$$

If one defines

$$\bar{\alpha}_r = [\alpha_r^2 + \beta_r^2]^{1/2} \quad \text{and} \quad \psi = \tan^{-1}(\beta_r/\alpha_r) \tag{3.43}$$

then the phase speed of the disturbance field is given by $c_{ph} = \omega_0/\bar{\alpha}_r$. Additionally, if one defines $\bar{\alpha}_i = [\alpha_i^2 + \beta_i^2]^{1/2}$ and $\bar{\psi} = \tan^{-1}(\beta_i/\alpha_i)$, then one can define two new directions, $\tilde{x}$ along $\bar{\alpha}_r$ and $\tilde{x}$ along $\bar{\alpha}_i$ and rewrite Eq. (3.42) as

$$q'(x, y, z, t) = \hat{q}(y)\, e^{-\bar{\alpha}_i \tilde{x}}\, e^{i(\bar{\alpha}_r \tilde{x} - \omega_0 t)} \tag{3.44}$$

Thus, one can similarly write a spatial amplification rate in the particular direction given by $\bar{\psi}$ as

$$\frac{1}{A}\frac{dA}{d\tilde{x}} = -\bar{\alpha}_i \tag{3.45}$$

The amplification rates are different in different directions, and of course, different $\bar{\alpha}_i$'s are functions of $\bar{\psi}$. For three-dimensional waves, spatial theory comes with a lot more complications as compared to temporal theory. In addition to the wave orientation angle ψ, the amplification direction $\bar{\psi}$ must also be specified before any calculation can be made. Once again, if

$$\bar{\alpha}_i > 0 \tag{3.46a}$$

then this corresponds to a *damped* solution.
On the other hand, if

$$\bar{\alpha}_i = 0 \tag{3.46b}$$

then this corresponds to *neutral stability*.
Finally, if

$$\bar{\alpha}_i < 0 \tag{3.46c}$$

then we have the situation of *instability*.

3.3.2.3 Relationship between Temporal and Spatial Theories

Consider the general dispersion relation

$$\omega_0 = \omega_0(\alpha, \beta, Re, ...) \tag{3.47}$$

From this, we can obtain the group velocity components for a three-dimensional disturbance field in the x- and z-directions, respectively, as

$$\vec{V}_g = \left(\frac{\partial \omega_0}{\partial \alpha}, \frac{\partial \omega_0}{\partial \beta}\right) \tag{3.48}$$

In *temporal theory*, one uses $\omega_0 = \omega_r$ and in *spatial theory*, one uses $\alpha = \alpha_r$ and $\beta = \beta_r$ in Eq. (3.48). The imaginary part of the group velocity is usually neglected. For parallel flows, one can form a spatial amplification rate by moving with the wave at the group velocity, i.e.,

$$\frac{d}{dt} = \vec{V}_g \frac{d}{d\tilde{x}} \tag{3.49}$$

where $\tilde{x}$ is chosen in the direction of $\vec{V}_g$. Consequently

$$\bar{\alpha}_i = -\frac{\omega_i}{|\vec{V}_g|^2}\,\vec{V}_g \tag{3.50}$$

and the direction of $\bar{\alpha}_i$ is obtained from

$$\bar{\psi} = \tan^{-1}\left[\frac{\partial \omega_r/\partial \beta}{\partial \omega_r/\partial \alpha}\right] \tag{3.51}$$

For two-dimensional flows, such a relation was used by [311] without any proof, which was later provided in [108]. But this can be shown for a general disturbance field by noting that ω_0 is an analytic complex function of α and β. Therefore, one can use the Cauchy–Riemann equations valid for complex analytic functions and here these are given by

$$\frac{\partial \omega_r}{\partial \alpha_r} = \frac{\partial \omega_i}{\partial \alpha_i}, \quad \frac{\partial \omega_r}{\partial \alpha_i} = -\frac{\partial \omega_i}{\partial \alpha_r} \tag{3.52}$$

and

$$\frac{\partial \omega_r}{\partial \beta_r} = \frac{\partial \omega_i}{\partial \beta_i}, \quad \frac{\partial \omega_r}{\partial \beta_i} = -\frac{\partial \omega_i}{\partial \beta_r} \tag{3.53}$$

For three-dimensional disturbances, the left-hand side of the first of Eqs. (3.52) and (3.53) are the group velocity components, as defined by Eq. (3.48). Also, $\frac{\partial \omega_i}{\partial \alpha_i}$ can be approximated by noting that ω_i decreases from its temporal value to zero in the spatial theory, as α_i goes from zero to its value obtained in the spatial theory. If the amplification rates are small, as they are in linear theory, then the above variations are linear and thus $\frac{\partial \omega_i}{\partial \alpha_i} \simeq -\frac{\omega_i}{\alpha_i}$.

Therefore

$$[V_g]_x = -\frac{\omega_i}{\alpha_i} \tag{3.54}$$

For three-dimensional disturbances, if $\bar{\psi}$ is specified arbitrarily and the x-axis is rotated to lie in the $\bar{\psi}$ direction, then Eq. (3.50) will apply with α_i replaced by $\bar{\alpha}_i$ and V_g by the component of group velocity in the $\bar{\psi}$ direction.

3.4 Properties of the Orr–Sommerfeld Equation and Boundary Conditions

For the stability analysis of a fluid dynamical system whose equilibrium solution is given by a parallel or quasi-parallel flow, one has to solve the Orr–Sommerfeld equation depending on whether the mean flow is two dimensional or three dimensional. For the two-dimensional problem, equivalent boundary conditions are given by

At y = 0:

$$f, \quad \phi = 0 \tag{3.55a}$$

And as

$$y \to \infty : \quad f, \quad \phi \to 0 \tag{3.55b}$$

To solve Eq. (3.35), the above boundary conditions are transformed in terms of ϕ for a two-dimensional disturbance field as

$$\phi' = -i\alpha f \tag{3.56}$$

Thus, one has to satisfy homogeneous boundary conditions for ϕ and ϕ' at both the boundaries for wall-bounded shear flows. The consequence of far stream boundary conditions, as given by Eq. (3.55b), is understood by using the mean flow information at

$y \to \infty:$ $U(y) = 1$ and $U''(y) \equiv 0$ in Eq. (3.35). One gets the following constant coefficient ODE at $y \to \infty$

$$\phi^{iv} - 2\alpha^2 \phi'' + \alpha^4 \phi = i\,Re\,[(\alpha - \omega_0)(\phi'' - \alpha^2 \phi)] \qquad (3.57)$$

The solution of the above can be obtained in the form $\phi \sim e^{\lambda y}$, such that one gets the characteristic roots as $\lambda_{1,2} = \mp\alpha$ and $\lambda_{3,4} = \mp Q$, where $Q = [\alpha^2 + i\alpha\,Re(1-c)]^{1/2}$. For boundary layer instability problems, $Re \to \infty$ and then $|Q| >> |\alpha|$. This is the source of stiffness, which makes obtaining a numerical solution of Eq. (3.35) a daunting task. This causes the fundamental solutions of the Orr–Sommerfeld equation to vary by different orders of magnitude, near and far away from the wall. This type of behavior makes the governing equation a *stiff differential equation* which suffers from the growth of *parasitic error* while numerically solving it.

For a pure hydrodynamics problem, the Orr–Sommerfeld equation is a fourth order ordinary differential equation. Thus, it will have four fundamental solutions whose asymptotic variation for $y \to \infty$ is given by the characteristic exponents of Eq. (3.57), i.e.,

$$\phi = a_1\phi_1 + a_2\phi_2 + a_3\phi_3 + a_4\phi_4 \qquad (3.58)$$

where at the free stream $(y \to \infty)$: $\phi_1 \sim e^{-\alpha y}$; $\phi_2 \sim e^{\alpha y}$; $\phi_3 \sim e^{-Qy}$ and $\phi_4 \sim e^{Qy}$.

To satisfy the boundary condition given in Eq. (3.55b), one must have $a_2 = a_4 = 0$ for real $(\alpha, Q) > 0$. Then the general solution is of the form

$$\phi = a_1\phi_1 + a_3\phi_3 \qquad (3.59)$$

This is an admissible solution of Eqs. (3.35) and/or (3.37) which satisfies one set of boundary conditions [Eq. (3.55b)] automatically. Two remaining constants in the solution of Eq. (3.59) can be fixed by satisfying the boundary conditions given by Eq. (3.55a) as

$$a_1\phi_1(y = 0,\ \alpha\ ;\ \omega_0,\ Re) + a_3\phi_3(y = 0,\ \alpha\ ;\ \omega_0,\ Re) = 0 \qquad (3.60)$$

$$a_1\phi_1'(y = 0,\ \alpha\ ;\ \omega_0,\ Re) + a_3\phi_3'(y = 0,\ \alpha\ ;\ \omega_0,\ Re) = 0 \qquad (3.61)$$

We will get a non-trivial solution for these if and only if the determinant of the associated matrix of the linear algebraic system given by above equations is zero, i.e.,

$$(\phi_1\phi_3' - \phi_1'\phi_3)_{y=0} = 0 \qquad (3.62)$$

This is the characteristic equation for the eigenvalues posed by the Orr–Sommerfeld equation, which can also be viewed as the dispersion relation of the problem. So the task at hand is to obtain a combination of α and ω_0 for a given Re such that the solution of OSE satisfies Eq. (3.62). The stiffness of OSE causes the numerical solution to lose the *linear independence* of different solution components corresponding to the different fundamental solutions. This is the source of *parasitic error* growth of any *stiff differential equation*. To remove this problem in a straightforward manner, one can use the compound matrix method (CMM).

As discussed in [91], there are three principal methods for solving linear stability problems and these are: (a) the matrix method based on finite difference or spectral discretizations; (b) shooting techniques along with orthogonalization of the fundamental solutions and (c) shooting with CMM. As discussed in [4], for eigenvalue problems in an infinite domain, matrix methods based on finite difference and spectral collocations produce spurious eigenvalues due to problems in satisfying boundary conditions correctly, due to fracturing of the continuous spectra. Shooting methods based on orthogonalization are cumbersome to program and require comparatively large computer memory, while producing non-analytic solutions. Considering these, CMM is the best available method for hydrodynamic stability problems. This method has been reformulated in [5] using exterior algebra to make it coordinate free. The advantages of CMM as compared to other methods have been described in [91, 248, 249, 320, 358]. There are other methods based on an ODE solver in the physical plane [318] or spectral collocation methods which involve Chebyshev discretization of the Orr–Sommerfeld equation as given in [314].

3.4.1 Compound Matrix Method

The compound matrix method (CMM) has been in use for stability calculations for over three decades, as the original method was reported in [248, 249]. Its main strength lies in its simplicity of application and interpretation of the results. CMM not only yields satisfactory results for the evaluation of eigenvalues and eigenfunctions of the Orr–Sommerfeld equation, but it also brings to the fore the analytic structure of the solution. For these reasons, CMM is recommended to readers. Moreover, because of the specific structure, it is possible to solve both the stability and the receptivity problems without a great deal of coding effort. Some essential modifications for appropriate choice of equation in CMM for eigenfunctions are reported in [320].

In this method, instead of working with ϕ, one works with a new set of variables which are combinations of the fundamental solutions ϕ_1 and ϕ_3. These new variables all vary with y at the identical rate, thereby removing the stiffness problem. For the Orr–Sommerfeld equation, these new variables are [91, 320]

$$y_1 = \phi_1 \phi_3' - \phi_1' \phi_3$$
$$y_2 = \phi_1 \phi_3'' - \phi_1'' \phi_3$$
$$y_3 = \phi_1 \phi_3''' - \phi_1''' \phi_3$$
$$y_4 = \phi_1' \phi_3'' - \phi_1'' \phi_3'$$
$$y_5 = \phi_1' \phi_3''' - \phi_1''' \phi_3'$$
$$y_6 = \phi_1'' \phi_3''' - \phi_1''' \phi_3''$$

$$(3.63)$$

It is easy to verify with the help of the solution of Eq. (3.57) in the free stream that y_1 to y_6 have identical growth rates as one integrates from the free stream to the wall. From the definition given above in Eq. (3.63), one gets the following

$$y_1' = \phi_1' \phi_3' + \phi_1 \phi_3'' - \phi_1'' \phi_3' - \phi_1'' \phi_3 = y_2 \qquad (3.64a)$$

$$y_2' = (\phi_1 \phi_3''' - \phi_1''' \phi_3) + (\phi_1' \phi_3'' - \phi_1'' \phi_3') = y_3 + y_4 \qquad (3.64b)$$

$$y_3' = \phi_1 \phi_3^{iv} + (\phi_1' \phi_3''' - \phi_1''' \phi_3') - \phi_3 \phi_1^{iv}$$

From Eq. (3.35)

$$\phi_1^{iv} = \{2\alpha^2 + i\alpha\, Re\{U - c\}\}\phi_1'' - \{\alpha^4 + i\alpha^3\, Re\{U - c\} + i\alpha\, Re\, U''\}\phi_1$$

or

$$\phi_1^{iv} = b_1 \phi_1'' - b_2 \phi_1$$

Similarly, a relation for ϕ_3^{iv} can be obtained and one can simplify to obtain $\phi_1 \phi_3^{iv} - \phi_1^{iv} \phi_3 = b_1 y_2$. Thus

$$y_3' = b_1 y_2 + y_5 \tag{3.64c}$$

$$y_4' = \phi_1' \phi_3''' + \phi_1'' \phi_3'' - \phi_1'' \phi_3'' - \phi_1''' \phi_3' = y_5 \tag{3.64d}$$

$$y_5' = \phi_1' \phi_3^{iv} + \phi_1'' \phi_3''' - \phi_1^{iv} \phi_3' - \phi_1''' \phi_3''$$

$$= [\phi_1'' \phi_3''' - \phi_1''' \phi_3''] + \phi_1'(b_1 \phi_3'' - b_2 \phi_3) - \phi_3'(b_1 \phi_1'' - b_2 \phi_1)$$

$$y_5' = y_6 + b_1 y_4 + b_2 y_1 \tag{3.64e}$$

$$y_6' = \phi_1'' \phi_3^{iv} + \phi_1''' \phi_3''' - \phi_1''' \phi_3''' - \phi_1^{iv} \phi_3'' = b_2 y_2 \tag{3.64f}$$

Equations (3.64a) to (3.64f) are six first order ordinary differential equations for the six unknown variables y_1 to y_6. Note that the order of the system is increased from four to six in CMM, while the governing equation is transformed from a boundary value problem to an initial value problem. To solve these six equations, we therefore need to generate initial conditions for the unknowns. As we know the property of the fundamental solutions in the free stream, we can use that information to generate the initial conditions for y_1 to y_6. As at $y \to \infty : \phi_1 \sim e^{-\alpha y}$ and $\phi_3 \sim e^{-Qy}$, therefore we can estimate the free stream values of the unknown as

$$y_1 \sim (-Q + \alpha) e^{-(\alpha+Q)y} \tag{3.65a}$$

$$y_2 \sim (Q^2 - \alpha^2) e^{-(\alpha+Q)y} \tag{3.65b}$$

$$y_3 \sim (-Q^3 + \alpha^3) e^{-(\alpha+Q)y} \tag{3.65c}$$

$$y_4 \sim (-\alpha Q^2 + \alpha^2 Q) e^{-(\alpha+Q)y} \tag{3.65d}$$

$$y_5 \sim (\alpha Q^3 - \alpha^3 Q) e^{-(\alpha+Q)y} \tag{3.65e}$$

$$y_6 \sim (-\alpha^2 Q^3 + \alpha^3 Q^2) e^{-(\alpha+Q)y} \tag{3.65f}$$

Note that in Eq. (3.65) all the variables have the same exponential rate of growth or attenuation — a special feature of CMM, where the problem of stiffness is removed by the specific choice of new variables. In integrating Eqs. (3.64), one can start off with the values given by Eqs. (3.65), with variables normalized with respect to one of these. Let us normalize every variable with respect to y_1, so that the initial conditions for solving Eqs. (3.64) are

$$y_1 = 1.0 \tag{3.66a}$$

$$y_2 = -(\alpha + Q) \tag{3.66b}$$

$$y_3 = \alpha^2 + \alpha Q + Q^2 \tag{3.66c}$$

$$y_4 = \alpha Q \tag{3.66d}$$

$$y_5 = -\alpha Q(\alpha + Q) \tag{3.66e}$$

$$y_6 = \alpha^2 Q^2 \tag{3.66f}$$

One solves Eqs. (3.64), starting off from the free stream to the wall, with the initial condition provided by Eqs. (3.66). After marching up to the wall, observe that the satisfaction of the characteristic equation, Eq. (3.62), is equivalent to locating (α, ω_0) combinations for a given Re. This is exactly equivalent to enforcing

$$y_1 = 0 \quad \text{at } y = 0 \tag{3.67}$$

Having recast the problem in CMM in terms of the new variables, one can calculate the eigenvalues very quickly. That leaves one with the task of finding out the corresponding eigenvector(s). This also can be done readily by noting that the eigenvector ϕ is a linear combination of ϕ_1 and ϕ_3 such that

$$\phi = a_1 \phi_1 + a_3 \phi_3 \tag{3.68a}$$

$$\phi' = a_1 \phi_1' + a_3 \phi_3' \tag{3.68b}$$

$$\phi'' = a_1 \phi_1'' + a_3 \phi_3'' \tag{3.68c}$$

$$\phi''' = a_1 \phi_1''' + a_3 \phi_3''' \tag{3.68d}$$

One can eliminate a_1 and a_3 from Eqs. (3.68) using Eqs. (3.64) in many ways. This leads to the following differential equations for ϕ

$$y_1 \phi'' - y_2 \phi' + y_4 \phi = 0 \tag{3.69a}$$

$$y_1 \phi''' - y_3 \phi' + y_5 \phi = 0 \tag{3.69b}$$

$$y_2 \phi''' - y_3 \phi'' + y_6 \phi = 0 \tag{3.69c}$$

$$y_4 \phi''' - y_5 \phi'' + y_6 \phi' = 0 \tag{3.69d}$$

In principle, having obtained y_1 to y_6 for all y's, it is possible to obtain ϕ by solving any one of the four equations listed above. It is noted that the retained fundamental solution automatically satisfies the far stream boundary conditions and thus one can solve any one of the equations in Eq. (3.69), starting from the wall and marching towards the free stream. The homogeneous boundary conditions given by Eq. (3.55a) cannot be used as the initial condition to march out from the wall to the free stream, as the differential equations in Eq. (3.69) are homogeneous. For stability problems, one should provide non-trivial normalized boundary conditions, as discussed in [358], instead of Eq. (3.55a). For receptivity problems, the wall boundary conditions are inhomogeneous and raise no problems at all. This is the benefit of CMM where a difficult boundary value problem is replaced by two relatively easier initial value problems.

The presence of four alternatives in Eqs. (3.69) raises the question of which of these equations is more accurate and consistent for the solution process. It was shown in [320] that the eigenvector cannot be obtained by all four equations for high Reynolds numbers and/or high wavenumbers. It was shown clearly in [320] that the third equation, Eq. (3.69c), produced divergent results for high wavenumbers due to the following reason.

All four equations, (3.69a) to (3.69d), reduce to constant coefficient ordinary differential equations at the free stream, for which one can use the values provided in Eqs. (3.66) for the compound matrix variables. It is easy to see that Eq. (3.69c) has three characteristic roots at the free stream given by $-\alpha$, $-Q$ and $\frac{\alpha Q}{\alpha + Q}$. The first two roots correspond to the fundamental solutions ϕ_1 and ϕ_3, but the third root is not only extraneous, but is also unstable! Thus, for the solution of Eq. (3.35) subject to boundary conditions in Eqs. (3.55a) and (3.55b), this spurious mode will not satisfy the free stream boundary condition. For moderate to high Reynolds numbers (as will be case for all post-critical, but pre-transitional flows), $|Q| >> |\alpha|$ and the third characteristic root can be simplified to $\frac{\alpha Q}{\alpha + Q} \simeq \alpha$. Thus, this third mode is a close approximation to ϕ_2 which violates the boundary condition of Eq. (3.55b). It is possible to miss this problem, as in [112, 78], where results were reported using this third equation. The problem associated with this growing mode was not observed because the leading eigenvalue of such problems is usually very small and if the far stream boundary is not taken far enough outside the shear layer, this problem remains masked. However, when one wants to solve a receptivity problem, one needs to solve the Orr–Sommerfeld equation for a large range of α's and there this problem becomes very apparent. This was the case investigated in [322] for the receptivity of the Blasius boundary layer.

The above investigation for the behavior of the solution at the free stream can be extended to other equations of Eq. (3.69) to check the usefulness of these in obtaining eigenfunctions. At the free stream, the characteristic roots for Eq. (3.69a) are $\{-\alpha, -Q\}$. Equation (3.69b), being a third order equation, has three roots given by $[-\alpha, -Q, (\alpha + Q)]$. Thus, this equation is also violently unstable, even for low wavenumbers and Reynolds numbers. Equation (3.69c) has asymptotic behavior for large y's, as dictated by the characteristic roots given by $[-\alpha, -Q, \frac{\alpha Q}{\alpha + Q}]$. Finally, characteristic roots for Eq. (3.69d) are given by $\{-\alpha, -Q, 0\}$, i.e., one mode is neutral for large values of y. One sees that the first and the fourth equations are compatible with the required formulation and boundary conditions, while the second and third equations are unstable (when the shear layer is excited at the wall) and need to be avoided. In [91] it is mentioned that *it is certainly not obvious which one is best for numerical purposes*, and this simple analysis tells us that the correct choice of equation is either Eq. (3.69a) or (3.69d). In obtaining eigenfunctions of the stability problem defined via the Orr–Sommerfeld equation, it is noted that Eq. (3.69a) is not suitable because y_1 at the wall is zero. It was suggested in [248] that this difficulty can be overcome by integrating Eq. (3.35) itself a few steps from the wall before switching over to Eq. (3.69a). At the wall, it is observed that while ϕ and ϕ' are zero, ϕ'' is indeterminate. The value of ϕ'' therefore can be arbitrarily set equal to any value and then either Eq. (3.69a) or Eq. (3.69d) can be solved, avoiding the above suggestion in [248]. This procedure has been followed in [358]. While solving receptivity problems, boundary conditions are the prescribed non-zero values and the problem of the starting solution for stability analyses does not arise. However, another problem may bedevil the CMM for obtaining eigenfunctions of Eq. (3.35) for high Reynolds numbers. This is due to the dynamic ranges of the compound matrix variables inside the shear layer. For example, in Eq. (3.69a) one would require accurate calculation of the ratio of $\frac{y_2}{y_1}$ and $\frac{y_4}{y_1}$. These ratios exhibit very large variations near the wall, where y_1 is close to zero. This causes the eigensolution to be highly oscillatory. For Eq. (3.69d) the ratios $\frac{y_5}{y_4}$ and $\frac{y_6}{y_4}$ are of order one and non-oscillatory. Thus, the eigenfunctions obtained by solving Eq. (3.69d) are quite regular, as seen from the results in [322] for a receptivity problem. Problems of solving the Orr–Sommerfeld equation analytically for very large values of α can be handled in a satisfactory way by looking for an asymptotic solution, which is dealt with later.

3.5 Instability Analysis from the Solution of the Orr–Sommerfeld Equation

The Orr–Sommerfeld equation can be solved either as a temporal or as a spatial instability problem. For a disturbance field created as a consequence of a localized excitation inside boundary layers, temporal growth of the disturbance field is not realistic. It has been observed phenomenologically that for attached flows, instability is usually of the convective type and obtaining a solution by spatial analysis is appropriate. In Chap. 5, we will note that even for such a problem there can be spatio-temporally growing wave-fronts which dominate in attached boundary layers, even when the shear layer is spatially stable.

Let us discuss the spatial instability of parallel flow past a flat plate at zero angle of attack — the problem with canonical status in stability analyses. For the spatial instability problem associated with a two-dimensional disturbance field of two-dimensional primary flows, disturbance quantities will have the appearance of Eq. (3.42) with $\beta = 0$. Thus, for a fixed Re, one would be looking for a complex α, when the shear layer is excited by a fixed frequency source, ω_0. If we define Re in terms of the displacement thickness δ^* as the length scale, then $Re = \frac{U_e \delta^*}{v}$. Results obtained are plotted as contours of constant amplification rates α_i in the (Re, ω_0)-plane, as shown in Fig. 3.2 for zero pressure gradient flow.

The solid line in the figure corresponds to $\alpha_i = 0$ and represents the condition of neutral stability and hence this is called the *neutral curve*. The area inside the neutral curve corresponds to $\alpha_i < 0$ and represents flow instability for these parameter combinations. Similarly, the area outside the *neutral curve* corresponds to stable disturbances in the shear layer. It is noted that the foremost position of the *neutral curve* occurs for $Re = 519.23$ and corresponds to the fact that the flow is stable below this value of Re for all circular frequencies. Hence, this value of Reynolds number is called the *critical Reynolds number*. There is also similarly an upper bound corresponding to the circular frequency above which the disturbance field is stable. It is customary to call the lower part of the *neutral curve branch I*, while the upper part of the neutral curve is called *branch II*. This picture corresponds to the stability portrait of the system, as obtained using a linear theory, making a parallel flow approximation.

Note that the circular frequency of Fig. 3.2 is non-dimensional, based on the length scale δ^* and velocity scale U_e. Thus

$$\omega_0 = 2\pi \, f \, \frac{\delta^*}{U_e} = \left(\frac{2\pi v \, f}{U_e^2} \right) \frac{U_e \delta^*}{v} = F \, Re \qquad (3.70)$$

In this equation, F is the non-dimensional physical frequency of excitation. Thus, a constant physical frequency line in the (Re, ω_0)-plane is a straight line passing through the origin and different constant-frequency lines have slopes that correspond to different values of F. Let us consider a physical frequency of excitation f_1, as shown in Fig. 3.2. If the point of origin of the disturbance (say, at the point A) is outside the neutral curve, then the created disturbances will be decaying to begin with, and one must follow the corresponding constant-frequency F-line, ABC. Once the disturbance crosses a location corresponding to B, at *branch I* of the neutral curve, from that point onwards the disturbance will amplify as it propagates downstream, until the line hits *branch II* of the neutral curve at C. Beyond that point, once again the disturbance will be damped. Now, let us say that the exciter is located at the same physical location, but the frequency of excitation is increased to f_2, marked by point P in the figure. Then, once again the disturbance will be damped in its early evolution until it reaches point Q. In between Q and R, the disturbance will amplify and beyond that it will again be damped. It is noted that the flow, with parallel flow approximation, readjusts itself to the local condition in fixing the growth or decay rate.

There are additional aspects that need recounting here for viscous flow instability problems, other than the zero pressure gradient flow given by the Blasius equation. If the *neutral curve* of Fig. 3.2 is extended to higher Reynolds numbers then *branch I* and *branch II* tends towards each other, i.e., the *neutral curve* closes on itself. In contrast, if the mean flow profile possesses an inflection point, the upper and the lower branches of the *neutral curve* remain distinct even when $Re \rightarrow \infty$. This is the case for wall-bounded flows with adverse pressure gradient, mixing layers, jets and wakes. Also for such flows, the critical Reynolds number (Re_{cr}) is significantly lower than that for the zero pressure gradient flat plate boundary layer. This is relevant for external flows past complex shapes, e.g., the flow over airfoils or wings.

Along with the constant-α_i contours, the constant-α_r contours can be plotted in the (Re, ω_0)-plane, as shown in Fig. 3.3. The *neutral curve* is included in this figure to provide a reference. From this figure, it is noted that the maximum wavenumber for instability of the Blasius profile is given by $\alpha_{max}\delta^* \approx 0.35$, which corresponds to the smallest unstable waves, such shear layer supports disturbance with wavelength given by $\lambda_{min} = \frac{2\pi}{\alpha_{max}} \approx 18\delta^*$. Thus, the TS waves are significantly longer than the boundary layer thickness, justifying the quasi-parallel flow approximation for stability analysis.

Early corroboration of the linear viscous stability theory came from experiments where background disturbances were minimized to extremely low values and simultaneously two-dimensional controlled disturbances were introduced in the flow. This was reported in the classic work in [316], with experiments performed in the NBS wind tunnel specially fitted

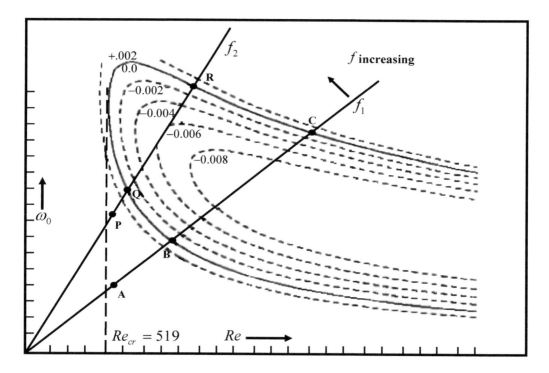

FIGURE 3.2
Contour plots of asymptotic growth rate in the Reynolds number-circular frequency plane.

with damping screens that reduced the turbulence intensity of the oncoming flow by increasing the solidity ratio or number of such screens. In the experiments without explicit excitation, the authors could progressively shift the location of transition by reducing the turbulence intensity all the way down to 0.08 percent. Further reduction in turbulence intensity had no effects on the position of the *natural transition*. Following Fig. 3.4, taken from [316], reveals the *natural transition*, as observed when no extraneous excitation was imposed on the flow, for a flow speed of 53 ft/sec. The sequences shown in the figure are a set of film records made by photographing the oscillograph screen with a moving film camera.

The first few frames have been amplified in the figure for ease of viewing, as indicated by the relative magnification on the right margin of the figure. While one can see sinusoidal oscillations at and up to 9 ft from the leading edge of the flat plate, it is also clear that the so-called *natural disturbances* are not purely *monochromatic*, i.e., the disturbances with the oncoming flow have many harmonic components. To detect TS waves and circumvent noticing multi-harmonic disturbances, the authors next introduced perfectly controlled two-dimensional disturbances by electromagnetically vibrating a phosphor-bronze ribbon (of 0.002 inch thickness and 0.1 inch wide) inside the shear layer, at a height of 0.006 inch from the plate. The positioning of this vibrator did not alter the mean flow, while it created pure sinusoidal oscillations starting from a little distance downstream from the ribbon. In this experiment, the *neutral curve* was charted out and compared with Schlichting's theoretical *neutral curve*. The agreement was remarkable and this led to complete justification of the viscous stability theory of the Göttingen-school. Historically, it was Heisenberg who actually led this type of analysis under the guidance of Sommerfeld.

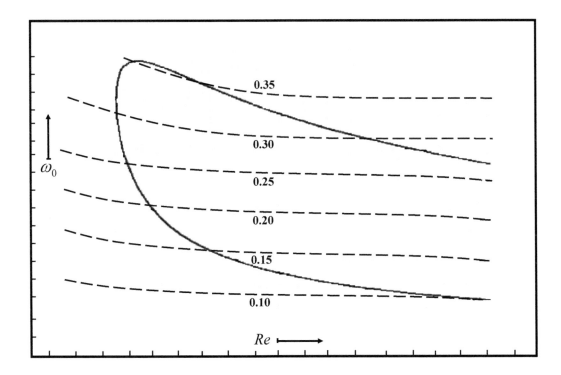

FIGURE 3.3
Contour plots of disturbance wavenumber in the Reynolds number-circular frequency plane.

The natural background disturbance must be viewed as an irregular pattern of two-dimensional and three-dimensional wave packets with non-uniform spectral contents. Also, study of an isolated spectral component is only a valid concept as long as non-linearity can be ignored and the superposition principle is valid. Even for linear problems, we have noted in Chap. 2 that the group around a wavenumber can interact to form packets which move with the group velocity. Furthermore, the stability analysis identifies a complex α for a given non-dimensional frequency ω_0 and Reynolds number under the assumption of parallel flow. However, for low speed incompressible flow, the shear layer thickness increases downstream even when the edge velocity is kept constant. Such growth is spectacular near the leading edge and the growth rate of the shear layer thickness is moderate to negligible at larger distances from the leading edge of the flat plate. Hence parallel flow analysis is more relevant for larger Reynolds numbers. For such flows, changes in boundary layer properties being very small, stability analysis works locally and the TS wave generated at a given location, in its movement downstream, will adjust itself to the local properties dictated by the dispersion relation. This local adjustment, as an assumption, is synonymous with what is known as the *quasi-parallel assumption*.

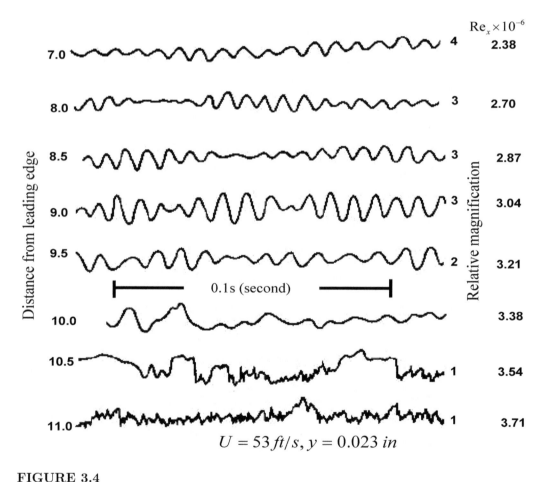

FIGURE 3.4
Hot-wire oscillogram traces showing natural transition from laminar to turbulent flow on a flat plate [316].

3.5.1 Local and Total Amplification of Disturbances

The Reynolds number at an observed transition location (defined as a location where the intermittency factor is about 0.1, i.e., the flow is 10% of the time turbulent and the rest of the time it is laminar) for a zero pressure gradient flat plate boundary layer is of the order of 3.5×10^5. This corresponds to $Re_\delta^* = 950$. The distance between the point of instability and the point of transition depends on the degree of amplification and the kind of disturbance present with the oncoming flow. This calls for a study of local and total amplification of disturbances. The following description is as developed in [10] for two-dimensional incompressible flows.

Consider the application of spatial theory for a wall-bounded external flow. On the top panel of Fig. 3.5, we once again show the *neutral curve* in the (Re, ω_0)-plane. If we focus our attention on a particular frequency f_1, then we will follow that ray emanating from the origin. According to this figure, this disturbance will decay up to $x = x_0$ from the leading edge. This, thereafter, will grow until $x = x_1$, i.e., up to the location where the ray intersects branch II.

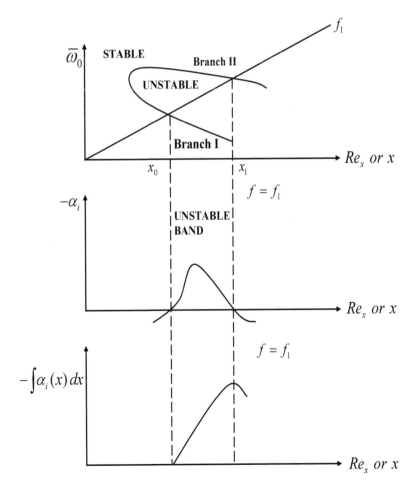

FIGURE 3.5
Sketch of local and total amplification of the disturbance field.

The amplification rate $(-\alpha_i)$ suffered by the disturbances within the neutral curve is shown in the middle panel of Fig. 3.5. Note the sign of the plotted rate, with negative values plotted along the positive ordinate direction. For two-dimensional disturbances in two-dimensional mean flow the amplification rate can be expressed as

$$\frac{1}{A}\frac{dA}{dx} = -\alpha_i \tag{3.71}$$

If the level of disturbance amplitude at $x = x_0$ is indicated by A_0, then the amplitude at any location downstream of x_0 is given by

$$A(x) = A_0 e^{-\int_{x_0}^{x} \alpha_i(x)\, dx} \tag{3.72}$$

In the lower panel of Fig. 3.5, the exponent n is shown as a function of x where

$$n = -\int_{x_0}^{x} \alpha_i(x)\, dx \tag{3.73}$$

This factor $n(x)$ is calculated for a particular frequency and the exercise can be repeated for a range of frequencies and a composite plot can be made, as shown in Fig. 3.6.

The total amplification suffered by individual frequencies is shown by the solid lines in Fig. 3.6. The envelope to these curves, shown by the dotted line, represents the maximum amplification suffered by different frequencies since their entry into the unstable zone. The envelope is generally designated by

$$N = \left[\mathrm{Ln}\frac{A}{A_0}\right]_{max} = \max_{f}\left[\int_{x_0}^{x} -\alpha_i(x)\, dx\right] \tag{3.74}$$

Note that x_0 is different for different frequencies as it starts from *branch I* where that particular frequency enters the unstable region. Second, the lower frequencies dwell inside the neutral loop for longer distances and these components tend to amplify the most.

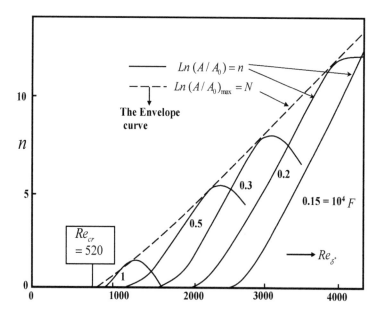

FIGURE 3.6
Total amplification rates for different frequencies for the Blasius flow [10].

3.5.2 Effects of the Mean Flow Pressure Gradient

So far, we have been talking about the stability of zero pressure gradient flows. It is possible to extend the studies to include flows with a pressure gradient using a quasi-parallel flow assumption. To study the effects in a systematic manner, one can also use the equilibrium solution provided by the self-similar velocity profiles of the Falkner-Skan family. These similarity profiles are for wedge flows, whose external velocity distribution is of the form $U_e = k\, x^m$. This family of similarity flows is characterized by the Hartree parameter $\beta_h = \frac{2m}{2m+1}$ and the shape factor $H = \frac{\delta^*}{\theta}$. Some typical non-dimensional flow profiles of this family are plotted against non-dimensional wall-normal coordinates in Fig. 3.7. The wall-normal distance is normalized by the boundary layer thickness of the parallel shear layer.

The flow profiles with $H > 2.591$ correspond to velocity distributions with inflection point and these are the decelerated flows or flows with an adverse pressure gradient. On the contrary, the flow profiles with $H < 2.591$ correspond to $\frac{dp}{dx} < 0$ (accelerated flows). The figure with $\beta_h = 0$ and $H = 2.59$ corresponds to the Blasius profile. The profile with $\beta_h = 1$ and $H = 2.22$ corresponds to stagnation point flow. The other two profiles in Fig. 3.7 are for flows with adverse pressure gradients and the crosses on the profile indicate the locations of the inflection points. The profile for $\beta_h = -0.1988$ ($H = 4.032$) corresponds to the case of incipient separation of the boundary layer.

In Fig. 3.8, the neutral curves for the above four velocity profiles of Fig. 3.7 are compared. The deciding trend is that as H increases, Re_{cr} decreases. And for flows with $\frac{dp}{dx} > 0$, the critical Reynolds numbers are significantly lower.

The fact that such profiles are prone to instability earlier is the reason for calling these adverse pressure gradient flows. Along with this, the growth rate of unstable waves becomes larger as the shape factor increases. Also, for decelerated flows, the neutral curve does not close, as seen for the $H = 2.80$ ($\beta_h = -0.10$) case. For separated profiles, there is no *branch II* of the neutral curve at all. The stagnation region on a body in the presence of fluid flow

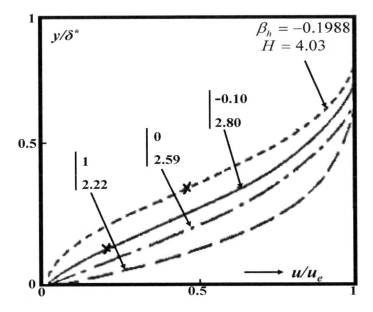

FIGURE 3.7
Typical mean velocity profiles of some thin shear layers [10].

is a site of high stability, as can be seen for the Hiemenz flow (stagnation point flow profile with $\beta_h = 1$) which is far more stable than the Blasius flow in Fig. 3.8.

The envelope curve (which represents the envelope of the maximum amplification suffered by all the frequencies considered present in the background disturbance), shown by the dotted line in Fig. 3.6, was drawn for a particular pressure gradient. This process can be repeated for different pressure gradient parameters. In Fig. 3.9, the envelope curves are shown for a few representative adverse pressure gradient parameters starting from the Blasius profile ($H = 2.59$) to the incipient separated flow ($H = 4.032$). As the shape factor H for the flow increases, Re_{cr} also decreases (indicated by the starting point of these curves). The slope of these envelope curves ($\frac{dN}{dRe_{cr}}$) also increases with H. One can use the local stability property of the flow by solving the OSE and characterize the effects of pressure gradient in the same way that we did for Blasius profiles. It is apparent that the shape factor of the shear layer H is a very important parameter which indicates flow separation and enhanced or reduced instability.

In Fig. 3.10, the Reynolds number based on momentum thickness at transition onset is shown plotted against the shape factor, as reported in [10].

The shape factor turns out to be an important parameter in deciding flow instabilities and transition. Thus, by plotting H variation in the streamwise direction it is possible to indicate the transition location in a given flow. It is adequate to define the transition onset location as the place where the shape factor variation exhibits a sudden negative slope. The following results in Fig. 3.11 [10] show that flow transition in an adverse pressure gradient shear layer always precedes theoretical location of laminar separation, indicated by vertical bars in the frames.

In the set of frames in Fig. 3.11, (A) represents a zero pressure gradient shear layer

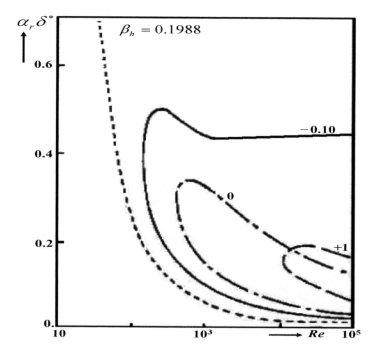

FIGURE 3.8
Neutral curves of some shear layer velocity profiles shown in Fig. 3.7.

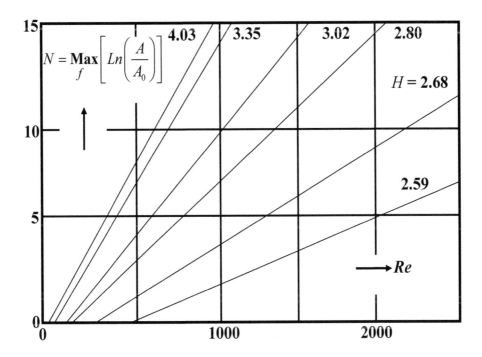

FIGURE 3.9
Envelope curves for total amplification for Falkner–Skan similarity profiles [10].

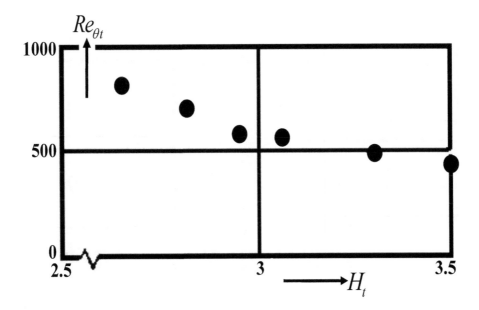

FIGURE 3.10
Momentum thickness Reynolds number and shape factor at transition onset [10].

for which the flow is not predicted to separate theoretically. The sudden decrease of H is the harbinger of flow transition. From figures (B) through (F), the pressure gradient is progressively made more adverse and consequently the theoretical location of the laminar separation point moves forward and the H variation becomes sharper with a larger fall in value at the location of transition. For severe adverse pressure gradients, the point of separation is almost coincident with the point of transition. This is often made use of for severe adverse pressure gradient flow calculations. However, this is not always the case for predicting transition location. For example, at low Reynolds numbers, flow often separates first and then transition occurs subsequently. The separated flow being very unstable, the following transition is very quick — one has practically a free shear layer with a point of inflection. As turbulent flows are capable of withstanding much higher adverse pressure gradients, such separated turbulent flows can reattach to the wall. It results in mean wall streamline forming a separation bubble. Beyond the reattachment point, the flow is turbulent and such flows are observed for uniform flow past circular cylinders in Reynolds number ranges of 3×10^5 to 3×10^6, on the leeward side. The flow after turbulent reattachment separates again and remains fully separated. Separation bubbles are also seen in flow past airfoils downstream of the suction peak at moderate Reynolds numbers and at non-zero angles of attack. There is an intense interaction of the viscous flow in the shear layer with the outer inviscid flow in such cases, and boundary layer approximation fails.

The importance of the shape factor H can also be gauged for its ability to correlate it with instability and transition by looking at Fig. 3.12, where the values of $Re_{\delta^* crit}$ are plotted versus H [258, 438], which shows a regular variation of data for different varieties of flows.

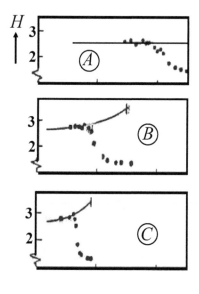
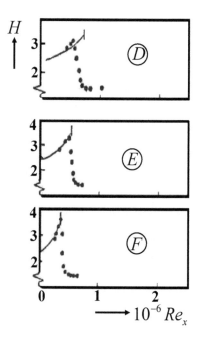

FIGURE 3.11
Effects of a positive pressure gradient on separation and transition. Discrete symbols are experimental data points and continuous lines are laminar calculations. The vertical bar in the laminar calculation indicates the location of the point of laminar separation [10].

This importance of H has been further utilized in predicting transition [435] by noting that similar correlations exist for parameters such as pressure gradient, suction/blowing, heating/cooling, etc., if $(Re_{\delta^*})_{crit}$ is plotted in terms of shape factor at transition. In Fig. 3.13, this correlation is shown by the upper curve that is premised on linear instability studies with $N = 9$. The lower curve gives Re_{crit} as computed from Fig. 3.12 by converting Re_{δ^*} to Re_x. To use Fig. 3.13, computed $H(x)$ by any laminar boundary layer code is used to predict transition, when the local $H(x)$ intersects the upper curve. This is also given by the following empirical fit

$$\log_{10}(Re_{x,tr}) = -40.4557 + 64.8066H - 26.7538H^2 + 3.3819H^3 \qquad (3.75)$$

This correlation is valid only in the range $2.1 < H < 2.8$. This method is promising,

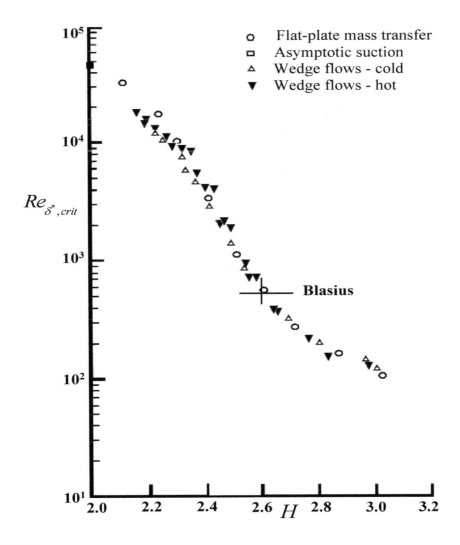

FIGURE 3.12
Collation of critical Reynolds number variation data with shape factor for different flow types [258].

as it is premised on the e^9 method (but with significantly lesser effort), as described in the next sub-section.

3.5.3   Transition Prediction Based on Stability Calculation

Following the path-breaking experiments of [316], there have been sustained efforts to link the stability theory in predicting transition. Michel [224] reported first that his compiled data showed the transition to be indicated when the total amplification of TS waves corresponded to $A/A_0 \approx 10^4$, where A_0 is the disturbance amplitude at the onset of instability. This motivated authors in [374, 421], to use temporal theory results to show that at transition the total amplification is given by

$$\frac{A}{A_0} = exp\left[\int_{x_{cr}}^{x_{tr}} \alpha \, c_i \, dt\right] \tag{3.76}$$

Note that the integration ranges from the critical point (x_{cr}) to the point of transition

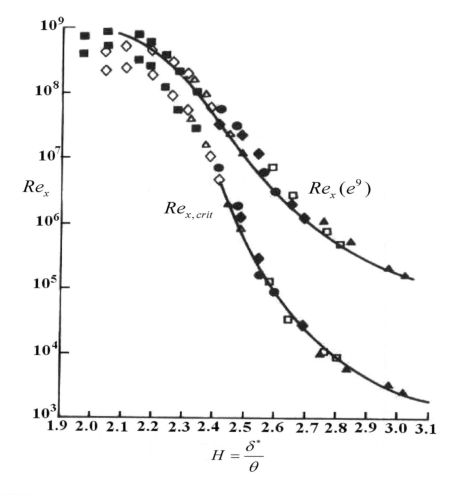

FIGURE 3.13
Correlation of critical and transitional Reynolds number versus boundary layer shape factor [438].

(x_{tr}). As the right-hand side roughly became equal to e^9 [374] at transition, this method has also acquired the name e^9 method. In [421], this value of the exponent was reported to be between 7 and 8. The disparity in the value of the exponent arises due to different levels and the spectrum of background disturbances. Currently, it is known as the e^N method, where the exponent has been given a general value depending on other factors affecting the transition process. Later on in [160], the authors performed spatial stability calculations and reported the exponent as equal to 10.

Experiments on transition for two-dimensional attached boundary layers have revealed that the onset process is dominated by TS wave creation and its evolution when the free stream turbulence level is low. Generally speaking, estimated quantities like frequency of most dominant disturbances, eigenvalues and eigenvectors matched quite well with experiments. It is also noted from experiments that the later stages of the transition process are dominated by non-linear events. However, this phase spans a very small streamwise stretch and therefore one can observe that the linear stability analysis more or less determines the extent of transitional flow. This is the reason for the success of all linear stability based transition prediction methods. However, it must be emphasized that nonlinear, nonparallel and multi-modal interaction processes are equally important in all cases.

Despite the reasons cited for the success of linear stability theory in predicting transition, it is important to underscore its limitation as well. This should help one to look for hitherto unknown mechanism(s) that may play a bigger role in transition prediction than might have been suspected. This is the scope of Chaps. 4 to 7 of this book. For example, the envelope method does not require any information about the frequency content of the background disturbances and always predicts transition based on lower frequency events. What if a particular disturbance environment precludes such components of disturbances? Will the non-linear process create those frequencies first and then these would amplify? In such a case, the transitional region will be prolonged as compared to other cases and transition prediction based on normal mode analysis will be in question. This possibility is not explored in the present form of linear instability theory. Also, it is stated that the above method works only when the free stream turbulence levels are low. It is legitimate to ask what happens when the turbulence intensities are higher. Would one see different transition scenarios? Would one see transition without even creating TS waves? This is the point of view for bypass transition research, discussed in Chap. 4. First, let us discuss incorporating semi-empirically the role of low-level free stream turbulence on transition along with the e^N method.

3.5.4 Effects of Free Stream Turbulence

It has been stated above that the success of the e^N method depends strongly on the fact that this method is based on experimental results obtained in low turbulence intensity tunnels. Here the turbulence intensity is defined as

$$Tu = \frac{1}{U_\infty} \left[\frac{(u'^2 + v'^2 + w'^2)}{3} \right]^{1/2} \tag{3.77}$$

Quantities indicated by primes are the fluctuating components and thus the numerator indicates the r.m.s. fluctuation level of the disturbance and is calculated taking a long time history. The effect of Tu is very strong on transition. For example, in the experiments of [316], the Reynolds number at the transition location dropped by 50% when Tu was increased to 0.35% from its highest value at 0.04%, as shown in Fig. 3.14 (a reproduction of Fig. 3.2 of [316]).

As seen here, the transition Reynolds number decreases rapidly for Tu greater than

0.10%, and this variation cannot be directly explained by the e^N method. Mack [212] has suggested the following empirical correlation, linking Tu with the exponent N of the e^N method

$$N = -8.43 - 2.4 \, Ln(Tu) \qquad (3.78)$$

The above correlation is applicable in the range $0.0007 \le Tu \le 0.00298$. This correlation is based on the experimental results of [92]. In [11], its successful application for flows with adverse pressure gradients is reported.

At the upper end of the range of application for $Tu = 0.00298$, one notes $N \approx 0$, implying that the transition would occur right at the location of Re_{cr}. In applying this correlation, the following has to be kept in mind:

- For $Tu < 0.10\%$, the transition location is insensitive to Tu. Such low levels of disturbances are typical of acoustic noise that controls transition, rather than vortical disturbances. This distinction is not made in the literature between the two and everything is included in the catch-all terminology of free stream turbulence (FST).

- For higher values of Tu (greater than 2 to 3%), transition often occurs without the appearance of TS waves at Reynolds numbers below Re_{cr} for wall-bounded flows. Thus the linear process of instability is *bypassed* and such transition processes are called *bypass transition*, as discussed in the next chapter.

If the excitation source is not inside the shear layer, then also a bypass transition can be caused by flows with low levels of FST embedding convected vortices in it — as is discussed in Sec. 3.9. Also, an important distinction must be made between periodic and aperiodic convecting vortices in the free stream. Following Fig. 3.15 [383], one clearly distinguishes causes which trigger transition due to FST. In this figure, generic causes shown are due to i) grid-generated turbulence, ii) acoustic noises and other vortical/acoustic sources which

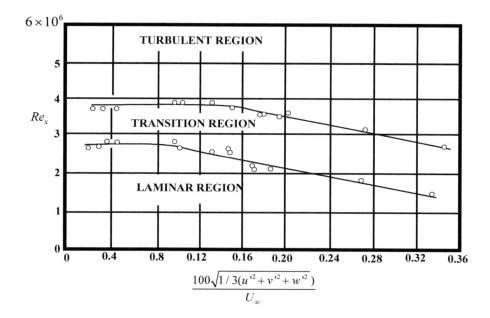

FIGURE 3.14
Effect of free stream turbulence on transition for flat plate boundary layers [316].

create iii) standing and iv) traveling waves. As compared to the "low noise" transition Reynolds number of 2.8×10^6 obtained in [316], the facility used in [436] produced transition at a Reynolds number equal to 4.9×10^6. The facility used for the experiment excluded traveling acoustic noise, but there were standing acoustic waves in the test section.

The results of [383] in Fig. 3.15 used acoustic signals of fixed frequency to trigger transition and these clearly demonstrate that transition is sensitively dependent on the frequency of excitation when Tu ranges from very small to significantly large values. Although the high frequency data in this figure (for 82 Hz) shows no variation at all in the plotted range — a value that is outside the neutral curve in the stable region. The grid data shows itself as a strong transition promoter the Re_{tr} value dropping sharply with a very small variation in Tu. Spangler and Wells [383] report that the spectral distribution of disturbances shows different amplitudes at different frequencies. To this one must also add that there can be significant interferences between acoustic and vortical excitation — as seen for the Tu dependence of transition for the 76 Hz data, with a gradual decrease of Re_{tr} with Tu.

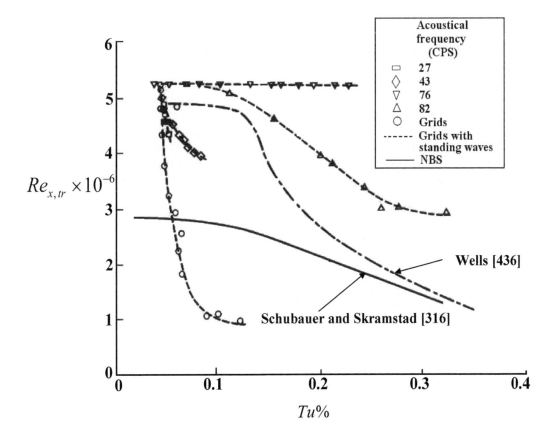

FIGURE 3.15
Correlation of transition Reynolds number with FST data from [383].

3.6 Receptivity Analysis of the Shear Layer

In Fig. 3.15, one notes the transitional Reynolds number plotted versus the disturbance amplitude for a flat plate boundary layer, excited by different means. It is evident that each case differs significantly from the others. The entrainment of different disturbance sources inside the shear layer to produce instability is at the core of receptivity studies, which relates the **cause** and its **effect(s)**. For example, from Fig. 3.15 it is apparent that the grid generated disturbances are very effective in triggering transition while free stream acoustic excitation is not so. Thus, one would state that shear layers are more receptive to vortical disturbances than acoustic disturbances. It is recognized that there are following sources which can trigger transition in laminar flows: (a) perturbations in the form of a vortical field, (b) an acoustic or isentropic weak pressure field and (c) an entropic field or temperature fluctuations. Note that the surface vibration or surface inhomogeneity also creates locally vortical perturbations, which are a potent trigger for transition in shear layers. It is the aim of receptivity studies to show the process of creation of instability in a shear layer by any one or a combination of these sources. What is essentially lacking so far is a proper mathematical framework which can deal with the problem of the processes for the initiation of instability − a topic known as receptivity. There have been some efforts on studying various aspects of receptivity, as given in [48, 62, 63, 70, 74, 75, 126, 195, 202, 232, 252, 296, 308, 327, 345, 362, 363, 398].

The experimental results of [316] were demonstrations of receptivity of a flat plate shear layer to induced vibration inside the shear layer. The authors could produce TS waves in a reproducible fashion only when they vibrated a metallic ribbon at a fixed frequency very close to the wall, with the help of an electromagnet. It is instructive to recall the following from [316]: *In the search for schemes to excite oscillations in the boundary layer, a number of devices were tried before completely satisfactory results were obtained. Methods using sound, both pure notes and random noise, were none too satisfactory because of resonance effects and the complexity of the wave pattern in the tunnel.* It is generally considered that the major problem with acoustic excitation from free stream is the problem of matching of scales between the acoustic excitation field and the TS waves. The wavelength and propagation speed of acoustic waves is almost two orders of magnitude larger than the unstable TS waves. Also, it is to be noted that the experiments of [316] was attempting to verify the theoretical developments for two-dimensional instability while the acoustic excitations are always three-dimensional! There is absolutely no scope of creating two-dimensional acoustic excitations! Since the instability of the shear layer is associated with spatial theory, Squire's theorem also does not help in looking for a two-dimensional disturbance field that is more unstable than the three-dimensional one. Schubauer & Skramstad [316] carefully devised the means of creating mainly a two-dimensional excitation field by vibrating a ribbon inside the shear layer that produced instability waves with properties predicted theoretically in [143, 311, 408].

However, one aspect of the theoretical developments of instability theory is quite easily overlooked. Instability theory requires neither the knowledge of excitation fields nor the wall-normal location, where it is applied. In eigenvalue analyses, excitation field information is obscured through the application of homogeneous conditions at the boundaries. It is even worse if the excitation is applied in the interior of the shear layer — there are no theories of instability at all! The instability theory developed in the previous sections relates to exciting the flow field by the boundary condition at the wall only, as explained next.

This is evident from the formulation as seen, for example, from the retention of two modes in Eq. (3.59), which only considers excitations at the boundary $y = 0$, those decay with height. It is therefore clear that the characteristic determinant of Eq. (3.62) will extract only those modes that are triggered by wall excitation and those decay with height. Thus in an experiment, TS waves are naturally going to be produced by excitation of the shear layer at the wall, as was demonstrated in Schubauer & Skramstad's experiment. When the frequency of the ribbon in the experiment is fixed, TS waves are predicted as a consequence of satisfying the dispersion relation of Eq. (3.62) via the spatial theory.

It is therefore also not directly apparent why TS waves will be created when the flow is excited at the free stream until and unless the coupling mechanism between free stream and wall excitation is established. It is not enlightening to read in the literature about the creation of TS waves in the shear layer by free stream excitation and to explain it all by theories that are developed exclusively for wall excitation. An exception to this is the work reported in [327], where the coupling mechanism was explained for the excitation by a train of convecting vortices moving at a constant speed at a fixed height outside the shear layer. It was pointed out that there is only a narrow convection speed range for which convected vortices can have very strong receptivity. This is discussed in detail later.

Furthermore, in the absence of any other mechanisms, most of the efforts so far have been in looking for TS waves as the harbinger of transition for any kind of excitation. This situation needs to be rectified, as suggested by Morkovin in various papers, including [235] and a schematic is shown here in Fig. 3.1. This relates to explaining bypass transition that causes transition without the appearance of TS waves. On bypass transition, one can look for additional discussions in [40, 159, 187, 188, 199, 327, 331, 349]. Having stated the inadequacies of the eigenvalue approach, in the following we describe the receptivity approach starting with the linear theory.

3.6.1 Receptivity Mechanism by a Linearized Approach: Connection to Stability Theory

Few articles on receptivity present a qualitative view of particular transition routes created by not so well-defined excitation fields [308]. Such approaches do not demonstrate complete theoretical and/or experimental evidence connecting the cause (excitation field) and its effect(s) (response field). Here, a model based on a linearized Navier–Stokes equation is presented to show the receptivity route for excitation applied at the wall. This requires a dynamical system approach to explain the response of the system with the help of Laplace–Fourier transforms.

3.6.1.1 A Brief Review of Laplace–Fourier Transforms

Fourier and Laplace transforms are linear transforms and are very often used for analyzing problems in various branches of science and engineering. Since receptivity is studied with respect to the onset of instability, it is quite natural that these transform techniques will be the tool of choice for such studies. Fourier transforms provide an approach wherein the differential equation of a time dependent system is solved in the transformed plane as

$$Y(\omega_0) = T(\omega_0)\, X(\omega_0) \tag{3.79}$$

where $X(\omega_0)$ and $Y(\omega_0)$ are the transform or spectrum of the input and output of the system, while $T(\omega_0)$ is called the *transfer function* of the system. If one focuses on an input of a particular frequency at $\omega_0 = \bar{\omega}_0$, then one can obtain the corresponding transfer function $T(\bar{\omega}_0)$ and the output of the system can be obtained using Eq. (3.79). Such a study for all possible frequencies will give us what is known as the *frequency response* of the system. In contrast, one can obtain the response of the system if the input to the system is instead a Dirac delta function in the physical plane; then the system will be excited simultaneously at all wavenumbers and the corresponding output (also in the physical plane) will be called the *impulse response*. Thus, the study of a linear system is as simple as finding the transfer function for the system and knowing the input spectrum. While the output in the transform plane is obtained by a simple multiplication, we will shortly see that the output can be obtained in the physical plane by performing a convolution integral between the impulse response and input spectrum. Use of Fourier transforms, as opposed to Fourier series, allows system analysis for aperiodic excitation. Almost all physical systems are *causal*, i.e., these are systems whose output does not anticipate input. For a linear and fixed system characterized by an *impulse response* $h(t)$ the causality requires

$$h(t) = 0 \quad \text{for } t < 0 \tag{3.80}$$

Thus, it becomes apparent that the output and the impulse response are one sided in the time domain and this property can be exploited in such studies. Solving linear system problems by Fourier transform is a convenient method. Unfortunately, there are many instances of input/output functions for which the Fourier transform does not exist. This necessitates developing a general transform procedure that would apply to a wider class of functions than the Fourier transform does. This is the subject area of one sided Laplace transforms that is being discussed here as well. The idea used here is to multiply the function by an exponentially convergent factor and then use the Fourier transform technique on this altered function. For causal functions, which are zero for $t < 0$, an appropriate factor turns out to be $e^{-\sigma t}$ where $\sigma > 0$. This is how a Laplace transform is constructed. However, there is another reason we use another variant of a Laplace transform, namely, the bilateral Laplace transform.

All fluid dynamical systems are continuous systems with *infinite* degrees of freedom and the governing equations depend continuously upon both space and time. While for any system the time-dependent signal cannot move back in time, the space-dependent signal can propagate in all directions with respect to the location of the source. This, therefore, requires that we develop a theory based on a bilateral Laplace transform – a topic described in greater detail in [264, 418].

3.6.1.2   Fourier and Laplace Transforms

If $F(\omega_0)$ is the Fourier transform of a time varying function $f(t)$

$$F(\omega_0) = \int_{-\infty}^{+\infty} f(t) \, e^{-i\omega_0 t} \, dt \tag{3.81}$$

then the above integral exists in the sense of having a Cauchy principal value for every value of ω_0. Conversely, if we have the Fourier transform, then we can construct the function in the time domain from

$$f(t) = \frac{1}{2\pi} \int_{-\infty}^{+\infty} F(\omega_0) \, e^{i\omega_0 t} \, d\omega_0 \tag{3.82}$$

Note that in Eqs. (3.81) and (3.82), ω_0 is real. As mentioned above, we will also use bilateral Laplace transform $F_{II}(\alpha)$ to describe the space dependence of the function by

$$F_{II}(\alpha) = \int_{-\infty}^{+\infty} f(x) \, e^{-i\alpha x} \, dx \tag{3.83}$$

Thus, $F_{II}(\alpha)$ and $f(x)$ form the transform-original pair for a space-dependent function. Here, α is a complex quantity, i.e., $\alpha = \alpha_r + i\alpha_i$, and represents a complex wavenumber.

Note that α being complex, $F_{II}(\alpha)$ is only defined in a limited region of the complex α plane, as shown in the sketch of Fig. 3.16.

In Fig. 3.16, a horizontal hatched strip is defined by $\gamma_2 < \alpha_i < \gamma_1$. Any contour chosen within this strip of convergence is called the *Bromwich contour*. On the right-hand side of the integral in Eq. (3.83), if we restrict the lower limit to $x = 0$, then we would retrieve the unilateral Laplace transform given by

$$F_I(\alpha) = \int_0^\infty f(x)\, e^{-i\alpha x}\, dx \qquad (3.84)$$

The right-hand side of the above can also be rewritten as

$$F_I(\alpha) = \int_0^\infty f(x)\, e^{\alpha_i x}\, e^{-i\alpha_r x}\, dx \qquad (3.85)$$

This shows that $F_I(\alpha)$ is the Fourier transform of the function $f(x)\, e^{\alpha_i x}$, if α_i is held constant. Applying the Fourier inversion formula Eq. (3.82), we obtain

$$f(x)\, e^{\alpha_i x} = \frac{1}{2\pi} \int_{-\infty+i\alpha_i}^{+\infty+i\alpha_i} F_I(\alpha)\, e^{i\alpha_r x}\, d\alpha_r \qquad (3.86)$$

Here, the integration contour is traced along an $\alpha_i = $ const. line, i.e., the integration contour is along a horizontal line, $\alpha_i = \gamma$. This is the Bromwich contour for a one sided Laplace transform, if the corresponding inverse Fourier transform in Eq. (3.86) exists. The integral in Eq. (3.85) will then converge for all other contours for which $\alpha_i < \gamma$, defining the region of convergence for the transform, where it is analytic. We can extrapolate this to define the strip of convergence for a bilateral Laplace transform.

The bilateral Laplace transform of Eq. (3.83) can be rewritten as

$$F_{II}(\alpha) = \int_0^\infty f(x)\, e^{-i\alpha x}\, dx + \int_{-\infty}^0 f(x)\, e^{-i\alpha x}\, dx \qquad (3.87)$$

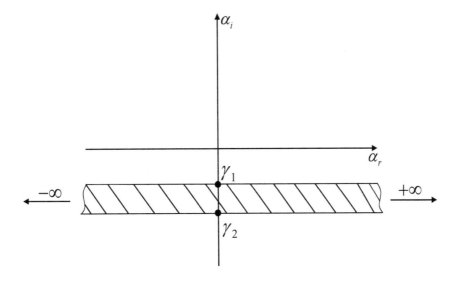

FIGURE 3.16
The strip of convergence for the Laplace–Fourier integral in the wavenumber plane.

The first integral in Eq. (3.87) is the unilateral Laplace transform of $f(x)\,U_1(x)$ where $U_1(x)$ is the unit step function or the Heaviside function. Therefore, it converges in a region $\alpha_i > \gamma_1$, as shown in Fig. 3.17.

Similarly, the second integral of Eq. (3.88) is analytic for $\alpha_i > \gamma_2$, as shown in Fig. 3.17. Therefore, $F_{II}(\alpha)$ is analytic in the region shown by cross-hatches in the figure, recovering the definition shown in Fig. 3.16.

3.6.1.3 Inversion Formula for Laplace Transforms

Once we identify the strip of convergence, we can write down the inversion formula by integrating along a Bromwich contour (here taken as an $\alpha_i = constant$ line for convenience) in the complex α-plane by

$$f(x) = \frac{1}{2\pi} \int_{Br} F_{II}(\alpha_r + i\alpha_i)\, e^{i\alpha_r x}\, e^{-\alpha_i x}\, d\alpha_r \qquad (3.88)$$

or

$$f(x)\, e^{\alpha_i x} = \frac{1}{2\pi} \int F_{II}(\alpha_r + i\alpha_i)\, e^{i\alpha_r x}\, d\alpha_r \qquad (3.89)$$

This is the equivalent inverse Fourier transform of the right-hand side. This integral can be evaluated by contour integration [264, 418]. Application of these in fluid flow instability studies is in [319, 322]. Suppose that the only singularities of $F_{II}(\alpha)$ are simple poles and

$$F_{II}(\alpha) \to 0 \qquad \text{for} \quad |\alpha| \to \infty \qquad (3.90)$$

Then one can invoke Jordan's lemma and Cauchy's theorem [441] for the line integral in Eq. (3.89) that can be converted to the contour integral, as shown in Fig. 3.18, with a single pole indicated at P_1. Let us also say that the disturbance corresponding to this pole

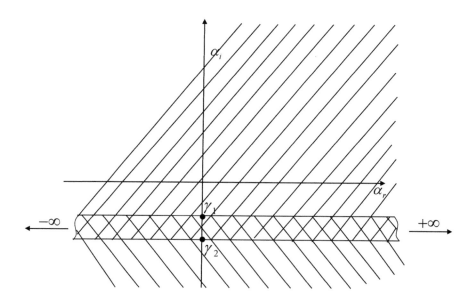

FIGURE 3.17
Fixing the Bromwich contour in the wavenumber plane inside the cross-hatched region.

has a positive group velocity, i.e., the associated disturbance propagates in the downstream direction.

We construct a closed contour C_u, which is the Bromwich contour plus the semi-circular arc, as indicated in the figure, with a small indented contour around the pole connected by a pair of vertical lines to this semi-circular arc. These vertical lines constitute a cut that links the small circular contour (C_1) to the semi-circle. Arrows show the direction along which the contour integral is taken for estimating the inverse transform corresponding to the downstream propagating signal. Because there are no other singularities, $F_{II}(\alpha)$ is analytic along and within C'_u, as indicated by hatching. Therefore

$$f(x) = \frac{1}{2\pi} \int_{C_u} F_{II}(\alpha) \, e^{i\alpha x} \, d\alpha \quad \text{for } x > 0 \tag{3.91}$$

where $C_u = C'_u + C_1$. For an analytic function $f(z)$ in a domain bounded by a closed contour C, *Cauchy's theorem* states that

$$\int_C f(z) \, dz = 0$$

Thus, we can apply this theorem to the integrand of Eq. (3.91) along the contour C'_u, i.e.,

$$\frac{1}{2\pi} \int_{C'_u} F_{II}(\alpha) \, e^{i\alpha x} \, d\alpha = 0$$

Hence

$$\frac{1}{2\pi} \int_{C'_u} F_{II}(\alpha) e^{i\alpha x} \, d\alpha = -\frac{1}{2\pi} \int_{C_1} F_{II}(\alpha) \, e^{i\alpha x} \, d\alpha \tag{3.92}$$

Note that while C_u is in the anti-clockwise direction and hence positive, the integral on the right-hand side of Eq. (3.92) is in the clockwise direction. If we perform this integral also in the anti-clockwise direction, then the negative sign can be removed. The integral on the right-hand side can be evaluated by calculus of residues, i.e.,

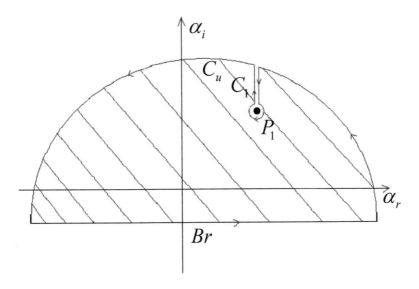

FIGURE 3.18
Bromwich contour and its closure for the integral in Eq. (3.91).

$$\int_{-C_1} F_{II}(\alpha)\, e^{i\alpha x}\, d\alpha = 2\pi\, i\, * \text{ Residue at } (\alpha = \alpha_{P_1}) \tag{3.93}$$

where the residue has to be calculated at the pole located at P_1. If the pole is of order m, then

$$\text{Residue}(\alpha|_{P_1}) = \frac{1}{(m-1)!} \frac{d^{m-1}}{d\alpha^{m-1}} \left[F_{II}(\alpha)\, e^{i\alpha x} \right] \tag{3.94}$$

If we had joined the Bromwich contour by a semi-circle in the lower part of the α-plane (indicated by a contour C_d), then

$$f(x) = \frac{1}{2\pi} \int_{C_d} F_{II}(\alpha)\, e^{i\alpha x}\, d\alpha \quad \text{for } x < 0 \tag{3.95}$$

However, one need not perform contour integrals to obtain $f(x)$. For example, if we perform the integral of Eq. (3.88) directly along the Bromwich contour, then we will be simultaneously getting both the downstream and upstream propagating solutions together. We advocate this procedure, because this also allows us freedom from finding out detailed information about all possible singularities (not only the simple poles) of the transform and their order. The only thing that has to be ensured is finding the correct strip of convergence for the Bromwich contour. This procedure has been followed in [319] in calculating the impulse response of the Blasius boundary layer in [114] for solving the signal problem for spatial growth and in [322, 349] for solving the full spatio-temporal problem, for a zero pressure gradient shear layer excited harmonically.

It is interesting to note that the response of a system to a harmonic input is itself harmonic at the same frequency under the twin conditions of linearity and time invariance of the system properties for stable systems. For instability and receptivity problems, there is no general proof of the same due to the nonlinear nature of the dispersion relation, despite the fact that one is studying the linearized Navier–Stokes equation. Thus it can at best be an assumption, which has been adopted in many analyses of this problem, except in [322, 349, 350], where the full time-dependent problem is solved as a general time-dependent problem by considering Bromwich contours in the α- and ω_0-planes simultaneously.

There is another reason for our preference in calculating the system response by integrating Eq. (3.88) directly and not using contour integral Eqs. (3.91) and (3.95). This is due to the restrictive condition of Eq. (3.90) needed to hold for Jordan's lemma to be used. We have shown here that the condition of Jordan's lemma does not hold for the Orr–Sommerfeld equation − a result that has not been used in stability studies of fluid dynamics.

3.6.1.4 A Short Tutorial on Fourier Integral and Transforms

Given a function of the real variable t, consider the integral

$$F(\omega_0) = \int_{-\infty}^{+\infty} f(t)\, e^{-i\omega_0 t}\, dt \tag{3.96}$$

The Fourier transform $F(\omega_0)$ is, in general, complex and can be expressed as

$$F(\omega_0) = R(\omega_0) + iI(\omega_0) \tag{3.97}$$

or

$$= A(\omega_0)\, e^{i\varphi} \tag{3.98}$$

where $A(\omega_0)$ is the amplitude or Fourier spectrum of $f(t)$, $A^2(\omega_0)$ its energy spectrum and φ its phase angle. The inversion formula

$$f(t) = \frac{1}{2\pi} \int_{-\infty}^{+\infty} F(\omega_0)\, e^{i\omega t}\, d\omega \qquad (3.99)$$

is valid at all continuous points. At discontinuities, one should take the average of its right and left limits, i.e.,

$$f(t) = \frac{1}{2}[f(t^+) + f(t^-)] \qquad (3.100)$$

If $f(t)$ is absolutely integrable, i.e., $\int_{-\infty}^{+\infty} |f(t)|\, dt < \infty$, then $F(\omega_0)$ exists. Let us now talk about some special forms of Fourier integrals. If $f(t) = f_1(t) + if_2(t)$, then

$$R(\omega_0) = \int_{-\infty}^{+\infty} [f_1\, \cos(\omega_0 t) + f_2\, \sin(\omega_0 t)]\, dt \qquad (3.101a)$$

$$I(\omega_0) = -\int_{-\infty}^{+\infty} [f_1\, \sin(\omega_0 t) - f_2\, \cos(\omega_0 t)]\, dt \qquad (3.101b)$$

And

$$f_1(t) = \frac{1}{2\pi} \int_{-\infty}^{+\infty} [R\, \cos(\omega_0 t) - I\, \sin(\omega_0 t)]\, d\omega_0 \qquad (3.102a)$$

$$f_2(t) = \frac{1}{2\pi} \int_{-\infty}^{+\infty} [R\, \sin(\omega_0 t) + I\, \cos(\omega_0 t)]\, d\omega_0 \qquad (3.102b)$$

Thus, if $f(t)$ is real, i.e., $f_2(t) \equiv 0$, then $R(\omega_0) = \int_{-\infty}^{+\infty} [f_1\, \cos(\omega_0 t)]\, dt$ and hence it is an even function. In the same way, $I(\omega_0)$ is an odd function. Furthermore, if $f(t)$ is real and even, then $f(t)\cos(\omega_0 t)$ is even and $f(t)\sin(\omega_0 t)$ is odd. Hence for such combinations

$$R(\omega_0) = 2 \int_0^{\infty} f(t)\cos(\omega_0 t)\, dt \quad \text{and} \quad I(\omega_0) \equiv 0 \qquad (3.103)$$

In contrast, if $f(t)$ is real and odd, then

$$R(\omega_0) \equiv 0 \quad \text{and} \quad I(\omega_0) = -2 \int_0^{\infty} f(t)\sin(\omega_0 t)\, dt \qquad (3.104)$$

For a causal function

$$f(t) = 0 \quad \text{for} \quad t < 0 \qquad (3.105)$$

Also from Eq. (3.96), it is easy to see that

$$F(-\omega_0) = \int_{-\infty}^{+\infty} f(t)\, e^{i\omega_0 t}\, dt = \int_{-\infty}^{+\infty} f(-t)\, e^{-i\omega_0 t}\, dt \qquad (3.106)$$

Thus, the Fourier transform of $f(-t)$ is given by $R(\omega_0) - iI(\omega_0)$. If we split $f(t)$ into an even and odd function, as given by

$$f_e(t) = \frac{1}{2}[f(t) + f(-t)] \quad \text{and} \quad f_0(t) = \frac{1}{2}[f(t) - f(-t)] \qquad (3.107)$$

Then it is clear that $R(\omega_0)$ is the Fourier transform of $f_e(t)$ and $iI(\omega_0)$ is the Fourier transform of $f_0(t)$. For a causal function, due to Eq. (3.105), one can see that $f(t) = 2f_e(t) =$

$2f_0(t)$ for $t > 0$. Therefore, a real causal function can be determined either in terms of $R(\omega_0)$ or in terms of $I(\omega_0)$ from

$$f(t) = \frac{2}{\pi} \int_0^\infty R(\omega_0) \cos(\omega_0 t) \, d\omega_0 \tag{3.108a}$$

or

$$= -\frac{2}{\pi} \int_0^\infty I(\omega_0) \sin(\omega_0 t) \, d\omega_0 \tag{3.108b}$$

In addition to the above simplifications for estimating the Fourier transforms and their inverses, the following properties are often used as further aids, which are stated as theorems [264]. For notational ease, let us use the following to indicate the connection between the original and its transform: $f(t) \Leftrightarrow F(\omega)_0$, as notation.

Linearity : If $\quad f_1(t) \Leftrightarrow F_1(\omega_0)$ and $f_2(t) \Leftrightarrow F_2(\omega_0)$

then $\quad f_1(t) + f_2(t) \Leftrightarrow F_1(\omega_0) + F_2(\omega_0)$ $\tag{3.109}$

This Fourier transform theorem directly applies to a Laplace transform without any further qualification.

Symmetry : If $\quad f(t) \Leftrightarrow F(\omega_0)$, then

$$F(t) \Leftrightarrow 2\pi \, f(-\omega_0) \tag{3.110}$$

Time Scaling: For any real $\bar{\alpha}$

$$f(\bar{\alpha}t) \Leftrightarrow \frac{1}{|\bar{\alpha}|} F\left(\frac{\omega_0}{\bar{\alpha}}\right) \tag{3.111}$$

This property also applies directly to a Laplace transform, for $\bar{\alpha} > 0$.

Time Shifting: For any real time t_0

$$f(t - t_0) \Leftrightarrow F(\omega_0) \, e^{-i\omega_0 t_o} \tag{3.112}$$

It is also the same for a Laplace transform.

Frequency Shifting: For any real frequency $\bar{\omega}_0$

$$e^{i\bar{\omega}_0 t} f(t) \Leftrightarrow F(\omega_0 - \bar{\omega}_0) \tag{3.113}$$

This theorem also applies directly to a Laplace transform.

Time Differentiation: The Fourier transform of the nth order derivative can be found in terms of the Fourier transform of the original function by

$$\frac{d^n f}{dt^n} \Leftrightarrow (i\omega_0)^n F(\omega_0) \tag{3.114}$$

For a Laplace transform, a more general expression is obtained using all the initial conditions as given by $(i\omega_0)^n F(\omega_0) - (i\omega_0)^{n-1} f(0^-) - \ldots\ldots\ldots - f^{(n-1)}(0^-)$.

Frequency Differentiation: In a similar fashion, one can relate the nth derivative in the spectral plane with the following function in the physical plane via

$$(-it)^n f(t) \Leftrightarrow \frac{d^n F(\omega_0)}{d\omega_0^n} \tag{3.115}$$

Moment Theorem: For the nth moment of a function, $m_n = \int_{-\infty}^{+\infty} t^n f(t)\, dt$, one has the following pair

$$(-i)^n m_n \Leftrightarrow \frac{d^n F(0)}{d\omega_0^n} \tag{3.116}$$

Next, we describe the important property of the convolution. Consider two functions $f_1(x)$ and $f_2(x)$ from which we can construct the following

$$f(x) = \int_{-\infty}^{+\infty} f_1(y)\, f_2(x-y)\, dy \tag{3.117}$$

The function $f(x)$ is called the convolution of $f_1(x)$ and $f_2(x)$. It is denoted symbolically as

$$f(x) = f_1(x) * f_2(x) \tag{3.118}$$

Let us now state the following convolution theorems.

Time convolution theorem: If $f_1(t) \Leftrightarrow F_1(\omega_0)$ and $f_2(t) \Leftrightarrow F_2(\omega_0)$ then

$$\int_{-\infty}^{\infty} f_1(\tau)\, f_2(t-\tau)\, d\tau \Leftrightarrow F_1(\omega_0)\, F_2(\omega_0) \tag{3.119}$$

Frequency convolution theorem: Similarly, one can state the following frequency convolution theorem

$$f_1(t)\, f_2(t) \Leftrightarrow \frac{1}{2\pi} \int_{-\infty}^{+\infty} F_1(y)\, F_2(\omega_0-y)\, dy \tag{3.120}$$

3.6.1.5 Some Useful Laplace–Fourier Transforms

Some useful transforms are summarized here, which will be of use in formulating and solving receptivity problems.

(a) **Dirac delta function:** The delta function treated as a distribution $f(t) = \delta(t)$ allows its Fourier transform to be obtained by

$$F(\omega_0) = \int_{-\infty}^{+\infty} \delta(t)\, e^{-i\omega_0 t} dt = 1 \tag{3.121}$$

Thus, applying the delta function as input is equivalent to exciting all circular frequencies/wavenumbers with equal emphasis. This is the basis of finding the natural frequency of any oscillator via impulse response. When the oscillator is subjected to an impulse, all frequencies are equally excited and the system dynamics picks out the natural frequency of vibration, leaving others to decay, in due course of time. It is noted that this result also applies to a Laplace transform and we are going to use it often by replacing time with space and circular frequency with wavenumbers.

Now using the time shift theorem (as given by Eq. (3.112)), it is immediately evident using Eq. (3.121) that

$$\delta(t-t_0) \Leftrightarrow e^{-i\omega_0 t_0} \tag{3.122}$$

Also, using the symmetry property of Eqs. (3.110) and (3.113), one gets

$$1 \Leftrightarrow 2\pi \, \delta(\omega_0) \tag{3.123}$$

Similarly, using the symmetry property of Eqs. (3.110) and (3.122), we get

$$e^{i\bar{\omega}_0 t} \Leftrightarrow 2\pi \, \delta(\omega_0 - \bar{\omega}_0) \tag{3.124}$$

Since $\cos \bar{\omega}_0 t = \frac{1}{2}(e^{i\bar{\omega}_0 t} + e^{-i\bar{\omega}_0 t})$, therefore

$$\cos \bar{\omega}_0 t \Leftrightarrow \pi[\delta(\omega_0 - \bar{\omega}_0) + \delta(\omega_0 + \bar{\omega}_0)] \tag{3.125}$$

In the same way, from $\sin \bar{\omega}_0 t = \frac{1}{2i}(e^{i\bar{\omega}_0 t} - e^{-i\bar{\omega}_0 t})$, one obtains the following transform pair

$$\sin \bar{\omega}_0 t \Leftrightarrow i\pi[\delta(\omega_0 + \bar{\omega}_0) - \delta(\omega_0 - \bar{\omega}_0)] \tag{3.126}$$

(b) **The sign function** (sgn (t)) is equal to $+1$ when $t > 0$ and is equal to -1 when t is negative. Its Fourier transform is then given by

$$\operatorname{sgn}(t) \Leftrightarrow \frac{2}{i\omega_0} \tag{3.127}$$

Proof of Eq. (3.127): For $F(\omega_0) = \frac{2}{i\omega_0}$, the original is given by

$$f(t) = \frac{1}{2\pi} \int_{-\infty}^{+\infty} \frac{2}{i\omega_0} e^{i\omega_0 t} \, d\omega_0 \quad = \frac{1}{\pi} \int_{-\infty}^{+\infty} \frac{\sin \omega_0 t}{\omega_0} d\omega_0$$

This integral on the right-hand side has the property to be equal to $+1$ when $t > 0$ and is equal to -1 when $t < 0$. Hence $f(t)$ is sgn(t).

(c) **Unit step function or Heaviside function** $U_1(t)$: This function is equal to zero for all negative values of the argument and for all positive values of the argument, it is equal to $+1$, taking a discontinuous jump at $t = 0$. Hence this function can also be written as

$$U_1(t) = \frac{1}{2} + \frac{1}{2} \operatorname{sgn}(t) \tag{3.128}$$

Hence, using the results of Eqs. (3.123) and (3.127) we get the following pair

$$U_1(t) \Leftrightarrow \pi \, \delta(\omega_0) + \frac{1}{i\omega_0} \tag{3.129}$$

Furthermore, using the frequency shift theorem of Eq. (3.113)

$$e^{i\bar{\omega}_0 t} U_1(t) \Leftrightarrow \pi \, \delta(\omega_0 - \bar{\omega}_0) + \frac{1}{i(\omega_0 - \bar{\omega}_0)} \tag{3.130}$$

The above result is of central importance for the study of harmonic excitation of a shear layer, with the excitation having a finite start-up time. Usage of the above allows one to distinguish between the transient and asymptotic part of the receptivity solution of a shear layer subjected to a harmonic excitation starting at a finite time. This has been used in [322, 349, 350] to study the receptivity of a zero pressure gradient shear layer to harmonic excitation.

It is now easy to show that a harmonic excitation starting at $t = 0$ has the transform pair given by

$$U_1(t) \, \cos(\bar{\omega}_0 t) \Leftrightarrow \frac{\pi}{2} \left[\delta(\omega_0 - \bar{\omega}_0) + \delta(\omega_0 + \bar{\omega}_0)\right] + \frac{i\omega_0}{\bar{\omega}_0^2 - \omega_0^2} \tag{3.131}$$

(d) **Derivatives of the delta function:** In this class of functions one can include the doublet, quadrupole, etc., which are often used as singularity functions in fluid mechanics and acoustics. It is worthwhile to point out that the potential flow results for source, sink, doublet, etc., can be obtained using the same procedure detailed here, using Laplace's equation as the governing differential equation.

It has been shown in [264]

$$\int_{-\infty}^{+\infty} \frac{d^n}{dt^n} \delta(t - t_0) \, \varphi(t) \, dt = (-1)^n \frac{d^n \varphi(t_0)}{dt^n}$$

Therefore

$$F(\omega_0) = \int_{-\infty}^{+\infty} \frac{d^n \delta}{dt^n} e^{-i\omega_0 t} \, dt = (-1)^n \frac{d^n}{dt^n} (e^{-i\omega_0 t})|_{t=0} = (i\omega_0)^n$$

Thus

$$\frac{d^n \delta}{dt^n} \Leftrightarrow (i\omega_0)^n \tag{3.132}$$

(e) **Gaussian function** $(e^{-t^2/2})$: This function belongs to the class of Hermitian functions and has the important self-reciprocity property with its transform. The Fourier transform of it is given by

$$F(\omega_0) = \int_{-\infty}^{+\infty} e^{-t^2/2} e^{-i\omega_0 t} \, dt$$

$$= \int_{-\infty}^{\infty} e^{-\frac{1}{2}(t^2 + 2i\omega_0 t)} \, dt$$

$$= e^{-\frac{\omega_0^2}{2}} \int_{-\infty}^{+\infty} e^{-\frac{(t + i\omega_0)^2}{2}} \, dt$$

$$= \sqrt{2} e^{-\frac{\omega_0^2}{2}} \int_{-\infty}^{+\infty} e^{-\frac{(t + i\omega_0)^2}{2}} \, d\left(\frac{(t + i\omega_0)}{\sqrt{2}}\right)$$

Since $\int_{-\infty}^{+\infty} e^{-x^2/2} dx = \sqrt{\pi}$
therefore

$$e^{-\frac{t^2}{2}} \Leftrightarrow \sqrt{2\pi} \, e^{-\frac{\omega_0^2}{2}} \tag{3.133}$$

This is an important property of the Gaussian function, namely, that its transform has the identical functional form as the original. This is called the property of *self-reciprocity*. All its derivatives also share the same property. The derivatives of the Gaussian function produce the well-known Hermite functions

$$h_n(t) = (-1)^n \, e^{t^2/2} \frac{d^n}{dt^n} (e^{-t^2}) \Leftrightarrow H_n(\omega_0)$$

Therefore $\sqrt{2\pi} \, h_n(t) = i^n \, H_n(t)$.

3.6.2   Receptivity to Wall Excitation and Impulse Response

This topic is described here based on the work reported in [114, 319, 322]. Consider the behavior of small amplitude disturbances in a given parallel mean flow, in response to a time harmonic localized disturbance source (with frequency $\bar{\omega}_0$). In the simplest form of a model for response, the perturbation stream function is defined in [114, 319]

$$\psi(x, y, t) = \frac{1}{2\pi} \int_{Br} \phi(y, \alpha; \bar{\omega}_0) \, e^{i(\alpha x - \bar{\omega}_0 t)} \, d\alpha \qquad (3.134)$$

A two-dimensional disturbance field arising in a two-dimensional parallel mean flow, described by the velocity profile $U(y)$, is considered first. The disturbance source is located at the origin of a Cartesian system and the circular frequency of the input excitation is given by $\bar{\omega}_0$. In writing the disturbance stream function above, one assumes the system to respond at the frequency of the excitation only − this is known as the **signal problem** [114, 319]. Also note that this represents a disturbance field which can move in both downstream and upstream directions. Hence, the above is a bilateral Laplace transform and the Bromwich contour is to be located in the strip of convergence for the transform amplitude, ϕ. Substituting Eq. (3.134) in the linearized Navier–Stokes equation and making the parallel flow assumption provides the Orr–Sommerfeld equation as given in Eq. (3.57). The difference in the receptivity from the stability approach is in boundary conditions, which are no longer homogeneous, unlike that given in Eq. (3.55a) for the stability problem.

For the case of receptivity to localized wall excitation, a delta function is used at the point of excitation; this gives rise to the *impulse response* of the shear layer [319]. Boundary conditions applied at the wall are given by

$$\text{at } y = 0: \quad u = 0 \quad \text{and} \quad \psi(x, 0, t) = \delta(x) \, e^{-i\bar{\omega}_0 t} \qquad (3.135)$$

And far from the wall ($y \to \infty$):

$$u, v \to 0 \qquad (3.136)$$

The decaying boundary condition of Eq. (3.136) at free stream excludes two fundamental modes of the Orr–Sommerfeld equation. With the other two retained modes one defines

$$\phi = a_1 \, \phi_1 + a_3 \, \phi_3 \qquad (3.137)$$

Constants a_1 and a_3 are fixed by the boundary conditions given by Eq. (3.135) as follows

$$a_1 \, \phi'_{10} + a_3 \, \phi'_{30} = 0 \qquad (3.138a)$$

$$a_1 \, \phi_{10} + a_3 \, \phi_{30} = 2\pi \qquad (3.138b)$$

The additional subscript 0 indicates quantities as evaluated at the wall. Equations (3.138) are solved for a_1 and a_3 and upon substitution simplifies Eq. (3.134) to

$$\psi(x, y, t) = \int_{Br} \frac{\phi_1(y, \alpha)\phi'_{30} - \phi'_{10}\phi_3(y, \alpha)}{\phi_{10}\phi'_{30} - \phi'_{10}\phi_{30}} \, e^{i(\alpha x - \bar{\omega}_0 t)} \, d\alpha \qquad (3.139)$$

The denominator in this expression is evaluated at the wall and is the characteristic determinant of the eigenvalue problem shown in Eq. (3.62). This connects the receptivity with the corresponding instability problem. It is also apparent that the eigenvalues are the zeros of the characteristic determinant in the denominator of Eq. (3.139) and constitute the poles for the receptivity problem.

The choice of Bromwich contour is made based on the qualitative knowledge of the eigen-spectra of the problem. For example, if one were to calculate the impulse response of the Blasius boundary layer, then one has to position this contour in such a way that all the downstream propagating modes (the group velocity of these are positive) lie above this contour and the upstream propagating modes lie below this contour in the wavenumber plane. A detailed discussion of this was given first in [322]. Response due to all the discrete modes, the essential singularities and the continuous spectra are built inside ϕ evaluated along the Bromwich contour itself. A possible Bromwich contour can be taken parallel and below the real axis, so that the downstream propagating modes stay above this contour. There is the added advantage for this choice of contour in applying DFFT along this contour − otherwise any other contour would be appropriate in the strip of its convergence.

A typical impulse response is shown in Fig. 3.19 for the Blasius boundary layer for $\bar{\omega}_0 = 0.1$ and $Re_{\delta^*} = 1000$, with the disturbance stream function shown at the inner and the outer maxima of the least stable mode.

About $30\delta^*$ from the exciter in the downstream direction, the computed disturbance profile matches the eigen-solution corresponding to the complex wavenumber value (0.2798261, −0.00728702), which is obtained by the stability analysis for the TS mode. It is interesting to note that there is a local component of the receptivity solution that decays rapidly in either direction. This is the near-field response or the local solution. Thus, the receptivity solution in this figure consists of the asymptotic solution (away from the exciter) and a local solution.

Descriptions of the local solution and other details of selecting the Bromwich contours

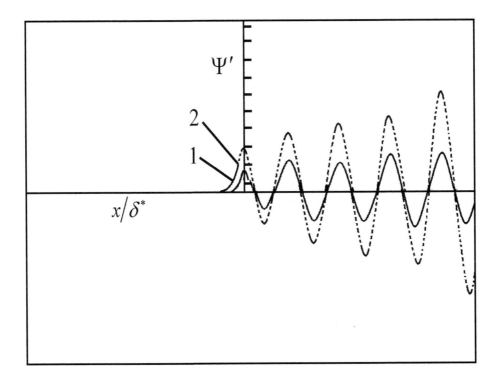

FIGURE 3.19
Disturbance stream function plotted versus streamwise distance for (1) the inner maximum at $y/\delta^* = 0.277$ and (2) the outer maximum at $y/\delta^* = 1.79$.

are given in [322, 348]. The near-field response created due to wall excitation is shown in these references as due to the essential singularity of the bilateral Laplace transform of the disturbance stream function. While the experiments of [316] verified spatial instability theory, instability theory is incapable of explaining all the aspects of the experiments or the solution obtained by the receptivity analysis, e.g., the near-field and the general time-dependent solution discussed in [350]. In this context, receptivity analysis is unique and assumes special importance in being able to explain the local component of the solution and the spatio-temporal growing wave front. In the following, the near-field of the receptivity solution is explained.

3.6.2.1　Near-Field Response Created by Localized Excitation

The near-field response created due to wall excitation is shown here as due to the essential singularity of the bilateral Laplace transform of the disturbance stream function. So far we have seen that the instability of the laminar boundary layer over the plane surface is manifested via the growth of spatially growing waves − as was theoretically shown in [143, 311, 408] and later verified experimentally in [316].

In [52, 109], some theoretical aspects of the receptivity of a fluid dynamic system from the initial-boundary value point of view were addressed. The first set of receptivity calculations were provided in [114, 319], under some restrictive conditions on time variation. In [322] this constraint was removed and the first time-dependent solution of the receptivity to wall excitation problem was provided. In [16], a "revised" formulation (with respect to [109]) was given following Briggs' method, originally developed to study plasma dynamics instability. The authors stated that the discrete spectrum and branch points constitute the time asymptotic solution. It was also conjectured that the near-field contribution comes from the branch cuts introduced from three fixed branch points of the governing Orr–Sommerfeld equation. Interestingly, they identified the branch cuts as the continuous spectral lines responsible for the local solution. This has not been proven since a one-to-one correspondence between the so-called continuous spectra and the near-field was established. On the other hand, the formal integral of the bilateral Laplace transform in space and the Fourier integral in time along carefully chosen Bromwich contours can automatically provide both the near- and far-field results; an example is shown in Fig. 3.19.

Here an explanation is provided on the structure of the near-field solution with the help of some fundamental theorems. These theorems provide the basis for interpreting both the near- and far-field solutions. These are due to Abel and Tauber and their utility was highlighted in [418] in connection with the properties of the bilateral Laplace transform. In exploring relationships between the original in the physical plane and the image or transform in the spectral plane, these two important theorems are used here.

　　If one is interested in the behavior of the original in the neighborhood of the exciter (i.e., near $x = 0$), then one needs to investigate the image at $\alpha \to \infty$. This is **Tauber's theorem**. Similarly, if one is interested in the solution far away from the exciter $x \to \infty$, then one needs to consider only the neighborhood of the origin in the α-plane. This is **Abel's theorem** and this simply points to the relevance of those poles very close to the origin in the α-plane, whose effects would be felt far away from the exciter.

In traditional stability analysis, not only is Abel's theorem implicitly used, but also attention is focused on the right half of the α-plane only.

Following the above, let us explore the relation between the original and image of bilateral Laplace transforms in relation to the evolution of small disturbances in boundary layers. In particular, one would be interested in the behavior of the original in the neighborhood of the exciter ($x = 0$) as determined by the image $\phi(y, \alpha)$ for $\alpha \to \infty$, according to Tauber's theorem.

In this sense, we will determine the contribution of $\alpha \to \infty$ for the original-image pair of Eq. (3.134). It is also relevant to discuss the role of Jordan's lemma [9], which shows the contour integral along the semi-circular arc approaching zero when the radius of the contour approaches infinity in Fig. 3.18. This can occur under some special condition satisfied by the integrand, as noted below. Let us recall that the integral along C, as shown in Fig. 3.18, is given by

$$I_C = \int \phi(\alpha)\, e^{i\alpha x}\, d\alpha \tag{3.140}$$

would vanish if and only if the degree of the denominator of Eq. (3.140) is at least two orders higher than the degree of the numerator, i.e.,

$$|\phi(\alpha)| < \frac{k}{|\alpha^2|} \tag{3.141}$$

for $\alpha \to \infty$. The main reason for the present discussion is to show that the image $\phi(\alpha)$ for this limit, as governed by the Orr–Sommerfeld equation, does not satisfy the condition given in Eq. (3.141). To show this, it is possible to expand $\phi(\alpha)$ as a function of ϵ_1 ($= \frac{1}{\alpha}$) by a singular perturbation analysis for $\alpha \to \infty$ and it is shown that such a series expansion starts with a $0(\epsilon_1)$ term, i.e., the condition given in Eq. (3.141) is not supported and hence Jordan's lemma is inapplicable here.

Thus, the next objective here is to estimate the contribution coming from the contour integral in Eq. (3.140) from the semi-circular arc in the α-plane, with the radius of the arc going to infinity. Along this arc one can represent

$$\alpha = \rho\ e^{i\theta} = \rho\beta \tag{3.142}$$

where ρ is the radius of the arc. To determine ϕ for large α, examine the asymptotic form of the Orr–Sommerfeld equation as an expansion in the small parameter $\epsilon_1 = \frac{1}{\rho}$, for $\rho \to \infty$. Equation (3.35) then takes the form

$$\epsilon_1^4\, \phi^{iv} - [2\epsilon_1^2\, \beta^2 + iRe\, \epsilon_1^3\, (\beta U - \epsilon_1\, \bar{\omega}_0)]\, \phi''$$
$$+ [\beta^4 + iRe\, \beta\epsilon_1^3\, U'' + iRe\, \beta^2\epsilon_1\, (\beta U - \epsilon_1\, \bar{\omega}_0)]\, \phi = 0 \tag{3.143}$$

Note that the image ϕ can also be expanded in a perturbation series and the analysis here is for the leading order term of such an expansion. The higher order terms of ϕ would produce trivial contributions to the contour integral, because each of the correction terms satisfies the condition for Jordan's lemma, as given above in Eq. (3.141).

Let us now discuss the case of wall excitation, where a localized delta function excites the flow. To simplify analysis, consider the following boundary conditions applied at the wall which is located at the origin of the coordinate system

$$y = 0: \quad u = 0 \quad \text{and} \quad \psi(x, 0, t) = \delta(x)\, e^{-i\bar{\omega}_0 t} \tag{3.144}$$

And far from the wall ($y \to \infty$):

$$u, v \to 0 \tag{3.145}$$

Boundary conditions given by Eqs. (3.144) and (3.145) of the impulse response problem can also be expressed as $\phi(0, \alpha) = 1$ and $\phi'(0, \alpha) = 0$ at $y = 0$ and as $y \to \infty$: $\phi(y, \alpha),\ \phi'(y, \alpha) \to 0$.

From Eq. (3.143), it is apparent that this is a singular perturbation problem (as the highest derivative term is multiplied by the small parameter) and then one can use matched asymptotic expansion to obtain ϕ by describing the solution in terms of outer and inner solutions.

3.6.2.2   Outer Solution

By definition, in the outer region, ϕ and all its derivatives are $0(1)$ and Eq. (3.143) simplifies to

$$\phi_0 = 0 \tag{3.146}$$

This solution is true up to any order and it automatically satisfies the outer boundary conditions. However, this does not satisfy the wall boundary condition and one must have a *boundary layer* or the inner layer of, say, thickness δ. The thickness δ can be obtained by the distinguished limits in the inner layer by following the method given in [27].

3.6.2.3   Inner Solution

In the inner layer, we define a new independent variable $Y = y/\delta$ and work with the dependent variable $\phi = \phi_i(Y)$. Then Eq. (3.143) takes the form

$$\left(\frac{\epsilon_1}{\delta}\right)^4 \phi_i^{iv} - \left[2\beta^2\left(\frac{\epsilon_1}{\delta}\right)^2 + iRe(\beta U - \epsilon_1\bar\omega_0)\left(\frac{\epsilon_1^3}{\delta^2}\right)\right]\phi_i''$$
$$+ \left[\beta^4 + iRe\beta\epsilon_1^3 U'' + iRe\beta^2\epsilon_1(\beta U - \epsilon_1\bar\omega_0)\right]\phi_i = 0 \tag{3.147}$$

Here the derivatives of ϕ_i are now with respect to Y. One can look at various distinguished limits by choosing terms pairwise in Eq. (3.147).

(i) For the distinguished limit $\delta << \epsilon_1$: Eq. (3.147) reduces to $\phi_i = 1$.

This is a non-trivial solution satisfying the wall boundary conditions. But it does not allow matching the inner and outer solutions and has to be discarded.

(ii) For the distinguished limit $\delta >> \epsilon_1$, one gets $\phi_i(Y) = 0$. However, this is not a valid solution as it fails to satisfy the wall boundary condition.

(iii) For the distinguished limit $\delta = \epsilon_1$: Eq. (3.147) takes the form
$$\phi_i^{iv} - 2\beta^2\phi_i'' + \beta^4\phi_i = 0$$
the solution of which is

$$\phi_i(Y) = A\,e^{\beta Y} + BY\,e^{\beta Y} + C\,e^{-\beta Y} + DY\,e^{-\beta Y}$$

Note that β is a complex constant and for $\beta_r > 0$, the inner solution that satisfies Eq. (3.144) is given by
$$\phi_i(Y) = (1 + \beta Y)\,e^{-\beta Y} \tag{3.148}$$

Similarly, one can obtain the inner solution for $\beta_r < 0$ as

$$\phi_i(Y) = (1 - \beta Y)\,e^{\beta Y} \tag{3.149}$$

It is easy to show that the other two distinguished limits, $\delta^2 = \epsilon_1^3$ and $\delta^2 = \epsilon_1^4$, produce only the trivial solution.

Thus, the only possible distinguished limit is $\delta = \epsilon_1$, implying that the inner layer of the Orr–Sommerfeld equation is of thickness $\delta = \frac{1}{|\alpha|}$. In terms of physical variables, the asymptotic value of ϕ is then given by

For $\alpha_r > 0$:

$$\phi = (1 + \alpha y)\, e^{-\alpha y} \tag{3.150}$$

And for $\alpha_r < 0$:

$$\phi = (1 - \alpha y)\, e^{\alpha y} \tag{3.151}$$

To evaluate the contribution of this solution to ψ, coming from the semicircular contour, we must consider three segments of the contour, as shown in Fig. 3.20, since ϕ is discontinuous across the α_i-axis. The imaginary axis demarcates the spectral plane in terms of the required sub-dominance of the fundamental solutions as given in Eq. (3.58) and a consequence of which is observed from Eqs. (3.148) and (3.149).

For the purpose of evaluating the contour integral, we choose the Bromwich contour along the real wavenumber axis, without any loss of analyticity of ϕ. Thus

$$\psi(x, y, t) = \frac{1}{2\pi} \int \phi(y, \alpha)\, e^{i(\alpha x - \bar{\omega}_0 t)}\, d\alpha \tag{3.152}$$

where

$$\text{Along} \quad C_1: \quad \phi(y, \alpha) = (1 + \alpha y)\, e^{-\alpha y} \tag{3.153}$$

$$\text{Along} \quad C_2: \quad \phi(y, \alpha) = (1 - \alpha y)\, e^{\alpha y} \tag{3.154}$$

$$\text{And at} \quad P: \quad \phi(y, i\rho) = \frac{1}{2}(e^{i\rho y} + e^{-i\rho y}) \tag{3.155}$$

The last contribution is required, as ϕ is discontinuous across P. To calculate the contribution coming from the neighborhood of point P, we consider a sector of the contour around P terminating at P_1 and P_2 in Fig. 3.20, defined by the small angle ϵ. One fixes the value of β at P_1 and P_2 in the following manner. The value of β corresponding to P_1 is
$\beta_1 = e^{i(\frac{\pi}{2} - \epsilon)} = i + \epsilon$, for small values of ϵ.
And the value of β corresponding to P_2 is
$\beta_2 = e^{i(\frac{\pi}{2} - \epsilon)} = i - \epsilon$, for small values of ϵ.
The contribution from C_1 is obtained as

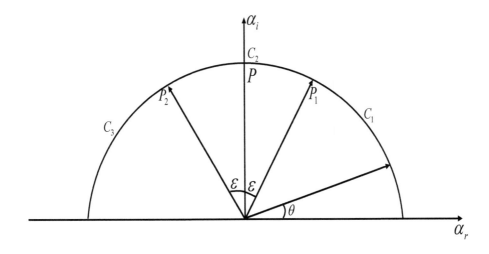

FIGURE 3.20
The contour used to evaluate the integral in Eq. (3.152).

$$I_1 = \frac{1}{2\pi} e^{-i\bar\omega_0 t} \int_0^{\beta_1} (1 + \rho\beta y) \, e^{i\beta\rho z} \rho \, d\beta$$

$$= \frac{e^{-i\bar\omega_0 t}}{2\pi} \left[e^{i\rho\beta_1 z} \frac{(1 + \rho\beta_1 z + \frac{iy}{z})}{iz} + e^{i\rho z} \frac{(1 + \rho y + \frac{iy}{z})}{z} \right]$$

The contribution I_2 coming from contour C_2, in the limit of $\epsilon \to 0$ is

$$I_2 = \frac{e^{-\rho x - i\bar\omega_0 t}}{2\pi} \cos\rho y$$

And finally the contribution I_3 from the contour C_3 is

$$I_3 = \frac{e^{-i\bar\omega_0 t}}{2\pi} \left[e^{-i\rho\bar z} \frac{(1 + \rho y - i\, y/\bar z)}{i\bar z} - e i\rho\beta_2\bar z \frac{(1 + \rho\beta_2 y - i\, y/\bar z)}{i\bar z} \right]$$

where $z = x + iy$ and $\bar z = x - iy$ is its complex conjugate.

Collecting various contributions, one obtains the perturbation stream function from the semicircular contour of radius ρ as

$$\psi(x, y, \rho, t) = \left[e^{-\rho x} \cos\rho y + \frac{ie^{i\rho z}}{z} \left(1 + \rho y + \frac{iy}{z} \right) - \frac{ie^{-i\rho\bar z}}{\bar z} \left(1 + \rho y - \frac{iy}{\bar z} \right) \right.$$
$$\left. - \frac{ie^{\rho z + iz}}{z} \left(1 + \frac{iy}{z} + i\rho y + y \right) + \frac{ie^{-\rho\bar z - i\bar z}}{\bar z} \left(1 - \frac{iy}{\bar z} - i\rho y + y \right) \right] \frac{e^{-i\bar\omega_0 t}}{2\pi} \quad (3.156)$$

To check the correctness of this result, let us investigate the solution at $y = 0$, where the wall excitation is applied to the shear layer. Here $\psi(x, 0, \rho, t)$ simplifies to

$$\psi(x, 0, \rho, t) = \frac{e^{-i\bar\omega_0 t}}{2\pi} \left\{ e^{-\rho x} - \frac{2\sin\rho x}{x} + 2e^{-\rho x} \left(\frac{\sin x}{x} \right) \right\} \quad (3.157)$$

In the limit, $\rho \to \infty$, the first and third terms in the above do not contribute. But the second term turns out to be the Dirichlet function, which is an approximation of the delta function, $\delta(x)$. Various approximate representations of the Dirac delta function are provided in [418] on pages 61−62. This clearly shows that we recover the applied boundary condition at $y = 0$. Therefore, the delta function is totally supported by the point at infinity in the wavenumber space (which is nothing but the circular arc of Fig. 3.20, i.e., the essential singularity of the kernel of the contour integral).

This result has the following consequence for the completeness of the basis function constructed from the eigenvectors obtained by stability analysis of external flows. It has been clearly shown in [211] that internal flows, like the channel flow, has denumerable infinite number of eigenmodes and any arbitrary applied disturbance can be expressed in terms of this complete basis set. However, for external flows, as we see for the Blasius flow in Table 3.1, there are only a few discrete eigenmodes and it is not possible to express any arbitrary functions in terms of these only, in the absence of any other singularities for this flow.

Theoretical analysis in the present section clearly indicates that the localized delta function excitation in the physical space is supported by the essential singularity ($\alpha \to \infty$) in the image plane. This is made possible because $\phi(y, \alpha)$ does not satisfy the condition required for the satisfaction of Jordan's lemma. Any arbitrary function can be shown as a convolution of the delta function with the function depicting the input to the dynamical system. The present analysis indicates that any arbitrary disturbances can be expressed in terms of a few discrete eigenvalues and the essential singularity. In any flow, in addition to these singularities there can be contributions from continuous spectra and branch points.

For the purpose of highlighting the above theoretical analysis explaining the near field of the response created by wall excitation, some numerical results are presented next, which have been obtained by solving Eq. (3.35) subject to the boundary conditions given by Eqs. (3.144) and (3.145). Cases have been considered where the harmonic source is excited with $\omega_0 = 0.1$, placed at three different locations with the Reynolds numbers, based on local displacement thickness, are 400, 1000 and 4000. As in calculating the impulse response case shown in Fig. 3.19, here also the Bromwich contour is located below and parallel to the real wavenumber axis, along $\alpha_i = -0.02$. Once again, CMM has been utilized to obtain solutions, as shown in Fig. 3.21.

The solutions presented in this figure demonstrate far fields corresponding to TS modes obtained by linear stability analysis. For $Re = 1000$ and $\bar{\omega}_0 = 0.1$, calculated impulse response displays TS wave with $\alpha_r = 0.279826$ and $\alpha_i = -0.007287$. Results are shown at a height of $y = 1.205\delta^*$, the location of the outer maximum of the eigenvector of the TS mode. Considering stability properties of the Blasius profile, one expects stable responses for $Re = 400$ and 4000 — with the latter case showing higher damping than the former, as clearly seen in Fig. 3.21.

Despite the differences in the values of the wavelength and the growth/decay rate of the three cases in this figure, one can notice the remarkable similarity of the near-field solution. While the upstream parts of all three solutions are exactly identical, minor differences on the downstream side of the near-field are due to the different wavelengths and growth (decay) rates of the asymptotic solutions. This is consistent with the observed properties of ϕ, for the noted behavior with $\alpha \to \infty$. Specifically, one can note that the near-field solutions obtained in Eqs. (3.156) and (3.157) are Reynolds number independent, as is also seen in the computed cases shown in Fig. 3.21. One also notes that the local solution originates in the inner layer whose governing differential equation is given by (obtained from the appropriate distinguished limit of Eq. (3.147))

$$\phi^{iv} - 2\beta^2 \phi'' + \beta^4 \phi = 0 \tag{3.158}$$

From this equation, it is possible to discuss further the general properties of the near-field solution. Notice that with the kinematic equation $\nabla^2 \psi = -\omega$ substituted in the governing

TABLE 3.1
Spatial modes and their group velocity for the impulse response analysis for $Re = 1000$ and $\bar{\omega}_0 = 0.10$ [322].

Mode Number	α_r	α_i	V_g (Group Velocity)
1	0.2798261	−0.00728702	0.4202
2	0.1380375	0.10991244	0.4174
3	0.1220209	0.17393307	0.8534

vorticity transport equation, one obtains

$$\frac{D}{Dt}\nabla^2\psi = \frac{1}{Re}\nabla^4\psi \tag{3.159}$$

It is easy to see that Eq. (3.158) is nothing but the right-hand side of Eq. (3.159), implying that the near field of the solution is given by the corresponding Stokes problem (in the limit of $Re \to 0$) as

$$\nabla^4\psi = 0 \tag{3.160}$$

Thus in the near field of the exciter, one observes a highly viscous flow and it is for the same reason the near-field does not penetrate very far upstream and downstream. For the cases shown in Fig. 3.21, the length of the near field is roughly about $20\delta^*$ at the height where the solution is displayed.

The above results also bring forth a very important aspect of real fluid flows. Despite the mathematical requirement of Jordan's lemma (that we must have $|\phi(\alpha)| \to 0$ for $\alpha \to \infty$), in real flows there always exists a cut-off wavenumber, due to the fact that smaller waves have larger strain-rates. Hence, an unlimited range of wavenumbers is not sustainable for an excitation problem driven by a finite source of energy. At the cut-off wavenumber, the kinetic energy of the wave would be converted to heat — an idea which has been traditionally employed to fix the cut-off wavenumber in Kolmogorov's theory of energy cascade for homogeneous turbulence. The present analysis also shows that the motion in the very small scale of any flow is governed by the Stokes equation given by Eq. (3.160).

So far, we have discussed very localized-in-space excitation cases. However, in an actual

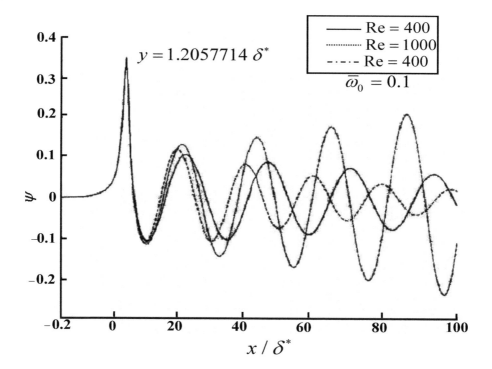

FIGURE 3.21
The solution of the signal problem at the indicated Reynolds numbers, for the same circular frequency. Note that the local solution components in all the cases are the same.

flow the exciter will be of finite width, as in the vibrating ribbon experiments of [316]. This is the topic of discussion in the following subsection.

3.6.3 Vibrating Ribbon at the Wall

If one wants to model the vibrating ribbon experiments of [316] by embedding the disturbance source on the surface of the plate, then the following is implied. Let the disturbance source be located at $x = x_0$ instead of the origin. Then instead of Eq. (3.134), one should rewrite the disturbance stream function as

$$\psi(x, x_0, y, t) = \frac{1}{2\pi} \int_{Br} \phi(y, \alpha; \bar{\omega}_0) \, e^{i[\alpha(x-x_0) - \bar{\omega}_0 t]} \, d\alpha \qquad (3.161)$$

It is seen that the governing equation for the bilateral Laplace amplitude is, once again, given by the Orr–Sommerfeld equation. To model this flow, we have to consider the width of the vibrating ribbon and not treat it as a simple line source. A finite width of the ribbon will excite a stream of contiguous finite wavenumbers. Such closely spaced wavenumbers will create groups of waves or a wave packet and in the response one would see modulated waves, as discussed and shown in [350].

Thus, for the finite-width ribbon located between $x = x_1$ and $x = x_2$ the disturbance stream function can be written as

$$\psi(x, y, t) = \frac{1}{2\pi} \int_{x_1}^{x_2} \left\{ \int_{Br} \phi(y, \alpha; \bar{\omega}_0) \, e^{i[\alpha(x-x_0) - \bar{\omega}_0 t]} \, d\alpha \right\} dx_0 \qquad (3.162)$$

Here it is implied that all the points between x_1 and x_2 are excited by the same amplitude. It need not necessarily be the case and a more general excitation would have the solution of the form written as

$$\psi(x, y, t) = \frac{1}{2\pi} \int_{x_1}^{x_2} \left[\int_{Br} W(x_0; x_1, x_2) \, \phi(y, \alpha; \bar{\omega}_0) \, e^{i[\alpha(x-x_0) - \bar{\omega}_0 t]} \, d\alpha \right] dx_0 \qquad (3.163)$$

where $W = W(x_0; x_1, x_2)$ is the prescribed weight function that fixes the type of excitation applied at the wall. This requires performing a weighted integral of the impulse response given by Eq. (3.161) with the prescribed weight function. Some typical weight functions are shown in Fig. 3.22, corresponding to different types of commonly used excitations.

These model excitations are commonly used in receptivity analysis for linear and nonlinear studies. For example, the combined blowing and suction excitation case shown in Fig. 3.22 has been used in few direct simulation attempts, as such excitation does not excite large bands of wavenumbers in the spectral plane to cause numerical instability, which is a common source of problems in DNS at high wavenumbers. Also, simultaneous blowing and suction excitation does not cause a numerical mass conservation problem at any time instant, as the amount of blowing is analytically counter-balanced by the amount of suction, always. Such an exercise for exciting TS waves was undertaken in [98]. Sometimes the Gaussian excitation shown in Fig. 3.22 is also preferred due to the advantage of the band-limited nature of the excitation field in the spectral plane as well, as a consequence of the *self-reciprocity* property of Hermite functions. The excitation shown in Fig. 3.22(a) is preferred for receptivity calculations based on the linearized Navier–Stokes equation, as it excites a large band of wavenumbers with equal emphasis.

Solutions for different excitations, at different x-locations, have simply to be added or convolved in linear analyses, as demonstrated in [350, 352], for the type of strip excitation shown in Fig. 3.22(a) applied on a flat plate surface.

To calculate the actual receptivity of a boundary layer in a correct time-accurate fashion, one should not start with the ansatz of the *signal problem*, as given by Eq. (3.134). Instead, one should define the disturbance stream function by

$$\psi(x,y,t) = \frac{1}{(2\pi)^2} \int \int_{Br} \phi(y,\alpha,\omega_0)\, e^{i(\alpha x - \omega_0 t)}\, d\alpha\, d\omega_0 \qquad (3.164)$$

For this case, Bromwich contours have to be traced simultaneously in the α- and ω_0-planes. The choice of the Bromwich contour in the ω_0-plane is not difficult, because it is strictly dictated by the causality requirement. However, the choice of the Bromwich contour in the wavenumber plane remains as difficult as before, for the *signal problem*.

Once again, when the ansatz of Eq. (3.164) is used in the linearized Navier–Stokes equation and the equilibrium solution is treated as parallel, one obtains the OSE as the governing equation. The boundary conditions applicable at the wall should now additionally incorporate the information of the finite start-up time of excitation as given below by

$$u = 0 \quad \text{and} \quad \psi(x,0,t) = U_1(t)\, \delta(x)\, e^{-i\bar{\omega}_0 t} \qquad (3.165)$$

At the free stream, disturbance quantities should decay to zero as before. Note that the presence of the Heaviside function $U_1(t)$ in Eq. (3.165) ensures that the excitation begins

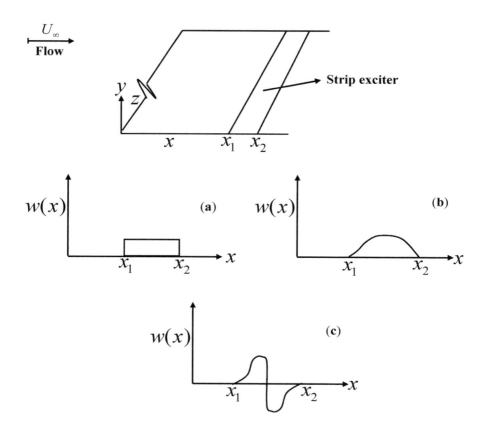

FIGURE 3.22

Typical time-harmonic excitations used on the surface for receptivity studies. (a) Strip excitation, (b) Gaussian amplitude distribution and (c) simultaneous blowing and suction excitation.

at $t = 0$, once again at the frequency $\bar{\omega}_0$. These boundary conditions in the physical plane translate in the spectral planes as

$$\phi'(\alpha, 0, \omega_0) = 0 \tag{3.166a}$$

and

$$\phi(\alpha, 0, \omega_0) = BC_w \tag{3.166b}$$

where $BC_w = [i(\bar{\omega}_0 - \omega_0)]^{-1}$ is the boundary condition for the full time-dependent problem. To satisfy the far-stream $(y \to \infty)$ conditions, solution of the OSE would be cast once again in the form

$$\phi(\alpha, y, \omega_0) = c_1 \phi_1 + c_3 \phi_3 \tag{3.167}$$

where ϕ_1 and ϕ_3 are the inviscid and the viscous fundamental decaying modes, as before. The constants c_1 and c_3 are fixed from the wall conditions given in Eqs. (3.166a) and (3.166b). This gives

$$\psi(x, y, t) = \frac{1}{(2\pi)^2} \int \int_{Br} \frac{\phi_1(\alpha, y, \omega_0)\phi'_{30} - \phi'_{10}\phi_3(\alpha, y, \omega_0)}{\phi_{10}\phi'_{30} - \phi_{30}\phi'_{10}}$$
$$BC_w \, e^{i(\alpha x - \omega_0 t)} \, d\alpha \, d\omega_0 \tag{3.168}$$

This is the problem solved for $Re = 1000$ and $\omega_0 = 0.1$ in [322], for which results of the corresponding *signal problem* were provided in [114, 319]. In Fig. 3.23, the full time-dependent solution is compared with the solution obtained for the corresponding *signal problem*, termed the time-asymptotic solution in the figure.

The presented time-accurate solution can be termed appropriately as the correct receptivity solution, as compared to its idealization in the *signal problem*. Later on, results obtained following this process are considered to look at cases of "spatially stable systems,"which actually admit spatio-temporally growing wave fronts which have been shown

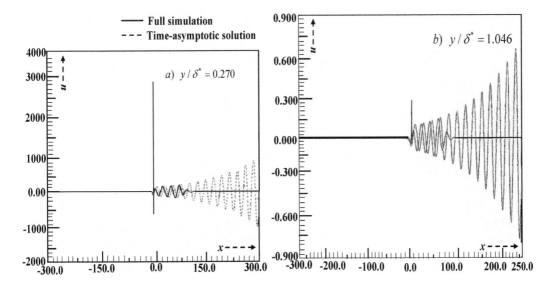

FIGURE 3.23
Comparison of solutions obtained by full receptivity analysis (solid line) and signal problem analysis (dotted line) at the indicated heights.

in [349, 350]. It is apparent for the spatially unstable cases that there are no differences between the solutions of the *signal problem* and the time-dependent problem, as these match in Fig. 3.23 up to a certain distance.

One of the reasons that the transition process is somewhat intractable is due to the fact that the receptivity process is inadequately understood for cases other than the wall excitation case, as discussed so far. This is specifically the case for shear layers excited by sources outside the viscous layer in the free stream, due to various reasons. First, unlike the wall excitation case, where the applied disturbance is uniquely located at the wall, for free stream excitation the response would depend upon the height of the input source in the free stream – a parameter that can continuously change with space and time, depending upon the dynamics of such disturbances interacting with other sources or by self-interaction. Second, the type of excitation that is present in the free stream also matters significantly. For example, acoustic, entropic and vortical disturbances will have significantly different receptivity. In the following, attention is focused upon the cases of vortical excitation only.

3.6.4   Receptivity to Free Stream Excitation

The main aspect of instability by free stream disturbances has not been investigated as successfully as has been the case of wall excitation. The reason has been indirectly explained earlier while discussing the observed results in the experiments of [316]. They did not detect TS waves when they irradiated the test section with acoustic waves. Of course, transition was noted without tracing the over-riding presence of monochromatic disturbances like the TS waves. Reasons for this are many fold: first, an acoustic wave is three dimensional and even when it creates TS waves, these would be due to streamwise and cross-flow instabilities. Moreover, it would create many interacting TS waves simultaneously, without showing monochromatic waves – as expected in normal mode analysis. Second, receptivity of the laminar boundary layer to this type of free stream disturbance shows very weak coupling. Some experimental efforts have been made starting with the interesting work of [400], who tried to estimate dependence of critical Reynolds number on free stream turbulence, treating the latter as vortical disturbances. In discussing Taylor's work in [229], the authors conjectured that the convected vortices embedded in the free stream cause a small adverse pressure gradient which gives rise to unsteady separation at multiple scales. Such unsteady separations can cause very rapid and catastrophic transition. The assumption implicit in this scenario is that the *effect is connected with the generation of fluctuations of longitudinal pressure gradient by these disturbances, leading to the random formation of individual spots of unstable S-shaped velocity profile* [229]. Thus, the effects of free stream disturbances are viewed as buffeting the shear layer by the ever-present disturbance sources and not necessarily due to instabilities.

In contrast to this earlier viewpoint, Morkovin [232] proposed that the response to free stream excitation occurs in two stages. In the first stage (termed in [232] the receptivity stage) the external perturbations are internalized as unsteady fluctuations giving rise to TS waves accompanying the equilibrium state. In the second stage, these internalized excitations have direct receptivity in causing transition – as described in previous sub-sections.

The problem of the excitation of the shear layer by disturbance sources convecting outside the shear layer has been experimentally investigated in [86, 172, 173]. Kendall [173], through experiments on jet-induced free stream turbulence (which has some inherent periodicity of excitation), has provided direct evidence of TS wave packets forming in a nominally flat plate boundary layer.

Dietz [86] successfully created a single frequency gust by using a vibrating ribbon in the free stream and provided quantitative data on TS wave amplitudes generated by the interaction of the gust with a surface roughness element. It is noted that the convected vortices traveling with free stream speed will not show any receptivity, as was also shown in [327] by linear and nonlinear studies. However, for free stream acoustic excitation, the receptivity problem is further complicated due to the problem of scale conversion, i.e., the sound and TS waves are of a different order of magnitude for wavelength and also phase/group velocity. Additionally, it was noted in the numerical calculations in [239] and by theoretical analysis in [125, 128] that the response field amplitude in the shear layer is an order of magnitude lower than the forcing disturbance amplitude in the free stream. However, the experimental results of [194] had clearly demonstrated earlier that this coupling is of order one. At the same time, the experiments of [3] demonstrated experimentally that an introduction of a thin surface roughness element increases the receptivity linearly with forcing amplitude and roughness height. Most of the theoretical developments using asymptotic analysis were based on this observation — see, e.g., the triple deck theory in [127, 304]. In [308], the authors noted that the *receptivity has many different paths through which to introduce disturbance into the boundary layer. They include the interaction of free stream turbulence and acoustic disturbances with model vibrations, leading-edge curvature, discontinuities in surface curvature, or surface inhomogeneities... The incoming free stream disturbance (sound or turbulence) at wave number α_{fs} interacts with a body in such a way (roughness, curvature, etc.) so as to broaden its spectrum to include the response wavenumber α_{TS}*. In discussing the receptivity to free stream excitation in [126], the authors noted that the scale adjustment mechanism can be attributed to *(i) rapid streamwise variations in the mean boundary-layer flow (which invalidates the parallel-flow assumption of Orr-Sommerfeld equation) and (ii) sudden changes in surface boundary conditions*. As a result, there is some region near the leading edge of the plate where the correct asymptotic approximation to the Navier–Stokes equations is the unsteady boundary-region equations, which are just the Navier–Stokes equations with the streamwise derivatives neglected in the viscous and pressure gradient terms. The former is used in leading-edge receptivity problems and the latter comes into view via the appearance of the triple-deck structure of the flow at downstream locations, in the large Reynolds number limit.

In particular, receptivity to vortical disturbances was investigated theoretically in [302], who modeled the free stream disturbance as a convected array of harmonic vortices. Highly damped near-wall disturbances were calculated from this model. Kerschen [174] used the asymptotic method to calculate vortical receptivity and showed it to vary with the convection speed of vortices. Other Orr–Sommerfeld based models in the literature also could not reveal the physical picture seen in the experiments of [86, 172], except in [327].

Kendall [172] performed experiments in which a circular cylinder was rotated in a circular trajectory above a flat plate shear layer to create a convecting periodic disturbance source. The speed of convection of these vortices was controlled and it was demonstrated that the underlying shear layer was strongly receptive to imposed disturbances in a narrow range of convecting speeds around $c = 0.3U_\infty$. In the receptive range, the response field consisted of wave packets composed of many TS waves. In this experiment the disturbance (cause) always stayed outside the shear layer.

Subsequently, Liu & Rodi [205] repeated Kendall's experiment, but now the periodic disturbance was directed towards the shear layer with a large wall-normal velocity component. This latter experiment was supposed to mimic the physical events in the flow inside turbomachinery. Experiments in [172, 205] showed such strong receptivity, while contemporary theoretical calculations by various mathematical models showed lower receptivity. This discrepancy was explained in [327] as due to the constraint imposed in the theoretical models on the speed of vortical disturbances in the free stream. It has been seen that in turbomachinery or in flows over helicopter rotor blades, flow in subsequent stages or blades is strongly influenced by vortices which travel over them at speeds much lower than the free stream speed. For example, experimental data and their correlation in [312] reveals that the vortices in the far wake of a single bluff body convect at 14% of the free stream speed. This convection speed is expected to be different in the near wake and when large ensembles of unconstrained vortices are present in the wake. In the experiments of [172], it was clearly shown that maximum receptivity of the shear layer occurred when the convection speed of the free stream vortices was around 30% of the free stream speed. In a three-dimensional DNS (where there was no need to prescribe any convection speed of vortices) in [446], such strong receptivity was clearly observed. Sengupta et al. [327] further showed this connection by producing results simultaneously by solving the Orr–Sommerfeld equation and two-dimensional direct simulation of receptivity. This is discussed further in Sec. 3.7.

3.6.5   General Excitation and Upstream Propagating Modes

Consider the following flow field over a flat plate which is excited simultaneously at the wall $y = 0$ and at the free stream $(y = Y)$, as shown in Fig. 3.24, where Y is significantly larger than the boundary layer thickness. At the wall, a time-periodic blowing-suction device is placed at $x = x_0$, defined in a coordinate system fixed at the leading edge of the plate. The circular frequency of excitation of the wall device is $\bar{\omega}_0$, such that the transverse velocity oscillation at the wall is set up as

$$u = 0 \quad v = v_w \, \delta(x - x_0) \, e^{-i\bar{\omega}_0 t} \tag{3.169}$$

The line vortex of strength Γ convects in the free stream with a constant speed c and at a constant height Y over the flat plate. Let the instantaneous location of this irrotational vortex be given by $\bar{x}$ from the leading edge. The corresponding stream function at any field point (x, y) created by this localized line vortex at Y is given by

$$\psi_\infty = \frac{\Gamma}{4\pi} \ln \frac{(x - \bar{x})^2 + (y + Y)^2}{(x - \bar{x})^2 + (y - Y)^2} \tag{3.170}$$

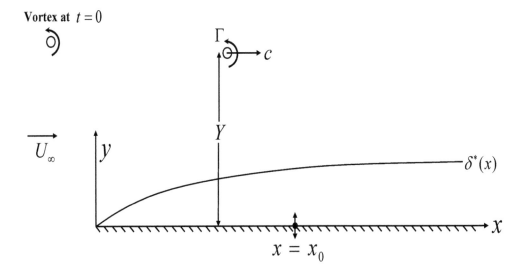

FIGURE 3.24
General excitation on a flat plate boundary layer by simultaneous time-periodic excitation at the wall and a convecting vortex at constant speed in the free stream.

where Γ is the strength of the potential line vortex located instantaneously at $\bar{x} = x_{v_0} - ct$, with x_{v_0} as the initial location of the vortex. The denominator accounts for the image system below the wall created by the reflection of the vortex in the free stream. One defines the full time-dependent perturbation stream function by

$$\psi(x, y, t) = \frac{1}{(2\pi)^2} \int \int_{Br} \phi(y, \alpha, \omega_0) \, e^{i(\alpha x - \omega_0 t)} \, d\alpha \, d\omega_0 \qquad (3.171)$$

One can write the Laplace–Fourier transform in terms of all the four fundamental solutions as

$$\phi(\alpha, y, \omega_0) = C_1 \, \phi_1 + C_2 \, \phi_2 + C_3 \, \phi_3 + C_4 \, \phi_4 \qquad (3.172)$$

Note that one has to retain all four modes for this general excitation case. To satisfy the first condition of Eq. (3.169), one must have the following satisfied

$$C_1 \, \phi'_{10} + C_2 \, \phi'_{20} + C_3 \, \phi'_{30} + C_4 \, \phi'_{40} = 0 \qquad (3.173)$$

with a prime indicating a derivative with respect to y. The subscript 0 refers to the condition at the wall. For the wall-normal velocity boundary condition of Eq. (3.169), one can write it using the "time-shift" theorem of the Laplace-Fourier transform as

$$v_w \, \delta(x - x_0) \, e^{-i\bar{\omega}_0 t} = \frac{1}{(2\pi)^2} \int \int_{Br} v_w \, e^{i[\alpha(x - x_0) - \omega_0 t]} \, \delta(\omega_0 - \bar{\omega}_0) \, d\alpha \, d\omega_0 \qquad (3.174)$$

Therefore

$$C_1 \, \phi_{10} + C_2 \, \phi_{20} + C_3 \, \phi_{30} + C_4 \, \phi_{40} = \frac{v_w}{i\alpha} \, e^{-i\alpha x_0} \, \delta(\omega_0 - \bar{\omega}_0) \qquad (3.175)$$

In the same way, one can convert the implied free stream condition of Eq. (3.170) as

$$C_1 \, \phi'_{1\infty} + C_2 \, \phi'_{2\infty} + C_3 \, \phi'_{3\infty} + C_4 \, \phi'_{4\infty} = B(\alpha, \omega_0) \qquad (3.176)$$

and

$$C_1 \, \phi_{1\infty} + C_2 \, \phi_{2\infty} + C_3 \, \phi_{3\infty} + C_4 \, \phi_{4\infty} = D(\alpha, \omega_0) \qquad (3.177)$$

The specific type of free stream condition can be represented by finding the appropriate functions, $B(\alpha, \omega_0)$ and $D(\alpha, \omega_0)$, defining the tangential and normal velocity components. The additional subscript (∞) refers to the conditions being evaluated at the free stream $(y = Y)$. Now one can solve for the constants C_1 to C_4 by simultaneously solving Eqs. (3.172) and (3.175) to (3.177). All these can also be written as the following linear algebraic equation

$$[\Phi]\{C_i\} = \{f_i\} \qquad (3.178)$$

where, $\{f_i\} = \lfloor 0 \quad \frac{v_w}{2\pi i \alpha} \, e^{-i\alpha x_0} \, \delta(\omega_0 - \bar{\omega}_0) \quad B(\alpha, \omega_0) \quad D(\alpha, \omega_0) \rfloor^T$ is the forcing, as applied through the boundary conditions. Thus, one can obtain the Laplace–Fourier transform with the constants C_i obtained from

$$\{C_i\} = [\Phi]^{-1}\{f_i\} \qquad (3.179)$$

with

$$\Phi = \begin{bmatrix} \phi'_{10} & \phi'_{20} & \phi'_{30} & \phi'_{40} \\ \phi_{10} & \phi_{20} & \phi_{30} & \phi_{40} \\ -\alpha e^{-\alpha Y} & \alpha e^{\alpha Y} & -Q e^{-QY} & Q e^{QY} \\ e^{-\alpha Y} & e^{\alpha Y} & e^{-QY} & e^{QY} \end{bmatrix}$$

where $Q^2 = \alpha^2 + i\alpha Re(1 - c)$ and $(\phi_{i\infty}, \phi'_{i\infty})$'s are obtained from the properties of the OSE given in Sec. 3.4. The constants obtained from Eq. (3.178) can be used in Eqs. (3.164), (3.171) and (3.172) to obtain the perturbation stream function for this excitation. However, it is also possible to obtain the eigenvalues by calculating these from the characteristic determinant of the corresponding stability problem obtained from

$$Det\ [\Phi] = 0 \qquad (3.180)$$

This locates the poles of the transfer function obtained from Eqs. (3.171) and (3.178). One feature that emerges from Eq. (3.179) is that for large Y, i.e., when the convecting vortices are far away from the plate, two sets of terms in the third and fourth rows are sub-dominant in $[\Phi]$.

This equation also holds promise to find eigenvalues in the left half of the complex wavenumber plane. It is noted that researchers have not been successful in establishing the existence of such modes; see, for example, figures in [16, 314], where the authors plot the so-called integration contours in the α-plane, indicating unknown eigenvalues on the left half of the plane with question marks. However, this has been detected and shown in [344] for the Blasius boundary layer.

Let us indicate how one can locate eigenvalues on the left half of the α-plane from the instability analysis point of view. Consider the case for which real $(\alpha) < 0$ and real $(Q) > 0$ and $Q^2 = \alpha^2 + i\alpha\ Re(1 - \frac{\omega_0}{\alpha}) = p + iq$. Now as $Y \to \infty$, $\phi_2(= e^{\alpha y})$ and $\phi_3(= e^{-Qy})$ are the modes which decay with height in the free stream. Applied boundary conditions in the free stream are supported there by ϕ_1 and ϕ_4.

Therefore, $Det[\Phi] = 0$ implies the following determinant to be equal to zero

$$Det[\Phi] = Det \begin{bmatrix} \phi'_{10} & \phi'_{20} & \phi'_{30} & \phi'_{40} \\ \phi_{10} & \phi_{20} & \phi_{30} & \phi_{40} \\ -\alpha e^{-\alpha Y} & 0 & 0 & Qe^{QY} \\ e^{-\alpha Y} & 0 & 0 & e^{QY} \end{bmatrix}$$

This implies that the following provides the dispersion relation

$$-[\alpha + Q]\ e^{(Q-\alpha)Y}\ [\phi'_{20}\phi_{30} - \phi_{20}\phi'_{30}] = 0 \qquad (3.181)$$

Thus, the characteristic determinant obtained from Eq. (3.180) for the eigenvalues in the left half plane are obtained by the decaying modes ϕ_2 and ϕ_3 only — as noted from Eq. (3.181). Thus, for the general case with free stream excitation as one of the components, while one needs to retain all four components of the fundamental solutions of the OSE, it is only the decaying modes which determine the dispersion relation. Therefore, it establishes that the general case of excitation requires modes both in the left half and the right half planes, simultaneously. Modes on the left half plane indicate solution components whose phase "move" upstream. However, whether such modes actually travel upstream or not, would be dictated by the group velocity of that mode. This possibility was investigated in [344] to detect upstream propagating modes for the Blasius profile. Indirect evidence of upstream propagating modes has been reported in [446], where the authors talked about upstream-facing turbulent spots from Navier–Stokes simulation of flow past a plate, in the presence of convecting vortices outside the shear layer which enter the computational domain through the inflow.

However, for the Blasius boundary layer, these upstream propagating waves are strongly decaying and thus do not give rise to linear instability [344]. In anticipation, one notes that for mean flows with an adverse pressure gradient, upstream propagating modes can become unstable. These upstream propagating modes were discovered as recounted below.

We have noted so far that the when Blasius boundary layer is excited by two-dimensional moderate frequency sources, one notices TS waves. However, at very low frequencies of excitation, TS waves are not seen and instead the whole boundary layer executes a heaving motion, which is now known as the Klebanoff mode of motion. Such a mode of motion was also seen experimentally in [113] when a diaphragm was vibrated at a frequency of 2 Hz on a flat plate with a two-dimensional Blasius boundary layer forming over it. TS waves were not seen as a consequence of this excitation. The mean flow was described adequately by the Blasius profile. This prompted the authors in [113] to comment that a proper mathematical account of these disturbances was not known until then. This was provided subsequently in [345]. To understand the physical nature of the Klebanoff mode, first of all we would investigate what happens to the disturbance field when the frequency of wall excitation decreases for the Blasius profile.

It is worth mentioning that similar experiments were performed earlier in [178, 401]. Klebanoff called this the *breathing mode of motion*. Apart from the fact that TS waves are not seen experimentally, corresponding two-dimensional linear instability studies also do not reveal the presence of any eigensolutions (TS waves) which decay with height up to the edge of the shear layer. As observed experimentally, the disturbance field remains three dimensional, but it propagates predominantly in the streamwise direction.

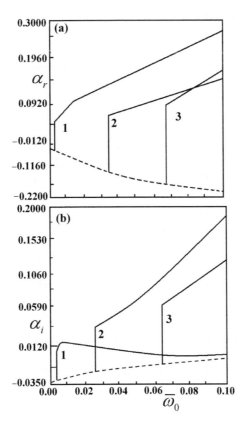

FIGURE 3.25

Variation of α_r and α_i with $\bar{\omega}_0$ for the Blasius boundary layer at $Re = 1000$. Note the downstream propagating modes (solid lines) and the upstream propagating mode (dotted line). Downstream propagating modes disappear as $\bar{\omega}_0$ decreases.

TABLE 3.2

Values of $\bar{\omega}_0$ for which two-dimensional modes of
the Blasius boundary layer disappear.

Mode Number	For $Re = 1000$	For $Re = 1196$
1	0.0026	0.0022
2	0.0663	0.0563
3	0.0276	0.0227

To show the relative roles of various two-dimensional modes at all frequencies, the real and imaginary parts of the wave numbers (α_r, α_i) are plotted in Fig. 3.25 as a function of non-dimensional circular frequency for $Re = 1000$. The three downstream propagating modes which are present at this Reynolds number for moderate frequencies are marked by numerals and are tracked in Fig. 3.25 for their variation with circular frequency. All these two-dimensional modes disappear abruptly one by one as the frequency is decreased, as depicted in Fig. 3.25. Values of circular frequencies at which these three modes disappear are listed in Table 3.2. Also shown in the second column are values of $\bar{\omega}_0$ for a Reynolds number of 1196, which corresponds to the experiments of [113].

If a fluid dynamical system is excited below the lowest of the three values of $\bar{\omega}_0$ in the above table, then CMM [91, 322, 4] detects an eigenvalue that is shown in Fig. 3.25 by a dotted line with its corresponding location on the left half of the α-plane. It is to be emphasized that finding the eigenvalues on the left half of the α-plane is possible by CMM due to its unique analytical features and it has not been reported by any other formulations and methods. From Table 3.2 for $Re = 1000$, this threshold value is given by $\bar{\omega}_0 = 0.0026$ and for $Re = 1196$, this value is slightly lower at $\bar{\omega}_0 = 0.0022$ and any excitation below this critical circular frequency shows the presence of only a single upstream propagating mode noted by the dotted line in Fig. 3.25.

One notes the interesting feature of this new mode, as both the values of α_r and α_i are negative. For TS waves, α_r is always positive, i.e., they are found in the right half of the wavenumber plane. This new upstream propagating mode is in the left half plane and one would be interested to find out the direction of propagation of such waves by finding numerically the group velocity of such disturbances from $V_g = \frac{\partial \omega_0}{\partial \alpha_r}$. One can readily testify that for the TS mode the group velocity is positive, showing the TS mode to propagate downstream. For this new mode, the corresponding variation of phase speed and group velocity with ω_0 are calculated numerically and shown in Fig. 3.26. This shows that the new mode travels upstream (as indicated by the negative value of the group velocity shown for this mode by the dotted line).

For the TS mode, a negative value of α_i implies the corresponding wave to be unstable. However, for this new mode, the disturbance travels upstream and a high negative value of α_i implies that this is a highly damped mode. One also notices that starting with the upstream propagating mode, upon increasing the frequency above the critical value, one can track this new mode, apart from the downstream propagating modes restored via the stability analysis. Thus, for a given Reynolds number at moderate frequencies, both types of modes are present. The upstream propagating modes also have the interesting property that their amplitude increases with y, i.e., this mode will be able to support any excitation at the free stream. Henceforth we will distinguish between these two classes of modes as the *wall mode* and the *free stream mode*. The *free stream mode* can support disturbances due to free stream turbulence or convected vortices in the free stream. There are also other interesting variations of properties of the upstream propagating modes with Reynolds number and these are demonstrated in Fig. 3.27, where $\alpha_r, \alpha_i, c_{ph}$ and V_g are plotted as a function of $\bar{\omega}_0$ for different Reynolds numbers. Chosen $\bar{\omega}_0$ varies over a large range and so do the Reynolds numbers.

For the wide range of $\bar{\omega}_0$ and Re in Fig. 3.27, all the properties seem to be a weak function of Reynolds number except for the value of α_i. The imaginary part of the wavenumber determines two aspects of the disturbance field in the boundary layer: (a) the extent of the upstream region over which disturbances can be found and (b) the disturbance flow structure in the wall-normal direction, within the boundary layer. As the magnitude of α_i reduces with increasing Reynolds number for a fixed $\bar{\omega}_0$, the region of upstream effect will increase with the excitation source convecting downstream in the free stream. As propagating downstream in a growing boundary layer is equivalent to increasing Reynolds number, for sufficiently high Reynolds numbers this will create disturbances that will affect the boundary layer over a larger portion of the flow field with a dominant high frequency component. This result establishes the fact that the upstream propagating mode is always damped for the

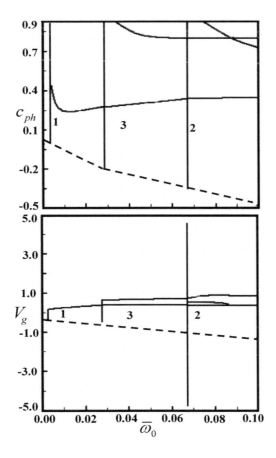

FIGURE 3.26
Variation of phase speed (top) and group velocity (bottom) of the modes shown in Fig. 3.25, with $\bar{\omega}_0$ for the Blasius boundary layer at $Re = 1000$.

Blasius boundary layer and would not be responsible for linear instability. There is another important aspect noted from Fig. 3.27, that the modes disappear with the circular frequency reduced, for which the phase speed of the eigensolution reaches the free stream speed ($c = U_\infty$).

3.6.6   Low Frequency Free Stream Excitation and the Klebanoff Mode

Low frequency excitation of a zero pressure gradient shear layer behaves qualitatively differently, as has been discussed before, with respect to the experimental results of [113, 401]. In these experiments, the zero pressure gradient shear layer was disturbed by a shallow oscillating bump on the wall, with the oscillation frequency very low. For example, in [113], the oscillation frequency was only 2 Hz and inside the shear layer no waves were seen; instead the whole boundary layer executed a heaving motion. Such low frequency fluctuations were also noticed in [178] earlier, and was called the *breathing mode*; now it is known as the *Klebanoff mode*.

There are two noticeable features of the experiments reported in [113]. First, the mean flow field is adequately represented by the Blasius profile and for the parameter ranges, two-dimensional instability studies did not reveal the presence of any eigensolutions which decay with height, as one approaches the edge of the shear layer. Second, the disturbance field was three dimensional, but this propagated predominantly in the streamwise direction. It is easy to see that if the disturbance field has very large wavelengths, then the experimental results will only indicate a heaving motion for relatively low to moderate length

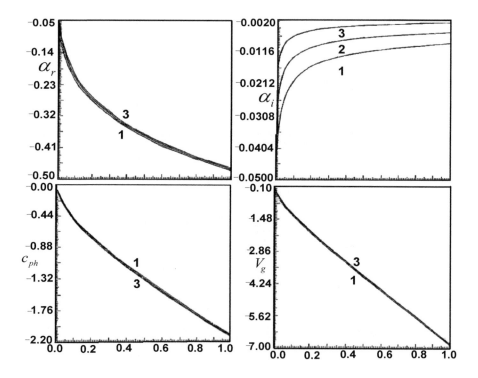

FIGURE 3.27
Variation of wave properties of the upstream propagating mode with $\bar{\omega}_0$ for the Reynolds numbers: (1) $Re = 400$; (2) $Re = 1000$ and (3) $Re = 4000$.

test-sections of tunnels (as was the case for these two experimental results in [113, 401]). In [113], measurement stations were located at 70 and 105 boundary layer thicknesses aft of the shallow bump. The most interesting aspect of the *Klebanoff mode* of motion is that although the mean flow is two dimensional the flow does not support a two-dimensional disturbance field. This led Gaster et al. [113] to comment that a *proper mathematical account of these disturbances has not yet appeared.* Subsequently, in [345] a rational explanation of the same phenomenon was provided and it is the subject of the present discussion.

In this reference, as the flow field was shown not to support any two-dimensional disturbances, three-dimensional disturbances were sought in the analysis. The wall-normal component of the disturbance field is therefore represented as

$$v'(x, y, z, t) = \frac{1}{4\pi^2} \int \int_{Br} \phi(\alpha, y, \beta, \bar{\omega}_0) \, e^{i(\alpha x + \beta z - \bar{\omega}_0 t)} \, d\alpha \, d\beta \qquad (3.182)$$

with α and β as the streamwise and spanwise wavenumbers, respectively. One is discussing the associated signal problem only, with the excitation frequency as $\bar{\omega}_0$. However, experiments conducted in closed tunnels, spanwise wavenumber will have a lower cut-off β_0, determined by the spanwise extent of the tunnel, i.e., $\beta_0 = 2\pi/\lambda_z$, where the spanwise wavelength (λ_z) is twice the tunnel width. For such cases, one can rewrite the disturbance field as

$$v'(x, y, z, t) = \frac{1}{2\pi} \sum_{n=1}^{\infty} \int_{Br} \phi(\alpha, y, \beta_0, \bar{\omega}_0) \, e^{i(\alpha x + n\beta_0 z - \bar{\omega}_0 t)} \, d\alpha \qquad (3.183)$$

The summation is over all spanwise modes. One can use the above ansatz in the three-dimensional Navier–Stokes equation and linearize the resultant equation after making a parallel flow approximation to get the following Orr–Sommerfeld equation for the Laplace–Fourier transform ϕ of v' as

$$\phi^{iv} - 2(\alpha^2 + n^2\beta_0^2)\,\phi'' + (\alpha^2 + n^2\beta_0^2)^2\,\phi$$
$$= iRe\{(\alpha U + n\beta_0 W - \bar{\omega}_0)[\phi'' - (\alpha^2 + n^2\beta_0^2)\,\phi] - [\alpha U'' + n\beta_0 W'']\,\phi\} \qquad (3.184)$$

In Eq. (3.184), $U(y)$ and $W(y)$ are the parallel mean flow components in the streamwise and spanwise directions, Re is the Reynolds number based on the displacement thickness of the boundary layer and primes indicate derivatives with respect to y.

To explain features of the experiments in [113], some details of the experimental conditions are recounted first. A flat plate was mounted in the test section of a low disturbance wind tunnel which is 3.5 m long and 0.91 m by 0.91 m in cross section. A three-dimensional velocity field was created by a circular bump of 20 mm diameter which was located 400 mm from the leading edge of the plate. At the location of the bump, the undisturbed boundary layer had a thickness of $\delta^* = 0.99$ mm. Based on this thickness and the free stream speed of 18.10 m/s, the Reynolds number is found to be 1196. The circular bump was oscillated at a frequency of 2Hz, which makes the non-dimensional circular frequency $\omega_0 = 6.248 \times 10^{-4}$. The span of the tunnel test section was 920 times δ^*(approx.) and thus the maximum spanwise wavelength of disturbances that can be supported is twice this dimension. This is the rationale for fixing the lower limit of spanwise wave number β_0 and the corresponding value is 3.41777×10^{-3}. It has been noted in [113] that most of the disturbance energy is carried by only the first ten modes and this corresponds to $n\beta_0 = 0.7854$.

For the above experimental conditions, it is apparent from Figs. 3.24 and 3.25 that the two-dimensional modes disappear as the value of $\bar{\omega}_0$ is lower than the critical value $(\bar{\omega}_0 = 6.248 \times 10^{-4})$ given in the last column of Table 3.2 for $Re = 1196$. This was taken as a cue to investigate if there are three-dimensional modes present for such low frequency excitation, in [345]. The spatial eigenvalues were located by the grid-search method of [211]. Once the eigenvalues were located, streamwise and spanwise components of group velocity were obtained numerically using the following

$$\vec{V}_g = \left(\frac{\partial \omega_0}{\partial \alpha_r}, \frac{\partial \omega_0}{\partial \beta_r} \right) \tag{3.185}$$

Numerical evaluation of the components of the group velocity required three eigenvalue evaluations. The wavenumber (α_r), damping rate (α_i), phase speed (c) and the x- and z-components of group velocity are shown in Figs. 3.28 and 3.29 for the first five modes for different spanwise wavenumbers. It is clearly evident that all the excited modes are damped. Far away from the exciter, only the effects of the least-damped mode (as indicated by the fourth and fifth modes in Fig. 3.28) will be felt. Also, the least damped modes have wavelengths which are a few thousand times the displacement thickness, δ^*. To detect such large wavelength disturbances, the test section of the tunnel in the experiments has to be long enough to accommodate at least a few wavelengths. In the experiments of [113], measurements were made only up to $500\delta^*$ downstream of the exciter and hence it appeared that the whole boundary layer was heaving. From Fig. 3.29, it is also clearly evident that the direction of propagation of the fourth and fifth modes was predominantly in the x-direction only, since the z-component of the group velocity is negligibly small, as compared to the x-component.

In a more general case, if the input represents wide-band disturbances, then the low frequency components will cause the shear layer to heave up and down, as the Klebanoff mode of motion. In many experiments involving FST, researchers have reported such heaving motion. This has led researchers to identify response to FST with the Klebanoff mode erroneously.

3.7   Direct Simulation of Receptivity to Free Stream Excitation

In general, transition problems are governed by the full Navier–Stokes equation and could be solved as receptivity problems by direct simulation, with a well-defined excitation field. However, unlike stability problems, receptivity to excitation has to be posed as initial-boundary value problems. Depending upon the nature of the equilibrium flow, various assumptions are made, including linearization, to arrive at different receptivity models. For example, in applying asymptotic theory [125, 128], authors have shown that the Navier–Stokes equation can be approximated by the unsteady boundary layer equation at the leading edge of a flat plate to study receptivity to acoustic waves. Murdock [239] numerically solved the incompressible Navier–Stokes equation for flow over a flat plate in a domain excluding the leading edge. The inflow boundary condition was obtained from the solution of the unsteady boundary layer equation. Similarly, Lin et al. [202] have modeled the distributed receptivity to free stream vortical disturbances by solving the *Parabolized Stability Equation* (PSE) for the Blasius boundary layer over a flat plate with small surface waviness. The authors in [201] numerically solved the full Navier–Stokes equation to study the role of discontinuity in curvature near the leading edge of a flat plate and found that the TS wave amplitudes were roughly halved when the curvature discontinuity was removed. Buter and Reed [48] have studied the receptivity to free stream vorticity by a boundary layer forming over a flat plate with an elliptic leading edge, by numerically solving the Navier–Stokes equation in stream function — vorticity formulation to track the formation of TS waves. In all cases considered, the first clear appearance of TS waves occurred at a location aft of the location where the pressure gradient is maximum. The receptivity increases with the magnitude of the pressure gradient maximum. Smoothing the discontinuity in curvature at the juncture of the elliptical leading edge and the flat plate shifted the pressure gradient maximum forward and increased its magnitude, resulting in a stronger TS response. This affirmed the importance of pressure gradient and showed that the continuous changes in surface curvature provide a receptivity source — as the case might be for flow past airfoils.

In Sec. 3.6.5, we have seen how a general excitation field can give rise to both downstream and upstream propagating disturbances, with the dispersion relation given by Eq. (3.181). Sengupta et al. [362] have demonstrated that there is a direct way of exciting TS modes by free stream excitation while solving the Orr–Sommerfeld equation — despite the contrary observations in the literature. This is explained next, where a train of periodic vortices is allowed to convect over a zero pressure gradient shear layer. The physical arrangement of the problem is shown in Fig. 3.30. For this problem, results were also obtained by solving

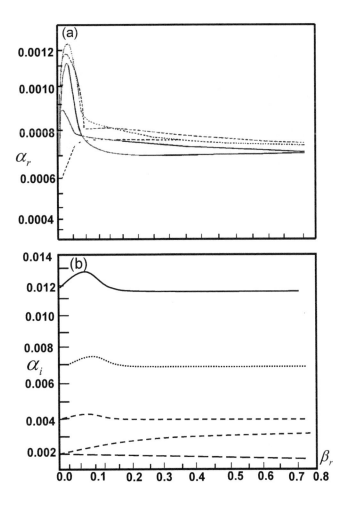

FIGURE 3.28
Variation of (a) α_r and (b) α_i of the first five modes with spanwise wave number (β_r) for the Klebanoff mode problem. Mode 1: solid line; Mode 2: sparse dotted line; Mode 3: short-chain line; Mode 4: dense dotted lines and Mode 5: long-chain line.

the Navier–Stokes equation, and the computational domain is also marked in the figure. An infinite array of irrotational vortices is seen to convect over a flat plate in the free stream. The presence of the wall gives rise to an image vortex system, as indicated also in the figure. The vortices of individual strength Γ are at a spacing of a and are convecting at a constant height of $b/2$ over the plate; the height is significantly higher than the shear layer thickness and thus the vortices are always in the inviscid part of the flow.

If these vortices are considered as potential vortices, then the induced perturbation velocity in the inviscid part of the flow is given by

$$u_\infty = \frac{\Gamma}{2aD}\left[\sin^2\frac{\pi\bar{x}}{a}\cosh\frac{2\pi y}{a} - \sinh\frac{\pi}{a}(y-\frac{b}{2})\sinh\frac{\pi}{a}(y+\frac{b}{2})\right]\sinh\frac{\pi b}{a} \qquad (3.186)$$

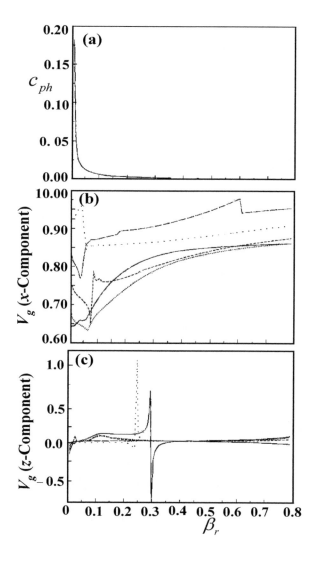

FIGURE 3.29
(a) Phase speed of the first five modes as functions of (β_r); (b) streamwise component of group velocity of the first five modes as functions of (β_r); (c) spanwise component of group velocity of the first five modes as functions of (β_r).

$$v_\infty = \frac{\Gamma}{4aD} \sinh \frac{\pi b}{a} \sin \frac{2\pi \bar{x}}{a} \sinh \frac{2\pi y}{a} \qquad (3.187)$$

where $\bar{x} = x - ct$, with c as the convection speed of the vortices and

$$D = \left[\sin^2 \frac{\pi \bar{x}}{a} \cosh \frac{2\pi y}{a} - \sinh \frac{\pi}{a}\left(y - \frac{b}{2}\right) \sinh \frac{\pi}{a}\left(y + \frac{b}{2}\right) \right]^2 + \frac{1}{4} \sin^2 \frac{2\pi \bar{x}}{a} \sinh \frac{2\pi y}{a}$$

If all lengths are non-dimensionalized by the displacement thickness and velocity by U_∞, then the periodic vortices impose a time scale on the flow given by $\bar{\omega}_0 = 2\pi c/a$. Periodicity of vortices excites the shear layer at circular frequencies $\bar{\omega}_0, 2\bar{\omega}_0, 3\bar{\omega}_0....etc.$ Thus, the disturbance stream function can be expressed as

$$\psi(x, y, t) = \frac{1}{2\pi} \sum_{n=1}^{\infty} \int_{Br} \phi(\alpha, y, n\bar{\omega}_0) \, e^{i(\alpha x - n\bar{\omega}_0 t)} \, d\alpha \qquad (3.188)$$

It has already been explained that the eigenvalues near $\alpha = 0$ give rise to the asymptotic solution, while the local solution is contributed by the point at infinity in the spectral plane ($\alpha \to \infty$) (in Sec. 3.6.2.1 and [348]). Also, for small values of Γ and large values of b, the receptivity problem can be solved by linearizing the Navier–Stokes equation.

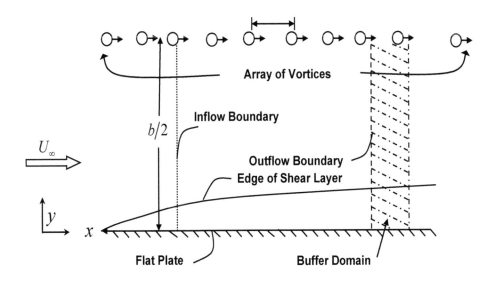

FIGURE 3.30
Physical arrangement for the flat plate receptivity problem excited by a periodic array of convecting vortices in the free stream.

Imposed Cartesian velocity components, u_∞ and v_∞, are plotted in Fig. 3.31 for the indicated parameters, over a single vortex pair spacing given by $a = 100\pi\delta^*$. Input disturbances are shown for a single period with $Y = 18\delta^*$ and these are calculated at $y = 16\delta^*$.

It is to be noted that the values of u_∞ and v_∞ obtained by an inviscid analysis are used here to calculate the impressed pressure over the shear layer. If Γ is considered to be small, then the impressed pressure gradient is going to be negligibly small and not cause flow separation. In the absence of separation, the boundary layer assumption holds and the impressed pressure remains the same across the shear layer and is the reason that the analysis results using inviscid pressure distribution provide a vital clue to the receptivity route.

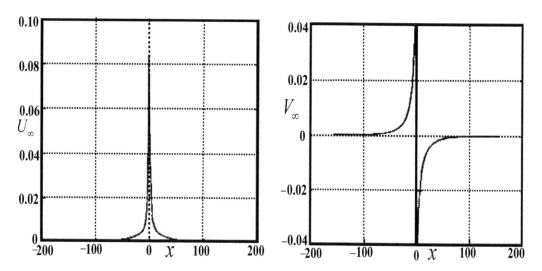

FIGURE 3.31
Imposed velocity perturbation by a periodically passing train of vortices in the free stream convecting with U_∞. Shown are the velocity components at $y = 16\delta^*$ when the array is convecting at $Y = 18\delta^*$ with vortices at a gap of $a = 100\pi\delta^*$.

In Fig. 3.32, the bilateral Laplace transform, ϕ_∞, and its normal derivative, ϕ'_∞, have been shown which correspond to the imposed velocity boundary conditions of Fig. 3.31. It is clearly evident that ϕ_∞ is much larger than ϕ'_∞ and this information is used to compare different types of free stream excitations in the next subsection.

3.7.1   Coupling between the Wall and Free Stream Modes

The solution of the Orr–Sommerfeld equation has four fundamental modes, as expressed below

$$\phi = C_1\,\phi_1 + C_2\,\phi_2 + C_3\,\phi_3 + C_4\,\phi_4 \tag{3.189}$$

These four fundamental modes have already been defined in Sec. 3.6; their asymptotic values in the free stream are given by $\phi_{1\infty} \sim e^{-\alpha y}$, $\phi_{2\infty} \sim e^{\alpha y}$, $\phi_{3\infty} \sim e^{-Qy}$, $\phi_{4\infty} \sim e^{Qy}$, where $|Q| = [\alpha^2 + i\alpha Re(1-c)]^{1/2}$. The first and the third modes decay, while the second and the fourth modes increase with y, whenever the real part of α and $\sqrt{Q^2}$ are positive. The decaying modes are required for pure wall excitation and we have formally defined these as the *wall mode* and note it as

$$\Phi_I = C_1\,\phi_1 + C_3\,\phi_3 \tag{3.190}$$

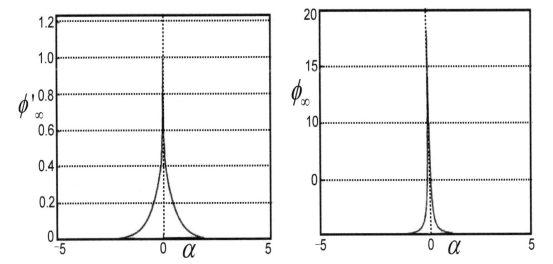

FIGURE 3.32
The bilateral Laplace transform of imposed velocity disturbance components shown in Fig. 3.31.

Similarly, the *free stream mode* is defined as

$$\Phi_{II} = C_2\, \phi_2 + C_4\, \phi_4 \tag{3.191}$$

Φ_{II} grows with y, to match the applied disturbance at the free stream. Now for a receptivity problem, where the shear layer is excited at the free stream, one can fix the values of C_2 and C_4 by matching u_∞ and v_∞ or ϕ_∞ and ϕ'_∞. The far field boundary for the problem is considered far out enough so that $\phi_{1\infty}$ and $\phi_{3\infty}$ are negligibly small and the disturbance solution is solely due to the *free stream mode*. Thus, fixing of the free stream boundary condition leaves the homogeneous boundary conditions at the wall to be satisfied. However, there are two qualitatively different ways for coupling between the two modes to be achieved: if the convected vortices move at free stream speed or at a speed different from the free stream speed.

Pure Convection Problem
When the convected vortices move at the free stream speed, notice that the fundamental solutions coalesce, i.e., $\phi_{1\infty} = \phi_{3\infty}$ and $\phi_{2\infty} = \phi_{4\infty}$. Thus, to satisfy the free stream boundary conditions, we fix C_2 and C_4 from

$$\Phi_{II\infty} = \phi_\infty = C_2\, e^{\alpha Y} + C_4 Y e^{\alpha Y} \tag{3.192a}$$

and

$$\Phi'_{II\infty} = \phi'_\infty = \alpha C_2\, e^{\alpha Y} + C_4\, (1 + \alpha Y)\, e^{\alpha Y} \tag{3.192b}$$

This occurs, as $\phi_{I\infty} = \phi'_{I\infty} \equiv 0$, for $Y \to \infty$.
Solution of these provides

$$C_2 = [(1 + \alpha Y)\phi_\infty - Y \phi'_\infty]\, e^{-\alpha Y}$$

and

$$C_4 = [\phi'_\infty - \alpha \phi_\infty]\, e^{-\alpha Y}$$

Now to satisfy the homogeneous boundary conditions at the wall, one must have

$$\phi(y = 0) = 0 = \Phi_{I0} + \Phi_{II0}$$

Thus, the wall boundary condition for the *wall mode* (noted as ϕ_{PC}) is given by

$$\phi_{PC} = \Phi_{I0} = - \Phi_{II0} = - e^{-\alpha Y}[\{\phi_\infty(1 + \alpha Y) - Y\phi'_\infty\}\bar{\phi}_{20} \\ + \{\phi'_\infty - \alpha\phi_\infty\}\bar{\phi}_{40}] \tag{3.193}$$

where the fundamental solutions, in this case, have been written with an overbar. Similarly, one can write an expression for Φ'_{I0}, providing two equations to solve for the other two unknowns, C_1 and C_3, from

$$C_1 \bar{\phi}_{10} + C_3 \bar{\phi}_{30} = -(C_2 \bar{\phi}_{20} + C_4 \bar{\phi}_{40})$$

$$C_1 \bar{\phi}'_{10} + C_3 \bar{\phi}'_{30} = -(C_2 \bar{\phi}'_{20} + C_4 \bar{\phi}'_{40})$$

Solution of these two equations provides

$$C_1 = (r_1 \bar{\phi}'_{10} - r_2 \bar{\phi}_{10})/D$$

and

$$C_3 = (r_2 \bar{\phi}_{30} - r_1 \bar{\phi}'_{30})/D)$$

where

$$D = (\bar{\phi}'_{10}\bar{\phi}_{30} - \bar{\phi}_{10}\bar{\phi}'_{30})$$

$$r_1 = e^{-\alpha Y}[\bar{\phi}_{20}\{\phi'_\infty Y - (1 + \alpha Y)\phi_\infty\} + \bar{\phi}_{40}(\alpha\phi_\infty - \phi'_\infty)]$$

and

$$r_2 = e^{-\alpha Y}[\bar{\phi}'_{20}\{\phi'_\infty Y - (1 + \alpha Y)\phi_\infty\} + \bar{\phi}'_{40}(\alpha\phi_\infty - \phi'_\infty)]$$

The non-zero values of C_1 and C_3 obtained in terms of C_2 and C_4 provide the coupling for this case.

Bypass problem

If the convected vortices do not move at the free stream speed, then the modes do not coalesce, as in the previous case, and one can simply calculate the wall boundary condition for the *wall mode* and call it as ϕ_{BP} to distinguish it from the previous case. Thus

$$\phi_{BP} = \Phi_{I0} = \frac{1}{p - \alpha}[e^{-\alpha Y}(\phi'_\infty - p\phi_\infty)\phi_{20} - e^{-pY}(\phi'_\infty - \alpha\phi_\infty)\phi_{40}] \tag{3.194}$$

where $p = Real(\sqrt{Q^2})$. Thus, Eqs. (3.193) and (3.194) represent the equivalent *wall mode* amplitudes calculated at the wall for pure free stream excitation, in pure convection and bypass modes, when the real part of α and p are positive. For instability problems, usually $|p| >> |\alpha|$. Also, note for the case of free stream vortical disturbances, $\phi_\infty > \phi'_\infty$, as shown in Fig. 3.31 and then

$$\frac{\phi_{PC}}{\phi_{BP}} = -(p - \alpha)\left[\frac{\bar{\phi}_{20}[\phi_\infty(1 + \alpha Y) - Y\phi'_\infty] + \bar{\phi}_{40}[\phi'_\infty - \alpha\phi_\infty]}{(\phi'_\infty - p\phi_\infty)\phi_{20}}\right] \tag{3.195}$$

This can be further simplified to

$$\frac{\phi_{PC}}{\phi_{BP}} = \left(\frac{\bar{\phi}_{20}}{\phi_{20}}\right)[1 + \alpha Y - Y\frac{\phi'_\infty}{\phi_\infty}] \qquad (3.196)$$

Hence, it is apparent that for the same level of excitation at the free stream, the above ratio indicates that the pure convection of vortices will be a far weaker mechanism for creating disturbances inside the shear layer as compared to the case when $c = \bar{\omega}_0/\alpha \neq 1$.

The above analysis is for the spatial stability problem associated with free stream excitation, where a time scale is imposed. Next, we consider the same case with explicit time scales imposed on the flow — the case of train of convective vortices — for the receptivity problem.

3.7.2   Receptivity to a Train of Convected Vortices in the Free Stream

The above distinction between the cases of pure convection with free stream speed and the bypass route where the train of vortices travels at speeds lower than the free stream speed can also be demonstrated by solving the Orr–Sommerfeld and Navier–Stokes equations, as shown in [327]. For the results shown, spacing between successive vortices was taken as $a = 100\pi\delta^*$ with $Y = 20\delta^*$ and $Re = 1000$. The Orr–Sommerfeld equation has been solved for the boundary conditions displayed in Fig. 3.32, by the Chebyshev collocation method with the weighting coefficients developed using the Generalized Differential Quadrature (GDQ) method in [368].

For the pure convection case, i.e., when $c = U_\infty$, excited modes are those for which the following dispersion relation must be satisfied: $\alpha_n = n\bar{\omega}_0$. The cause of the severely damped solution (not shown) can be ascribed to the stability property of the basic profile. This has been shown in Fig. 3.33. In this figure, the neutral curve is re-plotted along with $c = constant$ loci in the $(Re, \bar{\omega}_0)$-plane. The existence of a contour with $c = U_\infty$ (in the figure shown by the non-dimensional value of $c = 1$), far removed from the neutral curve in the stable zone, implies that such a convecting mode would decay very rapidly. Similarly, upstream propagating modes corresponding to $\bar{\omega}_0 = 0.02$ are also highly damped. Figure 3.33 also indicates the possibility by which wavy or oscillatory disturbances can be generated. Of particular significance are the properties of the upstream and downstream modes with $c < U\infty$. The real phase speed c indicates the speed at which a free stream disturbance convects, and for a vortex train that does not disperse, this would be the speed of the individual vortices. Figure 3.33 clearly indicates that instability can be very effectively triggered if the convection speed of the vortices lies within the narrow range of $0.26U_\infty$ to $0.32U_\infty$ that in this case corresponds to $n = 13$ to 16.

Convecting free stream disturbances within this speed range are likely to trigger strong sustained instability, because of the high amplification rate which these modes would experience. Also for $c > 0.4U_\infty$, convecting disturbances would create damped wave packets.

For a convected vortical disturbance field in the free stream, the growth process is also qualitatively different as compared to the growth of disturbances that are created at the wall or inside the shear layer at fixed frequency, as reported in [316]. For monochromatic wall excitation, the real frequency of the disturbance field is held fixed and the phase speed adjusts itself continually to the local stability property of the shear layer. Contrarily, for the disturbance field generated by convecting vortices, it is the phase speed or group velocity that is an invariant of the input disturbance field, while the corresponding real frequency will continuously vary, satisfying the dispersion relation $\alpha c = n\bar{\omega}_0$. In other words, for free stream excitation, the disturbance will follow a path of constant phase speed, while for localized wall excitation, this path will be along a straight line in the $(Re, \bar{\omega}_0)$-plane with a slope which denotes the physical frequency. Hence, if the convection speed of the vortex train is chosen to be between $0.26U_\infty$ to $0.32U_\infty$, then the created disturbance field inside the shear layer will experience a very high sustained growth, as the c = constant lines remain closer to the α_i maximum region over a longer streamwise stretch, as shown in Fig. 3.33. Moreover, the figure also indicates that the rate of growth will be much higher than that for disturbances which are excited from inside the shear layer at a constant physical frequency. There is clear experimental evidence of the presented analysis here with the results reported in [172]. In [172], a circular cylinder was rotated in a circular trajectory above a flat plate. During the motion of the cylinder in its trajectory, equivalent ejected disturbances over the flat plate were much more complicated, because the strength, location and migration speed of vortices are not controlled, as compared to the model problem discussed above. Despite these differences, Fig. 10 of [172] clearly showed that a rotation speed corresponding to $c = 0.3U_\infty$ showed maximum receptivity as compared to the $c = 0.23U_\infty$ and $c = 0.5U_\infty$ cases. The hot-wire oscillograms corresponding to the circumferential speed $c = 0.3U_\infty$ of the rotor indicated the presence of very large amplitude wave packets.

One notes here that the free stream convecting vortical excitation is not monochromatic. This is the fundamental difference between free stream excitation by convecting vortices

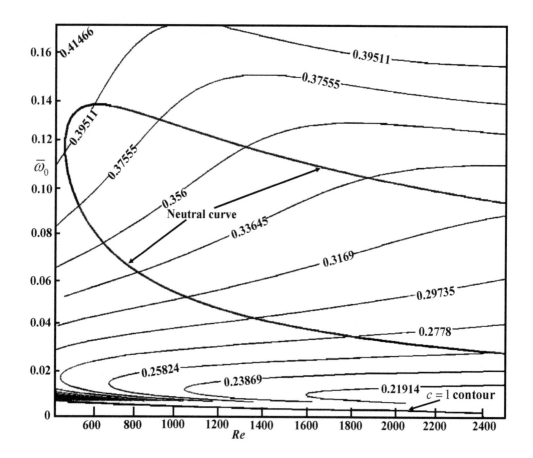

FIGURE 3.33
Stability diagram for a flat plate boundary layer showing the neutral curve superposed over $c = constant$ disturbance propagation contours shown by thin lines. Note that the line $c = 1$ corresponds to the case of pure convection of vortices.

and the vibrating ribbon excitation at a fixed frequency, discussed in [316]. If nonlinearity is important and one solves the full Navier–Stokes equation for disturbance quantities, then the wave packets can grow as they propagate downstream, while the growing disturbance within the packet can saturate in amplitude. Therefore, the solution of the Navier–Stokes equation will provide vital information about the role of nonlinearity for this problem, apart from including the effects of growth of the shear layer.

A direct simulation of the flow field was also attempted in [327], where the following stream function-vorticity formulation of the Navier–Stokes equation was used

$$\frac{\partial^2 \psi}{\partial x^2} + \frac{\partial^2 \psi}{\partial y^2} = -\omega \tag{3.197}$$

$$\frac{\partial \omega}{\partial t} + \frac{\partial}{\partial x}(u\omega + u_b\omega + u\omega_b) + \frac{\partial}{\partial y}(v\omega + v_b\omega + v\omega_b) = \frac{1}{Re_1}\left[\frac{\partial^2 \omega}{\partial x^2} + \frac{\partial^2 \omega}{\partial y^2}\right] \tag{3.198}$$

where u_b, v_b and ω_b are the undisturbed base flow quantities for the flat plate boundary layer flow as given by the Blasius similarity solution. Re_1 is the Reynolds number based on the free stream velocity, the displacement thickness at the inflow and ν, the kinematic viscosity.

To resolve flow gradients near the wall, the above equations were solved in a stretched co-ordinate system (in the wall-normal direction) via the transformations

$$x = \xi$$

$$y = \frac{y_{max}\, \sigma\, \eta}{\eta_{max}\sigma + y_{max}\,(\eta_{max} - \eta)} \tag{3.199}$$

where y_{max} is the height of the domain in the physical plane and η_{max} is the corresponding height in the computational plane; σ is a control parameter that clusters the grid points close to the wall.

At the inflow boundary and on top of the computational domain, the analytic solution for the disturbance velocity is used in accordance with Eqs. (3.186) and (3.187). On the flat plate, the no-slip condition simultaneously provides a Dirichlet boundary condition for the stream function and the wall vorticity at every instant of time. The computational domain is indicated in Fig. 3.30.

In order to eliminate any reflection of waves from the outflow boundary, the buffer domain technique developed in [204] has been used in these simulations. The buffer domain, as indicated in Fig. 3.30, is a narrow strip of the computational domain adjacent to the outflow boundary. A continuous buffer function $b(\xi)$ is introduced (in Eqs. (3.200) and (3.201)) which has a value of one in the main computational domain that decreased monotonically in the buffer domain to zero at the outflow boundary. To treat growing or unstable modes, a second buffer function $b_{Re}(\xi)$ was used (in Eq. (3.201)) to gradually reduce the Reynolds number in the buffer domain to a value below the critical Reynolds number. The buffer functions $b(\xi)$ and $b_{Re}(\xi)$ are as given in [204].

The transformed equations which have been actually solved are given by

$$b\frac{\partial^2 \psi}{\partial \xi^2} + \frac{\partial^2 \psi}{\partial \eta^2}\frac{1}{y_\eta^2} + \eta_{yy}\frac{\partial \psi}{\partial \eta} = -\omega \tag{3.200}$$

$$\frac{\partial \omega}{\partial t} + \frac{\partial}{\partial \xi}(u\omega + u_b\omega + u\omega_b) + \frac{1}{y_\eta}\frac{\partial}{\partial \eta}(v\omega + v_b\omega + v\omega_b)$$

$$= \frac{b_{Re}}{Re_1}\left[b\frac{\partial^2 \omega}{\partial \xi^2} + \frac{\partial^2 \omega}{\partial \eta^2}\frac{1}{y_\eta^2} + \eta_{yy}\frac{\partial \omega}{\partial \eta}\right] \tag{3.201}$$

At the outflow of the domain, extrapolation based on $\frac{\partial^2 \psi}{\partial \xi^2} = \frac{\partial^2 \omega}{\partial \xi^2} = 0$ has been applied for the stream function and vorticity transport equations. Equations (3.200) and (3.201) have been solved in [327] first for the case of $c = U_\infty$ in a domain for which the Reynolds number varied from 165 to 1900. Results are shown in Fig. 3.34 at different indicated times.

The abscissa in the figure is the Reynolds number based on local displacement thickness. In this case, only the local solution is predominant at early times, indicating the stable nature of the boundary layer to this free stream excitation. This local solution also disperses and decays, as can be seen from the solution at $t = 100$. Observed individual peaks at $t = 0$ due to the local solution disperse into multiple peaks at later times, as can be noted from the solutions at all subsequent times. This dispersion of solution is due to the presence of upstream propagating modes and the presence of multiple harmonics for the downstream propagating modes. The adjective upstream here is to be understood with respect to the local condition of the disturbance field. More details about this dispersion mechanism and tracking of the upstream propagating modes were first discussed in [362] and are discussed again in the next subsection.

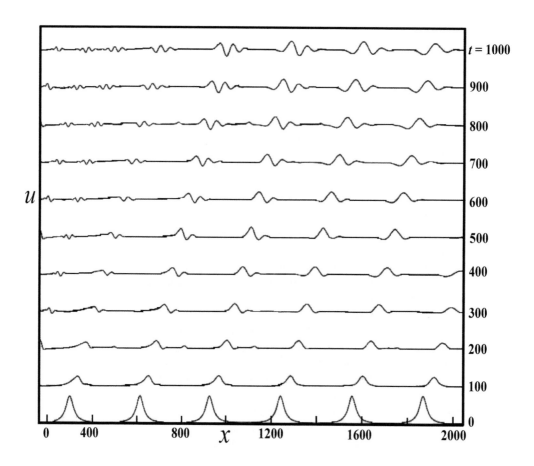

FIGURE 3.34
Streamwise disturbance velocity component plotted at a height $y = 0.3\delta^*$ for the pure convection case ($c = U_\infty$) at the indicated times, when solution is obtained by solving the Navier–Stokes equation for the problem shown in Fig. 3.30.

Next, the case for $c = 0.3U_\infty$ was investigated by solving the Navier–Stokes equation with the same formulation. As per the previous discussion based on stability properties of the Blasius profile depicted in Fig. 3.33 and the experimental observation of [172], one would expect to see a rapid growth of disturbances. The domain is the same that was considered for the previous case in Fig. 3.34 and other excitation field data are also the same, except the convection speed of the train. Results for this case are as shown in Figs. 3.35(a) and 3.35(b) for the locations at $y = 0.3\delta^*$ and $1.5\delta^*$, respectively. Here it is apparent that with time the shear layer displays supporting growing disturbances as it convects downstream. At early times up to $t = 400$, one notices the splitting of the full solution into upstream and downstream propagating components. While these results are for two-dimensional flows, similar computations have been performed in [446] in a three-dimensional box where free stream vortices were directed towards the base by imposing constant downward velocity at the inflow. The resultant trajectory of the vortex-induced disturbance is intuitively expected to lie in the southeast direction if the mean flow is from west to east. However, induced disturbances in [446] moved down and in the upstream direction instead, clearly indicating that the free stream excitation causes a disturbance field that propagates upstream, as predicted by two-dimensional calculation results. The existence of common features among the results from three-dimensional DNS, two-dimensional simulation and the solution of the linearized Navier–Stokes equation points to the common mechanism which is seen in all the cases.

3.7.3 Further Explanation of Free Stream Periodic Excitation

The physical problem described in Fig. 3.30 and its solution in Figs. 3.34 and 3.35 for $c = U_\infty$ and $c = 0.3U_\infty$, respectively, brings out some interesting features of the response field due to periodic free stream excitations. We have already noted that for such excitations one would get both downstream and upstream propagating disturbances, whose properties are as shown in Figs. 3.25 to 3.27. However, the upstream propagating mode will dominate for the case of $c = U_\infty$, as the downstream propagating mode corresponding to $c = U_\infty$ is highly attenuated, as seen from Fig. 3.34. The upstream propagating part will be less attenuated, as shown in Fig. 3.27 for the variation of α_i with increasing Reynolds number. The downstream mode is always excited for any free stream excitation which is introduced via the coupling mechanism described in Sec. 3.7.1. Thus, the response consists predominantly of the upstream propagating mode as it convects downstream.

The upstream propagating mode properties are indicated by the dotted line in Fig. 3.25 and this component of the solution can be seen in Figs. 3.34 and 3.35. Here we demonstrate these aspects further with respect to the case shown in Fig. 3.34, for which the free stream vortices travel with the free stream speed, $c = U_\infty$, for the pure convection problem. For the case of Figs. 3.34 and 3.35, the gap between successive vortices is taken as $a = 100\pi$ and corresponds to a time scale given by $\bar{\omega}_0 = \frac{2\pi c}{a} = 0.02$. Thus, the excitation creates a multi-periodic response field with $n\bar{\omega}_0$ (for $n = 1, 2, 3, ...$) as the frequencies in Eq. (3.188).

For the fundamental frequency $\bar{\omega}_0$ of the problem in Fig. 3.34, one notices a propagation speed of disturbance energy to be given by $V_{g0} \approx -0.5U_\infty$ (from Figs. 3.27 and 3.36), while for the next few harmonics the group velocity increases with circular frequency. In Fig. 3.36, variation of group velocity with circular frequency is shown for the range $0 \leq \bar{\omega}_0 \leq 0.2$ for different Reynolds numbers.

This figure shows the value of group velocity to increase in magnitude from $0.5U_\infty$ to $0.8U_\infty$ for the first and second harmonics, respectively. For higher harmonics, this further increases almost linearly. The initial solution created by the free stream vortices inside the shear layer appear as wave packets. At early times, the packets retain their coherence due to small dispersion of various harmonics. However, the free stream vortices convect at a speed given by $c = U_\infty$ and the vanishingly small downstream propagating disturbance field inside the shear layer convects with c. The upstream component would travel with respect to the downstream component with the group velocity $V_{g0} = -0.5U_\infty$. We note that the higher harmonics would be less damped as compared to the fundamental mode depending furthermore upon the Reynolds number, as shown in Figs. 3.25 and 3.27. Higher harmonics would dominate at later times, due to the fact that they have lower damping and as the packets move downstream, the Reynolds number also increases, for which the damping rate decays even further. This would therefore create a dispersion of the disturbance field. The effects of such dispersion would be equivalent to creating newer length scales. This is a

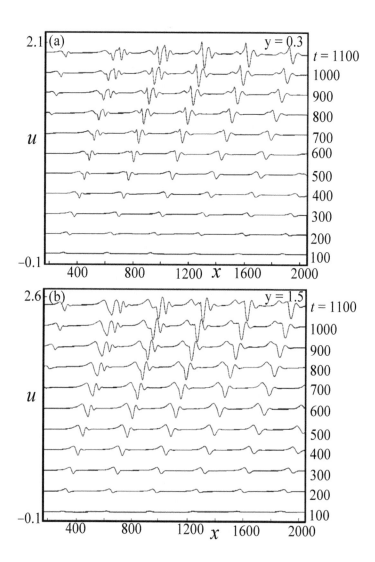

FIGURE 3.35
Streamwise velocity component plotted for the case $c = 0.3U\infty$ at the indicated times obtained by solving the Navier–Stokes equation for the problem shown in Fig. 3.30 for the heights (a) $y^* = 0.3\delta^*$ and (b) $y^* = 1.5\delta^*$.

different mechanism for the creation of different length scales (specifically, smaller length scales) which has been overlooked before for unstable flows.

To understand the creation of new length scales by the dispersive effect, it is explained clearly with an additional figure and a table. In Fig. 3.37, disturbance stream function is shown at two heights ($y = 0.333\delta^*$ and $1.0\delta^*$) at four distinct time instants, as indicated in the figure for the case of $c = U_\infty$.

Comparison of disturbance fields at these two heights clearly demonstrates a strong coherence of the data, while the disturbance increases with height — once again attesting to the properties explained before for the free stream mode. Dispersion of the solution field can be explained by looking at this figure and Table 3.3.

In the table, the initial condition ($t = 0$) identifies the packets at a non-dimensional distance of 100π, with packets identified by their location outside the computational domain by negative signs. Similarly, the time of entrance for different vortices is indicated by asterisks. Due to a higher damping rate of both upstream and downstream modes, the solution at very early times consists of only the local solution, as discussed with respect to the results in Fig. 3.19. Six disturbance packets are identified in the table at $t = 0$, those created exactly below the free stream vortices. Subsequently at $t = 100$, each of the two packets shown in Fig. 3.37 splits into two asymmetric clusters. If the disturbance field had moved with $c = U_\infty$, then these two packets would have been noted at $x = 400$ and at $x = 715$. One notes two smaller peaks at these two locations at $t = 100$. However, the major peaks are at the locations indicated by the quantities within parentheses in Table

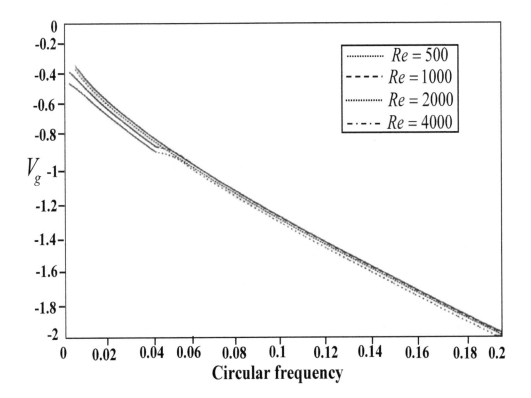

FIGURE 3.36
Variation of group velocity with circular frequency for the indicated Reynolds number for the upstream propagating mode.

3.3. These are exactly at locations if the disturbance clusters move with the group velocity $V_{g0} = 0.5U_\infty$ with respect to the corresponding free stream vortices. This is verified for the clusters' location at $t = 100$ and 410, from Table 3.3 and Fig. 3.37. At later times, higher harmonics move farther upstream due to their higher propagation speeds and lower damping rates, showing visible dispersion. This is evident from the results in the figure and the table at $t = 720$ and 1260.

Thus, one notes that pure convection of vortices in the free stream causes a dispersed disturbance field inside the shear layer which grows with height and those are not TS waves. Moreover, these damped disturbances are multi-periodic. In Sec. 3.3.2, we have classified instability belonging to either temporal or spatial theory, depending upon whether spatial or temporal scales are imposed. How does one view free stream periodic vortical excitation? Should it be viewed as a juxtaposition of spatial modes created by multi-periodic free stream excitation? Or should one interpret the creation of small length scales inside the shear layer as causing temporal instability? From a spatial theory perspective, we have noted how free stream excitation also causes an equivalent wall excitation in Sec. 3.7.1.

It has been shown in [349, 350] that spatially damped flows can display spatio-temporal growing wave fronts. While such a transient growth mechanism has been shown in these references for two-dimensional disturbance fields, many researchers have proposed algebraically growing transients for three-dimensional flows. This has been termed bypass transition by some authors, as in [40, 232, 235] and references in [349, 350]. We will discuss spatio-temporal growing wave fronts in Chap. 5. There still remain large numbers of other mechanisms which qualify as bypass transition. One such mechanism of bypass transition is

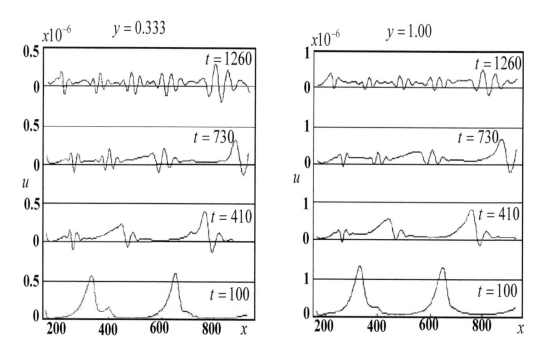

FIGURE 3.37
The disturbance stream function at two indicated heights, as functions of streamwise distance for the indicated times for the pure convection case.

introduced in the next chapter and termed vortex-induced instability caused by aperiodic free stream excitation.

3.8 Nonparallel and Nonlinear Effects on Instability and Receptivity

In this section, we discuss nonlinear and nonparallel receptivity of the zero pressure gradient boundary layer using the solution of the full two-dimensional Navier–Stokes equation with respect to wall excitation at fixed frequencies. The excitation is triggered through a simultaneous blowing-suction (SBS) strip at a location near the leading edge of the plate, as shown in Fig. 3.38. Attention is focused specifically for cases which are linearly stable in a spatial formulation, for different amplitudes of excitation of the strip. There has been lot of debate about these aspects of flow instability, in particular. While some rudimentary receptivity studies have been reported in [98] and earlier references contained therein, which supported mainly the basic findings of existing linear stability theory. However, in recent times, use of DRP schemes in computing unstable flows has revealed many other aspects of flow instability and receptivity which are worth recounting. However, in this section, we discuss the main prototypical example of zero pressure gradient flow past a flat plate. The results presented reveal many new findings from [325] which not only show distinctly nonlinear, nonparallel effects, but also display early markers of bypass transition.

The linear instability theory depends on three main assumptions/restrictions: (a) for its sensitivity to imperceptible ambient disturbances, it was developed for small perturbations, which allows studying linearized governing equations for the evolution of disturbances; (b) also to make the problem tractable theoretically, a parallel flow assumption is made and (c) further to simplify the spatial stability analysis, it is assumed that the flow responds at the same frequency of the exciter itself — the so-called *signal problem*. Experimental verifi-

TABLE 3.3
Position of response field with respect to free stream vortices and their time of entry in the computational domain.

Vortex No. → Time ↓	1	2	3	4	5	6
$t = 0$	−960	−645	−330	−15	300	615
$t = 100$	−860	−545	−230	85	400	715
				(35)	(350)	(665)
$t = 410$	−550	−235	80	395	710	1025
				(270)	(505)	(820)
				$t_1 = 165*$		
$t = 720$	−230	85	400	715	1030	1345
			(280)	(450)	(665)	(980)
		$t_1 = 810*$	$t_1 = 490*$			
$t = 1260$	300	615	930	1245	1560	1875
	(230)	(390)	(550)	(695)	(945)	(1260)
	$t_1 = 1120*$					

cation of TS waves predicted by linear stability theory in [316] was a path breaking event. This experiment identified the region where the zero pressure gradient boundary layer is unstable for high to moderate frequencies in Fig. 3.39 by discrete symbols in following the theoretically neutral curve shown by the continuous curve in the (Re, ω_0)-plane. Experimental data points in the figure show good agreement at low frequency and high Reynolds number combinations. As the theoretical analysis is based on the parallel flow assumption, its results require interpretation of propagating disturbances in a growing boundary layer. A constant physical frequency excitation follows a ray starting from the origin. For example, if one focuses attention on the single frequency f_0 and the excitation source is placed anywhere between O and S, the disturbance will travel along OR and will decay up to S, i.e., until the ray meets the point P. Between the locations S and T, disturbances grow in space. Again beyond Q, the constant frequency disturbance decays. However, it is noted in Fig. 3.39 that experimental results reveal discrepancies for low Re-high ω_0 combinations. The experimental critical Reynolds number is significantly lower than the value obtained by linear instability theory. In all cases in the experiment, the boundary layer was excited at the same physical location and thus the effects of the local solution are not same for all the frequencies. According to linear instability theory, three frequencies corresponding to OA, OB and OC in Fig. 3.39 create disturbances which are stable, while the experimental results indicate a finite length where the flow is unstable.

The experiment verified certain features of instability theory, yet there were many features in the experiment which instability theory is simply incapable of explaining. One realizes that the linear stability theory for a parallel flow predicts system response with respect to small perturbations which is only valid at a distance removed from the leading edge of the plate. The created waves (either spatially stable or unstable) are with small wavenumbers — as explained with the help of Tauber's theorem. This relates poles (eigenvalues) near the origin of the complex wavenumber plane with the asymptotic behavior in the physical plane $(x \to \infty)$. The essential singularity of the Laplace–Fourier transform in the wavenumber plane determines the local solution, as described in Sec. 3.6.2.1.

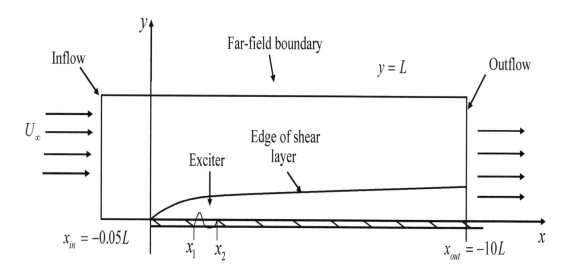

FIGURE 3.38
Schematic of the flow and the computational domain including the simultaneous blowing and suction (SBS) strip.

A typical receptivity solution to localized wall excitation for a fixed frequency is shown in Fig. 3.40 at $t = 8$, with the solution obtained by solving the full Navier–Stokes equation for the excitation frequency corresponding to OA in Fig. 3.39. The governing two-dimensional Navier–Stokes equation is solved using the stream function vorticity (ψ, ω) formulation in the transformed plane. The flow is computed using an analytical uniform grid with 1600 points in the streamwise (ξ) and 600 grid points in the wall-normal (η) directions. These equations are written in a transformed (ξ, η)-plane as

$$\frac{\partial}{\partial \xi}\left(\frac{h_2}{h_1}\frac{\partial \psi}{\partial \xi}\right) + \frac{\partial}{\partial \eta}\left(\frac{h_1}{h_2}\frac{\partial \psi}{\partial \eta}\right) = -h_1 h_2\, \omega \tag{3.202}$$

$$h_1 h_2 \frac{\partial \omega}{\partial t} + h_2 u \frac{\partial \omega}{\partial \xi} + h_1 v \frac{\partial \omega}{\partial \eta} = \frac{1}{Re}\left[\frac{\partial}{\partial \xi}\left(\frac{h_2}{h_1}\frac{\partial \omega}{\partial \xi}\right) + \frac{\partial}{\partial \eta}\left(\frac{h_1}{h_2}\frac{\partial \omega}{\partial \eta}\right)\right] \tag{3.203}$$

where h_1 and h_2 are the scale factors of transformation given by $h_1^2 = x_\xi^2 + y_\xi^2$ and $h_2^2 = x_\eta^2 + y_\eta^2$. These equations are non-dimensionalized with L as the length scale and U_∞ as the velocity scale. We choose L in such a way that $Re = \frac{U_\infty L}{\nu} = 10^5$. The other essential details of the numerical methods and boundary and initial conditions are given in [324]. All the distances are referred with respect to this length scale, L, and this distance is approximately chosen as $L \equiv 60\delta_D^*$, where δ_D^* is the displacement thickness at the outflow of the computational domain.

Here $t = 0$ corresponds to the onset time of the excitation. This solution should be completely damped according to linear theory, unlike what is shown in Fig. 3.40, which indicates a stretch where the solution grows first, followed by subsequent decay. The solution corresponds to $y = 0.008$, a point very close to the wall.

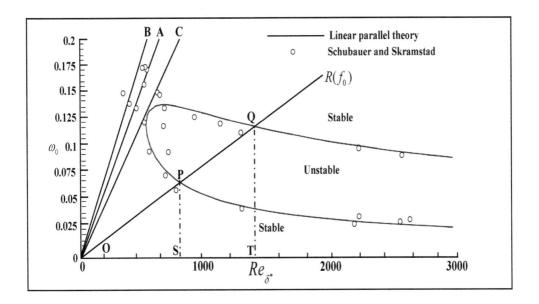

FIGURE 3.39
Theoretical neutral curve in the (Re, ω_0)-plane with stable and unstable parts identified. According to linear theory with the parallel flow assumption, three frequencies shown corresponding to OA, OB and OC are all stable. Discrete symbols are experimental data from [316].

This structure of the solution in Fig. 3.40 is generic and has features shared by the solution of the problem obtained using the linearized governing equation, as will be discussed in Chap. 5. A distinct leading wave front different from the following TS wave packet in Fig. 3.40. The exciter is located near the leading edge of the flat plate and the computed solution shows large peaks in the physical plane for the local solution. When such an excitation is applied, the local and the asymptotic solution remain stationary in space at the location depicted in Fig. 3.40. Results here are based on an accurate solution of the full Navier–Stokes equation, without the restrictive assumptions of (i) the signal problem (ii) nonparallelism and (iii) nonlinearity.

3.8.1 Time Varying Receptivity Problem vis-a-vis the Signal Problem

To study the receptivity to wall excitation following time-dependent normal velocity is applied, in between $x = x_1$ and $x = x_2$ of Fig. 3.38 and is given by

$$v_{wall}(x) = A_m(x)\,\sin(\omega_0 t) \qquad \text{for } x_1 \leq x \leq x_2 \qquad (3.204)$$

where x_1 and x_2 represent the beginning and the end of the streamwise extent of the strip exciter placed along the wall. On rest of the plate, no transpiration velocity is imposed. The amplitude function $A_m(x)$ is defined in terms of the amplitude used in the SBS strip in [98] given by

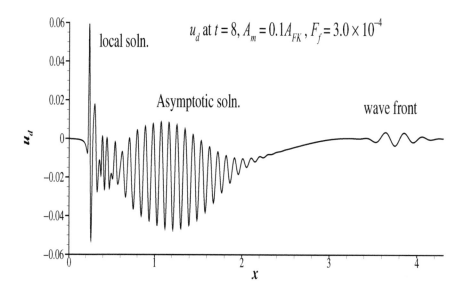

FIGURE 3.40
Streamwise disturbance velocity obtained as a receptivity solution at $t = 8$ inside a boundary layer (at $y = 0.008$) identifying the local, asymptotic and wave front components due to a strip excitation near the leading edge. The frequency of excitation of the blowing-suction strip on the wall is $F_f = 3.0 \times 10^{-4}$.

$$A_{FK}(x) = a \left(\frac{x - x_1}{x_{st} - x_1} \right)^5 - b \left(\frac{x - x_1}{x_{st} - x_1} \right)^4 + c \left(\frac{x - x_1}{x_{st} - x_1} \right)^3 \qquad (3.205)$$

for $x_1 \leq x \leq x_{st}$, and

$$A_{FK}(x) = -a \left(\frac{x - x_1}{x_{st} - x_1} \right)^5 + b \left(\frac{x - x_1}{x_{st} - x_1} \right)^4 - c \left(\frac{x - x_1}{x_{st} - x_1} \right)^3 \qquad (3.206)$$

for $x_{st} \leq x \leq x_2$.

Here $a = 15.1875$, $b = 35.4375$ and $c = 20.25$ and $x_{st} = (x_1 + x_2)/2$. Using the high accuracy DRP methods of Chap. 2, one requires only a small fraction of A_{FK} by taking $A_m = \alpha A_{FK}$. In Fig. 3.41, results are shown for the case where the SBS strip is defined by $x_1 = 0.22$ and $x_2 = 0.264$; the frequency of excitation is defined by $\omega_0 = 3.0 \times 10^{-4} * Re$ and the amplitude in Eq. (3.204) is taken as $A_m = 0.1A_{FK}$.

On the left column of the figure, a time series for disturbance vorticity is shown at a distance of $x = 1.34$ from the leading edge of the plate for the indicated heights very close to the wall. The time series indicates the variations to be different at different heights and it is also seen from the Fourier transforms shown on the right of Fig. 3.41 that the first superharmonic of the excitation frequency is very clear and a weak second superharmonic is noted for heights closest to the wall. When the time variations are recorded closer to the exciter ($x = 0.27$), as shown in Fig. 3.42, then the results are seen to be significantly different for same heights over the plate.

For the lowest height in Fig. 3.42, the amplitude of the fundamental is seen to be three times higher at this location from the exciter and the first superharmonic is of comparable magnitude, with more harmonics seen in the spectrum. For the other two heights also, one notices the clear presence of higher harmonics. These results clearly show that the assumption of treating it as a signal problem is wrong. This assumption is always made in all spatial instability studies and its limitations are noted here - which will be even more for higher amplitude of excitation. These results also indicate that effects due to shear layer growth are visible near the exciter due to the local solution component. However, the signal problem assumption can be relaxed within the linear receptivity analysis framework, as performed in [322, 349], where the authors obtained the full spatio-temporal dynamics. Thus, the receptivity analysis is capable of providing more details of the results as compared to those provided by the stability analysis.

3.8.2   Evidence of Nonparallel and Nonlinear Effects

The amplitude of the blowing-suction excitation case displayed in Figs. 3.41 and 3.42 is one-tenth of that used in [98]. These authors did not report any nonlinear and nonparallel effects for the moderate frequency excitation cases considered. They did not report any cases for higher frequencies of excitation. Here, specifically high frequency excitations at lower Re are considered for those parameter combinations which continue to be plagued by uncertainty, with conflicting claims made by various researchers. We note that the method in [98] was of lower accuracy as compared to the present one, which significantly filters the computed results. The presented results in [325] were obtained using a very high accuracy compact difference scheme, OUCS3, developed in [336], which has resolution an order of magnitude higher than the method of [98]. In fact, one can compute receptivity cases with amplitude of excitation reduced further by a factor of ten using the OUCS3 scheme for nonlinear convection terms.

To emphasize that the noted features in Figs. 3.41 and 3.42 are not due to nonlinearity, another case is computed with the amplitude of excitation further reduced tenfold, keeping the exciter at the same location. In Fig. 3.43, the disturbance amplitude of streamwise velocity is plotted at $t = 15$ for a height $y = 0.00662$, as a function of a streamwise coordinate for the three frequencies noted in Fig. 3.39.

All three cases show clearly the growth of disturbance velocity over a limited stretch, followed by decay downstream. In the frames of Fig. 3.43, a vertical dotted line indicates the location up to which the solution grows; beyond that line the solution decays. An interesting point emerges when the disturbance vorticity is plotted against x for the same height and at the same time in Fig. 3.44. For all three frequencies, the disturbance vorticity decays, in contrast to the growing disturbance velocity shown in Fig. 3.43. It is worthwhile remembering that the vorticity represents the rotational part of the flow field, while the velocity also contains an irrotational part of the flow field.

However, a completely contrasting scenario is noted at a higher height, $y = 0.0112$, in Figs. 3.45 and 3.46. At this height, the vorticity field indicates instability in Fig. 3.45, while

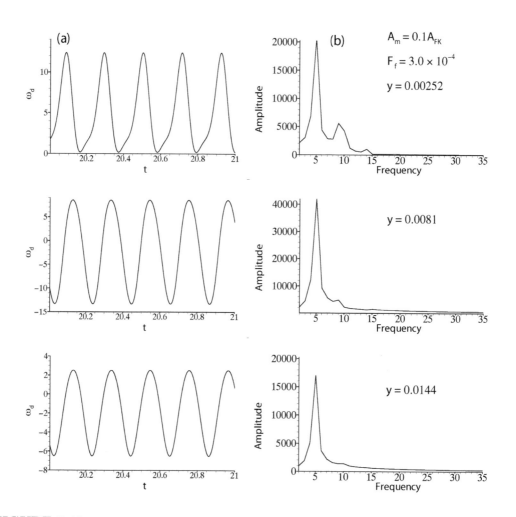

FIGURE 3.41
Disturbance vorticity time series (left) for the indicated amplitude and frequency of excitation at $x = 1.34$ for three indicated heights close to the plate and the corresponding Fourier transforms shown on the right.

the velocity field is stable in Fig. 3.46. Note the absence of local solution and the growth in the disturbance vorticity in Fig. 3.45. In Fig. 3.46, the disturbance velocity shows monotonic decay for the two higher frequencies, while for the lowest frequency, a very small growth of the signal is noted between two vertical lines in the frame.

Thus, the perceived anomaly between the noted instability at higher frequencies is neither due to nonlinearity nor due to any differences of the observed quantity. This implies incompleteness of the linear stability theory, developed with the rotational part of the flow field, via the Orr–Sommerfeld equation.

Amplitude effects are noted in Fig. 3.47 at $t = 15$ for results obtained with three levels of excitation, as indicated by the amplitude of the SBS strip. The disturbance streamwise velocity at $y = 0.000252$ is plotted as a function of x for these three cases. This height corresponds to the first grid line above the plate. Results have been normalized with respect to the amplitude, i.e., for the case of $A_m = 0.01A_{FK}$, amplitude is multiplied by ten and the

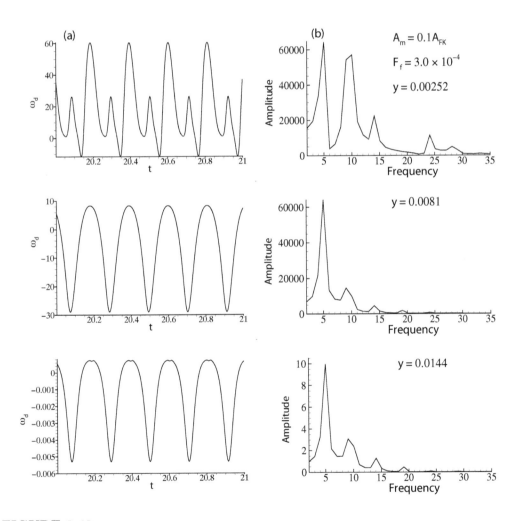

FIGURE 3.42
Disturbance vorticity time series (left) for the indicated amplitude and frequency of excitation at $x = 0.27$ for the same three heights shown in Fig. 3.41 and the corresponding Fourier transforms shown on the right.

disturbance velocity is multiplied by two for the results of $A_m = 0.05A_{FK}$ to meaningfully compare with the case of $A_m = 0.1A_{FK}$.

In Fig. 3.47 the following distinct differences are noted. The lowest amplitude case shows stable variation, while higher amplitude cases show unstable growth. The disturbance u velocity for the higher amplitude cases shows the presence of superharmonics. A vertical line indicates the streamwise location up to which near-neutral behavior of the disturbance velocity is observed for this height. This also shows that the boundary layer displays instability or stability, depending on the quantity observed and also the height where it is noted. The last aspect is important, as in traditional stability study of parallel flows, one considers the growth/decay properties to be height independent. However, the eigenfunctions are distinctly dependent upon height. To understand the sensitive dependence on height over the plate, the disturbance u velocity variation with x for the same three cases is shown in Fig. 3.48 for a different and higher height, $y = 0.0081$, as compared to that shown in Fig. 3.47.

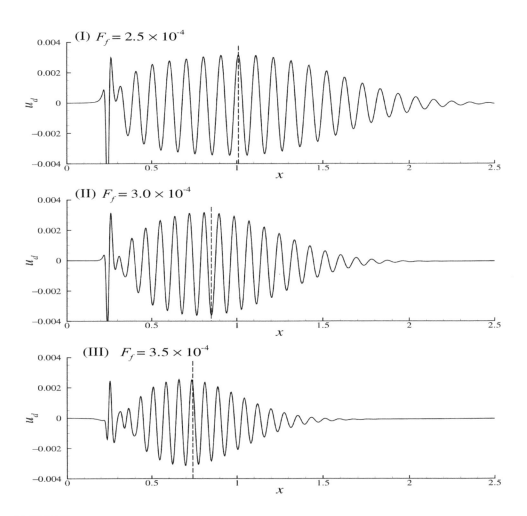

FIGURE 3.43

Disturbance streamwise velocity plotted against streamwise distance at $t = 15$ for $y = 0.00662$. The three indicated frequency cases have an amplitude of excitation which is one-tenth of that for the case in Figs. 3.41 and 3.42.

Results in Fig. 3.48 are for the same excitation parameters and at the same location of the exciter, but recorded at a higher height (the thirtieth grid line from the plate), after normalizing for amplitude in the same way performed in Fig. 3.47. The qualitative nature of disturbance velocity is now completely different and one can see pronounced growth of the signal for the asymptotic component. This predominant dependence on height of the solution is a typical attribute of nonparallel effects. For all the amplitude cases, one observes significant growth and subsequent decay of the disturbance along with the local solution. Having talked about dependence of solution with height, one must also note the essential difference between linear stability results with the present receptivity results, even when the excitation amplitude is very small and linearization appears as a valid option. For one reason, TS waves predicted by linear stability theory are not synonymous with the asymptotic solution obtained by receptivity analysis. In normal mode analysis, only the least stable eigenmode is tracked, ignoring the presence and contribution of other modes. Receptivity analysis accounts for the presence of all discrete modes and singularities.

Receptivity of the same boundary layer is studied for a higher amplitude of $A_m =$

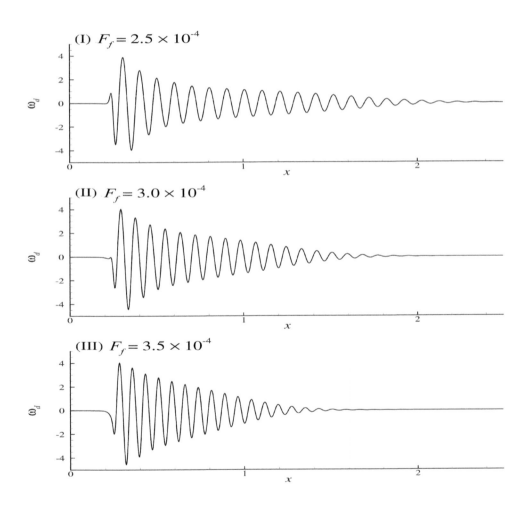

FIGURE 3.44

Disturbance vorticity plotted as a function of streamwise distance at $t = 15$ for $y = 0.0066$ with $A_m = 0.01 A_{FK}$ for the three displayed frequencies.

0.05 A_{FK} for the same excitation frequency of $F_f = \omega_0/Re = 3.0 \times 10^{-4}$ and the streamwise disturbance velocity component is plotted at the indicated heights in Fig. 3.49. The displayed variations of the streamwise disturbance velocity with the streamwise coordinate at different heights over the plate indicate an interesting phenomenon. This solution corresponds to events at $t = 20$. While the linear normal mode analysis predicts this flow field to be stable, one notices that barring the top- and bottom-most frames on the left frames in Fig. 3.49a, disturbances actually grow spatially at intermediate heights. All frames clearly indicate the presence of multiple modes, which is directly evident from the multiple peaks of the Fourier transform of these signals shown in the right frames of Fig. 3.49b. The Fourier amplitudes show the dominant normal mode near the wall, while multiple dominant modes are seen in the third and fourth frames showing the transform. Also, one notices the contribution responsible for the local solution as the distant low peak at high wavenumbers in Fig. 3.49b. The second to fourth frames in Fig. 3.49a show different streamwise stretches at different heights over which the disturbance actually grows with x, corresponding to a particular mode that moves downstream at a faster speed. The presence of this mode

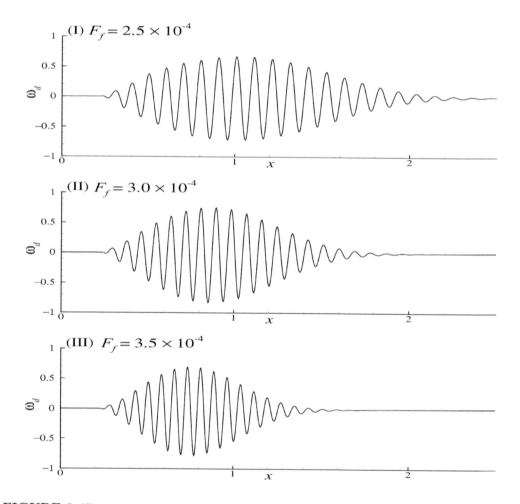

FIGURE 3.45
Disturbance vorticity plotted against x at $t = 15$ for the indicated frequencies of excitation at $y = 0.0112$ with $A_m = 0.01 A_{FK}$.

becomes weaker as the height increases. However, there is another damped mode that dominates with increasing height over the plate. It is noted that in earlier nonparallel studies [111], investigators have tried to explain experimental data which indicated instability at particular heights for streamwise disturbance velocity components.

Another interesting aspect of the computed results is noted with respect to time variation, as shown in Fig. 3.50 for signals collected at a distance of $x = 0.3$ from the leading edge at the indicated heights. Despite the fact that the fluid dynamic system is excited at a single constant frequency (ω_0), the Fourier transform in Fig. 3.50b indicates presence of superharmonics at all heights with different proportions. From Fig. 3.38 one clearly notes the corresponding normal modes to be more stable from a linear stability point of view for these additional superharmonics. Of specific interest is the data for $y = 0.00662$ and 0.00808, which show the presence of strong second and third harmonics. In Fig. 3.49 also, we noted a most pronounced instability at these heights. Thus, disturbance growth at these heights is related to nonlinear effects and is experienced at intermediate heights only.

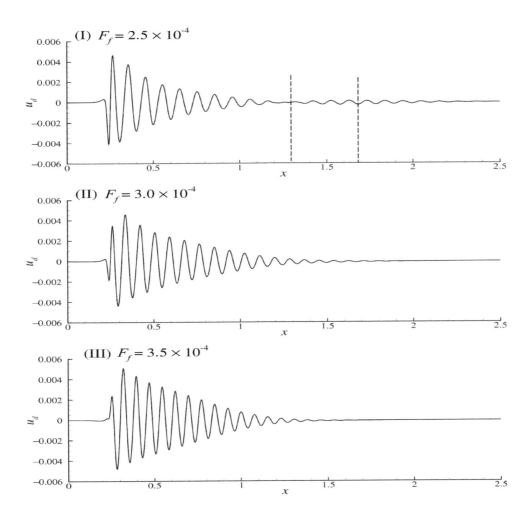

FIGURE 3.46
Disturbance streamwise velocity plotted against x at $t = 15$ for the indicated frequencies of excitation at $y = 0.0112$ with $A_m = 0.01 A_{FK}$.

3.8.3   Limitations of Linearized Nonparallel Theories

One notices that in [111], the author introduced height dependence as a nonparallel effect in a linear theory, showing it as a weak function due to variation of the eigenfunction with streamwise distance. In [111], the author followed the earlier work of [36], estimated these effects through a *kinetic energy integral* (averaging the growth with height); the integral $\int u_d^2 \, dy$ and streamwise disturbance velocity at the inner maximum; at a distance $y/\delta^* = 0.15$ and at fixed values of y below the inner maximum to compare with the measurements of [316]. Despite showing some dependence of growth rate with height, comparison with [316] was not good. These estimates using a linear nonparallel flow model depended on simply scaling the parallel flow results by a function $A(x)$. However, in creating a mathematical framework for receptivity of a growing shear layer to wall excitation, it is shown below why this scaling function has no role to play if the assumptions in [111] hold.

In these linearized nonparallel stability theories, perturbation stream function was re-

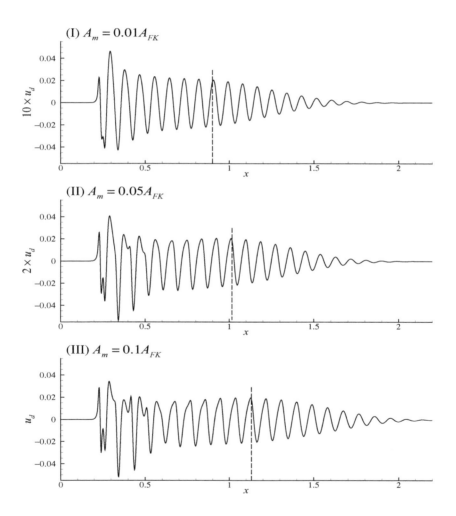

FIGURE 3.47
Normalized disturbance velocity plotted against x at $t = 15$ for the indicated amplitudes of excitation at $y = 0.000252$, the first grid line above the plate.

defined in terms of an additional slow scale $X = \epsilon x$, with ϵ as the small perturbation parameter. The perturbation stream function is given by

$$\psi_d(x, y, t) = \int \left(A(X) \, \phi_0 + \epsilon\phi_1 \right) e^{i(\alpha x - \beta_0 t)} d\alpha \tag{3.207}$$

Substituting Eq. (3.207) in the linearized Navier–Stokes equation and making a parallel flow approximation, the governing equations for ϕ_0 and ϕ_1 are obtained following the notations of [111] as

$$L_2(\Phi_0) = \left(U - \frac{\omega_0}{\alpha} \right) \left(\Phi_0'' - \alpha^2 \Phi_0 \right) - U'' \Phi_0 - \left(\Phi_0^{iv} - 2\alpha^2 \Phi_0'' + \alpha^4 \Phi_0 \right) \frac{1}{i\alpha Re} = 0 \tag{3.208}$$

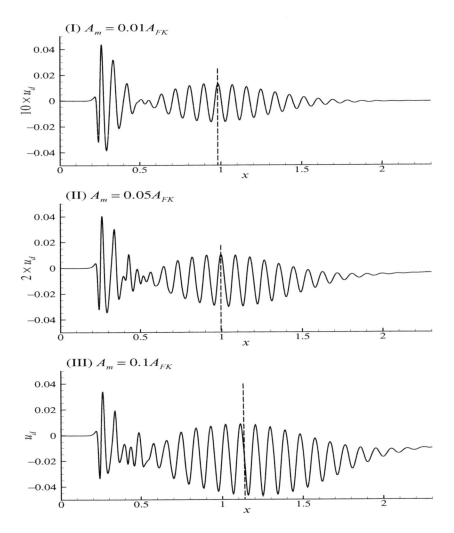

FIGURE 3.48

Normalized disturbance velocity plotted against x at $t = 15$ for the indicated amplitudes of excitation at $y = 0.0081$, the thirtieth grid line above the plate.

$$L_2(\phi_1) = -\left(F_0 A + F_1 \frac{dA}{dX}\right) \tag{3.209}$$

where the solution of Eq. (3.208) is the same as the parallel flow case, as L_2 is the familiar Orr–Sommerfeld operator. The detailed expressions for F_0 and F_1 can be seen in [111], which are in terms of the eigenfunctions of the parallel flow solution of Eq. (3.208). However, we would not recommend solving these, as Eq. (3.209) has conceptual difficulties.

A higher order correction term ϕ_1 is given by the differential equation, which is *driven* by terms dependent on ϕ_0 through the various terms on the right-hand side. This indicates

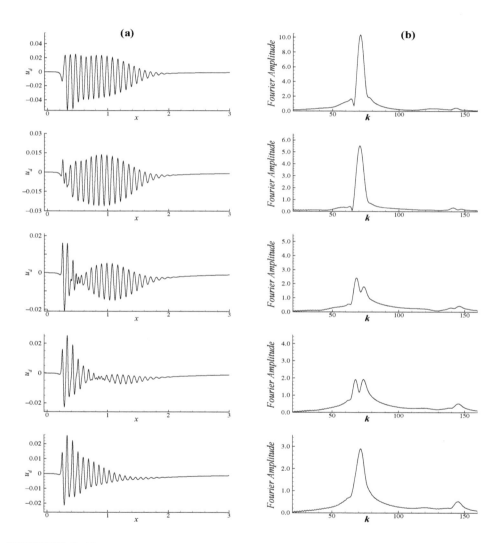

FIGURE 3.49

u_d plotted as a function of x on the left for $t = 20$ at the heights 0.00385, 0.00521, 0.00662, 0.00808, 0.00958 and 0.0144, respectively, from top to bottom. Signals are for the SBS strip excitation amplitude of $A_m = 0.05 A_{FK}$, $F_f = 3.0 \times 10^{-4}$ and exciter location between $x_1 = 0.22$ and $x_2 = 0.264$. (b) Fourier transform of the signals are shown on the right.

secular growth for ϕ_1. Without addressing this problem, the *solvability condition* was used to obtain an equation for $A(X)$ in [111], which was used as the correcting factor at the lowest order and using it as the linear nonparallel result.

Even if one restricts oneself to the lowest order term and follows the assumption in [111] (namely, the eigenvalues of the parallel flow analysis remain unchanged in the non-

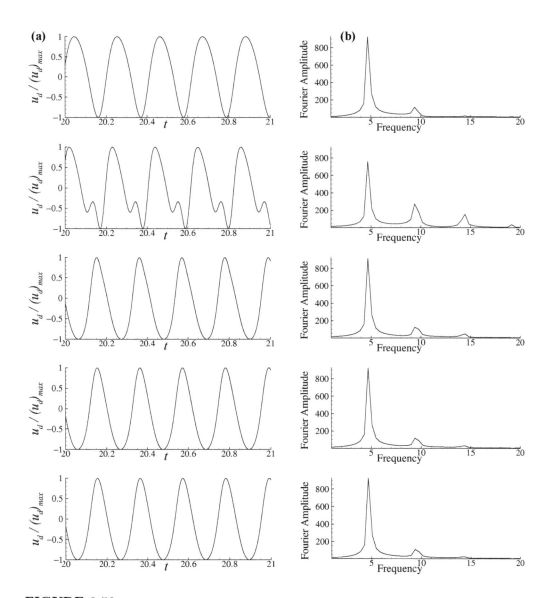

FIGURE 3.50

(a) Normalized streamwise disturbance velocity time series plotted for a distance of $0.3L$ from the exciter at the heights 0.00385, 0.00662, 0.00808, 0.00958 and 0.0144, respectively, from top to bottom. The figure is shown for the SBS strip excitation amplitude of $0.05A_{FK}$; $F_f = 3.0 \times 10^{-4}$ and the exciter location between $x_1 = 0.22$ and $x_2 = 0.264$. (b) Fourier transform of the signal shown in (a) which shows the presence of superharmonics.

parallel framework), then also one can show that there are no nonparallel corrections in the receptivity framework. This is shown below, by writing the disturbance stream function as

$$\psi_d(x, y, t) = \int A(X)\, \phi(y, \alpha, \omega_0, X)\, e^{i(\alpha x - \omega_0 t)} d\alpha \qquad (3.210)$$

where ϕ is once again given by the Orr–Sommerfeld equation. Looking at the receptivity problem for the wall excitation given by

$$u = 0 \quad \text{and} \quad v(x, t) = \delta(x)\, e^{-i\omega_0 t} \qquad (3.211)$$

for a localized excitation provided by the delta function and the time period given by ω_0, one can construct a leading order solution for the wall excitation case. Disturbances decay in the far field and wall excitation conditions in Eq. (3.211) can be expressed in terms of the decaying modes (φ_1, φ_3) of the Orr–Sommerfeld equation given by

$$a_1 \varphi'_{10} + a_3 \varphi'_{30} = 0 \qquad (3.212)$$

$$(i\alpha A + \dot{A})(a_1 \varphi_1 + a_3 \varphi_3) + A(a_1 \dot{\varphi}_{10} + a_3 \dot{\varphi}_{30}) = 1 \qquad (3.213)$$

where the additional subscript 0 indicates any quantity evaluated at the wall and the dot indicates a derivative with respect to the slow variable X. For the parallel flow, the dispersion relation is given by

$$D_p = (\varphi_{10} \varphi'_{30} - \varphi'_{10} \varphi_{30})$$

Solving Eqs. (3.212) and (3.213) one gets

$$a_1 = -\varphi'_{30}/D$$

and

$$a_3 = \varphi'_{10}/D$$

with

$$D = (i\alpha A + \dot{A}) D_p + A(\dot{\varphi}_{10} \varphi'_{30} - \varphi'_{10} \dot{\varphi}_{30})$$

This receptivity analysis clearly reveals that inclusion of the variation of the fundamental solutions with streamwise distance alters eigenvalues through the dependence of D on the second set of terms. However, if one assumes the dispersion relation to remain unchanged, as assumed in Gaster [111], then the disturbance stream function is given by

$$\psi_d = \int \left[\frac{A(X)}{(i\alpha A + \dot{A})} \right] \frac{(\varphi'_{10} \varphi_3 - \varphi_1 \varphi'_{30})}{D_p} e^{i(\alpha x - \omega_0 t)} d\alpha \qquad (3.214)$$

In the denominator of this equation, the first factor within the square bracket is due to nonparallelism following the assumption of [111], while the second factor is due to receptivity of the parallel boundary layer following the *signal problem* model. The above equation can also be rewritten in the form

$$\psi_d(x, y, t) = \int \frac{(\varphi'_{10} \varphi_3 - \varphi_1 \varphi'_{30})}{i(\alpha - \alpha_1) D_p} e^{i(\alpha x - \omega_0 t)} d\alpha \qquad (3.215)$$

where, $i\alpha_1 = -\dot{A}/A$ is used, which represents an additional pole introduced due to the slow growth of the shear layer. Thus, in Eq. (3.215), one notes the absence of the scaling function $A(X)$, which is supposed to account for nonparallelism. If one ignores the additional pole due to slow growth, then there are indeed no nonparallel effects in this formulation. Moreover, such an approach does not remove the secular growth of ϕ_1. A solution to this can be achieved by using a new slow scale in the wall-normal direction in a multiple scale theoretical approach. However, this will not be discussed any further here.

4

Bypass Transition: Theory, Computations, and Experiments

4.1 Introduction

In Fig. 3.1, a system portrait for the road to turbulence provided, taken from a review written in [235] highlighting certain conjectures by the author. In talking about receptivities of any basic equilibrium flow, the author presented two tracks following which flow transition occurs from laminar to turbulent flows: (i) the classical primary instability route and (ii) the bypass route. There are two troublesome aspects of this classification scheme. First, it is not necessary that all primary instabilities be based on the Orr–Sommerfeld equation (as indicated in Fig. 3.1), as one must pay equal attention to inviscid instability mechanism(s) following the necessary condition given by Rayleigh-Fjørtoft's theorem, which is due to temporal instability for velocity profiles displaying an inflection point inside the shear layer, following Fjørtoft's criterion. Second, there is the artificial segregation of the transition problem into an eigenvalue approach (instability theories which do not require explicit excitation) and a forced motion approach (termed bypass route of receptivity). This classification implies the inevitability of loss of stability of an equilibrium state if primary instability is indicated, without the need to quantify the input. This is unfortunate, when one recalls that in the original investigation of Osborne Reynolds, the role of background disturbance was considered very important. Also, the bypass route is considered to be different in essence from the instability problem. This has not been established at all. In contrast, we will note in subsequent chapters that the applied excitation causes flow instability, as was also done in [316].

It is a common belief that canonical flows, such as flow past a flat plate under the action of a pressure gradient, experience disturbance growth via spatial viscous instability. Primary instability marked by a viscous mechanism is manifested by the appearance of Tollmien-Schlichting (TS) waves and has become the centerpiece of instability research for external flows past streamlined bodies, as discussed in the previous chapter. It was also noted that in the presence of an adverse pressure gradient, such spatial instabilities worsen in terms of lower critical Reynolds number and higher growth rate. However, there are only a handful of theoretical/computational efforts where simultaneous spatial and temporal instabilities have been studied. These aspects are discussed here using the Bromwich contour integral method in Sec. 3.6.3 and DNS of receptivity to wall excitation in Sec. 3.8.

In [235] it has been conjectured that there are other mechanism(s) which do not display TS waves as a marker of instability − all of which are collectively termed bypass transition. A similar point of view was also advanced in [40]. There are many such flows, where transition has been caused without the presence of monochromatic unstable TS waves. Prime examples of bypass transition are flows past bluff bodies, Couette and pipes flows, leading edge contamination on swept wings, two-dimensional and three-dimensional roughness effects, etc. In some of these flows unstable TS waves are not noted; instead disturbances are seen to grow with time due to the presence of inflection point(s) for the velocity profile.

Flow past a bluff body is a typical example. In contrast, plane Poisseuille flow displays a critical Reynolds number in excess of 5770, while the flow is noted experimentally to be turbulent at significantly lower Reynolds numbers (around 1000 in [80]). This flow is also not characterized by an inflectional velocity profile, yet it displays subcritical transition. Another example of aeronautical interest is the flow undergoing transition, right at the leading edge of a swept wing. There are no definitive physical mechanisms identified in the system portrait of Fig. 3.1 in [235] for any of the bypass transition examples. Instead, question marks have been raised related to the possibility of nonlinear, nonparallel or some unknown mechanisms as potential reasons. In this chapter, we focus upon the physical mechanism for a few prototypical examples of flow transition with the help of results from carefully designed receptivity experiments and accurate flow computations.

In this chapter, a known TS wave mechanism for small amplitude excitation is shown to modify under the actions of nonlinearity and nonparallelism, showing one bypass route for flow past a flat plate under a zero pressure gradient. This is shown by solving the full Navier–Stokes equation using a high accuracy method; typical results are reported in Figs. 3.49 and 3.50 for $A_m = 0.05\ A_{FK}$ for this numerical method. It will be noted that even when a bypass route is activated, TS waves are also present and this raises questions about the definition proposed by various researchers for bypass transition to imply the absence of TS waves.

To understand the effects of nonlinearity via the role of the amplitude of excitation through a blowing-suction strip, results for the case of $A_m = 0.10A_{FK}$ are discussed which has identical excitation parameters of the case shown in Figs. 3.49 and 3.50, except the amplitude of excitation doubled. Still this larger amplitude case corresponds to an amplitude which is one-tenth of the amplitude considered in [98]. This case of $A_m = 0.10A_{FK}$ causes unsteady separation near the exciter on the plate and resultant separation bubbles are confined to a small height near the wall, as shown in the stream function contours in Fig. 4.1 at $t = 20$. Only a few representative contours have been plotted in this figure which clearly show the micro-bubbles near the exciter seen up to $x \leq 1.6$ only. Should this be construed as instability following the bypass route?

To understand the dynamics of this higher amplitude case better, an instantaneous snapshot at $t = 20$ is shown for the streamwise disturbance velocity variation with x at three heights in Fig. 4.2a. In the figure, the corresponding Fourier transform is shown in right column in Fig. 4.2b, which helps identify length scales of various components of the solution noted in the physical space.

One can compare the middle frame of Fig. 4.2 (at $y = 0.00662$ for $A_m = 0.1A_{FK}$) with the second frame from the top of Fig. 3.49 (at the same y for $A_m = 0.05A_{FK}$) to check for nonlinearity effects. This is evident from the Fourier amplitude that shows the increase is by more than two and a half times for the Fourier amplitude, while the input excitation amplitude is increased two-fold. However, the length scale selected is almost identical for these two amplitude cases, despite the presence of separation bubbles for the higher amplitude excitation. Having the same wavenumber for the two cases indicates that both the cases suffer instability by the same physical mechanism. The manifestation of micro-bubbles in Fig. 4.1 is for the total flow field. For points very close to the wall, the disturbance flow is larger in magnitude as compared to the mean flow. When the flow is excited harmonically by the SBS strip at twice the amplitude, then during the blowing phase, flow separates locally. This separation bubble pulsates and becomes extinct during the suction phase of excitation. These bubbles convect and an instantaneous snapshot is shown in Fig. 4.1. We conjecture that with increasing the width of the exciter and/or increasing the amplitude of excitation further, separation bubbles will be more pronounced, which could interfere destructively with TS waves, as happened in this case with low intensity. There is another difference for the higher amplitude excitation case, noted in the outer part of the boundary layer. For the $A_m = 0.05A_{FK}$ case, the solution shown at $y = 0.0113725$ in Fig. 3.49 did not show any instability. In contrast, for the case of $A_m = 0.10A_{FK}$, results shown for $y = 0.035$ distinctly indicate a streamwise stretch over which the disturbance actually grows. This is essentially a combination of nonlinear and nonparallel effects.

However, the presence of TS waves is unmistakable for this case of $A_m = 0.1A_{FK}$, as further noted in Fig. 4.3, from the time series for streamwise disturbance velocity shown at $x = 0.3$ (which is in the neighborhood of the exciter) for the three indicated heights. These time series are also multi-periodic, as is evident from the Fourier transform shown in Fig. 4.3b. The distinct presence of second and third harmonics is noticeable in the interior of the shear layer (at around $y = 0.00662$), similar to the case shown in Fig. 3.50, while the superharmonics are not seen for $y = 0.034$ and above, for the higher amplitude excitation case.

Next, we investigate a case where the exciter width is increased by considering the strip between $x_1 = 0.2$ to $x_2 = 0.29$. Having extended it further towards the leading edge of the plate, more nonparallel effects are incorporated and by extending it in the downstream direction also, we are imparting more energy to the shear layer. In Fig. 4.4, stream function and vorticity contours are compared for the two cases of different exciter widths. For both the cases, the amplitude of excitation is taken as $A_m = 0.1A_{FK}$ and the frequency of excitation given by $F_f = 3.0 \times 10^{-4}$.

From the ψ contours in Fig. 4.4a, we note the presence of a few small bubbles in the immediate downstream of the exciter spaced at a fixed gap similar to the wavelength of the TS wave for this ω_0. The wall-normal direction is stretched in all the frames shown in

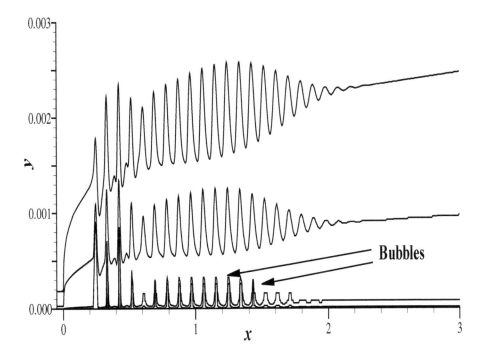

FIGURE 4.1
Streamline contours shown at $t = 20$ for the case of $A_m = 0.1A_{FK}$, $F_f = 3 \times 10^{-4}$ for the exciter located between 0.22 and 0.264. Note the micro-bubbles near the exciter on the plate.

Fig. 4.4. The vorticity contours exhibit a two-tier structure with the lower layer showing an upstream propagating tendency, while the upper tier shows vertical direction of the ejected vortices (those are slightly tilted towards the downstream direction). For the case of the wider excitation strip in Fig. 4.4c, the stream function contours display bubbles over the streamwise stretch up to $x = 3.3$, as compared to the case of the shorter exciter, which shows visible undulation up to $x = 1.7$ only. Also, the bubbles are of significantly larger dimension in the wall-normal direction for the wider extent exciter. Despite this, the physical mechanism of disturbance generation is as in creating TS waves by harmonic excitation shown in [322]. This once again establishes that despite the appearance of convecting separation bubbles for higher amplitude cases of $A_m = 0.1A_{FK}$, the classical receptivity route described

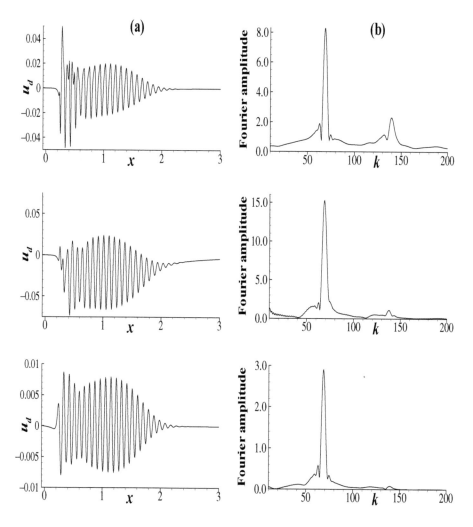

FIGURE 4.2

(a) Disturbance streamwise velocity component plotted for the heights $y = 0.00245$, 0.00662 and 0.035, respectively, from top to bottom at $t = 20$, for the case of $A_m = 0.1A_{FK}$, $F_f = 3 \times 10^{-4}$ for the exciter located between 0.22 and 0.264. (b) Fourier transform of the signals shown in (a).

in Sec. 3.8 is seen here for the primary instability and these wall excitation cases should not be said to create bypass transition, as the space-time scale selection is still governed by the same dispersion relation which produces TS waves. We add here that with further increase in excitation amplitude of harmonic wall excitation, one would note secondary and higher order instabilities which can cause premature transition, without going through the exponential growth of TS modes, as conjectured in the literature. We would note in Chaps. 5 to 7 that the TS wave packet does not cause transition for flow past a flat plate, with and without heat transfer and for transition over NLF airfoils in Chap. 9.

The above discussion relates to receptivity of flow to disturbances created at the wall by imposed time scale(s). It is shown in the previous chapter that the correct way to treat the flow is by a spatio-temporal theory using Bromwich contour integrals, as in [322], and not by spatial theory. This will be further emphasized in Chaps. 5 to 7, where combinations

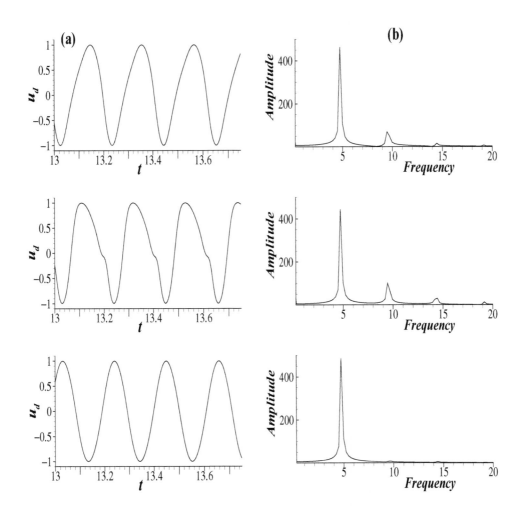

FIGURE 4.3
Normalized streamwise disturbance velocity time series plotted at a distance $x = 0.3$ from the leading edge at the heights 0.00124, 0.00662 and 0.034 for the SBS exciter with $A_m = 0.1A_{FK}$, $F_f = 3 \times 10^{-4}$ for the exciter located between 0.22 and 0.264. (b) Fourier transform of the time series shown in (a).

of linear and nonlinear approaches are used. It is shown that by direct simulation of the Navier–Stokes equation, harmonic wall excitation on a zero pressure gradient boundary layer creates TS wave packets, even when the response displays micro-bubbles in the vicinity of the exciter. Next, we want to study cases where the excitation is applied at the free stream without introducing any time scale(s), unlike the multi-periodic cases discussed in the previous chapter.

Thus, we focus upon the problem of vortex-induced instability which creates large perturbations, without the appearance of TS waves, as reported in [199, 331, 342]. A shear layer

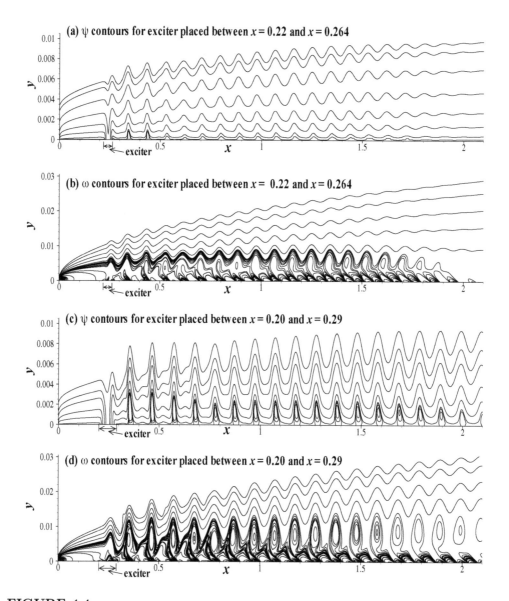

FIGURE 4.4
Stream function and vorticity contours for the two different strip exciter cases having different width. (a) and (b) correspond to exciter located between 0.22 and 0.264; (c) and (d) correspond to exciter located between 0.2 and 0.29.

can interact with a finite-core vortex convecting far outside itself leading to unsteady separation in the shear layer and this was noted in [90] *as one of the most important unsolved problems of fluid dynamics.* Unsteady separation as such has been noted to be present in: (i) flow past surface-mounted obstacles, (ii) dynamic stall and blade vortex interaction (BVI), (iii) impulsive motion of bluff bodies, (iv) in the near-wall region of a fully turbulent flow and (v) bypass transition initiated by aperiodic convecting vortices. Here, focus is retained on the last topic, as an example through which the physical mechanism during this bypass transition is explained. This prototypical mechanism is seen to be also present in the other examples cited above. Unsteady flow separation has also been the topic of discussion in [82] and other references contained therein. Brinckman & Walker [44] have studied near-wall eddy formation in turbulent boundary layers. In this last work, a late stage of transition, whose origin is completely different from the one created by linear instability, was studied in a quasi two-dimensional framework. In [301, 376], the authors have also discussed formation of hairpin vortices in the near-wall region of turbulent flows. Another study [267, 268] considered the scenario where a vortex placed above a plane wall caused the vortex to move and thereby create a thin unsteady boundary layer over the wall. Thus, the problem considered in this chapter has many variations, but we will only consider the case where primary instability is triggered by a convecting vortex over a wall-bounded shear layer, as described in [199, 331, 334].

4.2   Transition via Growing Waves and Bypass Transition

In the previous chapter, we noted that a train of periodic convecting vortices moving at a constant speed in the free stream produces wave packets formed by TS waves, whose dynamics are different from those produced by wall excitation or were reported via the vibrating ribbon experiment of [316]. There are few reasons for qualitative differences between free stream and wall excitations. In the previous chapter, we identified *wall* and *free stream modes* as two different types of ensuing motions seen for these two types of excitations. First, in the vibrating ribbon experiment a predominantly monochromatic TS wave is created by constant frequency excitation at or very near the wall. For free stream excitation by a train of vortices, the response field is multi-periodic in time, as expressed in Eq. (3.188) for the stream function. Typical results are shown by numerical results in Figs. 3.34, 3.35 and 3.37. The convecting train of vortices causes an excitation field that is essentially broad-band, but multi-periodic. Second, free stream excitation can create both downstream and upstream propagating disturbance fields. Finally, in contrast to the vibrating ribbon case, free stream excitation creates disturbance wave packets, which can remain within the unstable range over a longer stretch, and also suffers much higher growth rates, as reasoned while discussing the results in Fig. 3.33.

For the case of a periodic train of vortices convecting in the free stream at a constant speed, one follows the c = constant line in Fig. 3.33. For monochromatic wall excitation, waves initially grow upon entering the unstable region, in the (Re, ω_0)-plane, monotonically up to a maximum and thereafter the growth rate decays, until it falls to zero when it exits the neutral curve. Outside the neutral curve, disturbances actually decay while propagating, according to linear stability theory results. In contrast, for the free stream excitation case, disturbances are noted as forming wave packets which experience explosive growth if the excitation field convects at a speed in a narrow range (as shown to be between $0.26U_\infty$ and $0.32U_\infty$) for the zero pressure gradient case shown in Fig. 3.33.

The most important difference between wall and free stream excitations is the direction of propagation of disturbance field with respect to the local mean flow. For low amplitude wall excitation, disturbances propagate downstream only for unseparated flows. In contrast, when the shear layer is excited by a train of vortices in the free stream, then a part of the disturbance field travels upstream, in addition to the downstream propagating *wall mode*. While these observations are for the two-dimensional flow field, it is noted that very low frequency wall disturbances create only three-dimensional disturbances in the form of the Klebanoff mode, as discussed in Sec. 3.6.6. Thus, when a shear layer is excited by periodic free stream sources, the resultant flow field is a mixture of two-dimensional and three-dimensional disturbances and interactions between these two types of disturbances, as these travel upstream and downstream, with respect to the excitation source. This type of behavior was noted in the DNS in [446]. In this exercise, periodic free stream excitation is created by a periodic wake entering the computational domain through the inflow. In the computed results, *transition precursors of long backward jets* are noted in the outer part of the shear layer which propagate downstream to form turbulent spots and very long *elongated streaks of velocity fluctuations*. The latter are the three-dimensional Klebanoff modes, while the former are associated with the upstream propagating modes of the two-dimensional component of the disturbance field. While DNS can produce results for the imposed disturbance field when performed carefully without spurious dispersion, still it does not provide a clear picture of the physical mechanism(s) involved. The shear layer properties of different modes and simulations in [362] and [327] provide vital clues to free stream excitation problems by periodic vortices validating itself by providing results which were noted in early observations of the control experiments in [172].

Rapid transition caused by a periodic train of vortices follows a maximum growth path when the convection speed is within $0.26U_\infty$ to $0.33U_\infty$ if Re is restricted to a maximum of 1900 at the outflow. Instability is initiated by TS wave packets as one component of disturbance field if the applied perturbation is not too large − as will be explained in subsequent chapters. This route of instability is not the bypass transition. However, when the free stream turbulence intensity exceeds 1%, it has been observed experimentally that transition occurs rapidly, bypassing the TS route. It is noted in [159] that *inertial-time-scale processes apparently come into play, but their origin and nature are not at present known*. These large amplitude perturbations are noted to cause bypass transition and the authors in [159] have stated that bypass transition is *stochastic by nature, and so lies within the province of statistical fluid dynamics*, which is truly speculative in nature, as we will note in later chapters. Progress has been limited in understanding bypass transition due to the lack of definitive controlled experimental observation and data in identifying physical mechanism(s) in the spirit of the experiment in [316]. The design of control-experiments to identify unit processes has been advocated in [373] for transitional and turbulent flows; the author notes that one must excite the fluid dynamic system with non-negligible input while reducing the background disturbances to a minimum, such that the causality of inputs can be unambiguously established in the same way the classical experiments of [316] proved the viscous instability theory and existence of TS waves.

Such an experiment was performed and reported in [61, 199, 342]. In this experiment, the aim was to cause subcritical violent instability of the zero pressure gradient shear layer by a convecting single vortex moving at a fixed height and speed. Thus, no time scale was enforced for the disturbance field. In the experiment, violent breakdown during bypass transition was shown to be initiated by a two-dimensional mechanism and the resultant broadband energy spectrum of the corresponding turbulent flow is excited at an early receptivity stage itself, immediately following the onset of instability. The experiment was conducted in a water tunnel, using a dye visualization technique, where the bypass transition was triggered by the controlled motion of a captive vortex in the streamwise direction, at a speed lower than the free stream speed. The experimental results are discussed in detail in the next section.

4.3 Visualization Study of Vortex-Induced Instability as Bypass Transition

In this experiment, stability of the zero pressure gradient laminar boundary layer is examined when the subcritical flow (with respect to the critical Reynolds number, at which the growing TS wave first makes its appearance) is affected by convecting a single vortex outside the shear layer. Experiments on vortex-induced instability are challenging, as it is difficult to control and quantify the strength and the propagation speed of the vortex causing the instability. This experiment was designed to achieve complete control over these parameters by creating the captive convecting vortex by a rotating cylinder, at a high rotation rate, to create a stable captive vortex of constant strength and by constraining it to translate at a fixed height with constant convection speed. It is to be emphasized that shed vortices in the wake of the rotating cylinder, if there are any at all, have strengths an order of magnitude lower than that of the captive vortex created due to rotation by the Robins–Magnus effect (as discussed in [356]). The role of a captive translating vortex with a finite-core size is to destabilize the shear layer by creating a longitudinal adverse secondary pressure gradient. This was also the scenario proposed in [229], where the process was thought to arise due to constant buffeting by convected vortices associated with free stream turbulence. In the reference, the authors were attempting to explain a similar scenario proposed in [400].

The experiment was conducted in the re-circulating water channel in NUS, Singapore. Figure 4.5 shows the schematic of the experimental set-up. The test section measured 400 mm × 400 mm in cross section and was 1800 mm long, made of transparent Plexiglass. The flow before entering the test-section passes through a honeycomb, three fine screens and a 5:1 contraction section. The flow velocity in the test section is controlled by a variable-speed A.C. motor driving a centrifugal pump on the return leg of the flow circuit.

The boundary layer under investigation was allowed to form over a 1100 mm long flat plate held vertically on its longer edge, on the base of the test section. The plate had a rounded leading edge and an adjustable trailing edge flap which is used to create a stable boundary layer in the absence of excitation. The single captive vortex was created by rotating a circular cylinder of diameter $d = 15$ mm, whose axis was parallel and along the spanwise direction of the plate. The cylinder was attached to a stepper motor, whose rotation rate was controlled and varied from 3.5 to 6 revolutions per second, in either a clockwise or anticlockwise direction on demand. The distance between the flat plate and the rotating cylinder was kept in most of the cases equal to 90 mm, but was also varied between 60 and 120 mm for some other cases. For the purpose of visualization, dye was released from six dye ports located 88 mm downstream from the leading edge of the plate.

In Fig. 4.5a, the side view showing a rotating cylinder translating over a flat plate is shown. In Fig. 4.5b, the same set-up is viewed from the top of the tunnel. H_1 is the adjustable distance of the cylinder from the plate; U_∞ is the oncoming flow speed; c indicates the translational speed of the cylinder (not to be confused with the phase speed) and Ω is the angular velocity of the cylinder. The distance of the cylinder from the leading edge of the plate is indicated by x. This arrangement creates a captive vortex (at the center of the cylinder), which can be made to travel at a predetermined speed. The Reynolds number of the rotating cylinder was 2975, based on the diameter and the free stream speed, U_∞. The displacement thickness (δ^*) at the location of the dye port was calculated as 3.27 mm and thus the distance between the cylinder and plate is equivalent to 27.52 δ^*. Therefore, the disturbance source was, by design, kept significantly outside the shear layer to mimic the unit process of free stream turbulence effects. The cylinder was constrained to travel at the fixed height (H_1) over the flat plate and the translational speed c was also controlled. All of these parameters were used to control the events in a reproducible fashion.

Eight significant cases were reported in [199] (as given in Table 4.1). The first two cases correspond to conditions when the cylinder rotates in the counterclockwise direction and translates slower than the free stream speed. However, the translation speed of the cylinder in Case 2 is almost double that of Case 1. For Case 3 the cylinder was not rotated and

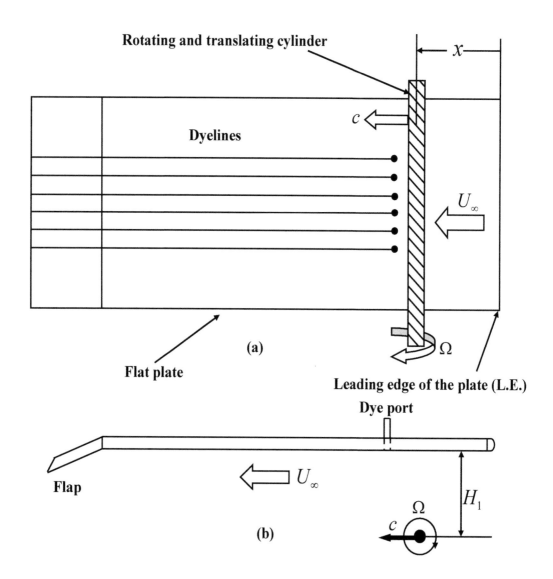

FIGURE 4.5
Schematic of the experimental set-up. (a) Side view showing a rotating cylinder translating over a flat plate. (b) The same set-up as viewed from the top of the tunnel. H_1= adjustable distance of the cylinder from the plate, U_∞ = flow speed, c = translational speed of the cylinder and Ω = angular velocity of the cylinder, x = distance of the cylinder from the leading edge of the plate. This arrangement creates a captive vortex (at the center of the cylinder), which can be made to travel at a predetermined speed.

TABLE 4.1

Cases of vortex-induced instability experiments.

	c/U_∞	$\Omega_{(rps)}$	H_1/δ^*	$U_s/(U_\infty - c)$
Case 1	0.386	+5	27.52	2.360
Case 2	0.772	+5	27.52	6.364
Case 3	0.386	0	27.52	0
Case 4	0.386	−5	27.52	2.360
Case 5	0.237	−5	27.52	2.324
Case 6	0.386	+5	18.35	2.360
Case 7	0.386	+5	24.45	2.360
Case 8	0.386	+5	36.70	2.360

this was a case where the shear layer on the flat plate would have been perturbed by the shed vortices which are significantly weaker and periodic in the near wake of the cylinder. Also, such weak shed vortices would not move at constant height and speed. This case also demonstrated the importance of controlled over uncontrolled disturbances in the free stream. For Cases 4 and 5, the cylinder rotated in the opposite direction to that of Cases 1 and 2.

In Cases 6 to 8, the rotation and translation velocities of the cylinder are the same, but the cylinder is located at different heights above the boundary layer. The last column of the table shows the ratio of surface speed ($U_s = \Omega d/2$) to the relative free stream speed, ($U_\infty - c$), for the cylinder. Except for Case 3, where the cylinder is not rotating, values given in the last column for all other cases are greater than 2. This parameter value is known to cause limited or no Karman vortex shedding, as noted experimentally in [407] and [85].

In Fig. 4.6, flow visualization sequences for Case 1 are shown which indicate vortex-induced instability as a localized increase in mixing, diffusion and irregularity of the dye filaments. The dye filaments were initially parallel, implying two dimensionality of the primary flow. As the vortex moved in the flow direction from right to left, each dye filament split into two, with one part lifting off with little or no spanwise spreading at the early stages and the other part staying in its course. Below each frame, the instantaneous location of the free stream vortex is noted with respect to the leading edge of the plate. Higher order instability is noted to have occurred some time between the third and fourth frame. The direct consequence of this is in the incipient formation of disturbance packets which keep lengthening due to higher *front speed*. Disturbances amplify ahead of the cylinder, and as time progresses it grows while convecting faster than the cylinder and hence affecting a larger part of the flow with time. Violent breakdown of dye filaments indicates strong unsteadiness due to an instability caused by the translating vortex (for $c = 0.386U_\infty$).

When the translation speed of the cylinder was increased to ($c = 0.772U_\infty$) in Case 2, there was no violent breakdown of the flow, as shown in Fig. 4.7. This indicates that the boundary layer is insensitive to the vortex convecting at higher speeds. For the range of translation speeds investigated, it was found that the slower the translation speed of the vortex, the greater is the propensity for boundary layer instability, when other parameters were kept the same.

The above aspect of lesser instability with increased speed of the rotating cylinder can be explained by noting that a rotating and translating cylinder of diameter d induces an additional disturbance stream function (ψ) in the inviscid irrotational part of the flow field (free stream disturbances) as given in [299] by

$$\psi = (U_\infty - c)\left[2y - \frac{(y - H_1)(d/2)^2}{\bar{x}^2 + (y - H_1)^2} - \frac{(y + H_1)(d/2)^2}{\bar{x}^2 + (y + H_1)^2}\right] + \frac{\Gamma}{4\pi}\mathrm{Ln}\frac{\bar{x}^2 + (y + H_1)^2}{\bar{x}^2 + (y - H_1)^2} \quad (4.1)$$

with $\bar{x} = x - x_v$, $x_v(= x_0 - ct)$ is the current location of the convecting vortex with x_0 its initial location (at $t = 0$) and Γ is the circulation of the vortex. This expression is the input to the underlying shear layer acting as the dynamical system and it takes into account the image system due to the presence of the wall. Thus, Eq. (4.1) is an expression that is the cause for the fluid dynamical system to be destabilized inside the shear layer in a receptivity scenario. The imposed disturbance, as given by Eq. (4.1), consists of two

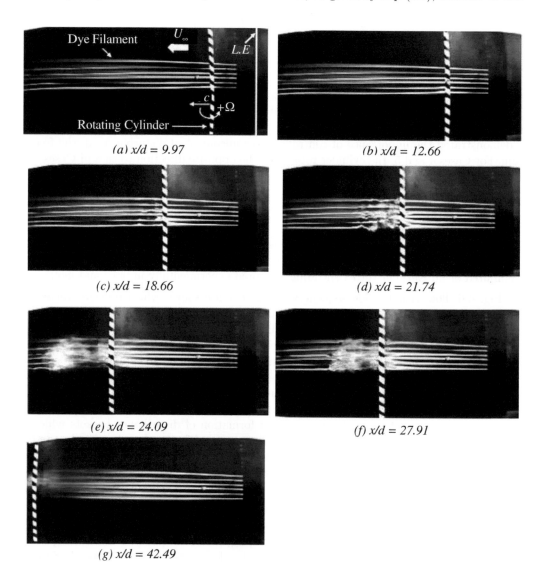

(a) x/d = 9.97

(b) x/d = 12.66

(c) x/d = 18.66

(d) x/d = 21.74

(e) x/d = 24.09

(f) x/d = 27.91

(g) x/d = 42.49

FIGURE 4.6
Visualization pictures for Case 1: $c = 0.386$, $H_1/\delta^* = 27.52$, $\Omega = +5$. Note: x is measured from the leading edge of the plate and d is the diameter of the cylinder.

parts: (i) the displacement effect of the finite-core vortex given by the first term with c, d and H_1 as the defining parameters and (ii) the circulatory effect caused by the Biot–Savart interaction between the lumped free stream vortex with its image system given by the last term. This circulatory effect depends upon c, H_1 and Γ. In Figs. 4.8a and 4.8b, these effects are shown separately for the parameters given by $H_1 = 6d$, which is also equal to $27.52\delta^*$ and two strengths of vortex, $\Gamma = 14$ and 30.

In Fig. 4.8a(i), the disturbance stream function induced by a positive (counter-clockwise) spanwise vortex is shown for $\Gamma = 14$, which establishes that the circulation effect is subdominant to the displacement effect for the values of H_1 and Γ chosen. This disturbance stream function is evaluated for the vortex location given at $x_v = 5$ by the dotted vertical line. As the displacement effect does not vary with x, so it does not induce any wall-normal velocity components, as compared to the circulation effect, which induces a blowing to the right and suction to the left of the vortex in the free stream. In Fig. 4.8a(ii), the imposed pressure gradient is calculated and shown as a function of x for three values of non-dimensional c for the same case. This is the induced pressure gradient by the convecting vortex over an equilibrium zero pressure gradient flow. It is seen that the imposed pressure gradient has similar excursion about the zero value, showing variation with c. For all the three c cases, maximum favorable and adverse pressure gradients occur at the same upstream and downstream locations, with the magnitude decreasing with increasing c. Thus, the adverse pressure gradient is mostly determined by the circulation effect and the induced adverse pressure gradient can destabilize the flow for the same Γ when c is reduced. This in effect explains the flow receptivity to vortical excitation for Cases 1 and 2 in Figs. 4.6 and 4.7. The difference can be further understood when one looks at Fig. 4.8a(iii), where the induced pressure gradient is plotted as a function of time at $x = 5$ for the indicated five values of c.

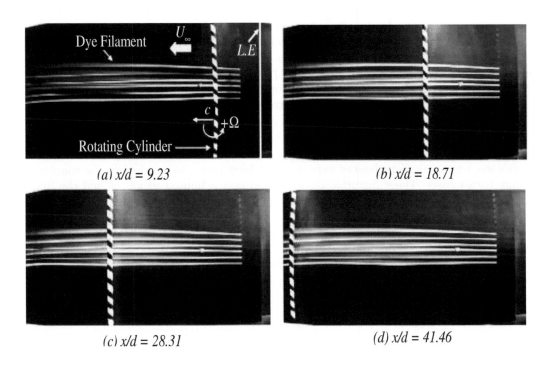

(a) $x/d = 9.23$

(b) $x/d = 18.71$

(c) $x/d = 28.31$

(d) $x/d = 41.46$

FIGURE 4.7

Visualized flow for Case 2 with $c = 0.772$, $H_1/\delta^* = 27.52$, $\Omega = +5$ showing lower receptivity.

For this figure, the vortex is moved from $x_o = 1$ at $t = 0$ to $t = 65$. For the cases considered in this figure, the stationary vortex case ($c = 0$) shows the induced pressure gradient to remain the same at all times. However, for $c \neq 0$ cases, for higher c's, the boundary layer experiences an adverse pressure gradient quicker, for a shorter duration and subsequently it asymptotes towards a zero pressure gradient. For $c = 0.8$, an adverse gradient is created as early as $t = 5$ and the duration over which the adverse pressure gradient is experienced is $5 \leq t \leq 30$; for $c = 0.4$ the duration of the adverse pressure gradient is $10 \leq t \leq 60$ and for $c = 0.2$ the duration starts at $t = 20$ and continues to remain adverse for the displayed range. As c is reduced further down to 0.1, the imposed adverse pressure gradient onset time is $t = 40$. But the subsequent pressure gradient is exceedingly adverse.

One of the important non-dimensional pressure gradient parameters is the Falkner–Skan pressure gradient parameter, $m = \frac{x}{U_e} \frac{dU_e}{dx}$. This parameter is shown as a function of x for the two indicated time instants, for $c = 0.2$ in frame (iv) of Fig. 4.8a. Apart from the initial time instant $t = 0$, another large time is considered for plotting m in this frame. This latter time is chosen when the solution of the Navier–Stokes equation indicated unsteady separation at $x_{MP} = 6.4$ at $t = 23$ for the $c = 0.1548$ case shown in [331]. The horizontal dotted line in the figure is drawn at $m = -0.19884$, for which the similarity flow suffers steady separation. This figure clearly shows that the considered cases represent truly unsteady flows which can sustain a much larger adverse pressure gradient for a long time before it shows unsteady separation, as compared to steady flows.

In Fig. 4.8b, corresponding induced effects are shown when the vortex strength is increased to $\Gamma = 30$. One distinguishing feature of this case is that the disturbance stream function is of the same order for displacement and circulatory effects. One also notes that the onset time for the adverse pressure gradient remains the same for both the cases of different values of Γ. Subsequent maximum adverse pressure gradient scales with the amplitude of the free stream vortex strength. Also, the induced pressure gradients for the displayed cases are proportionately higher for this higher vortex strength case.

In Fig. 4.9, visualization sequences are shown for Case 3. In this case of a non-rotating translating cylinder, no violent instability was seen to occur for two reasons. First, the imposed disturbance field as given by Eq. (4.1) has no captive vortex (*i.e.*, $\Gamma \equiv 0$) as the cylinder does not rotate while translating. Second, if there are shed vortices present these are very weak in the Benard–Karman vortex street, they are seen to affect the flow weakly at later times, far downstream of the translating cylinder.

Figure 4.10 shows visualized frames for Case 5, when the translating cylinder rotates in the opposite direction at the same rotation rate, while it translates with a speed $c = 0.237$. In this case, very weak disturbances were noted at later times, as opposed to the counter-rotating vortex cases. These disturbances are noted over a limited region behind the free stream vortex, do not show much spanwise mixing and do not lead to complete transition to turbulence, even for the last frame. For this case, while there is an adverse pressure gradient behind the cylinder, the motion of the captive vortex in the free stream stabilizes the flow ahead of the rotating cylinder and this is clearly evident in the figure.

For Cases 4 and 5, the main distinguishing features are 1) the disturbances trail behind the free stream vortex and 2) the disturbance growth is far less than that for Cases 1 and 2. Both the observations are due to the input adverse pressure gradient imposed on the shear layer. In the case of clockwise circulation, the adverse pressure gradient trailed behind the vortex, while in cases of counter-clockwise rotation, the adverse pressure gradient is created ahead of the vortex. This is explained in Fig. 4.11 with the help of a sketch of streamline patterns in the inviscid part of the flow which would be created by a rotating and translating cylinder for both the clockwise and the counter-clockwise supercritical rotation cases.

Streamline curvature in the vicinity of the flat plate would be an indicator of the secondary imposed pressure gradient by the vortex. This vortex convects far outside the shear layer forming over the plate and can be idealized by an equivalent lumped vortex at the

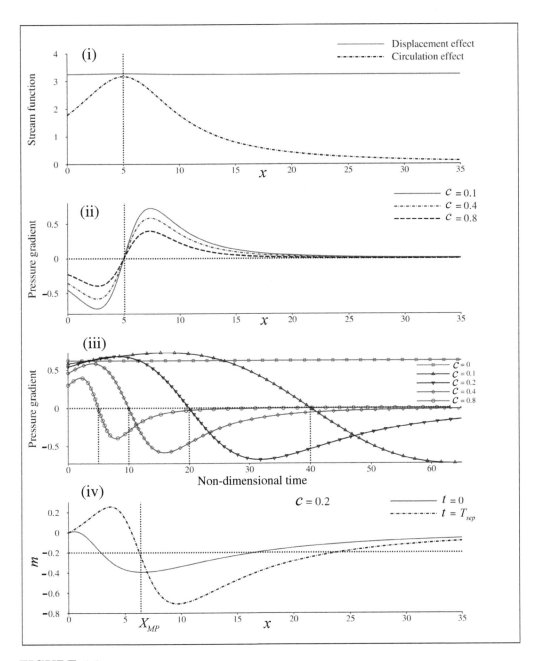

FIGURE 4.8a
Disturbance stream function versus x caused by circulatory and displacement effects at $y = 1.92$, when the free stream vortex is at $x_c = 5$, $H_1 = 6$, moving with $c = 0.1545$ and a strength of $\Gamma = 14$. (ii) Induced pressure gradient at the same height for the case (i). (iii) Pressure gradient as a function of time at $x = 5$ and the same height when the vortex is started from $x_0 = 1$. (iv) Variation of the Falkner–Skan parameter m with x at the indicated time instants.

center of the cylinder. We have already noted that the imposed disturbance stream func-

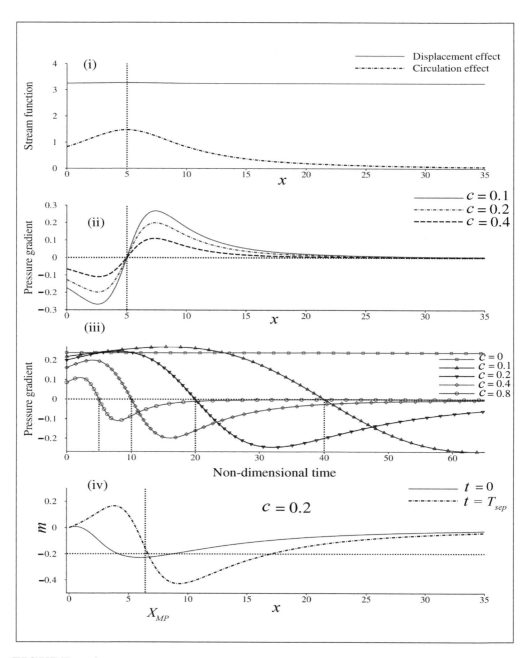

FIGURE 4.8b
Disturbance stream function versus x caused by circulatory and displacement effects at $y = 1.92$, when the free stream vortex is at $x_c = 5$, $H_1 = 6$, moving with $c = 0.1545$ and a strength of $\Gamma = 30$. (ii) Induced pressure gradient at the same height for the case of (i). (iii) Pressure gradient as a function of time at $x = 5$ and the same height when the vortex is started from $x_0 = 1$. (iv) Variation of the Falkner–Skan parameter m with x at the indicated time instants.

tion causes predominantly lesser circulatory effects than displacement effects, given by Eq. (4.1) and shown in Figs. 4.8a and 4.8b. Thus, in essence, the drawn streamlines in Fig. 4.11 resemble that seen past a translating and rotating cylinder. For ease of understanding, we have displayed the case of supercritical rotation rate with distinct saddle points displaced along the vertical line appropriately due to the lift experienced by the cylinder. The actual lift experienced by a translating and rotating cylinder for viscous flow is far lower than that given by the Kutta-Jukowski theorem (see the results in [439] based on the work in [407] and [243, 339]). Thus, the displayed streamlines are somewhat exaggerated, but qualitatively correct in explaining the perceived effects.

For the counter-clockwise rotating case in Fig. 4.11, the streamline divergence ahead of the cylinder is more pronounced than that behind the cylinder for the clockwise rotating cylinder cases. This explains why the counter-rotating vortex will create a stronger effect ahead of the cylinder than for the location behind the clockwise rotating vortex. This is consistent with computational and experimental results reported in [331, 342] and [61]. In closing this section, we note that the displayed evidence is for an instability of wall-bounded shear layer at a streamwise location where theoretical investigation reveals that there would not be instability leading to formation of TS waves. For this reason, in [199] this is referred to as subcritical instability. There is a very good reason for the receptivity experiment to have been conducted for subcritical conditions of a zero pressure gradient boundary layer. For supercritical conditions, instability can be triggered by residual background disturbances; these background disturbances are difficult to measure and categorize. Also, at higher Reynolds numbers, the transfer function of the dynamical system is very receptive to trace magnitude of disturbances. In contrast, for subcritical conditions the transfer function

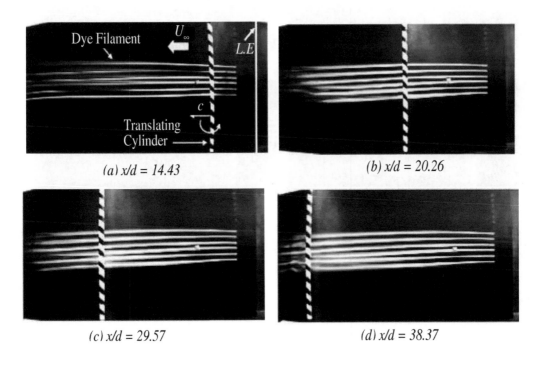

(a) x/d = 14.43

(b) x/d = 20.26

(c) x/d = 29.57

(d) x/d = 38.37

FIGURE 4.9
Visualized flow field for Case 3: $c = 0.386$, $H_1/\delta^* = 27.52$, $\Omega = 0$ showing negligible receptivity for this parameter combination.

is less receptive to small amplitude disturbances and one needs definitive stronger input to destabilize the shear layer. Thus, at subcritical conditions one has more control in studying the receptivity problem with the help of carefully designed experiments, such as this.

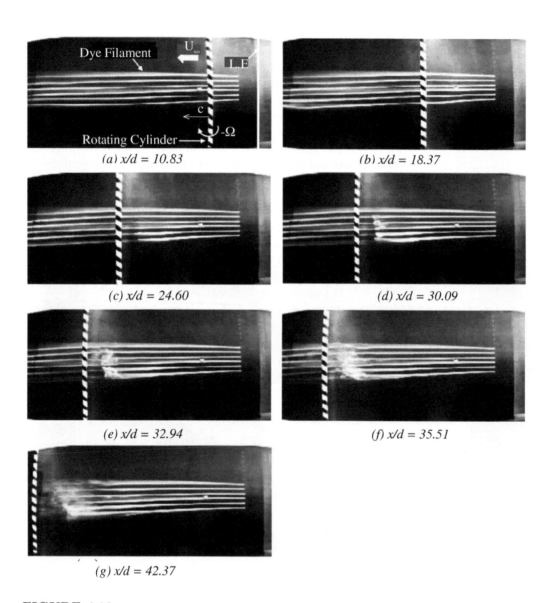

(a) x/d = 10.83

(b) x/d = 18.37

(c) x/d = 24.60

(d) x/d = 30.09

(e) x/d = 32.94

(f) x/d = 35.51

(g) x/d = 42.37

FIGURE 4.10
Visualized flow field for Case 5 with the vortex having opposite circulation for $c = 0.237$, $H_1/\delta^* = 27.52$, $\Omega = -5$. Note the weak disturbances behind the free stream vortex.

4.4 Computations of Vortex-Induced Instability as a Precursor to Bypass Transition

For the subcritical instability of the zero pressure gradient boundary layer shown in the experimental results of the previous section, the mean flow was given by the Blasius bound-

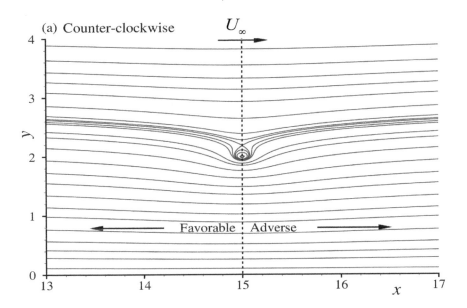

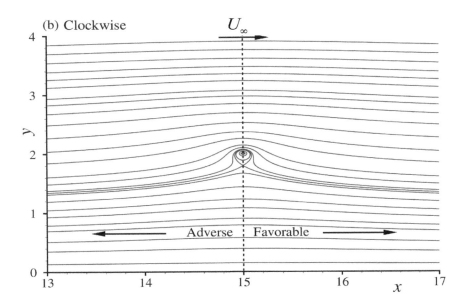

FIGURE 4.11

Sketches showing ideal streamline patterns created by a rotating and translating cylinder for both clockwise and counter-clockwise rotation. In the real situation for viscous flows, the streamlines are somewhat different near the cylinder surface, especially for supercritical rotation rates.

ary layer that was destabilized by a convecting captive vortex. In [61, 342, 331] the early stages of the bypass transition were computed by solving the full Navier–Stokes equation in two dimensions. We also note that a demonstration of a physical event in two dimensions is more general than that shown for three-dimensional flows. In recent years, some researchers have reported transient growth as an alternative to the primary instability of Fig. 3.1, with an implication that one is looking at a bypass route of transition. However, such growth is only relevant for three-dimensional flows and makes that a less generic a mechanism than what one would like to call a general bypass mechanism.

The two-dimensional Navier–Stokes equation is solved in the stream function vorticity formulation, as reported in Chap. 2, based on various reported work in [331, 334, 342]. Brinckman and Walker [44] the burst sequence of turbulent boundary layer excited by streamwise vortices (in the x-direction) was simulated using an identical formulation for which a stream function was defined in the (y, z)-plane only. Ideally, one should work with vector potential for three-dimensional flows, instead of a decoupled "stream function" in one plane alone. To resolve various small scale events inside the shear layer, the vorticity transport equation (VTE) and the stream function equation (SFE) are solved in the transformed (ξ, η)-plane given by

$$h_i h_2 \frac{\partial \omega}{\partial t} + h_2 u \frac{\partial \omega}{\partial \xi} + h_1 v \frac{\partial \omega}{\partial \eta} = \frac{1}{Re}\left[\frac{\partial}{\partial \xi}\left(\frac{h_2}{h_1}\frac{\partial \omega}{\partial \xi}\right) + \frac{\partial}{\partial \eta}\left(\frac{h_1}{h_2}\frac{\partial \omega}{\partial \eta}\right)\right] \tag{4.2}$$

and

$$\frac{\partial}{\partial \xi}\left[\frac{h_2}{h_1}\frac{\partial \psi}{\partial \xi}\right] + \frac{\partial}{\partial \eta}\left[\frac{h_1}{h_2}\frac{\partial \psi}{\partial \eta}\right] = -h_1 h_2\, \omega \tag{4.3}$$

where h_1 and h_2 are the scale factors of transformation defined by

$$h_1^2 = \left(\frac{\partial x}{\partial \xi}\right)^2 + \left(\frac{\partial y}{\partial \xi}\right)^2 \text{ and } h_2^2 = \left(\frac{\partial x}{\partial \eta}\right)^2 + \left(\frac{\partial y}{\partial \eta}\right)^2 \tag{4.4}$$

The above equations are solved in a domain $-1 \leq x \leq 25$ and $0 \leq y \leq 1.92$, with all lengths non-dimensionalized by the core size of the convecting vortex, which in this case is identified with the diameter of the convecting cylinder. A uniform grid spacing of $\Delta x = 0.04$ is used in the streamwise direction and a stretched grid is used in the wall-normal direction with the wall resolution given by $\Delta y = 0.000780$. This grid has been used along with the high accuracy compact schemes of [336, 348] for resolving the nonlinear convection terms. From visual experimental observation, one cannot obtain the strength of the captive convecting vortex and in the computation this has been treated as a parameter. For the results shown in Fig. 4.12, a value of $\Gamma = 9.1$ has been taken.

The Reynolds number based on the displacement thickness of the undisturbed flow at the outflow of the computational domain is 472, which is lower than 519.7 − the critical Reynolds number obtained for the Blasius boundary layer by spatial stability theory. Thus, the flow is fully subcritical in the computational domain.

At the inflow and at the top of the computational domain, one calculates the flow variables as induced by the free stream vortex via the Biot–Savart interaction rule. At the outflow, the fully developed condition is applied for the wall-normal component of the velocity ($\frac{\partial v}{\partial \xi} = 0$) that is used for the SFE. One obtains the vorticity boundary condition at the outflow from Eq.(4.3). In the top frame of Fig. 4.12, one sees incipient unsteady separation on the wall. In subsequent frames, one notices secondary and tertiary events induced by the rotating and translating cylinder in rapid succession, which is more than the maximum growth rate that is predicted by the exponential rate of primary linear spatial instability theory. In these computed cases, one does not notice TS waves and the vortices formed on the wall are essentially due to unsteady separation, which is initiated by the free stream convecting vortex. This ensemble of events has been noted as the vortex-induced instability or bypass transition in [331, 342, 356] and in [334].

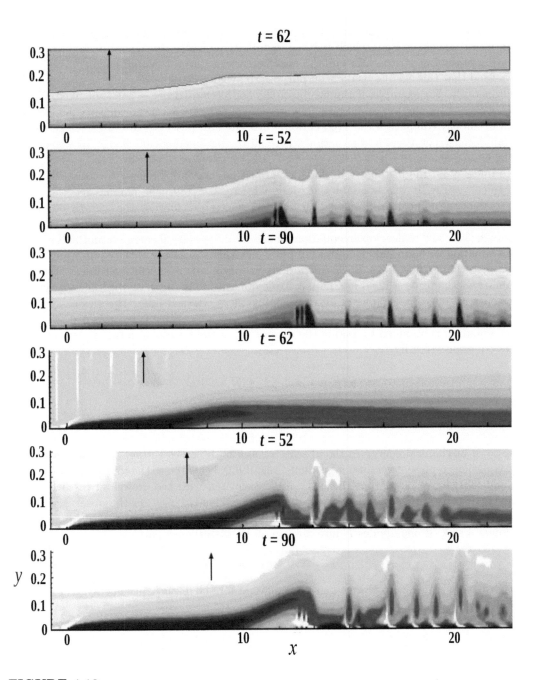

FIGURE 4.12
Stream function (top three panels) and vorticity contours plotted at the indicated times. The same contour values are plotted for each quantity. Arrowheads at the top of each frame indicate the instantaneous streamwise location of the free stream vortex.

4.5   Instability Mechanism in Vortex-Induced Instability

The mechanism at play during vortex-induced instability has been explained in [331, 342] and compiled here in explaining one of the bypass transition mechanisms. Experimental and computational results shown so far clearly reveal the existence of a receptivity mechanism inside the shear layer which induces instability as a consequence of a single vortex migrating in the free stream at a uniform speed convecting for a constant height. It is to be understood that this exercise is done on purpose to avoid variability of multiple parameters simultaneously. For example, if the height of the convecting vortex is not kept fixed, then the Biot-Savart interaction would provide different coupling between the vorticity fields and would needlessly complicate the explanation. Once the unit processes are understood, then one can deconstruct a real flow situation where all the parameters can change, as was done for the Case 3 shown in Fig. 4.9.

To understand the instability in flows, one must understand the growth of disturbance mechanical energy. It was Landahl and Mollo-Christensen [189] who emphasized the fact that in understanding transition and turbulence, one must consider the growth of total mechanical energy and not just simply the disturbance kinetic energy. This prompted the authors in [331, 342] to develop an equation for the total mechanical energy from the Navier–Stokes equation without making any assumption on either the nature of the equilibrium flow or the disturbance flow field. Such restrictions are imposed in all classical linear viscous instability mechanisms studied in Chap. 3. We emphasize that the developed new equation for receptivity must be capable of also explaining these classical linear theories as special cases. This has been demonstrated in [349, 350]. For incompressible flows, this disturbance energy equation is obtained by taking the divergence of the rotational form of the Navier–Stokes equation given by

$$\nabla^2 E = \vec{\omega} \bullet \vec{\omega} - \vec{V} \bullet (\nabla \times \vec{\omega}) \tag{4.5}$$

where

$$E = \frac{p}{\rho} + \frac{V \bullet V}{2}$$

is the total mechanical energy in the absence of any body forces acting. The solid dots in these equations represent vector dot products. Instability is related to the rotationality of the flow and [331] have shown that the instability associated with the growth of disturbance energy is driven by the right-hand side of Eq. (4.6). This is based on the observation in [381] that a negative right-hand side indicates a source of energy while a positive quantity represents a sink. If one divides E into a mean and a disturbance part via

$$E = E_m + \varepsilon E_d$$

and substitutes it in Eq. (4.5), one gets the equation for the disturbance energy given by

$$\nabla^2 E_d = 2\vec{\omega}_m \bullet \vec{\omega}_d + \vec{\omega}_d \bullet \vec{\omega}_d - \vec{V}_m \bullet (\nabla \times \vec{\omega}_d) - \vec{V}_d \bullet (\nabla \times \vec{\omega}_m) - \vec{V}_d \bullet (\nabla \times \vec{\omega}_d) \tag{4.6}$$

This equation can be used to describe the onset of instability when a suitable mean flow is defined. We note that this equation is very generic for all incompressible flows (steady or unsteady flows), as it is based on the full Navier–Stokes equation without making any assumptions. In [350], this equation has been used to explain the classical linear instability theory for parallel flows showing exactly identical TS waves obtained from the Orr–Sommerfeld equation. In Sec. 4.3, this is fully explained with the development of the actual equations and results. For the computational data, a mean flow was taken at $t = 20$

as representative undisturbed flow and the right-hand side of Eq. (4.6) was calculated and plotted, as shown in Fig. 4.13, at some representative times.

In the figure, regions with darker shades of blue indicate the presence of disturbance

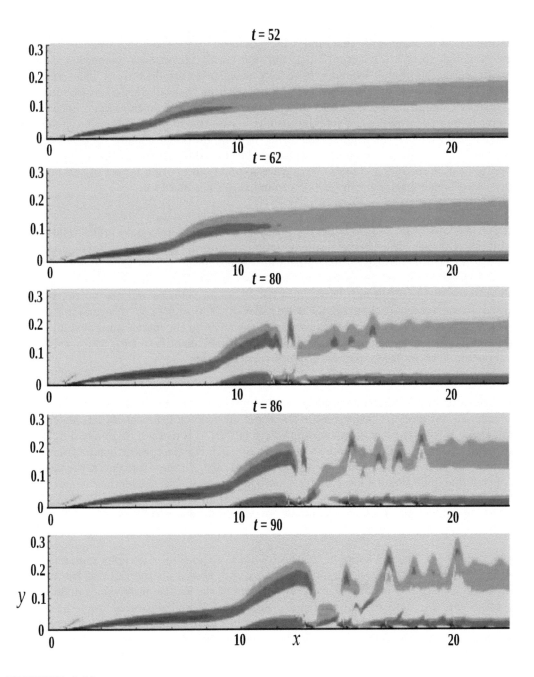

FIGURE 4.13
Contours of the right-hand side of the disturbance energy Eq. (4.5). The negative contours are shown by a darker blue color in the figure, indicating the energy source for disturbance quantities.

energy sources. At early times, one notes two sites from which disturbances originate. These are from the leading edge of the plate and from a location more than $5d$ from the leading edge of the plate. It is seen that the disturbances from the leading edge are weaker and do not enter within the shear layer convecting downstream along the edge of the shear layer. Thus, the inability of the disturbances to entrain within the shear layer do not make these disturbances responsible for the onset of subcritical instability. The disturbance from the other site grows stronger with time and a little before $t = 80$, these disturbances become stronger enough to create an eruption in the wall-normal direction — somewhat similar to events during the burst sequence in a fully developed turbulent flow. This eruption connects these disturbances via formation of the spike from the wall. In the stream function and vorticity contour plots of Fig. 4.12 the spikes are also evident, in the form of secondary and tertiary bubble formations. Thus, the ensuing bypass transition would not be simulated properly if one excludes the leading edge from the computational domain. Otherwise, one would compute an unimpeded spike stage from a location downstream of the leading edge, as reported earlier in [267] and [256]. We must also point out another aspect of this present energy-based instability theory developed around Eqs. (4.5) and (4.6).

It is to be noted that the corresponding homogeneous equation would indicate intrinsic instabilities. However, the homogeneous equation is governed by the Laplacian operator which has no known intrinsic primary instability mechanism at this time except for the side-band instability noted for surface gravity waves in [28]. Thus, in the present case, the noted primary disturbances are driven by interactions of velocity and the vorticity field acting as source terms on the right-hand side of Eqs. (4.5) and (4.6). The major issue is about how the energy is initially exchanged from the mean to the disturbance field and this is clearly brought out by the first term on the right-hand side of Eq. (4.6), which indicates an interaction between mean and disturbance vorticity fields.

During later stages of bypass transition caused by a convecting vortex in the free stream, large coherent vortices are formed inside the shear layer. Such coherent structures can be characterized by proper orthogonal decomposition (POD) — a method originally developed in [181] and later used in fluid dynamics in [371] and [151]. For the present bypass transition problem, POD analysis results have been shown in [331, 356]. In these references, the method of snapshot due to [371] and a local analysis method were used. For the method of snapshot, 21 frames have been used in a non-dimensional time-span of 10 to perform POD. In Fig. 4.14, the eigenvalues and leading eigenmodes are shown for the vorticity data.

The ordinate in Fig. 4.14 represents the fraction of the total enstrophy contained by a specific number of leading eigenmodes and these are shown on the right of the figure. The fractional enstrophy content is given by the sum of the leading eigenvalues divided by their total sum. One notices that the largest eigenvalue is well separated from the other eigenvalues. Also, up to around $t = 50$, five eigenmodes capture 99% of the total disturbance enstrophy. This number increases to 14 during $t = 80$ to 90. Thus, POD analysis provides a very good basis for reduced order modeling (ROM) of fluid dynamical systems during bypass transition, as discussed in [356].

The prototypical vortex-induced instability described in this chapter so far also makes its appearance in other flows. In the following, we describe a special flow that depends on this mechanism to trigger instability that otherwise would have remained unexplained.

4.6   Instability at the Attachment Line of Swept Wings

The leading edge of a sweptback wing in contact with the fuselage is noted to experience abrupt flow transition due to the convection of continuous turbulent puffs along the attachment line plane whenever the Reynolds number based on momentum thickness Re_θ is greater than 100 or $Re_{\delta*} \geq 245$, as noted in [276] and [11]. Various investigations based on linear and weakly nonlinear theories have shown the lowest critical Reynolds number as

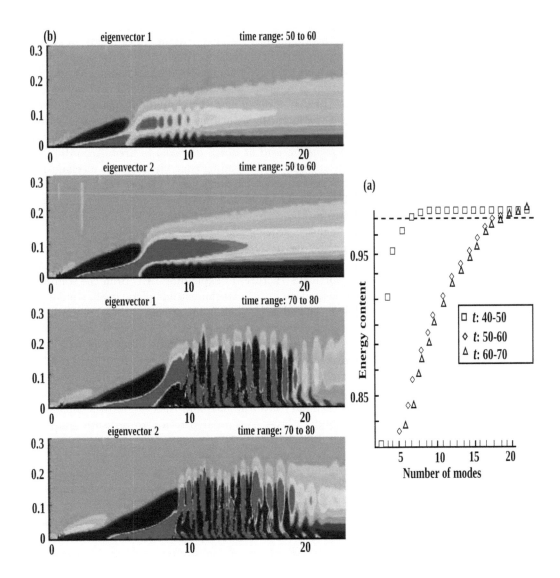

FIGURE 4.14
Sum of a specific number of eigenvalues divided by their total sum, indicating disturbance enstrophy content. The dotted line indicates the 99% level. (b) The first two eigenvectors of disturbance vorticity during the indicated time ranges are shown.

$Re_{\delta^*} = 535$, indicating the actual transition to be a subcritical phenomenon. This is pointed out in [334], who put forward the viewpoint that flow rapidly turning to the turbulent stage right at the leading edge is due to vortex-induced instability described in the previous sections of this chapter. Otherwise, it is well known that the attachment line boundary layer is very stable for the streamwise direction and the critical Reynolds number is orders of magnitude larger as compared to a zero pressure gradient boundary layer. There were two major issues that remained unanswered and were pointed out in [405]. The first is related to the subcritical instability and the second issue is that of relating the instability at the attachment line to the events downstream in the chordwise direction. The first issue was tackled in [356] and in [334] and the second issue follows from the first as self-explanatory, if the flow becomes turbulent at the leading edge itself, as stated in these references.

It has been demonstrated that attachment line boundary layer supports instability waves, the kind predicted in linear theory (shown by very careful experiments in [272], [276], [10], [138], [279]). However, they do not cause transition occurring at the attachment line, as no linear or nonlinear theories have explained transition at the attachment line. In the literature, this premature transition is referred to as the leading edge contamination (LEC) problem. In [356] and in [334], LEC was posed as a vortex-induced instability problem, where the free stream convecting vortex in the attachment line plane is created at the wing-body junction, as has been suggested in [11] that *the leading edge is contaminated by large turbulent structures coming from the wall at which the wing is fixed.* According to Arnal [11], the flow undergoes transition right at the leading edge via a bypass mechanism. Gaster [110] specifically noted that vortices associated with junction flow are fed into the attachment line boundary layer. The mean flow field and coordinate systems for flow past a sweptback wing schematically shown in Fig. 4.15. On the attachment line plane itself, the flow is essentially two dimensional, as noted in [283] and [71].

In all early experiments, including the one in [276], the existence of attachment line vortical structures is well established. It is thus natural to investigate the subcritical instability by looking at the role of convecting vortical structures in explaining LEC from the solution of the two-dimensional Navier–Stokes equation in the attachment line plane itself, similar to the vortex-induced instability problem studied in [199] and [331], for zero pressure gradient flow.

The Navier–Stokes equation has been solved in [334], using a 501×101 grid with 501 points in the streamwise direction distributed uniformly. In Fig. 4.16, computational results are shown for the case of a counter-clockwise circulating vortex convecting at a speed of $0.2U_\infty$ at a height of $30\delta^*$ of the attachment line boundary layer. This height is more than that considered for the Blasius boundary layer in [331] and thus establishes enhanced receptivity of the attachment line boundary layer as compared to a zero pressure gradient boundary layer.

In the figure, the first frame corresponds to an early time when the convecting vortex is to the left and out of the computational domain and its influence is seen as an upwelling of the shear layer near the inflow of the domain. By $t = 748$, the convecting vortex appears over the computational domain and a very intense sequence of instabilities is seen over the attachment line. At $t = 2001$, the convecting vortex is located around $x = 460$ and the leading coherent structures are out of the domain. As explained before, the shear layer aft of the convecting vortex is always stabilized due to the impressed favorable pressure gradient to the extent that the shear layer thins down, and this is clearly evident in the frames at $t = 1500$ and 2001. It has been shown in [334] that for a particular case with $\Gamma/v = 211.063$ and the non-dimensional core size of the vortex (with respect to δ^* of the shear layer at inflow) as 6, the first bubble occurs at a location where the Reynolds number based on displacement thickness was $Re_{\delta^*} = 190$. If the appearance of the first bubble is taken as an indication of flow criticality, then this is a case of subcritical instability.

Once the first bubble forms on the wall, a succession of others follows due to a local strong adverse pressure gradient being created ahead of the primary bubble. Also, this cascading phenomenon occurs at a very high speed and the flow contaminates very rapidly. As the attachment line shear layer is more stable with respect to the linear mechanism, it is of more interest to investigate this bypass mechanism for LEC.

It is interesting to note that the instability is only feasible when the speed of the convecting vortex is far below the free stream speed — as was explained in the last chapter via Eqs. (3.193) to (3.196), following the work reported in [327]. However, when we consider an aperiodic vortex convecting in the free stream, then the mechanism of instability is completely different, as there are no imposed time scales. In Sec. 4.3, we have seen the visual signature of such events that we termed bypass transition. In this mode also, a lower

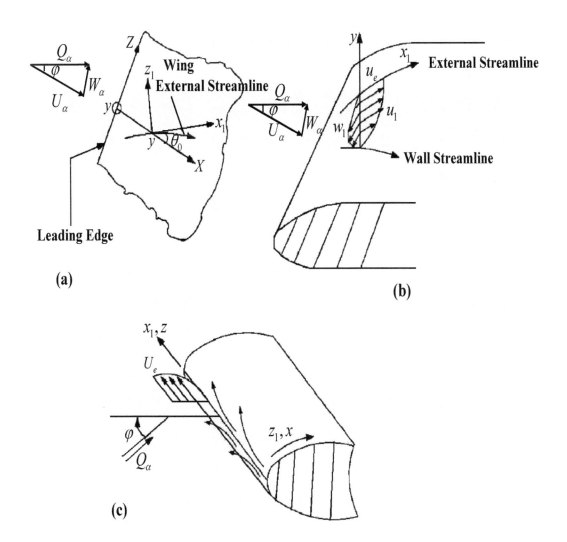

FIGURE 4.15
Flow profiles and schematic of coordinate systems for flow past a sweptback wing. (a) Notation and coordinate system; (b) streamwise and cross flow mean velocity profiles and (c) attachment line flow.

speed of convection has higher receptivity, as explained with the help of Eq. (4.1) and the associated discussion with the help of results shown in Fig. 4.8. For the same reason, the present computations of LEC show strong bypass transition as compared to that shown in [259], where the computations displayed lower (exponential) growth rates for the introduced vortex moving at free stream speed. In [334], this bypass mechanism was explained with respect to the disturbance energy equation, Eq. (4.6). Furthermore, it is worth quoting the following from this reference in explaining this bypass mechanism: *The disturbance energy equation arises by taking the divergence of the Navier–Stokes equation in the rotational form, representing the irrotational component of the disturbance field. The rotational field of the Navier–Stokes equation as governed by the vorticity transport equation yields the Orr–Sommerfeld equation obtained by linearization and making parallel flow approximation. Although [235] suggested that unsteadiness during bypass transition is due to the shear noise term in the Poisson equation for the static pressure.* What is more important to realize is that the Poisson equation for disturbance energy actually shows coupling between the rotational part of the flow field (through velocity and vorticity terms) with the irrotational part

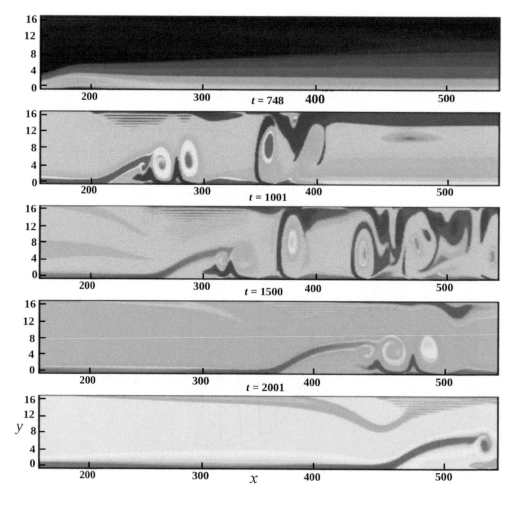

FIGURE 4.16
Vorticity contours for the LEC by a counter-clockwise rotating vortex.

(given by the Laplacian on the left-hand side). Furthermore, this mechanism is equally valid for both two-dimensional and three-dimensional flow fields. Thus, significant unsteadiness with large spectral bandwidth disturbances can be created without the vortex stretching mechanism to explain the excited small scales in transitional and turbulent flows.

The effects of a clockwise convecting vortex in the free stream are studied for this case too in [356] and [334]. A typical set of results is shown in the vorticity contour plots in Fig. 4.17. As noted in [199] and [331], the role of a negative vortex is to create an adverse pressure gradient behind it and hence instability onset is noted upstream of the convecting vortex. It has been noted and explained in [342] and also here that there prevails a favorable pressure gradient region ahead of the clockwise convecting vortex. Thus, the cascade of secondary and higher instabilities seen for the counter-clockwise vortex ahead of it is completely absent here. First and foremost, the absence of an adverse pressure gradient does not promote disturbance growth at all. Even if there is some disturbance present, an induced favorable pressure gradient attenuates it strongly. This feature of the flow is clearly visible in Fig. 4.17, which shows the existence of virtually a single coherent vortex.

We have already commented about the DNS of attachment line instability study performed in [259] and explained the reason for their failure to explain LEC due to the wrong choice of convection speed of the vortex. In the following, we also make some general remarks about various other such efforts towards DNS. When the full Navier–Stokes equation is solved to reproduce experimentally observed LEC effects, there also appears to be no consensus among reported results. Spalart's [382] three-dimensional DNS could not reproduce even the nonlinear equilibrium solution reported in [137], but it did produce the correct experimental transitional Reynolds number in [276]! Spalart [382] used white noise to trigger instability for spatial DNS, where spanwise periodicity and a buffer domain in the chordwise direction were used additionally. Two-dimensional DNS results, however, produced conflicting results, with [404] predicting the wrong frequency of disturbance as compared to the experimental value of [279]. Joslin [165], in reporting DNS results from a formulation that does not make spanwise periodicity assumption, showed the existence of the subcritical two-dimensional equilibrium solution in [137]. Subsequently, Joslin [166] postulated that interactions of multiple three-dimensional modes lead to observed computational bypass transition. Thus, it appears that these simulations suffered due to multiple reasons, the primary ones being (i) inappropriate modeling of the problem through boundary conditions, see, e.g., the failure of [382] to capture even the mean flow; (ii) inappropriate excitation field, as seems to be the case in all these simulations except in [259] and (iii) inappropriate convection speed of the free stream vortex, as was the case in [259]. Finally, it is also required that the adopted numerical methods must be dispersion relation preserving (DRP), which most of these methods are not for the chosen numerical methods and the computational parameters, barring the spectral methods.

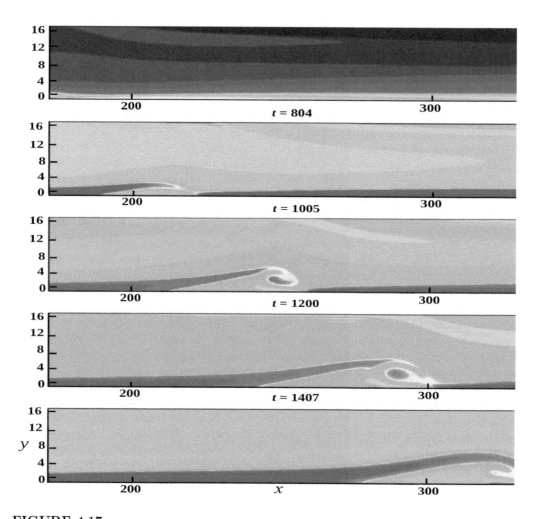

FIGURE 4.17
Vorticity contours for the case of a convecting negative vortex with strength given by $|\Gamma|/v =$ 126.638 and the other parameters are the same as in the previous case. Details of this are available in [334] and [356].

5

Spatio-Temporal Wave Front and Transition

5.1 Introduction

In Chap. 3, we familiarized ourselves with different aspects of spatial instability theory. We specifically noted that flow instability depends intimately upon the way the flow is excited. A distinction was made between wall and free stream modes for the solution of the Orr–Sommerfeld equation in Secs. 3.6.5 and 3.7.1. Unfortunately, classical stability theory approaches are incapable of making the connection between instability and receptivity of flows. To circumvent this, in many recent studies, fluid flow transition from the laminar to the turbulent state has been viewed as a consequence of phase transitions and/or bifurcation(s) of a dynamical system [353, 360]. The response of a dynamical system is a convolution of transfer function with the input to the system, in the spectral plane. In stability theories, one is only interested in interpreting system dynamics in terms of the transfer function of the dynamical system. Furthermore, focus on instability implicitly considers only small perturbations. This aids in linearizing the system dynamics and one can superpose solutions. This is the basis of normal mode analysis, and flow instability is often interpreted in terms of the least stable mode only. However, this approach does not produce information about the response field in the neighborhood of the exciter. This has been rectified in receptivity approaches in Sec. 3.6, where one correlates the input with the transfer function for disturbances which evolve in space. The role of disturbance amplitude was also earlier highlighted in the pioneering pipe flow experiments of [297].

Transition from laminar to turbulent flow can take many routes, depending on physical parameters relating the transfer function with the input spectrum. Initially, it was thought that viscous actions are not central in predicting flow instability and this was the basis of the inviscid instability approaches pioneered by Kelvin, Rayleigh and Helmholtz — as recounted in Sec. 3.2. When flows were found to be unstable due to inflectional velocity profile, then the Rayleigh–Fjørtoft theorem produced a necessary condition for temporal growth. For shear driven external flow problems, it is found that small disturbances grow in space and not in time. This prompted Orr and Sommerfeld to propose viscous instability theories with the Orr–Sommerfeld equation, which was analytically solved by asymptotic methods in [143, 310, 408]. In this formalism, disturbance quantities are expressed as

$$q(x, y, t) = \int \hat{q}(y; \beta_0, Re) \, e^{i(kx - \beta_0 t)} \, dk$$

with β_0 as the fixed frequency of the imposed disturbance. The complex wavenumber $k = k_r + i k_{imag}$ is sought as an eigenvalue in spatial stability analyses. Instability analysis uses a local parallel flow model to predict TS waves for flow over a flat plate with a zero pressure gradient (ZPG).

The prevalent idea that viscous actions damp disturbances was the main obstacle in accepting predicted TS waves from the solution of the Orr–Sommerfeld equation. These waves were also not detected experimentally, until the results of the famous experiment in [316] were announced much later. Earlier experiments suffered from experimental device noise and due to lack of appropriate design of experiments which follow the theoretically predicted structure of the disturbance field. This last aspect relates to the eigenvalue problem which describes disturbances as waves, excited by a time-harmonic localized source at the wall. In this formulation, there is also an implicit view that the disturbance created at the wall decays with height. In noisy experiments, distributed disturbance sources in the flow field are not monochromatic, and decay of disturbance with height is irrelevant, as the flow is excited at the free stream. Even when these were accounted for by placing a vibrating diaphragm at the wall in [401], it was not successful because the frequency of vibration was too low, as compared to that used in theoretical instability prediction.

Instead of studying stability using mass and momentum conservation, alternate approaches based on energy considerations had been initiated early, leading to the well known Reynolds–Orr equation given in [260]. This has been further explained in the literature in [200, 314, 396]. This equation deals with evolution of disturbances in terms of kinetic energy. If one writes the Navier–Stokes equation in the indicial notation for the disturbance quantities (in lower case) and takes a dot product of it with the velocity vector, one gets the following equation

$$
\begin{aligned}
u_i \frac{\partial u_i}{\partial t} = &- u_i u_j \frac{\partial U_i}{\partial x_j} - \frac{1}{Re} \frac{\partial u_i}{\partial x_j} \frac{\partial u_i}{\partial x_j} \\
&+ \frac{\partial}{\partial x_j} \left[-\frac{u_i u_i}{2} U_j - \frac{u_i u_i u_j}{2} - u_i p \delta_{ij} + \frac{u_i}{Re} \frac{\partial u_i}{\partial x_j} \right]
\end{aligned}
\tag{5.1}
$$

Here, the upper case quantities represent the mean flow. If one defines the disturbance kinetic energy of the full domain as

$$
E_v = 1/2 \int_v u_i u_i \, dV
$$

then the above can be integrated over the whole domain to give rise to the Reynolds–Orr equation as

$$
\frac{dE_v}{dt} = - \int_v u_i u_j \frac{\partial U_i}{\partial x_j} dV - \frac{1}{Re} \int_v \frac{\partial u_i}{\partial x_j} \frac{\partial u_i}{\partial x_j} \, dV
\tag{5.2}
$$

However, this equation is derived subject to the assumption that the disturbance field is localized and/or spatially periodic. This assumption removes any contribution coming from nonlinear convection terms. The authors in [200, 396] pointed out that various estimates of critical Reynolds number obtained by this approach are erroneously too low due to the above mentioned assumption that leads to the elimination of nonlinear terms appearing in the divergence form of the equations of motion. In contrast, the disturbance mechanical energy concept of Eqs. (5.1) and (5.2) retain the nonlinear term contributions, which are central to any instability studies. This energy-based receptivity analysis is all-inclusive, as it is based on the full Navier–Stokes equation without any assumption.

Traditional eigenvalue analyses seek a disturbance field to grow either in space or in time. We emphasize that this approach is strictly for ease of analysis and there are no general proofs which inform us which growth rate to investigate. In [153], the authors have used a complex dispersion relation, termed spatio-temporal theory (developed originally to study plasma instabilities in [29]), to a family of mixing layers with the goal to determine a criterion for flows to be analyzed by either a spatial or a temporal theory by inspecting the complex wave number-frequency plane. One should note that this does not treat the growth in the spatio-temporal framework, but simply tells one which of the two theories may be appropriate for its study. In contrast, the spatio-temporal approach based on the Bromwich contour integral method of Chap. 3 enables one to perform an analysis without any presumption of adopting either a spatial or temporal approach. In fact, there must be many cases of instability where both spatial and temporal growths are present simultaneously.

Classical approaches to instability studies require an equilibrium state, whose stability is studied by eigenvalue analysis by linearizing the governing equations. Results obtained by this approach match controlled laboratory experiments for thermal and centrifugal instabilities. But instabilities dictated by shear force do not match so well, e.g., (i) Couette and pipe flows are found to be linearly stable for all Reynolds numbers, while the former was found to suffer transition in a computational exercise at $Re = 350$ [208] and the latter found to be unstable experimentally at $Re \geq 1950$ [297], with the exact value dependent upon facilities and background noise level; (ii) plane Poiseuille flow has $Re_{cr} = 5772$, while in the laboratory experiment transition was shown to occur even at $Re = 1000$ [80]. Interestingly, according to [411], the other example for which *eigenvalue analysis fails include to a lesser degree, Blasius boundary layer flow*. This is the flow which many cite as the success story of linear stability theory.

In this context, one should evaluate the contribution made by the receptivity analysis in [319, 322]. In [16] and references contained therein, various researchers have attempted to provide a mathematical framework of receptivity analysis to overcome the above shortcoming of instability theory to reconcile with real flows. The first set of results for linear receptivity of the ZPG boundary layer to wall excitation was reported in [319]. An assumption was made that the response is at the frequency of excitation — as in spatial stability analysis. Results obtained using parallel flow approximation identified a local solution in the immediate neighborhood and the TS wave as an asymptotic part of the response field. This is the impulse response of the Blasius boundary layer reported for the first time in [319] by using this signal problem assumption.

In [322], this barrier was removed by treating receptivity as a spatio-temporal problem, while retaining the parallel flow assumption, to demonstrate creation of a spatially unstable TS wave by time harmonic wall excitation. These linearized receptivity analyses are based on the Bromwich contour integral method. Despite the absence of nonlinear and nonparallel effects in these studies, some new advances were reported. First, the local solution in the immediate neighborhood of the exciter is clearly brought out by receptivity analysis, which was shown to be due to the essential singularity of the solution of the Orr–Sommerfeld equation, which does not satisfy Jordan's lemma, as explained in Sec. 3.6.2. Second, even with the signal problem assumption, the response field is different from results of normal mode eigenvalue analysis, for which response is assumed to be by the least stable mode. In receptivity analysis, effects of all modes present are obtained simultaneously in the response field. More importantly, the effort in [322] was special, as the artificial division of spatial and temporal theories was removed by performing a spatio-temporal analysis. Results in [322] confirmed that the unstable disturbances in the Blasius boundary layer can be equivalently obtained either by spatial instability theory or the Bromwich contour integral method. While the latter approach provides the complete disturbance field, instability theory provides the asymptotic part of the least stable mode only.

A further advance of linear spatio-temporal receptivity analysis by the Bromwich contour integral method was made in [349, 350] with the finding that the spatially stable Blasius boundary layer for a fixed frequency displays a spatio-temporal growing wave front. A spatially stable system displaying a spatio-temporally growing wave front was not known before, although some attempts have been recorded in detecting transient growth in channel flow (see, e.g., [314]). The detected frontrunner in the Blasius boundary layer was novel, but forerunners have been predicted in [43] in the context of electromagnetic waves. The propagation speed of the forerunner has also been variously termed group velocity (by

Rayleigh), signal velocity (by Sommerfeld) and velocity of energy transfer (Brillouin). For unstable systems all these definitions coincide, as shown in [349]. For stable systems, signal and energy propagation speed coincide, which is in general not the same as the group velocity of the leading eigenmode. The difference of speed between the signal speed with group velocity of eigenmodes creates a detached wave front from an asymptotic TS wave. It is also noted that the spatio-temporal front has a smaller wavenumber as compared to the asymptotic TS wave packet. Without the spatio-temporal approach, it was not possible to predict the existence of a spatio-temporal front, as its existence is inherent in the system dynamics. This is where spatial and temporal theories failed.

Apart from these receptivity approaches, other efforts have been reported in trying to understand the process of transition, as distinct from flow instability. We have already noted one such example in the previous chapter, where vortex-induced instability took a subcritical flow from an equilibrium state to the evolved turbulent-like state, as long as the source of the disturbance persisted. As the source of disturbances in the free stream convected away, the underlying shear layer reverted back to the original equilibrium flow state. In addition to the above examples, it is also noted that three-dimensional disturbances like a surface roughness element or equivalent localized disturbances can lead directly to turbulence without any observed TS wave growth. The authors in [40] mention that *these kinds of transitions are collectively known as 'bypass mechanisms'*. However, in Chap. 4, we have also noted that in [159], considered bypass transition as stochastic and not deterministic, even in the presence of large amplitude disturbances. Instead of classifying these as bypass transition, as discussed in Chap. 4, such instabilities have also been attempted to be described in terms of transient growth of kinetic energy. This kinetic energy approach is distinctly different from the developed theory in [331, 342] of disturbance mechanical energy growth described in Sec. 4.5. It is to be noted that such a transient kinetic energy growth scenario was sought to be explained in the literature only for the three-dimensional disturbance field, as the corresponding analysis for two-dimensional flows showed a meager growth rate, as described in [411].

Landahl [187] showed that, subject to certain general constraints, three-dimensional disturbances grow algebraically in a mean shear via an inviscid mechanism. This instability is not related to the Orr–Sommerfeld equation solely and is related to tilting of vorticity by mean shear. However, in [38], it was revealed that the algebraic instability is due to a coupling between the instability modes obtained by the solution of the Orr–Sommerfeld and Squire equations. The latter equation is nothing but the linearized vorticity transport equation for the vertical component of vorticity. Thus, this pertains once again to a three-dimensional disturbance field. The above coupling inclines the shear layer, which intensifies with time. Furthermore, the authors in [39] investigated corresponding nonlinear growth and found a secondary instability that leads to direct breakdown to turbulence. However, in a DNS, the authors in [145] have reported bypass transition occurring through a different mechanism.

The focus of a mechanism by algebraic growth of transient energy is two-fold, as conjectured in [40]. First, transient energy growth can occur for linearly stable systems by two to three orders of magnitude at which *nonlinear effects might push the disturbance into turbulence* directly. This has been termed the subcritical route of bypass transition [40]. Second, such transient energy growth in a supercritical flow can boost relatively slow TS waves, with the former providing the initial algebraic growth that will be followed by the exponential growth of the linear mechanism of TS waves. However, none of these mechanisms has been experimentally verified in a controlled and unambiguous manner.

There is an alternate viewpoint that treats the transient growth as due to non-normal modes. It has been shown in [314] that regular and adjoint solutions of the Orr–Sommerfeld equation are orthogonal, while Orr–Sommerfeld eigenfunctions are not orthogonal to each other. This leads to the sensitivity of the eigenvalues to perturbations to the underlying stability operator, which was shown for a channel flow. It has been pointed out in [314] that the *non-normality of the Orr–Sommerfeld operator not only has consequences for the sensitivity, of the spectrum, but it also influences the dynamics of disturbances governed by the linearized Navier–Stokes equations.* Note in Fig. 4.6 of [314] transient growth of the Blasius boundary layer for $Re = 1000$ was shown to be thousandfold for three-dimensional disturbance fields. However, for two-dimensional disturbance fields, corresponding maximum transient growth rates are very small, as clearly stated in [411] as the *essential feature of this non modal amplification is that it applies to three-dimensional perturbation of the laminar flow field ... When only two-dimensional perturbations are considered, some amplification can still occur, but it is far weaker.* In general, this should have been just the opposite. Apart from these, there are other asymptotic stability studies involving algebraic or transient growth of three-dimensional disturbance fields, as in [449].

5.2   Transient Energy Growth

The above discussion reveals that the linearized Navier–Stokes equation supports transient energy growth for three-dimensional disturbance fields. However, a different route of spatio-temporal growth via linearized receptivity analysis has been reported in [349, 350] for the Blasius boundary layer using the Bromwich contour integral method [322]. Additional results have been also reported in [348]. In these papers, a curious property of the shear layer has been reported. Here, one performs the Bromwich contour integrals simultaneously in wavenumber and circular frequency planes, rather than performing integrals only along the Bromwich contour in the wavenumber plane for spatial stability analysis, which is the practice for wall-bounded unseparated flows, posed as a signal problem.

In performing spatio-temporal analysis, it is shown that for spatially stable systems with multiple modes, one notices spatio-temporally growing wave packets. The existence of a spatio-temporal wave front raises a question on the validity of normal mode eigenvalue analysis following either spatial or temporal theories. Creation of the growing spatio-temporal wave front indicates that it is natural to adopt receptivity analysis using the Bromwich contour integral method described in [322, 348] as a spatio-temporal problem. The moot point to note about this method and associated results is that one does not require any restrictions on the dimensionality of the problem in explaining large spatio-temporal growth. In [349, 350], the authors explained transient energy growth from a mechanical energy perspective which has been developed in explaining bypass transition, as explained in the previous chapter. Furthermore, the equivalence of viscous instability theory based on the Orr–Sommerfeld equation (when TS waves are generated) with an energy-based receptivity approach was shown in these papers. Due to the unique and important perspective that the spatio-temporal approach brings in explaining viscous instability and wave fronts for spatially stable systems, the Bromwich contour integral method is discussed in the next section.

5.3   Bromwich Contour Integral Method and Energy-Based Receptivity Analysis

An explanation of this mechanism is presented here from the solution obtained by the Bromwich contour integral method used in [322, 348] for two-dimensional disturbances in two-dimensional mean flows, with the disturbance stream function given by

$$\psi(x,y,t) = \frac{1}{(2\pi)^2} \int \int_{Br} \phi(y,\alpha,\beta) \, e^{i(\alpha x - \beta t)} d\alpha \, d\beta \qquad (5.3)$$

The Blasius boundary layer problem was solved in these references for a parallel mean flow for $Re = 1000$, excited at the wall. Here, Re is defined in terms of the displacement thickness of the boundary layer at the location of the exciter. In terms of wall modes, ϕ_1 and ϕ_3, the disturbance stream function can also be written as

$$\psi(x,y,t) = \frac{1}{(2\pi)^2} \int \int_{Br} \frac{\phi_1(\alpha,y,\beta)\phi'_{30} - \phi'_{10}\phi_3(\alpha,y,\beta)}{\phi_{10}\phi'_{30} - \phi_{30}\phi'_{10}} BC_w \, e^{i(\alpha x - \beta t)} d\alpha \, d\beta \qquad (5.4)$$

where BC_w was defined in Eq. (3.168) for the harmonic excitation at the wall, as shown in Fig. 5.1. The excitation field is defined by the condition given at $y = 0$ by

$$u = 0$$

$$\psi(x,0,t) = U_1(t) \, \delta(x) \, e^{-i\beta_0 t}$$

Here $U_1(t)$ is the Heaviside function, which represents the finite start-up of the exciter placed at the origin of the co-ordinate system.

The governing equation for the Fourier–Laplace transform is given by the following Orr–Sommerfeld equation

$$\phi^{iv} - 2\alpha^2 \phi'' + \alpha^4 \phi = iRe\{(\alpha U - \beta)[\phi'' - \alpha^2 \phi] - \alpha U'' \phi\} \qquad (5.5)$$

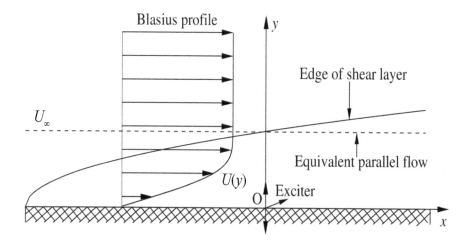

FIGURE 5.1
Harmonic excitation of a parallel boundary layer corresponding to the location of the exciter.

5.4 Spatio-Temporal Wave Front Obtained by the Bromwich Contour Integral Method

To understand spatio-temporal growth, a few cases were considered in [349] for parameter conditions corresponding to that marked as A, B, C and D in Fig. 5.2, shown with respect to the neutral curve in the (Re, β_0)-plane for the leading eigenmode.

The Bromwich contour for point A was chosen in the α-plane on a line extending from -20 to $+20$, which is below and parallel to the α_{real} axis, at a distance of 0.009 and in the β-plane it extended from -1 to $+1$, above and parallel to the β_{real} axis, at a distance of 0.02. For other points, the Bromwich contour in the α-plane was located at a distance of 0.001 below the α_{real} axis. The choice of the Bromwich contour in the α-plane is such that all the eigenvalues corresponding to downstream propagating disturbances lie above it. The Orr–Sommerfeld equation was solved along these contours with 8192 equidistant points in the α-plane and 512 points in the β-plane. The Orr–Sommerfeld equation was solved taking equidistant 2400 points across the shear layer in the range $0 \leq y \leq 6.97$. Spatial stability analysis produced waves for the four points of Fig. 5.2 with properties shown in Table 5.1.

For the point A, receptivity analysis produced streamwise perturbation velocity (u) which is shown in the bottom frame of Fig. 5.3 at $t = 801.1$. In this figure, the top two frames show solutions for the case of point B at the indicated times. The result obtained for point A by a spatio-temporal approach is indistinguishable for the growing asymptotic solution obtained by treating this as a signal problem, barring one major point. For the signal problem, the unstable wave continues to grow in space, while in the spatio-temporal approach the streamwise extent is a function of time and the response is more appropriately termed a wave packet. Hence, in the receptivity framework, one does not talk of TS waves. Instead, it is appropriately termed the TS wave packet which lengthens at early time instants and after some time the length of the packet saturates, as determined by the entry and exit

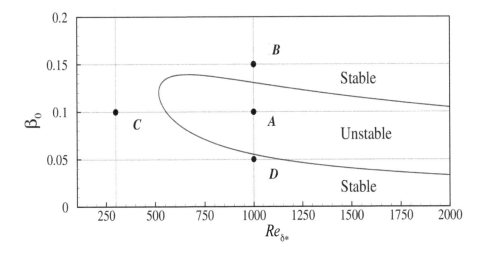

FIGURE 5.2
Neutral curve for the Blasius boundary layer identifying stable and unstable regions, with representative marked points.

points of the constant frequency line intercepting the neutral curve. Comparisons of results by the signal and spatio-temporal approaches were made in [322], as shown in Fig. 3.23.

Both types of receptivity analyses (signal problem and full time dependent problem) provide the local solution in the neighborhood of the exciter, but a forerunner preceding the asymptotic solution is distinctly visible for the full time dependent analysis of the spatially stable case. For a spatially unstable system one cannot see a clear demarcation line between the asymptotic solution from the forerunner for the indicated solution at $t = 801.1$, with one merging smoothly into the other. This terminal time was dictated by the number of points taken in the β-plane. If one would take more points, then one would be able to compute for a longer duration and the wave front will detach from the asymptotic part of the solution.

For the point A, the receptivity solution is dominated by the leading eigenmode, without any effects coming from the second and third modes of Table 5.1. In contrast, for point B the asymptotic solution is due to the first mode of Table 5.1 (in terms of wavelength and decay rate) and the growing wave front properties coincide with the second mode in terms of the wavelength. Effects of the third mode are not seen to contribute to the overall solution for point B. It is noted that the leading edge of the asymptotic solution continues to decay at a rate similar to that predicted by spatial stability analysis, while the forerunner continues to grow spatio-temporally.

Additional conditions for the creation of a forerunner are found by looking at the receptivity solutions for points C and D, with the former having a single stable mode and the latter having two damped modes. Results are shown in Fig. 5.4 for the streamwise perturbation velocity, at the indicated times. The essential difference between these and the cases shown in Fig. 5.3 is that the latter have three modes, while C possesses a single mode and D possesses two modes. The frontrunner in Fig. 5.3 is present only for those cases with multiple modes. In the absence of multiple modes, as for point C, no such forerunner is seen in Fig. 5.4. Again for point D, there are only two stable modes which create a spatio-temporally growing wave front. Thus, it appears that for fluid dynamic systems, the presence of a minimum of two stable modes seems necessary to produce a spatio-temporally growing wave front, even when the least stable mode is spatially damped. This explanation was advocated in [349].

There is an alternate possibility for the results shown in Table 5.1 and Figs. 5.3 and 5.4. The spatio-temporal wave front for the case of B has a wavelength of $B2$ in Table 5.1. However, a superposition of all the spatio-temporal fronts (in terms of wavelength and propagation speed) shows similar properties, including what is shown for the spatially unstable point A in Fig. 5.3. This raises the possibility that the forerunner is a wave packet

TABLE 5.1

Wave properties of the selected points in Fig. 5.2.

Mode	α_r	α_i	V_g	V_s	V_e
$A1$	0.279826	−0.007287	0.4202	0.42	0.42
$A2$	0.138037	0.109912	0.4174	....	....
$A3$	0.122020	0.173933	0.8534	....	....
$B1$	0.394003	0.010493	0.4267	0.352	0.352
$B2$	0.272870	0.167558	0.2912		
$B3$	0.189425	0.322635	0.1159		
$C1$	0.246666	0.013668	0.5026	....	....
$D1$	0.160767	0.001520	0.3908	0.33	0.33
$D2$	0.062141	0.069659	0.2762		

centered around one of the unstable wavenumbers corresponding to the fixed Reynolds number of $Re_{\delta*} = 1000$ for all three points A, B and D. There is a band of circular frequencies which are unstable for all $Re > Re_{cr}$. However, this possibility is ruled out for the following reasons.

When the Bromwich contour integral method is used for points B and D, the Fourier transform of the solution at $t = 801.1$ indicates different wave properties for the wave fronts in Figs. 5.9 and 5.11, in terms of the amplitude. The wave front corresponds to $B2$ of Table 5.1 and a similar wavenumber band for the case of D. Also, one must also look at the important difference between linear and nonlinear receptivity results. Here for point C, the exciter is located at a place where $Re_{\delta*} = 300 < Re_{cr}$, and thus, there is no single frequency for which there is an unstable wave solution. However in Fig. 3.40, we have noted from the Navier–Stokes solution that the exciter is located at a place for which $Re_{\delta*} < Re_{cr}$, yet the computed Navier–Stokes solution indicates the existence of a wave front. It has to be emphasized that normal modes obtained by spatial instability theory for harmonic wall excitation are qualitatively different from the spatio-temporal wave front noted here. One is

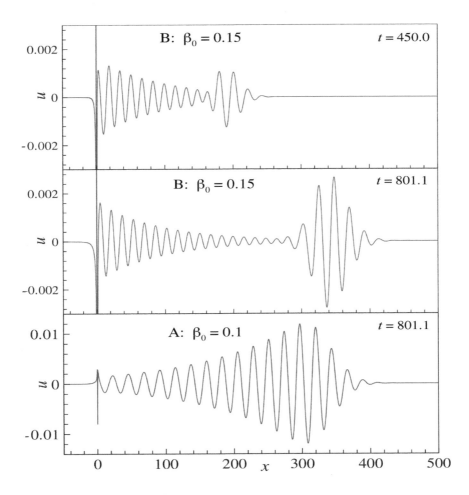

FIGURE 5.3

The streamwise disturbance velocity component plotted as a function of x at the indicated times for $Re_{\delta*} = 1000$ at $y = 0.278$.

not the consequence of the other. Thus, the nonlinear and nonparallel effects are significant and no generalization with solely linear results is definitive.

Also, to highlight the shortcomings of normal mode analysis and to emphasize the role of mode interactions, another case was considered in [350], where the Blasius boundary layer was excited by a wide-band excitation given by $0.08 \leq \beta_0 \leq 0.12$. If the response field for an excitation at β_0 is given by

$$\psi(x,y,t;\beta_0) = \frac{1}{(2\pi)^2} \int \int_{Br} \frac{\phi_1(\alpha,y,\beta)\phi'_{30} - \phi'_{10}\phi_3(\alpha,y,\beta)}{i(\phi_{10}\phi'_{30} - \phi_{30}\phi'_{10})(\beta - \beta_0)} e^{i(\alpha x - \beta t)} \, d\alpha \, d\beta \qquad (5.6)$$

then the solution for the banded excitation case is obtained from the convolution of the above given by

$$\psi(x,y,t,\beta_1,\beta_2) = \int_{\beta_1}^{\beta_2} \psi(x,y,t,\beta_0) \, d\beta_0 \qquad (5.7)$$

where β_1 and β_2 define the bandwidth of excitation. The chosen bandwidth was such that the fundamental mode remained unstable for all frequencies. Thus, for banded excitation, each unstable mode will interact with others to create wave packets. The presented results in [350] for the banded excitation case are reproduced here, in Fig. 5.5, for the indicated time instants.

From the figure one notices that up to some early times ($t = 110$), the effects of banded excitation are not perceptible, as the evanescent waves are dominated by the local solution.

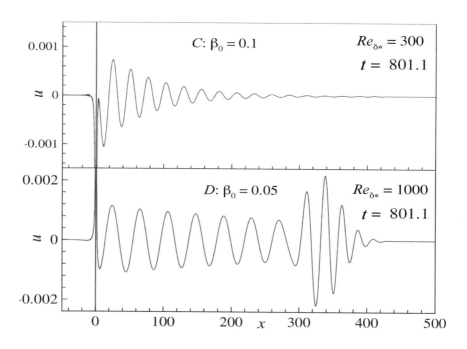

FIGURE 5.4

Streamwise velocity plotted as a function of x at the indicated times for $Re_{\delta*} = 300$ and 1000 for the indicated circular frequencies. Results are shown for non-dimensional height $y = 0.278$, over the plate.

However, at later times, this is not the case (seen at $t = 185.4$), as one can see the presence of multiple frequencies behind the leading wave front. There appear to be significant cancellations among different unstable waves at the back of the front. As a result of such cancellations, one notices attenuated multiple packets traveling downstream, except the leading wave front that continues to grow. This is an example of multiple unstable waves annihilating each other (including the local solution), while leaving the growing wave front.

The feature of signal cancellation for unstable wave systems can be better understood if one looks at some representative harmonic components of the response field. In Fig. 5.6a, the streamwise disturbance velocity component of various circular frequencies as a function of space is shown for $t = 411.5$.

Here, only seven distinct circular frequency solutions are shown for ease of explanation. Significant phase shifts among these solutions for $x \leq 120$ are responsible for mutual cancellations within the signal. The zoomed part in Fig. 5.6b shows more phase shift among the displayed components at the left of the frame, resulting in cancellation of these components in antiphase. In contrast, there is lesser phase shift among the signal components seen in the right of the frame in Fig. 5.6a. This will lead to reinforcing of the components in this part of the domain.

It is also noted that the spatio-temporal front has a smaller wavenumber as compared to the asymptotic TS wave packet. With the spatial stability approach, it is not possible to predict the spatio-temporal front. Group velocity (V_g) for the presented problem is also given in Table 5.1, obtained by an eigenvalue analysis. From Figs. 5.3 and 5.4, one can directly estimate the signal speed (V_s) by tracking the crests and this information is also given in Table 5.1. An estimate of energy propagation speed (V_e) can also be obtained from the equation for the disturbance energy given in Eq. (4.6). If one represents (E_d) in terms of its Fourier–Laplace transform as

$$E_d(x, y, t) = \frac{1}{(2\pi)^2} \int \int_{Br} \hat{E}_d(y, \alpha, \beta) \, e^{i(\alpha x - \beta t)} d\alpha \, d\beta$$

then the governing equation for $\hat{E}_d$ is given by

$$\hat{E}_d'' - \alpha^2 \hat{E}_d = \phi''' U + 2\phi'' U' + \phi'(U'' - \alpha^2 U) - 2\alpha^2 \phi U' \tag{5.8}$$

Equation (5.8) was solved in [350] as a function of α and β and the solution was reconstructed as a function of x and t by performing inverse Bromwich integrals successively. Results for E_d are shown as a function of x for points A and B in Fig. 5.7. This governing equation clearly shows that this is not a propagation equation − as β is absent in Eq. (5.8). At the same time, this equation shows that the spatial diffusion of the energy is forced by the right-hand side via the mean and disturbance flow velocities.

Here the plot shows the variation of E_d to be smoother than u. Once again, for point A there is no detached forerunner from the asymptotic solution, while point B displays the same detached forerunner as before. The rate at which E_d propagates can be estimated roughly from the figures. This is shown in Table 5.1 as V_e. We note that the system dynamics are determined by the least stable mode ($A1$) for the spatially unstable case, with all three definitions of propagation speed producing identical results. In contrast, for stable systems with multiple modes, the forerunner has identical V_e and V_s, which lie between the group velocity values of the leading two modes.

Further understanding of the role played by different stable modes, when significant transient growth is present, was studied in [348]. Two specific cases of $\beta_0 = 0.05$ and 0.15 considered in Table 5.1 are compared. The complex wave numbers are listed again in Table 5.2.

For these excitation parameters, the existing spatial modes are all damped: One set corresponding to the lower frequency (identified as 1 and 2 in Table 5.2) is below the neutral curve and the other set (B) corresponding to the higher frequency (identified as

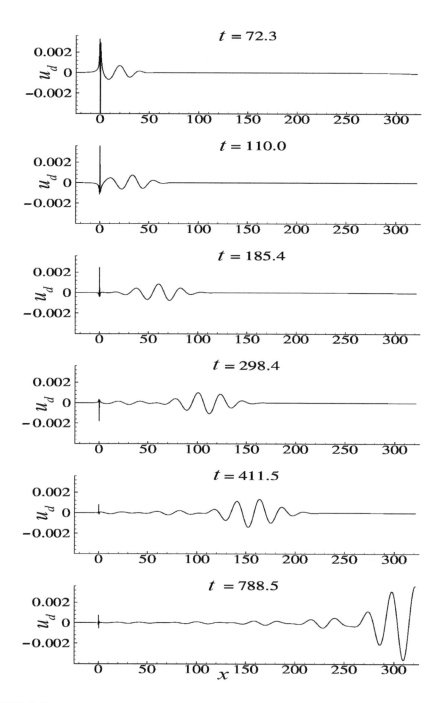

FIGURE 5.5

Streamwise disturbance velocity plotted as a function of x at the indicated times for banded excitation centered around $\beta_0 = 0.1, \delta\beta = \pm 0.02$ for $Re_{\delta^*} = 1000$ and $y = 0.278$.

3, 4 and 5 in the table) is above the neutral curve. In Fig. 5.8, the computed disturbance velocity in the streamwise direction, obtained by the Bromwich contour integral method, is shown for the case of point B ($\beta_0 = 0.15$).

Solutions at the indicated time instants show (i) the near field given by the local solution and the far field consisting of (ii) a decaying wave and (iii) a spatio-temporally growing wave front. The Fourier–Laplace transform of the solution at $t = 801.1$ is shown in Fig. 5.9 with

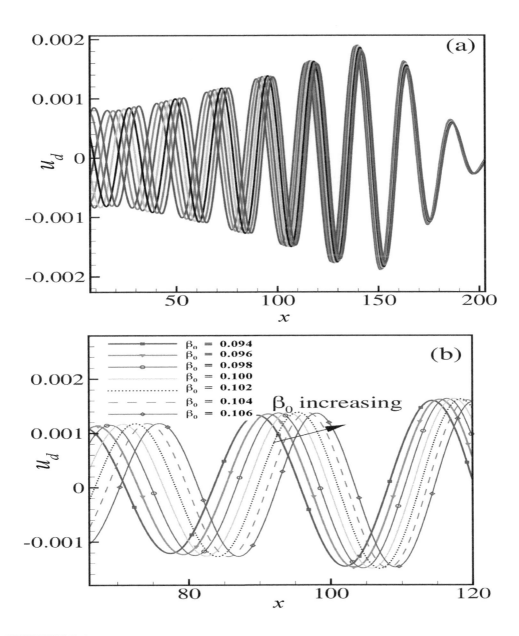

FIGURE 5.6
(a) Streamwise disturbance velocity plotted as a function of x at $t = 411.5$ for different β_0 with $Re_{\delta^*} = 1000$ and results shown at $y = 0.278$. (b) Enlarged view of (a) shown for selected frequencies.

vertical lines indicating the location of the three modes. This result indicates that the

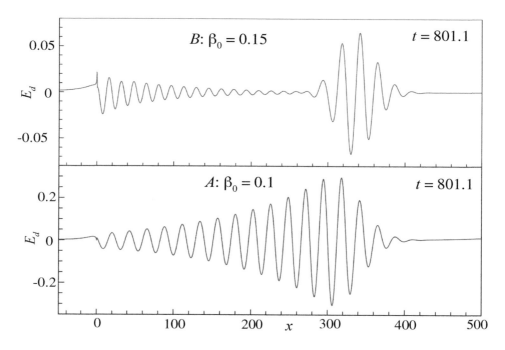

FIGURE 5.7
Disturbance energy E_d plotted as a function of x at the indicated times for $Re_{\delta*} = 1000$ at $y = 0.278$.

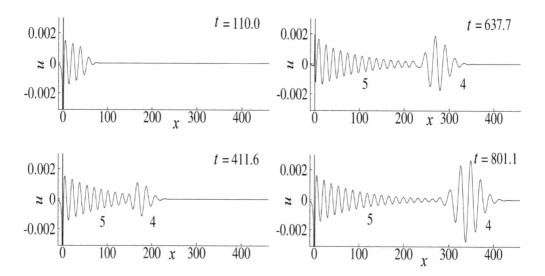

FIGURE 5.8
Streamwise disturbance velocity plotted as a function of x at different times for $Re_{\delta*} = 1000$, $\beta_0 = 0.15$ with results shown for $y = 0.278$.

displayed decaying wave corresponds to Mode 5 and the growing wave front is centered around Mode 4 of Table 5.2. The effects of Mode 3 are not visible due to its extremely large damping rate.

In Fig. 5.10, the computed disturbance velocity in the streamwise direction obtained by the Bromwich contour integral method is shown for the case of point D ($\beta_0 = 0.05$). Results at the indicated time instants show similar structures at near and far fields of the exciter, as was the case for point B.

In Fig. 5.11, the Fourier–Laplace transform of the solution at $t = 788.5$ is shown for the solution of Fig. 5.10. In this case, the asymptotic decaying signal corresponds to Mode 2, while the effect of Mode 1 is not visible here due to its high decay rate. The growing wave front corresponds to the packet to the right of Mode 2.

In summarizing the discussion on the spatio-temporal growth shown here to occur for viscous flows, we note that this is in sharp contrast to what has been written about algebraic growth in [187, 449] and the non-normal modes discussed in [313, 411] for more than one reason. All these works are premised on the basis of the inadequacy of the Orr–Sommerfeld equation and are valid for a three-dimensional disturbance field only. In [348, 349, 350], the spatio-temporal growth is embedded into system dynamics which is affected also due to mutual interactions of multiple modes and is equally valid for both two- and three-dimensional flows. We note the spatio-temporal growing front recorded here is not transient in nature, while other works report transient energy growth.

TABLE 5.2
Wave properties of indicated circular
frequency disturbances.

ω_0	α_r	α_i
0.05 (D)	1) 0.0621413	1) 0.0696594
	2) 0.1607670	2) 0.0015206
0.15 (B)	3) 0.1894256	3) 0.3226357
	4) 0.2722870	4) 0.1675585
	5) 0.3940036	5) 0.0104936

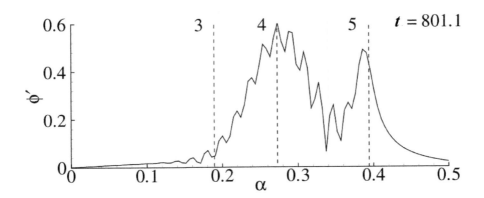

FIGURE 5.9
Fast Fourier transform (FFT) of streamwise velocity data of Fig. 5.8, shown as a function of wavenumber for point B.

There is also the distinction between the two approaches in the way the mechanism occurs. As stated in [40], transient energy growth can occur for subcritical systems by two to three orders of magnitude at which *nonlinear effects might push the disturbance into turbulence.* Second, in supercritical flows transient growth can boost existing TS waves with the initial algebraic growth followed by the exponential growth of TS waves. Neither of these two mechanisms for transient growth has been demonstrated experimentally yet. In contrast, the spatio-temporal growth shown in [348, 349, 350] is similar to what is seen

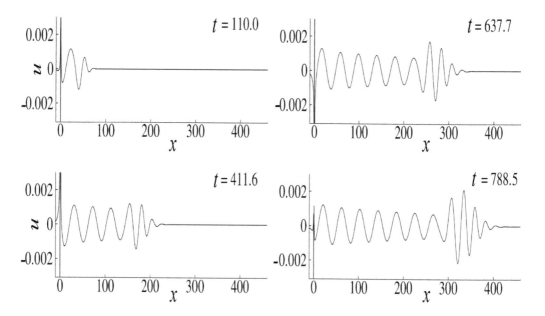

FIGURE 5.10
Streamwise disturbance velocity plotted as a function of x at different times for ($\beta_0 = 0.05$), $Re_{\delta^*} = 1000$ and results shown for $y = 0.278$.

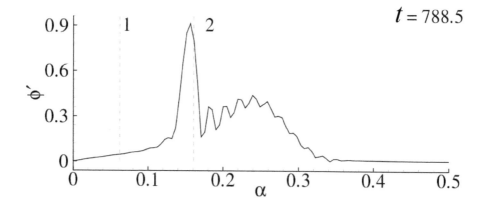

FIGURE 5.11
Fast Fourier transform (FFT) of the data shown in Fig. 5.10, shown as a function of wavenumber for point D.

in the ocean surface due to localized disturbances in the seabed; although the governing equations and flow topology of the equilibrium flow do not bear an exact resemblance.

5.5   Nonlinear Receptivity Analysis: Transition by the Spatio-Temporal Front and Bypass Route

Linear receptivity theory using the parallel flow model in [349] showed the amplitude of the wave front to increase with space and time continuously. Also at later times, the amplitude of the front is noted to be significantly higher than the TS wave packet. Whether this amplitude continues to grow further in the presence of nonlinearity is also the reason for undertaking a nonlinear receptivity study via solution of the Navier–Stokes equation, which also relaxes the parallel flow approximation of linear theory.

The existence of a spatio-temporal front was also demonstrated in [349] by numerical solution of the Navier–Stokes equation, with definitive excitation created on the plate. The simultaneous blowing-suction (SBS) strip exciter used is the same as was used in [98]. Such an exercise was repeated in [325], where nonlinear, nonparallel aspects of boundary layer receptivity have been investigated by solving the Navier–Stokes equation using the stream function vorticity formulation, with attention focused only on the local and TS wave packet part of the solution. Inclusion of the growth of the shear layer for monochromatic excitation leads to a limited extent of unstable region which leads to the TS wave packet, instead of progressive TS waves. Thus, instead of the propagating TS wave of linear spatial theory, one notices a packet, which is almost fixed in space and is of finite width, in [325] for high frequency excitation. This also suggests a need to perform receptivity calculations by solving the full time-dependent Navier–Stokes equation incorporating nonparallel and nonlinear effects.

Analysis performed by solving the full Navier–Stokes equation for a zero pressure gradient (ZPG) flow is shown schematically in Fig. 5.12, identifying the wall exciter inside a computational domain marked with an inflow and an outflow. For this problem, Reynolds number based on displacement thickness ($Re_{\delta*}$) and non-dimensional frequency of the exciter (β_0) are the relevant parameters. By definition, a constant physical frequency disturbance is marked by a ray given by $F = \beta_0/Re_{\delta*}$. The parameter F is related to physical frequency f in Hz by $F = 2\pi\nu f/U_\infty^2$. The neutral curve is considered important, as it can also be obtained from solution of the linearized Navier–Stokes equation in the spectral plane for fixed frequency excitation. Having established the onset of instability via TS waves, it has been ever since believed that such wavy disturbances are responsible for transition to turbulence when the background disturbances are low. The experiments of [316] validated linear stability analysis applied to a viscous flow, yet did not elucidate the complete process of transition. We note disturbances from receptivity to the intermittent turbulent stage via an accurate simulation of the two-dimensional Navier–Stokes equation in [323].

The failure of linear stability theory in predicting subcritical transition has prompted researchers to look at alternate nonlinear theories as in [191] and/or secondary instabilities, as in [24]. Such failure has been attributed to *non-normal* eigenvectors causing large transient energy growth for stable systems, but applicable to three-dimensional perturbation only, as reported in [411]. However, for two-dimensional perturbations, spatio-temporal growth has been reported in [349] by receptivity analysis, and in [331] subcritical bypass transition is explained. Here, the dynamics of the wave front are reevaluated with the Navier–Stokes solution obtained over a longer domain and time duration for different amplitudes of input excitation. However, the input excitation amplitudes considered are still less in one case than that used in [98]. In [101], it is noted that the *common feature of these two very different transition scenarios occurs when the primary instability reaches a certain amplitude and breaks down to turbulence, probably through a secondary instability mechanism.* This speculation is often used to explain differences between onset of instability and eventual transition process. Instead, in [323], it is shown that the TS wave packet remains subdominant, while the spatio-temporal wave fronts create secondary and higher order instabilities, leading to flow intermittency and widening of the spectrum.

Explaining the genesis and leading features of fluid flow turbulence has remained one of the unsolved problems of classical physics, and despite innumerable studies, no definitive explanations existed which show its creation from the onset stage (receptivity) of disturbances

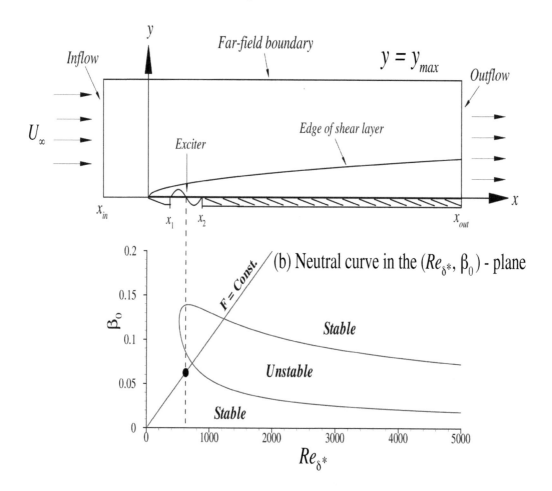

FIGURE 5.12
Schematic diagram of the problem with the exciter shown in frame (a) along with the neutral curve in the $(Re_{\delta*}, \beta_0)$-plane in frame (b). Constant physical frequency ray F and the location of the exciter in the $(Re_{\delta*}, \beta_0)$-plane are shown in frame (b).

in fluid flow. Recent studies at IIT Kanpur have attempted to bridge the gap and explain flow turbulence starting from laminar flow [323]. In this reference, one notes mechanisms of turbulence creation by three distinct routes in a nonlinear framework by solving the Navier–Stokes equation with an extremely accurate numerical methods developed. Despite varieties of mechanisms, the final turbulent state shows universal feature of the energy spectrum for two-dimensional turbulence [323]. Also, by demonstrating this for two-dimensional flow, the generic nature of turbulence is displayed which is not as restrictive as any effort showing it for three-dimensional flow only, as has been noted in some published literature.

5.5.1 Governing Equations and Boundary Condition

The physical problem considered in this study is a two-dimensional flow over a semi-infinite flat plate with the leading edge as the stagnation point. The free stream condition is specified as U_∞ for velocity. Study of nonlinear receptivity of an equilibrium state is described here. Both the equilibrium and the disturbance flows are obtained by solving the conservation equations in the derived variable formulation. Once again, the (ψ, ω) formulation is preferred due to its accuracy and efficiency in satisfying mass conservation exactly and computed in the transformed (ξ, η)-plane, which is utilized for grid refinement in selective parts of the domain. In the transformed plane, the vorticity-transport equation (VTE) and the stream function equation (SFE) are re-written in non-dimensional form. An identical formulation and methodologies were used in [325] for the problem shown in Fig. 5.12

$$h_1 h_2 \frac{\partial \omega}{\partial t} + h_2 u \frac{\partial \omega}{\partial \xi} + h_1 v \frac{\partial \omega}{\partial \eta} = \frac{1}{Re} \left[\frac{\partial}{\partial \xi} \left(\frac{h_2}{h_1} \frac{\partial \omega}{\partial \xi} \right) + \frac{\partial}{\partial \eta} \left(\frac{h_1}{h_2} \frac{\partial \omega}{\partial \eta} \right) \right] \qquad (5.9)$$

$$\frac{\partial}{\partial \xi} \left(\frac{h_2}{h_1} \frac{\partial \psi}{\partial \xi} \right) + \frac{\partial}{\partial \eta} \left(\frac{h_1}{h_2} \frac{\partial \psi}{\partial \eta} \right) = -h_1 h_2\, \omega \qquad (5.10)$$

with the contra-variant components of the velocity vector given by

$$u = \frac{1}{h_2} \frac{\partial \psi}{\partial \eta}, \qquad v = -\frac{1}{h_1} \frac{\partial \psi}{\partial \xi}$$

where h_1 and h_2 are the scale factors of the transformation given by $h_1 = (x_\xi^2 + y_\xi^2)^{\frac{1}{2}}$ and $h_2 = (x_\eta^2 + y_\eta^2)^{\frac{1}{2}}$, with the subscripts indicating partial derivatives. In the transformed grid, ξ is in a direction parallel to the wall and η is in the wall-normal direction are related to physical coordinates by $x = x(\xi)$ and $y = y(\eta)$, so that $h_1 = x_\xi$ and $h_2 = y_\eta$. The parameter Re has been used as $\frac{U_\infty L}{\nu} = 10^5$ in all the reported computations in this section. Thus, all coordinates are non-dimensionalized by L in this section.

Having obtained the equilibrium flow, vortical excitation is imposed through appropriate wall boundary conditions for receptivity calculations. Here we discuss necessary boundary conditions enforced for both equilibrium flow and for the excitation field. For receptivity calculations, dependent variables in Eqs. (5.9) and (5.10) are composed of a primary quantity (denoted by an overbar) and a disturbance quantity (denoted by a subscript d), i.e.,

$$\omega = \bar{\omega} + \omega_d$$

The governing equations, Eqs. (5.9) and (5.10), and boundary conditions used here are applicable for total quantities.

For calculating the equilibrium flow, at the inlet and the top boundary of the domain, free stream conditions are prescribed which yield

$$\frac{\partial \bar{\psi}}{\partial \eta} = h_2 \qquad (5.11)$$

$$\bar{\omega} = 0 \qquad (5.12)$$

At the wall, appropriate boundary conditions for primary quantities are given by

$$\bar{\psi}_w = \bar{\psi}_o = \text{constant} \qquad (5.13)$$

$$\bar{\omega}_w = -\frac{1}{h_2^2} \frac{\partial^2 \bar{\psi}}{\partial \eta^2} \qquad (5.14)$$

Receptivity analysis by vortical wall excitation is carried out by a simultaneous blowing-suction (SBS) strip at the wall. Inlet and top boundary conditions remain the same, but wall boundary conditions for stream function and vorticity change as follows for the excitation problem

$$\psi_w = \bar{\psi}_o + \psi_{wp} \qquad (5.15)$$

$$\omega_w = -\frac{1}{h_1 h_2} \frac{\partial}{\partial \xi} \left(\frac{h_2}{h_1} \frac{\partial \psi}{\partial \xi} \right) - \frac{1}{h_2^2} \frac{\partial^2 \psi}{\partial \eta^2} \qquad (5.16)$$

In writing Eq. (5.16), one notices that the wall is no longer a $\psi = constant$ line, due to mass transfer through the SBS strip, although the net mass transfer at any instant of time is zero. Consider the strip to be placed between x_1 and x_2, as shown in the schematic diagram of Fig. 5.12. The wall perturbation stream function value (ψ_{wp}) in Eq. (5.15) can be obtained from the no-slip condition and an expression for wall-normal velocity given by

$$u_d = 0, \quad v_d = A(x) \sin(\beta_L t) \qquad (5.17)$$

where β_L is the non-dimensional disturbance frequency based on length scale L and $A(x)$ is the amplitude of the disturbance velocity. The amplitude function $A(x)$ is related to $A_m(x)$ as $A(x) = \alpha_1 A_m(x)$, where α_1 is the amplitude control parameter. In [98], the wall-normal component of velocity was scaled by $\sqrt{Re}$, i.e., $\alpha_1 = Re^{1/2}$ and cases were considered with $Re = 10^5$. $A_m(x)$ is defined in [98, 325] as:
for $x_1 \leq x \leq x_{st}$

$$A_m = 15.1875 \left(\frac{x - x_1}{x_{st} - x_1} \right)^5 - 35.4375 \left(\frac{x - x_1}{x_{st} - x_1} \right)^4 + 20.25 \left(\frac{x - x_1}{x_{st} - x1} \right)^3 \qquad (5.18)$$

and for $x_{st} \leq x \leq x_2$

$$A_m = -15.1875 \left(\frac{x_2 - x}{x_2 - x_{st}} \right)^5 + 35.4375 \left(\frac{x_2 - x}{x_2 - x_{st}} \right)^4 - 20.25 \left(\frac{x_2 - x}{x_2 - x_{st}} \right)^3 \qquad (5.19)$$

where $x_{st} = (x_1 + x_2)/2$. This excitation produces vorticity disturbances which create experimentally observed downstream propagating waves in the flow, at early times. The implication of the value of α_1 is in the magnitude of the input disturbance amplitude. For example, $\alpha_1 = 0.01$ implies an input harmonic disturbance amplitude to be 1% of the free stream speed. In [98], a wall-normal velocity is imposed that is equal to $\alpha_1 = 0.00316$, i.e., a fixed 0.316% excitation case used to show the mere existence of TS waves. It was also not clarified whether the Navier–Stokes solution creates a wave packet or a constantly growing TS wave. This is due to the fact that the streamwise extent of computational domain taken in [98] is very small, restricted very closely to the lower branch of the neutral curve. In fact, it is so small that the outflow of the domain remains within the neutral curve. This provides computational advantage for the outflow boundary condition, which corresponds to the smooth passage of the TS wave and in reality it works as the Sommerfeld radiative boundary condition for a monochromatic wave. We will discuss similarities and dissimilarities with the boundary condition used here with that used in [98] and the buffer domain technique used in [204, 327].

In [325], small amplitude excitation cases have been considered to show the existence of all three components of the solution, i.e., the local solution, the TS wave packet and the spatio-temporal wave front, as described in Chap. 3. The domain size taken in [349, 325] is large enough for the outflow to be placed far outside the neutral curve. One must also note that the results in [416] were obtained with an extremely high resolution numerical method to solve the vorticity transport equation, with a numerical dispersion relation which mimics the physical dispersion relation faithfully. This is practically impossible with a fourth order explicit spatial discretization scheme, which has been used in [98]. Also in [98], velocity vorticity formulation of the Navier–Stokes equation has been used with the lower accuracy numerical method which does not ensure solenoidality of the velocity field, i.e., mass conservation is not strictly enforced.

At the outflow, radiative boundary conditions are enforced for both mean and perturbed flows as shown below

$$\frac{\partial v}{\partial x} = 0 \tag{5.20}$$

$$\frac{\partial \omega}{\partial t} + U_c \frac{\partial \omega}{\partial x} = 0 \tag{5.21}$$

The convective speed of disturbances through the outflow, U_c, in the Sommerfeld radiative outflow condition is taken as the free stream speed. This type of boundary condition is well suited to approximate better the flow over a semi-infinite plate to obtain a velocity profile which is identical to the Blasius profile away from the leading edge of the plate. Direct simulation of transitional flows is not straightforward due to the formation of large vortical structures and migration of these through the outflow boundary. Such vortical structures act like fronts, exciting a large range of spatial scales. Different wavenumber components travel at different group velocities and thus a single value of U_c for the structure often leads to numerical instabilities. This problem was not faced in [98], as the authors dealt only with a monochromatic wave in its linear state of evolution. The problem is solved by using a one-dimensional filter developed in [326] applied over the last ten points for all y, out of 4501 points in the streamwise direction spanning the region $-0.05 \le x \le 120$. This filter is much narrower than the usual buffer domain used in [204, 327].

The initial condition for obtaining the equilibrium flow is given by the inviscid solution as

$$\frac{\partial \bar{\psi}}{\partial \eta} = h_2 \tag{5.22}$$

$$\bar{\omega} = 0 \tag{5.23}$$

Computations have been performed in the domain in the wall-normal direction defined by $0 \le y \le 1.5$. In the wall-normal direction, a stretched tangent hyperbolic grid is used, given by the following, for the jth point, as

$$y_j = y_{max} \left[1 - \frac{\tanh[b_y(1 - (\eta)_j)]}{\tanh b_y} \right]$$

Here $y_{max} = 1.5$, $b_y = 2$ have been used for appropriate grid clustering and a total of 401 points have been taken in this direction. In the streamwise direction, the domain is divided in to two segments: the first segment extends from $x_{in} = -0.05$ to $x_m = 10$, which contains points distributed by a tangent hyperbolic function, and for the ith point it is obtained as

$$x_i = x_{in} + (x_m - x_{in}) \left[1 - \frac{\tanh[b_x(1 - (\xi)_i)]}{\tanh b_x} \right]$$

In the second segment, from $x_m = 10$ to $x_{out} = 120$, a uniform distribution of points is used. For the presented calculations, identical stretching is used in both directions, i.e., $b_x = b_y$ for the stretched segments.

Numerical methods for solving Eqs. (5.9) and (5.10) are given in [325] and are based on a compact scheme for the solution of the vorticity transport equation and a stabilized conjugate gradient (Bi-CGSTAB) method as given in [419] for the solution of the Poisson equation. Time integration is performed by the optimized Runge–Kutta (ORK_3) method of Rajpoot *et al.* [285], with $\Delta t = 8 \times 10^{-5}$.

The equilibrium solution is obtained when the velocity and vorticity profiles at all the

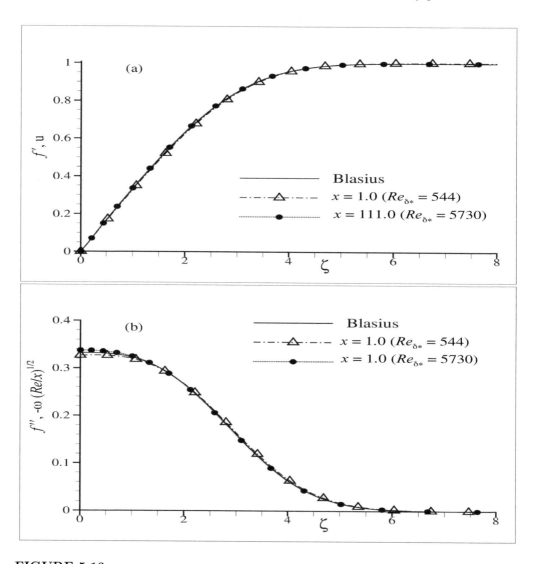

FIGURE 5.13
(a) f' obtained by solving the Blasius equation and u obtained from DNS results for the ZPG boundary layer at $t = 406$ and the indicated streamwise locations shown plotted as a function of similarity variable ζ. (b) f'' obtained by solving the Blasius equation and $-\omega\sqrt{Re/x}$ obtained from DNS results for the ZPG boundary layer at $t = 406$ and the indicated streamwise locations shown plotted as a function of similarity variable ζ.

stations match satisfactorily with the Blasius similarity profiles, except very near the leading edge. The Blasius profile can be obtained from the solution of

$$f''' + \frac{1}{2} f f'' = 0$$

Here a prime refers to the derivative with respect to the similarity variable

$$\zeta = y \sqrt{Re/x}$$

This equation is solved subject to the boundary conditions

$$f = f' = 0 \quad \text{at} \quad \zeta = 0$$

$$f' \longrightarrow 1 \quad \text{as} \quad \zeta \longrightarrow \infty$$

We note that the Blasius solution is used here for reference only, as the unexcited equilibrium flow and the perturbed flow due to time harmonic perturbation at the wall are obtained from the solution of the Navier–Stokes equation itself, with appropriate boundary conditions defined above. This is the DNS of the flow under consideration. For the DNS of equilibrium flow, filtering of the solution at the outflow is not required, while it is used for the receptivity solution. In addition to this, an adaptive filter developed in [33] is used in the interior of the domain whenever the severity of the nonlinear growth of disturbances inside created numerical instability. By its very nomenclature, such adaptive filters are used in a very localized manner as and when required for a few time steps only.

In Fig. 5.13, f' and f'' are nothing but u and $-\omega \sqrt{x/Re}$, respectively, plotted as a function of ζ, obtained by solving the Blasius equation. The DNS of the equilibrium flow is compared in this figure at $x = 1$ ($Re_{\delta*} = 544$) and $x = 111$ ($Re_{\delta*} = 5730$), where $Re_{\delta*}$ is the Reynolds number based on local displacement thickness δ^*. The first location is very near the plate leading edge and the second one is near the outflow of the computational domain. Results correspond to that obtained at $t = 406$ for the mean flow obtained by solving Eqs. (5.9) and (5.10), started with uniform flow as the initial condition given by Eqs. (5.22)and (5.23). Results obtained at $t = 406$ show an excellent match with the Blasius similarity result. However, the direct simulation result show a slight mismatch near the wall for f'' with that obtained from the Blasius equation, at $x = 1$ only. This mismatch at points near the leading edge is due to the nonparallelism of the flow. The mean flow profile obtained by solving Eqs. (5.9) and (5.10) is representative of that shown in Fig. 5.13.

5.5.2 Nonlinear Receptivity to Vortical Wall Excitation

We noted in Fig. 5.13 that one solves the Navier–Stokes equation to obtain an equilibrium solution, which is almost the same as the similarity solution of Blasius, except near the leading edge of the plate. To initiate the receptivity calculations, solution of the Navier–Stokes equation at $t = 406$ has been taken as the equilibrium solution, whose receptivity to vortical excitation given by Eqs. (5.17) to (5.19) is studied for different values of amplitude control parameter α_1 and exciter frequency β_0 or F.

Times noted in all subsequent discussions and in figures refer to time instants after the exciter at the wall is switched on, which is at $t = 406$. In Eq. (5.17), the amplitude of excitation for the blowing-suction strip is taken as a small fraction (α_1) of the value taken in [98]. A typical response field for a low excitation amplitude characterized by $\alpha_1 = 0.01$ (as compared to the amplitude excitation case in [98], with $\alpha_1 = 0.00316$) is shown in Fig. 5.14, at $t = 40$, to display various components of the response field for the streamwise disturbance velocity for $F = 10^{-4}$. In the figure, the local solution is marked by an asterisk, and the exciter begins at $x_1 = 1.445$ and ends at $x_2 = 1.545$. In all the computed cases, the width of the exciter is kept fixed as 0.09 and the center of the strip is indicated as x_{ex}. For the case in Fig. 5.14, the beginning and the end of the exciter correspond to Reynolds numbers based on displacement thickness 653.9 and 676.25, respectively. For the chosen frequency, the exciter is positioned below the neutral curve. The result is shown for the streamwise disturbance velocity plotted for the first η-line from the plate at a non-dimensional height of $y = 0.000426$. The disturbance component of the solution is obtained by subtracting the equilibrium solution from the instantaneous solution of Eqs. (5.9) and (5.10). In Fig. 5.14, the extent over which linear spatial theory indicates instability is marked by two vertical dashed lines.

Due to growth of the shear layer and changing stability properties, one notices a wave packet consisting of TS waves. We note that this is, in effect, a progressive wave train whose amplitude grows and decays due to stability properties, which is different due to the growth of the mean flow and nonlinearity of the disturbance flow. Thus, the TS wave packet should not be considered to be stationary, although the envelope might appear to be stationary in space and time. It is noted that nonparallel and nonlinear effects indicate instability over a longer stretch for this height in Fig. 5.14, as compared to that predicted by linear theory. However, when the solution is noted for $y = 0.0057$, as shown in Fig. 5.15, the unstable region actually is shorter for nonlinear computations, as compared to that predicted by linear theory. This height dependence of instability of the disturbance field is typical of nonparallel effects, as was also reported in [325] for higher frequency cases. Such differences between linear and nonlinear approaches have been shown earlier in [325] for flow without heat transfer and in [423] for flow past a heated vertical plate. In Fig. 5.14, one also notes the local component of the solution in the immediate neighborhood of the exciter. Such a local component is noted in all receptivity solutions following linear and nonlinear approaches reported in [325, 349]. The spatio-temporal wave front shown

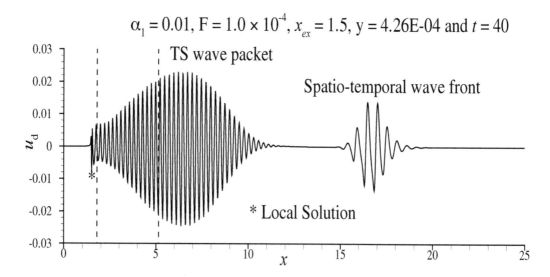

FIGURE 5.14

Disturbance streamwise velocity u_d plotted as a function of streamwise distance x at $t = 40$ and $y = 4.26 \times 10^{-4}$ for $\alpha_1 = 0.01$, $F = 1.0 \times 10^{-4}$ and $x_{ex} = 1.5$. Two dashed vertical lines indicate the entry and exit location of the $F = 1.0 \times 10^{-4}$ ray into and from the neutral curve shown in Fig. 5.12b.

in Fig. 5.14 on the right is centered around $x \approx 17$. Such a spatio-temporal front has been shown before in [349], using the linearized Navier–Stokes equation for spatio-temporal analysis using the Bromwich contour integral method. In the same reference, solution of the Navier–Stokes equation revealed the existence of a spatio-temporal wave front for the flow. Differences between experimental results and results obtained using linear approaches are well-known for higher frequency excitations. One can also note (not reported) a spatially growing vorticity field at lower heights, while the velocity field indicated growth at higher heights, where the vorticity displays spatial decay.

In Fig. 5.15, amplitude effects on the TS wave packet are shown for four different values of α_1. Plotted amplitudes have been normalized by taking the solution for $\alpha_1 = 0.1$ as a reference. That is, the solution for $\alpha_1 = 0.002$ is multiplied by a factor of 50, the solution for $\alpha_1 = 0.01$ by a factor of 10 and so on. These results plotted for $y = 0.0057$ at $t = 94$ indicate nonlinearity of the response field. It is evident that $\alpha_1 = 0.01$ shows maximum normalized instability, as compared to $\alpha_1 = 0.002$. The extent of instability predicted for $\alpha_1 = 0.01$ matches quite well with the linear theory prediction. However, the higher amplitude cases of $\alpha_1 = 0.05$ and $\alpha_1 = 0.1$ show nonlinear amplitude saturation. Also, these cases indicate

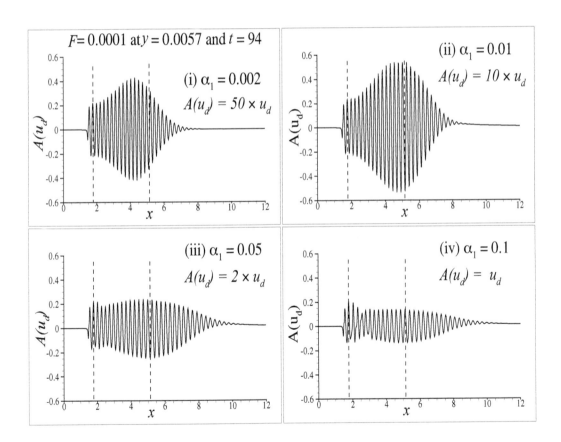

FIGURE 5.15
Normalized u_d plotted as a function of x at $t = 94$ and $y = 0.0057$ for $F = 1.0 \times 10^{-4}$, $x_{ex} = 1.5$ and indicated amplitudes of excitation α_1. Two dashed vertical lines shown in each frame indicate the entry and exit location of the $F = 1.0 \times 10^{-4}$ ray into and from the neutral curve shown in Fig. 5.12b.

steady separation bubbles at the wall near the exciter, as seen clearly for $\alpha_1 = 0.1$ with two different length scales present in the linearly unstable zone with a non-zero mean of u_d.

The information in Fig. 5.15 is only for the TS wave packet and one should instead interpret receptivity by looking at the total solution in the computational domain, as different components of the solution would be predominant for different input parameters. This has been performed by plotting contours for ψ and ω in Figs. 5.16a and 5.16b, respectively, at $t = 194$ for three different amplitudes of excitation. In these figures, the edge of the shear layer is also marked by a dash-dotted line. For $\alpha_1 = 0.002$, one can see a weak convecting spatio-temporal wave front. Yet the amplitude of the spatio-temporal wave front is larger than the TS wave packet for this input excitation. However, this wave front is not large enough to cause transition at this instant of time. For the $\alpha_1 = 0.01$ case, one can note two highly perturbed zones. These perturbations are due to the breakdown of the growing spatio-temporal wave front and disturbances induced upstream of it. Similarly, for $\alpha_1 = 0.05$, the extent of the perturbed zone is noted to be larger and continuous, as compared to the $\alpha_1 = 0.01$ case. One can also note in Fig. 5.16a induction of distinct unsteady separation bubbles for the $\alpha_1 = 0.01$ and $\alpha_1 = 0.05$ cases. The field indicated by vorticity contours in Fig. 5.16b also indicates similar patterns of disturbances, caused by these vortical eruptions piercing through the shear layer. Causes for these massive perturbations are categorized in the next subsections. The displayed components of the solution in Fig. 5.14 influence the transition process differently for different input parameters.

5.5.3 Low Amplitude, Moderate Frequency Excitation

In Fig. 5.17, evolution and transport of the spatio-temporal wave front is shown by tracking u_d at a height close to the wall (at $y = 0.0057$) as a function of streamwise distance at the indicated times, for the case of $\alpha_1 = 0.002$ and $F = 10^{-4}$. Here the exciter is located at $x_{ex} = 1.5$ and the TS wave packet is noted at the left of the frames. The top four frames show very slow variation of the TS wave packet amplitude with time, while the spatio-temporal front continues to grow at a much higher rate. At $t = 20$ and 100, the wave front is significantly smaller than the TS wave packet. At $t = 150$, the maximum amplitude of the spatio-temporal wave front is of the order of the TS wave packet amplitude. By $t = 200$, the wave front has grown almost two times the TS wave packet. Later stage growth rates are significantly higher than the initial growth rate of the wave front.

The last frame of Fig. 5.17 is shown for $t = 260$, when the spatio-temporal front amplifies significantly to induce unsteady separation locally, and the maximum peak-to-peak variation of the wave front is of the order of free stream speed. Comparatively, the TS wave packet amplitude remains the same and appears very small compared to the wave front. Hence, it is marked with an asterisk in the last frame of Fig. 5.17 for ease of identification. Thus, the instability process is definitively seen to be dominated by the spatio-temporal wave front and not by the TS wave for this low amplitude case. However, the traditional viewpoint is that the transition caused by low amplitude, moderate frequency inputs is solely dominated by TS wave. The presented results show instead that the TS wave packet amplitude envelope virtually remains stationary in space and time, while the transition is determined by the spatio-temporal wave front. However, a closer inspection reveals that the envelope of the TS waves also grows with time very slowly.

The evolution of the spatio-temporal wave front for the disturbance field is shown in Fig. 5.18 in terms of the maximum peak-to-peak amplitude of the spatio-temporal front and the variation of the transform of this amplitude with time. In Fig. 5.18(i), variation of peak-to-peak maximum amplitude of the wave front with time is shown for u_d, where one notes four distinct phases of growth. In the earliest Stage I, the wave front evolves through the local solution and the TS wave packet and during this process the maximum amplitude decreases due to dispersion, as the front widens rapidly in the wavenumber plane. This phase is followed by the log-linear growth in stage II. This growth Stage II is similar to that predicted by linear receptivity analysis in [349]. The secondary growth Stage III beginning at $t = 100$ is characterized by rapid growth of the wave front with respect to the TS wave packet, although the absolute growth rate seems a little lower in the initial phase as compared to Stage II. Beyond $t = 200$, the growth rate increases very rapidly due to mean flow distortion by wave-induced stresses. This is followed by bubble formation on the wall. This case suffers unsteady separation at around $t = 260$. Each growing bubble creates

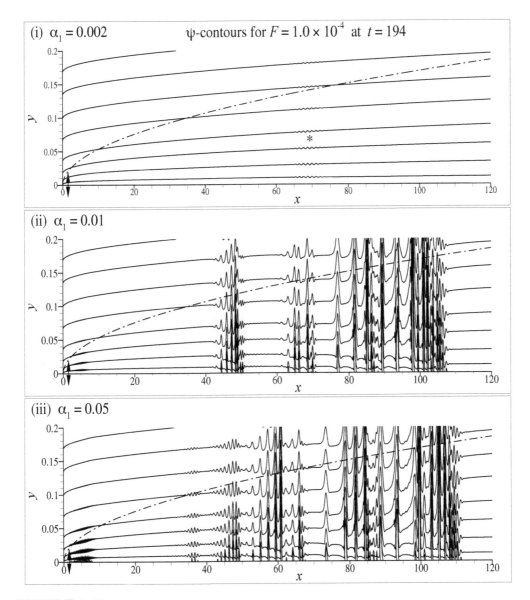

FIGURE 5.16a

ψ-contours shown in the (x, y)-plane at $t = 194$ for $F = 1.0 \times 10^{-4}$, $x_{ex} = 1.5$ and α_1 values. The shear layer edge is marked by a broken line, and the asterisk in the top frame indicates the spatio-temporal front.

an adverse pressure gradient upstream of it, which in turn creates another bubble, and this cascading effect is responsible for the eventual bypass transition. However, as bubbles keep appearing, these also convect downstream, creating a dynamical equilibrium whereby no further upstream penetration of disturbances takes place. This type of saturation of growth rate for the spatio-temporal front due to nonlinear interaction in Stage IV is shown to be universal for all input amplitude cases considered here. This type of unsteady bubble formation leading to transition is similar to the bypass route shown by the free stream convecting vortex in [331]. Once the bubbles form, u_d amplitude saturates to the level of the

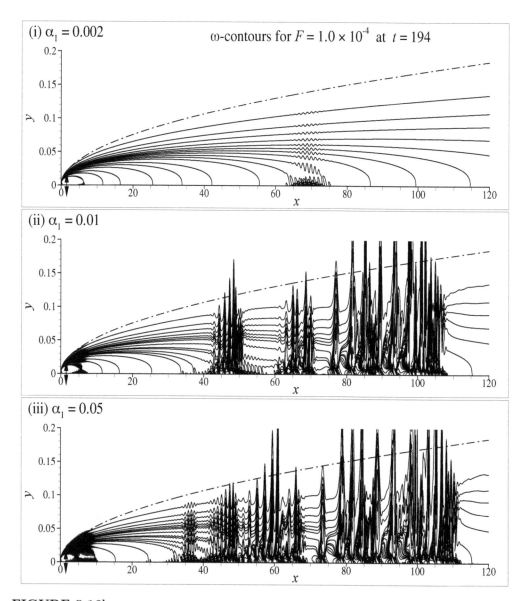

FIGURE 5.16b

ω-contours shown in the (x, y)-plane at $t = 194$ for $F = 1.0 \times 10^{-4}$, $x_{ex} = 1.5$ and α_1 values. The shear layer edge is marked by a broken line, and the asterisk in the top frame indicates the spatio-temporal front.

free stream speed for all input amplitude cases shown subsequently. The highest disturbance growth rate observed during the later part of Stage III is also seen in Fig. 5.18(ii), where the streamwise position of the maximum amplitude of the wave front is plotted as a function of time, which displays a kink marked by an arrow.

Scale selection during bypass transition is noted through the evolving spectrum in frames 5.18(iii) to 5.18(x), where the Fourier transform of u_d is plotted at the indicated times for the case of Fig. 5.18(i). The Fourier transform of u_d of the wave front alone is obtained and shown in the figure. Initiation of the wave front is through the linear mechanism of [349], while the subsequent propagation and growth are strongly affected by nonlinear mechanisms. At $t = 19$, the spectrum is characterized by two dominant peaks, with additional weak side-lobes. The front cleanly detaches from the TS wave packet and is dilated in the front, which is responsible for the lower wavenumber peak. Thereafter, the amplitude decreases due to dispersion up to $t = 60$, while widening of the front shifts the wavenumber of the packet to lower values. Then dispersion weakens, showing exponential growth of peak-to-peak amplitude, noted from the frames at $t = 34$ and 94. This growth continues to $t \simeq 230$, as noted in frame 5.18(i). The mean flow distorts by wave-induced stresses during this stage, shown originally in [408], as the action of the disturbance field on the mean field. This mean flow distortion intensifies with the appearance of multiple unsteady separation bubbles. This suggests that a linear receptivity analysis will hold up to stage II, beyond which one needs to solve the full nonlinear Navier–Stokes equation. We also note that the linear mechanism in [349] did not report stage I, due to the parallel flow approximation in the linear model. The results in this subsection explain how intermittent turbulent spots are formed by the spatio-temporal wave front for low input amplitude excitation cases at moderate to high frequencies.

5.5.4   High Amplitude Cases and Spot Regeneration Mechanism

First, a case is considered for which nonlinearity plays a significant role in accelerating the growth of the spatio-temporal wave front. Thus, the time evolution shown in Figs. 5.17 and 5.18 will happen quicker and/or may bypass the linear growth of disturbances. In other words, results obtained for the high amplitude case are equivalent to events which would be noted at a later time for low amplitude cases. Results for a typical high amplitude case are shown in Fig. 5.19a, for which $\alpha_1 = 0.01$ and $F = 10^{-4}$ and u_d is plotted as function of x at the indicated times for $y = 0.0057$. In the top frame of Fig. 5.19a, the result shown at $t = 94$ displays the TS wave packet and the spatio-temporal front, marked as A. In the frame at $t = 164$, one notes nonlinear saturation and widening of the disturbance packet caused by unsteady separation making the flow intermittent, as also noted in the bottom frame of Fig. 5.17. However, in this high amplitude case, a convecting intermittent spot induces another spatio-temporal front, marked as B in this frame. The wave front B also grows rapidly, as noted at $t = 194$. Also, one notes the trailing part of A elongates with time reinforcing the intermittent nature of the disturbance front. At $t = 244$, the fronts A and B approach each other, while a third spatio-temporal front is initiated as C. The bottom frame of Fig. 5.19a at $t = 324$ shows subsequent flow evolution, where A, B and C have merged into a single intermittent patch, while two other spatio-temporal fronts marked as D and E have appeared. However, the upstream point remains in the vicinity of $x \simeq 40$ for this amplitude case for $t = 194$ onwards. This location is a function of the excitation amplitude for moderate frequencies.

In Fig. 5.19b, u_d at the same height is plotted as a function of x for a higher amplitude case. When α_1 is increased to 0.05 for the same frequency of excitation, the response field shows nonlinear distortion for both the TS wave packet and the spatio-temporal front, even at early times. At $t = 34$, one notes the spatio-temporal front to be distorted, with separation

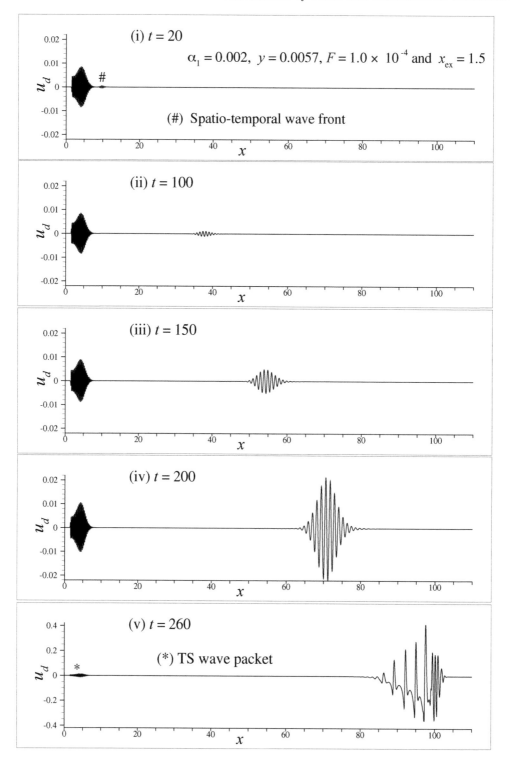

FIGURE 5.17
u_d plotted versus x at $y = 0.0057$ at the indicated times for $\alpha_1 = 0.002$, $F = 1.0 \times 10^{-4}$ and $x_{ex} = 1.5$. The spatio-temporal front in the top frame and the TS wave packet in the last frame are indicated by # and ∗, respectively.

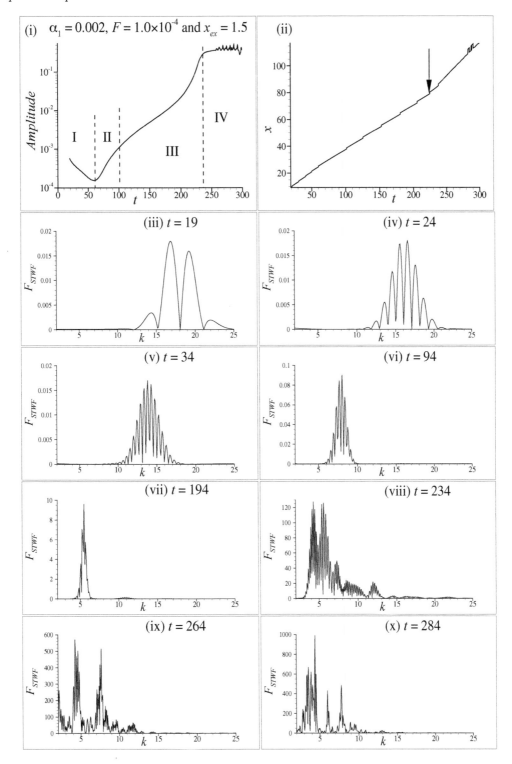

FIGURE 5.18

(i) Maximum peak-to-peak amplitude variation of the spatio-temporal front for u_d with time shown for $\alpha_1 = 0.002$, $F = 1.0 \times 10^{-4}$ and $x_{ex} = 1.5$. Growth stages of the front are identified. (ii) Location of the maximum of u_d plotted as a function of time. An arrow points to a kink, which signifies the onset of mean flow distortion. (iii) − (x) Fourier transform of the spatio-temporal front shown at the indicated times.

bubbles forming underneath. Also, the TS wave packet displays a non-zero mean for all the indicated times. By $t = 184$, the spatio-temporal front shows intermittent structures due to the first two spots A and B. At the same time, leading spots induce two trailing spatio-temporal fronts marked as C and D. All these four structures merge together by $t = 239$, while two more induced structures, E and F, are noted. With the passage of time, additional

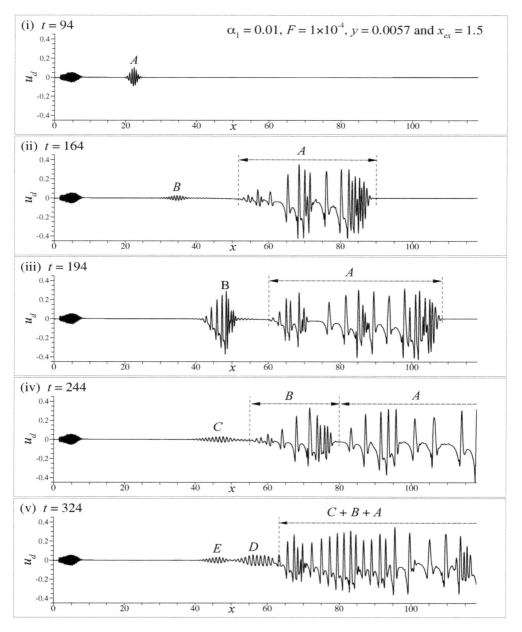

FIGURE 5.19a

Disturbance streamwise velocity u_d plotted as a function of x at $y = 0.0057$ and the indicated times for $F = 1.0 \times 10^{-4}$, $x_{ex} = 1.5$ and $\alpha_1 = 0.01$. Various disturbance packets have been identified and marked in the figure.

fronts are induced farther upstream, approaching the TS wave packet. In the last frame at $t = 289$, one notes that the leading front A has left the computational domain, while the upstream end of the intermittent packet has reached $x \simeq 20$. This should be contrasted with the upstream end of the packet in Fig. 5.19a, where the upstream end of the intermittent disturbance penetrated only up to $x \simeq 40$, at a later time of $t = 324$. Thus, increasing

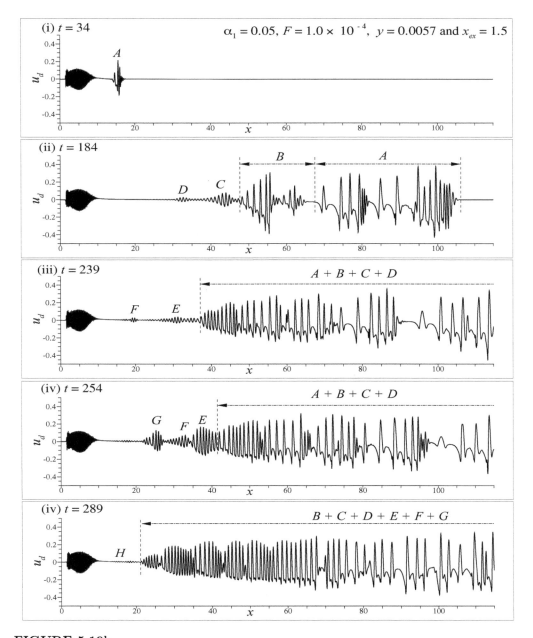

FIGURE 5.19b

u_d plotted as a function of x at $y = 0.0057$ and the indicated times for $F = 1.0 \times 10^{-4}$, $x_{ex} = 1.5$ and $\alpha_1 = 0.05$. Various disturbance packets have been identified and marked in the figure.

the amplitude of excitation causes earlier intermittency of the flow, which also penetrates upstream towards the TS wave packet. Interestingly, the TS wave packet for the cases shown in Figs. 5.19a and 5.19b remains rooted at the same spatial location with negligible temporal variation.

Peak-to-peak maximum amplitudes of spatio-temporal fronts are plotted in Fig. 5.20 as a function of time for different amplitudes, frequencies and locations of the exciter. In Fig. 5.20(i), effects of frequency of excitation are studied for the lowest amplitude case ($\alpha_1 = 0.002$) with the exciter located at $x_{ex} = 1.5$. The observed response field shows maximum growth for $F = 0.75 \times 10^{-4}$ at any time, as compared to the other two cases, while the exciter is located at the same position for all the cases. The disturbance for $F = 1.0 \times 10^{-4}$ enters the neutral curve the earliest, while its amplitude decreases initially due to dispersion. The subsequent growth rate is higher during the linear phase, while the distortion of mean flow begins after $t = 220$, which leads to unsteady separation and nonlinear saturation. In contrast, for $F = 0.75 \times 10^{-4}$ the spatio-temporal front makes its appearance with a higher amplitude. This also shows reduction in peak-to-peak amplitude by dispersion, as shown in Fig. 5.19(i) during Stage I. However, the growth takes u_d to its maximum value by $t = 180$, which thereafter induces unsteady separation and nonlinear saturation. For the lowest frequency of $F = 0.5 \times 10^{-4}$, the spatio-temporal front appears even later with an intermediate amplitude. This case does not exhibit sharp dispersion and reaches the terminal state around the same time as the case of $F = 1.0 \times 10^{-4}$. We see a reduction in the time span of Stages I and II (linear stage) with the reduction in frequency from $F = 1.0 \times 10^{-4}$ to $F = 0.75 \times 10^{-4}$. Stage II is completely absent in the case of $F = 0.5 \times 10^{-4}$, which directly exhibits secondary and other instabilities due to a totally different mechanism described in the next subsection. This explains the apparent lesser growth of the front for $F = 0.5 \times 10^{-4}$ than the other two frequency cases displayed in Fig. 5.20(i).

Five cases with different amplitudes, exciter locations and frequencies are shown in Fig. 5.20(ii). Case I corresponds to the $\alpha_1 = 0.002$ case of Fig. 5.20(i), used as a reference for comparison. For Case III, the amplitude is increased fivefold and the frequency is increased by a factor of 1.5, for which the transition onset time is $t = 150$. In comparison, when the exciter location is pushed upstream at $x_{ex} = 0.5$, while keeping $F = 1.0 \times 10^{-4}$ and $\alpha_1 = 0.01$, disturbance growth follows curve II, which is almost coincidental with curve III. This suggests that placing the exciter far from the neutral curve shows effects which are equivalently caused by exciting the flow at a higher input frequency. Case II should be contrasted with Case IV, with a difference noted for exciter location, and the transition onset time for the latter case occurs earlier, at $t = 70$. Case V corresponds to the highest amplitude case ($\alpha_1 = 0.05$) and here one notices unsteady separation to occur right from the beginning. We note that all the cases shown in Fig. 5.20 suffer bypass transition — a signature of transition occurring due to growth of the spatio-temporal front.

5.5.5   Low Frequency Excitation Cases: Different Route of Transition

For very low frequency cases, it is noted that spatio-temporal fronts are created from the exciter through the TS wave and in this respect this route of transition is different from what has been noted so far for moderate to high frequency of the exciter. For moderate to higher frequency cases, one notes the creation of the wave front following the spatio-temporal linear instability described in [349]. With the passage of time, this pierces through the TS wave packet and travels downstream while growing in amplitude. When the amplitude of the front saturates, it induces additional fronts upstream of it. This process of induction of fronts takes the disturbance field progressively upstream towards the TS wave packet. This scenario is clearly evident from Figs. 5.19a and 5.19b for $F = 1.0 \times 10^{-4}$ for two different amplitudes of excitation cases.

In Fig. 5.21, u_d is plotted at $y = 0.0057$ for the case of $\alpha_1 = 0.002$, for the lower frequency case of $F = 0.5 \times 10^{-4}$, with the exciter located at $x_{ex} = 1.5$. Here the spatio-temporal front grows slowly in the initial stages up to $t = 200$, without showing a significant linear dispersion phase. Thereafter the front grows very rapidly as compared to higher frequency cases, as comparatively shown in Fig. 5.20(i). This very rapid growth is seen by comparing

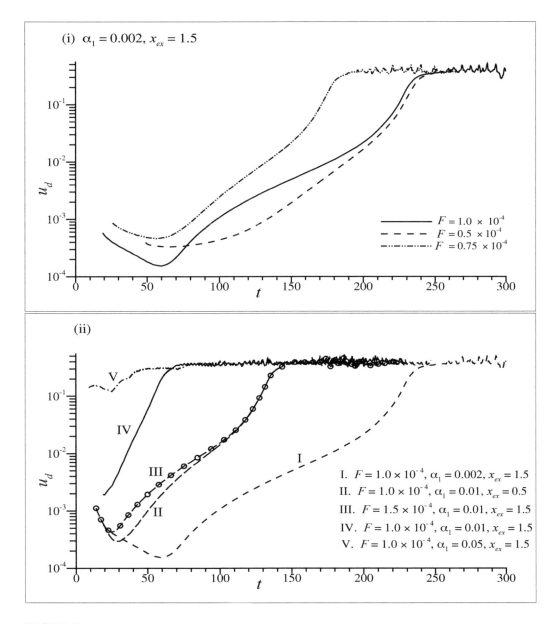

FIGURE 5.20
Maximum peak-to-peak amplitude of the spatio-temporal wave front plotted as a function of time for the indicated parameters of excitation.

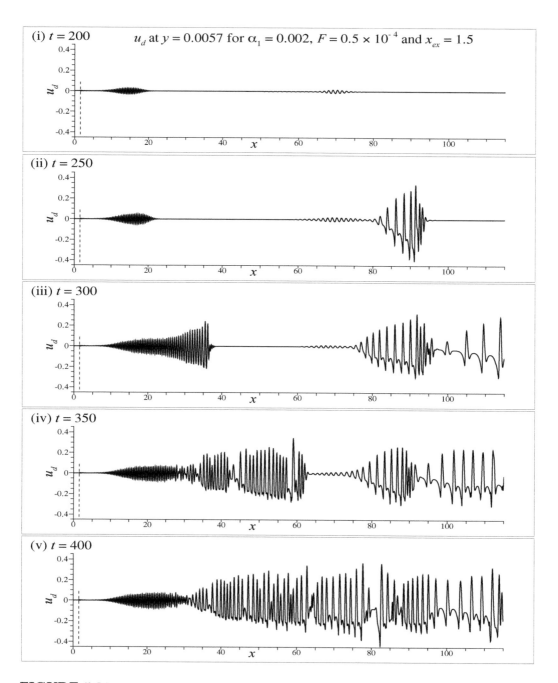

FIGURE 5.21

u_d plotted as a function of x at $y = 0.0057$ and the indicated times for $\alpha_1 = 0.002$, $F = 0.5 \times 10^{-4}$ and $x_{ex} = 1.5$. The location of the exciter is identified by a broken line in the figure.

the top two frames of Fig. 5.21. Such high growth causes nonlinear saturation of amplitude, which induces upstream elongated fronts. Upstream effects are also noted in the form of a new spatio-temporal front trailing behind the leading front at $t = 250$. Additionally, the TS

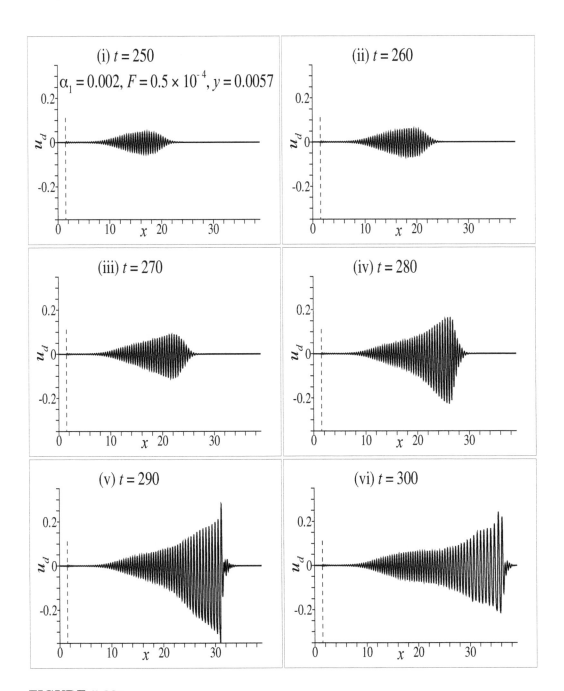

FIGURE 5.22a

u_d plotted as a function of x at $y = 0.0057$ and the indicated times for $\alpha_1 = 0.002$, $F = 0.5 \times 10^{-4}$ and $x_{ex} = 1.5$. In this figure, only the TS wave packet is highlighted. The location of the exciter is marked by a broken line.

wave packet is seen to distort too. Between $t = 250$ and $t = 300$, both the wave front and the TS wave packet evolve rapidly with significant unsteady separation associated with the

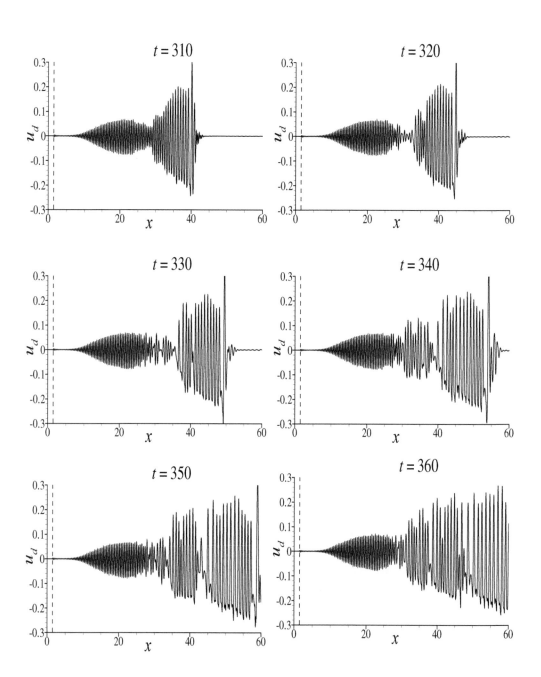

FIGURE 5.22b

u_d plotted as a function of x at $y = 0.0057$ and the indicated times for $\alpha_1 = 0.002$, $F = 0.5 \times 10^{-4}$ and $x_{ex} = 1.5$. In this figure, only the TS wave packet is highlighted. The location of the exciter is marked by a broken line.

spatio-temporal fronts and the TS wave packet amplifies and extends downstream, with the upstream part fixed at the same location.

Transformation of only the TS wave packet is shown during $250 \leq t \leq 360$ in Figs. 5.22a and 5.22b. One notices the formation of the second spatio-temporal front originating

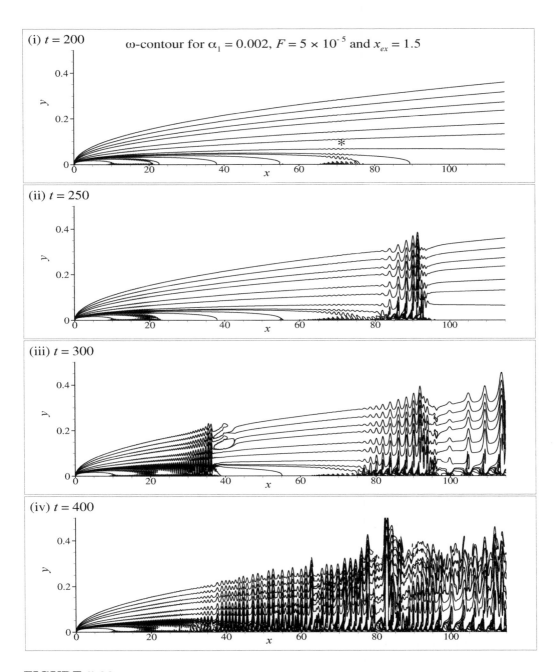

FIGURE 5.23
Vorticity contours shown plotted in the (x, y)-plane at the indicated times for $\alpha_1 = 0.002$, $F = 0.5 \times 10^{-4}$ and $x_{ex} = 1.5$. The spatio-temporal wave front is marked by an asterisk in the top frame.

from the exciter through the TS wave packet in these two figures. For low frequency of input excitation, the unstable range for the TS wave is elongated and during the passage of the front through the TS wave packet, its growth is very rapid. The effect of finite length of the unstable TS wave also affects the growth of the spatio-temporal front through the formation of a steep front, as noted in the frame at $t = 290$. This steep front induces secondary instability ahead of it via the formation of a small packet, as noted at $t = 290$ and 300. Thus, the steepening of the spatio-temporal front is a direct consequence of the finite length of the unstable region of the TS wave packet.

In Fig. 5.22b, detachment of the second wave front from the TS wave packet is shown during $t = 310$ and 360. By $t = 310$, the steepened wave front has come out of the TS wave packet, while maintaining the steepened front. One also notes the slight appearance of a third front between the second spatio-temporal front and the TS wave packet. This is prominently noted in the frame at $t = 320$, where one can also notice a fourth front forming. With the passage of time, sequence of wave fronts keeps forming, as noted in the subsequent frames in Fig. 5.22b. This detailed description of the succession of fronts forming between the TS wave packet and the second spatio-temporal front explains the flow evolution in Fig. 5.21. In Fig. 5.21, by $t = 400$ one notices the TS wave packet to be bridged up to the leading wave front, although the difference in length scale is clearly visible. This low frequency case clearly shows upstream and downstream propagation and formation of fronts from the first and second wave fronts emanating from the exciter. All these events tell us that the parabolized stability equation should be viewed as an engineering tool, with limited ability to describe the sequence of physical events during flow transition.

In Fig. 5.23, vorticity contours are plotted in the computational domain at the indicated times for the case of Figs. 5.21 and 5.22. This clearly reveals the traveling characteristics of the front created by the TS wave packet, as noted at $t = 250$ and $t = 300$. The leading edge of the TS wave packet displays a very sharp front. However, by $t = 400$ one notes the complete merger between the TS wave packet and the spatio-temporal front, with the former propagating downstream and the latter propagating in both directions. Between $t = 300$ and $t = 400$, one notices shrinking of the streamwise stretch between these two structures. In Fig. 5.24, plotted ψ-contours also confirm the above observations, specifically the leading edge of the TS wave packet forming a sharp front, like a hydraulic jump, visible across a large wall-normal distance.

5.6   Calculation of the N Factor

For spatial instability, the amplitude of disturbances at a station x, given by $A(x)$, is related to its initial amplitude, A_0 at $x = x_0$, by

$$A(x) = A_0 \, e^{-\int_{x_0}^{x} \alpha_{imag} \, dx} \tag{5.24}$$

where α_{imag} is the imaginary part of the complex wavenumber α, which signifies exponential growth or decay rate of disturbances. In classical stability analysis, the cumulative effects of this growth rate exponent in Eq. (5.24) is termed the N factor. This factor is empirically correlated for transition prediction, where N factors are obtained from the solution of the Orr–Sommerfeld equation in the presence of different strain rates. The emphasis of all such transition prediction models is in tracking the growth of TS wave packets. For a ZPG boundary layer, it is assumed that transition occurs for $N = 9$, in the absence of free stream turbulence and other disturbances [54]. The N factor for a general case can also be obtained as

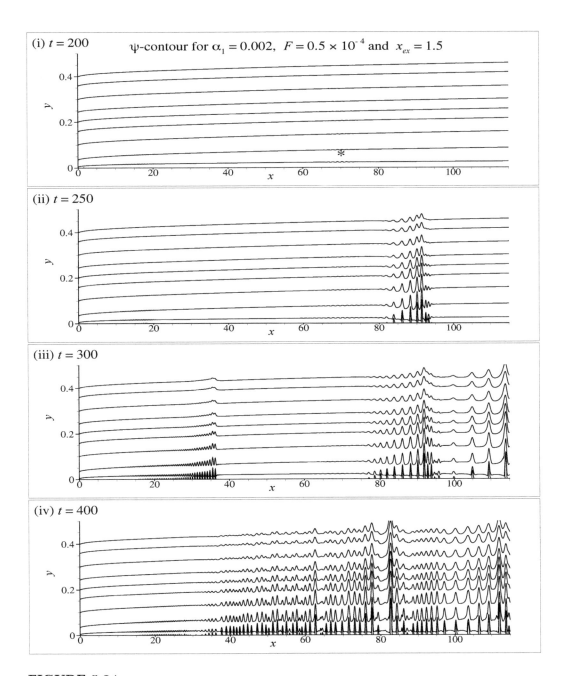

FIGURE 5.24

ψ-contours shown plotted in the (x, y)-plane at the indicated times for $\alpha_1 = 0.002$, $F = 0.5 \times 10^{-4}$ and $x_{ex} = 1.5$. The spatio-temporal wave front is marked by an asterisk in the top frame.

$$N = \ln\left[\frac{A(x)}{A_0}\right] \tag{5.25}$$

In Fig. 5.25, we have compared the cumulative growth rate factor (N) for the spatio-temporal wave front as a function of time (or equivalently with space) for the cases with maximum peak-to-peak u_d (at a particular height) shown in Fig. 5.20. In frame (a) of

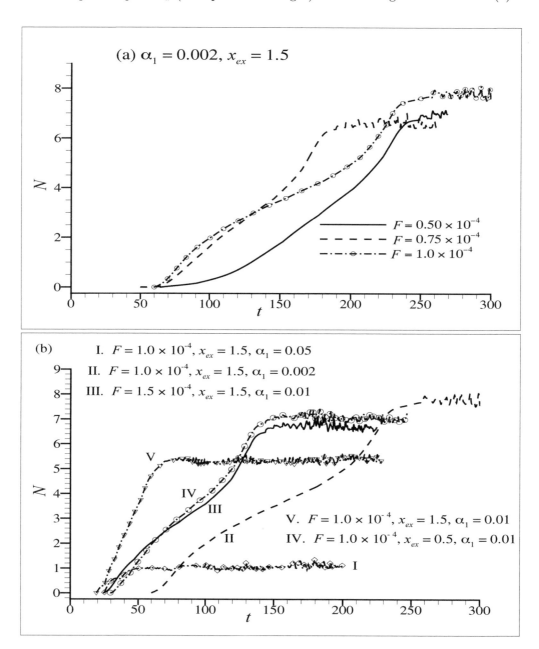

FIGURE 5.25

Equivalent growth rate factor (N) for the spatio-temporal wave front as a function of time shown plotted for the indicated parameters of excitation.

Fig. 5.25, the N factors for three different frequency cases are shown for $\alpha_1 = 0.002$ and $x_{ex} = 1.5$. We emphasize that the N factor shown here relates to the growing spatio-temporal wave front. To calculate the N factor for any case, we have to choose a reference location x_0 with amplitude A_0. We have considered the lowest amplitude of the wave front in Fig. 5.20, corresponding to the case under consideration. We see from this figure that the displayed cases show different values of N, when the nonlinear saturation occurs, despite u_d having similar saturated values at the time of turbulent spot formation, as seen from Fig. 5.20. It is interesting to note that the computed N factors for the low amplitude excitation case ($\alpha_1 = 0.002$) varied between 5.18 and 5.19 for low to moderate frequencies. Despite the near constant value of the N factor at transition, the transition zone is dictated by the onset of secondary instabilities. For example, for $F = 1.0 \times 10^{-4}$, mean flow distortion begins at $t \simeq 200$, while for $F = 0.75 \times 10^{-4}$, this is noted to begin at $t \simeq 150$. However, the lowest frequency case ($F = 0.5 \times 10^{-4}$) suffers transition due to a totally different physical mechanism for which one cannot distinguish between primary and secondary instabilities. We have already noted that in this case, the TS wave packet interacts significantly with the spatio-temporal wave front.

In Fig. 5.25(b), all the cases studied have been summarized, to show nonlinear effects and the position of the exciter. This last aspect is important, yet in many experimental studies, the location of the exciter is not mentioned. This can be of significant value for higher frequency excitation cases where published results have shown significant *nonlinear, nonparallel effects*. Those results could also be significantly contaminated by local solution behavior. The presented results in Fig. 5.20(ii) show strong nonlinear effects. For example, the curve corresponding to the case of $\alpha_1 = 0.05$, $F = 1.0 \times 10^{-4}$ and $x_{ex} = 1.5$ shows a terminal N value of around one at transition. All the cases eventually display unsteady separation due to nonlinear action. However, Case I shows bypass transition right from the onset. The wide discrepancy in the terminal value of the N factor can be attributed to different routes of transition suffered by the flow, which in turn depends upon the initial amplitude and frequency of excitation. Empirical transition prediction models do not take care of initial amplitude and frequency of background disturbances. But for free stream effects, turbulence intensity is empirically correlated, although without any consideration for the spectrum of FST.

6

Nonlinear Effects: Multiple Hopf Bifurcations and Proper Orthogonal Decomposition

6.1 Introduction

There are many internal and external flows which display exponential temporal growth of very small omnipresent disturbances by a linear mechanism. As these disturbances grow in amplitude, nonlinearity intervenes decisively in taking the system from one equilibrium state to another. Vortex shedding behind a circular cylinder represents this scenario for which flow instability begins with the growth of disturbances by a linear temporal mechanism followed by nonlinear saturation. This flow has been identified as an example of a nonlinear dynamical system representing phase transition and instabilities for external flows in [284, 353, 387]. For this dynamical system, it was considered until very recently that the transfer function is central, with the input spectrum playing only a residual role at the equilibrium stage (as incidental for systems representing intrinsic dynamics). Such flows are also referred to as hydrodynamic oscillators at any post-critical Reynolds number. Here the criticality refers to primary temporal instability, which is also described mathematically as a Hopf bifurcation in the parameter space in [129].

The authors in [353] have noted that *the idea that the flow represents intrinsic dynamics is also bolstered by the evidential success of Landau's model in explaining super-critical amplitude saturation that is independent of the initial condition. This idea seems to be supported by various numerical studies of the flow in [18, 156, 236, 447] who have predicted a critical Reynolds number in the range $45 \leq Re_{cr} \leq 47$. We note that all these simulations by different methods are for a uniform flow over a smooth cylinder.*

The authors further contested the above in [353] by establishing that the computationally predicted critical Reynolds number is dependent upon the numerical method and that their high accuracy method predicted a higher critical value, which is more consistent with the experimental results of [393] and the unexplained unique results of [152]. This was additionally supported by providing a solution of the Navier–Stokes equation for noisy flow past a cylinder which displayed a lower critical Reynolds number. This was established with the help of proper orthogonal decomposition (POD) of the high accuracy solution of the Navier–Stokes equation and the role of multiple modes has also been shown. POD modes have been used to classify flow instability modes. This approach has been further advanced in [360], where such instability modes have been shown to be universal by locating and comparing these modes for an internal and an external flow. The main theme of this chapter is explaining this nonlinear instability theory, as opposed to the viewpoint proposed in Landau's equation, which is based on the role of a single dominant mode. In the following, the classical viewpoint is first presented, before the newly developed approach is explained.

6.2   Receptivity of Bluff-Body Flows to Background Disturbances

In real flows, bifurcation initiates with linear temporal instability and thus the possibility for the flow to be receptive to background disturbances is expected. The dynamical system, therefore, must act as an amplifier during the receptivity stage. Hence, the input disturbance spectrum is as important as the transfer function. The dependence of criticality on noise is often ignored in the description of theoretical results for stability. However, this becomes evident, as shown in [353] for flow past circular cylinder, by collating experimental evidence gathered over decades. For example, Batchelor [22] conjectured Re_{cr} to be between 30 and 40; in [192] this is quoted as 34, based on some unreported experimental observation. In [182] it is reported as $Re_{cr} = 40$, while in [303] a value of 50 was reported for the same. In [175, 409] this is obtained as 52 and 53, respectively. The sensitive dependence of Hopf bifurcation or instability on facility-dependent disturbances becomes even more evident from the experimental results in [152], which showed $Re_{cr} \simeq 65$. These interesting results are also featured in Plate 2 of Batchelor [22] and in Schlichting [312] on page 18. Such a high value of Re_{cr} has not been satisfactorily explained until the explanation of it was provided in [346]. The title of Homann's paper [152] describes these experiments as *high viscosity flow past cylinder*. In the experiments two oils were used: For Re less than 30, machine oil was used, while for the experiments above Re equal to 30, a mixture of machine oil and gasoline was used. At 20°C, the former has a dynamic viscosity of 0.010135 kgs/m^2 and the latter has the value of 0.001549 kgs/m^2, which should be compared with the value of 0.0001021 kgs/m^2 for water at the same temperature. Both the working fluids have a density similar to that of water. The essential idea behind using the working fluids is to damp out vortical disturbances created by *kapselpumpe* or a rotary vane pump used to drive the fluid.

In Fig. 3 of [284] and Fig. 6 of [387], different values of Re_{cr} are reported for cylinders with different length (L) and diameter (D). Similar variations of Re_{cr} were also reported earlier in [251] and these authors attributed this to the flow three dimensionality indicated by the different aspect ratios of the models. This observation is proposed as the same tunnel was used for individual experimental cases. However, that the variations are due to three dimensionality of the flow field is questionable, as the experimental models were fixed from wall-to-wall in the tunnel, specifically to preclude flow three dimensionality. This also contradicts the observation in [443] that flow past a cylinder remains two dimensional for $Re \leq 180$. However, in all these explanations one aspect remained obscured by a tacit admission that the background disturbances in the tunnel do not affect the shedding phenomenon for a given tunnel operated at different speeds, a requirement to fix a Reynolds number while using cylinders with different diameters.

This idea was seized upon in [329, 353], where the authors argued and experimentally demonstrated that the background disturbances in a given tunnel are a strong function of tunnel speed. Specifically, two cases of $Re = 45$ and 53 were demonstrated using two cylinders with different diameters for each Re (5 mm and 1.8 mm for $Re = 53$; 2.6 mm and 1.8 mm for $Re = 45$) at two different speeds (17 cm/sec and 46.9 cm/sec for $Re = 53$; 26.1 cm/sec and 37.8 cm/sec for $Re = 45$). The phenomenon of flow instability is facility dependent. This is due to the fact that operation of a tunnel at different flow speeds creates disturbances of differing amplitude at the same frequencies. Performing experiments with different diameter cylinders at the same Reynolds number causes the actual physical frequencies for the same Strouhal number to be different. To explain this dependence unambiguously, experiments have been performed in a wind tunnel with a test section of dimension $30cm \times 40cm$ and a contraction ratio of 8:1. Visualization pictures and other relevant data are shown in Fig. 6.1 for $Re = 53$. Below the visualization picture, FFT of streamwise disturbance velocity (u') are plotted for the empty tunnel at the same speed, U_∞. For the experiments performed at different U_∞, disturbance spectra are similar, displaying a constant level tail, but with different discrete peaks at low frequencies. The turbulence intensity ($Tu = \sqrt{u'}/U_\infty$) is found to be 0.7% for $U_\infty = 17$ cm/sec, which is of the same order (1.0%) for $U_\infty = 46.9$ cm/sec. As the visualization pictures display significantly different instabilities, dependence of it on parameters other than Tu and Re is obvious. This can be detected from the detailed amplitude distribution shown at the lower frequency range for these two cases. The detailed spectrum shows that the amplitudes are an order of magnitude higher for the lower speed, which display well-formed vortex street as compared to the higher speed case at lower frequencies. If the vortex shedding behind the circular cylinder

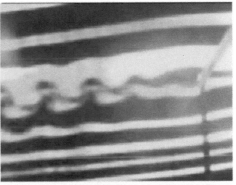

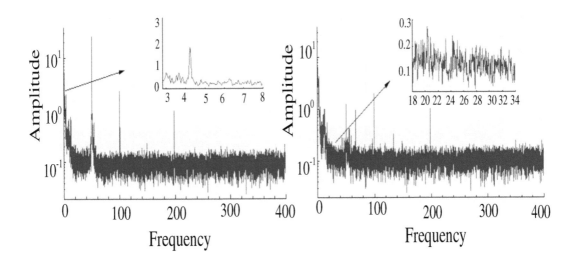

FIGURE 6.1
Experimental observation of flow past a cylinder for $Re = 53$ with $D = 5$ mm; $U_\infty = 17$ cm/s; $Tu = 0.007$ (left) and $D = 1.8$ mm; $U_\infty = 46.9$ cm/s; $Tu = 0.01$ (right). FFT of the empty tunnel disturbances at the same speed is shown below the visualization pictures. Insets show disturbance amplitudes near the Strouhal frequency.

is characterized by the Strouhal number, then for $Re = 53$ this is given by $St = 0.124$ [312] and the physical frequencies for the cases in Fig. 6.1 work out as $f_0 = 4$ Hz for $U_\infty = 17$ cm/sec and $f_0 = 31$ Hz for $U_\infty = 46.9$ cm/sec. From the detailed spectra at the corresponding f_0, one notes that the amplitude of u' is ten times more than for the $U_\infty = 46.9$ cm/sec case. This figure establishes the relevance of background disturbances on flow instability and shedding at natural frequency. While Re and f_0 determine the transfer function of the fluid dynamical system, the detailed distribution of Tu at different frequencies defines the input spectrum for the dynamical system. A spectrum-rich input, as in the case of $U_\infty = 17$ cm/sec, exhibits stronger receptivity and resultant vortex shedding.

A similar conclusion can also be drawn from Fig. 6.2 for $Re = 45$, which is below the

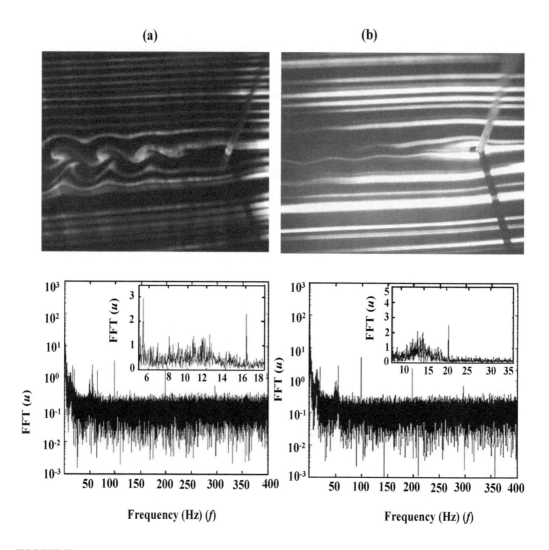

FIGURE 6.2
(a) Experimental observation of flow past a cylinder for $Re = 45$ with $D = 2.6$ mm, $U_\infty = 26.1$ cm/s (left) and (b) $D = 1.8$ mm, $U_\infty = 37.8$ cm/s (right). FFTs of the empty tunnel disturbances at the same speed are shown below the visualization pictures. Insets show disturbance amplitudes near the Strouhal frequency.

first Hopf bifurcation Reynolds number, reported in the literature to be between 45 and 47. For the two models used for this Reynolds number, flow speeds are not too different and corresponding Tu are also similar, with some fine distinction at low frequencies. For the case shown in Fig. 6.2a, for $U_\infty = 26.1$ cm/sec, the Strouhal number is $St = 0.11$ and the corresponding physical frequency is $f_0 = 11.8$ Hz, for which the input disturbance has the amplitude of more than 1. For the case shown in Fig. 6.2b with $U_\infty = 37.8$ cm/sec, one has $f_0 = 24.8$ Hz and an input disturbance amplitude of less than 0.25. The difference of input amplitude at the lower speed explains the corresponding higher receptivity.

This points to the receptivity aspect of the flow field to FST and is the likely explanation for the different Re_{cr} observed in earlier experiments [387, 284, 251], where different aspect ratio cylinders were used. Variability and coupling between FST and tunnel speed have not been explored systematically in the literature. Variation of equilibrium amplitude of the disturbance field in the near wake with FST was presented in Fig. 14 of [387], which clearly shows a large scatter in the growth rate for Re in the range between 50 and 60. Results for higher FST levels clearly reveal Re_{cr} to be significantly higher than 47. This conundrum of higher Re_{cr} for higher FST levels seems non-intuitive, but was discussed in detail in [353] with respect to calculations for FST cases. This is to do with smaller vortices being stripped from the edges of shed eddies in the near wake, by FST. The computational domain used along with inflow and outflow are shown in Fig. 6.3a. Note that the FST is introduced continuously through the inflow, using the parametrized components u' and v', whose details can be obtained from [329, 353].

The role of FST was shown in [353] with the help of the vorticity contour plots shown in Fig. 6.3b and 6.3c for two sets of calculations: zero FST computations, shown in Fig. 6.3b,

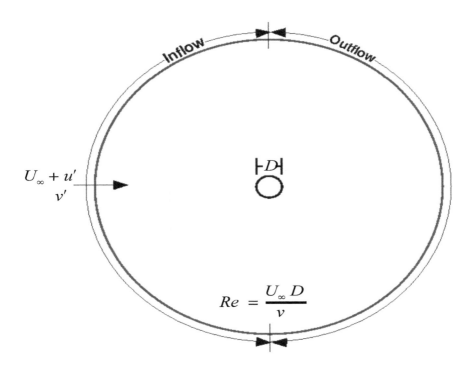

FIGURE 6.3a
Schematic of the physical and computational flow fields, showing the inflow, the outflow and FST introduced through the inflow.

and with an FST level of 0.06%, shown in Fig. 6.3c. Details of FST models and their uses in computing are described briefly later. In Figs. 6.3b and 6.3c, the same contour levels are plotted for the cases shown, which indicate marginal differences in contour values. The main effect of FST is seen in the slightly lower values of vorticity at the core of shed vortices, as compared to the no-FST case. This slight variation shows up in early onset time of primary instability and lower values of Re_{cr}, when the FST levels are low. The effects of higher FST levels are not considered here – those are significantly different, as seen in Chaps. 4 and 5 for flows past streamlined bodies.

So far, the highest value of $Re_{cr} = 65$ has been noted in the experiment in [152] and it was not certain if this relates to the level of FST that led to higher Re_{cr}. However, this experiment was unique, with the following particular features. The working medium used in the tunnel was a highly viscous oil which can dissipate background disturbances by molecular diffusion at a higher rate, compared to commonly used fluids like air or water used in other tunnels. Does it imply that the Hopf bifurcation at $Re \simeq 45$ to 47 is bypassed in the experiments of Homann, while it is not prevented above $Re = 65$? Interestingly, experimental results in [393] also showed qualitative change in flow between $Re = 60$ and 90. One of the aims in [353] was to provide explanations for these experimental observations, which also provided insight on the existence of multiple modes and their interactions during instability.

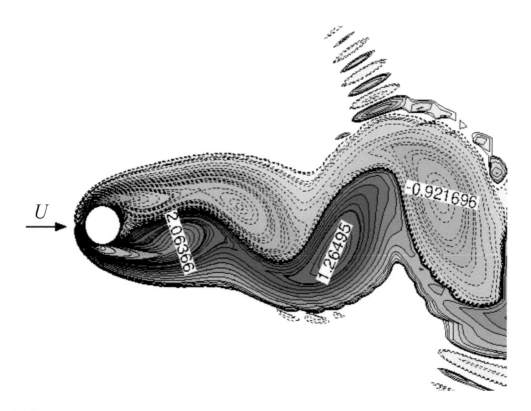

FIGURE 6.3b
Computed vorticity contours for the flow at $Re = 75$ when no FST is introduced through the inflow.

6.2.1 Numerical Simulation of Flow Past a Cylinder

For the numerical simulation of the governing two-dimensional Navier–Stokes equation, (ψ, ω)-formulation is once again preferred due to its ability to preserve solenoidality of both the unknowns. Divergence-free condition combined with absence of the pressure-velocity coupling problem ensures the success of DNS for instability studies of this formulation, as opposed to using primitive variables. Flow is computed in an O-grid with 153 points in the azimuthal (ξ) and 400 points in the radial (η) direction, with the outer boundary placed at a distance of 20 diameters from the cylinder surface. The computational domain with inflow and outflow boundaries is shown in Fig. 6.3a. Details of the formulation and adopted numerical methods are provided in [353, 360], with a brief description here. Governing equations written for a transformed (ξ, η)-plane are the same as given in Eqs. (3.202) and (3.203) reproduced.

$$\frac{\partial}{\partial \xi}\left(\frac{h_2}{h_1}\frac{\partial \psi}{\partial \xi}\right) + \frac{\partial}{\partial \eta}\left(\frac{h_1}{h_2}\frac{\partial \psi}{\partial \eta}\right) = -h_1 h_2\, \omega \tag{6.1}$$

$$h_1 h_2 \frac{\partial \omega}{\partial t} + h_2 u \frac{\partial \omega}{\partial \xi} + h_1 v \frac{\partial \omega}{\partial \eta} = \frac{1}{Re}\left[\frac{\partial}{\partial \xi}\left(\frac{h_2}{h_1}\frac{\partial \omega}{\partial \xi}\right) + \frac{\partial}{\partial \eta}\left(\frac{h_1}{h_2}\frac{\partial \omega}{\partial \eta}\right)\right] \tag{6.2}$$

Here h_1, h_2 are the scale factors of transformation given by $h_1^2 = x_\xi^2 + y_\xi^2$ and $h_2^2 = x_\eta^2 + y_\eta^2$. The diameter ($D$) of the cylinder is taken as the length scale and free stream velocity as

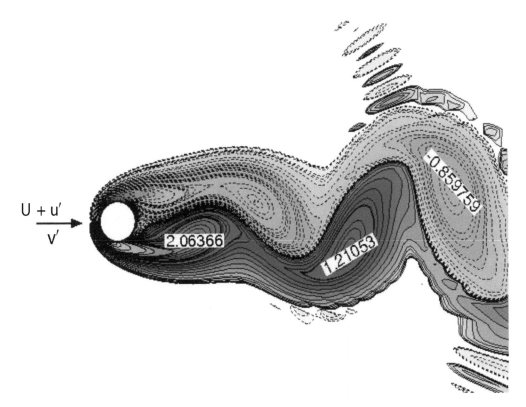

FIGURE 6.3c

Computed vorticity contours for the flow at $Re = 75$ with an FST level given by 0.06% corresponding to the data of [254].

the velocity scale for non-dimensionalization. The time scale is constructed from these two scales.

The loads (lift and drag) are calculated by solving the pressure Poisson equation (PPE) for the total pressure (P_t). The PPE is obtained by taking the divergence of the momentum equation expressed in primitive variables. For the orthogonal curvilinear coordinate system this equation is given by

$$\frac{\partial}{\partial \xi}\left(\frac{h_2}{h_1}\frac{\partial P_t}{\partial \xi}\right) + \frac{\partial}{\partial \eta}\left(\frac{h_1}{h_2}\frac{\partial P_t}{\partial \eta}\right) = \frac{\partial}{\partial \xi}\left(h_2 v\omega\right) - \frac{\partial}{\partial \eta}\left(h_1 u\omega\right) \tag{6.3}$$

The no-slip boundary condition on the cylinder wall is satisfied by

$$\left(\frac{\partial \psi}{\partial \eta}\right)_{body} = 0 \tag{6.4}$$

An additional condition arising out of the no-slip condition is used as

$$\psi = \text{constant} \tag{6.5}$$

This condition is used to solve Eq. (6.1), while both Eqs. (6.4) and (6.5) are used to evaluate the wall vorticity (ω_b) that provides the boundary condition to solve Eq. (6.2). At the outer boundary (i) the uniform flow condition is applied at the inflow (Dirichlet condition on ψ for the upstream part of the outer boundary) and (ii) the convective boundary condition is applied for the radial velocity at the outflow. Through the outflow part of the boundary, shed vortices depart the computational domain smoothly, without affecting the flow inside the domain. The radiative condition applied at the outflow is given by

$$\frac{\partial u_r}{\partial t} + u_c(r,t)\frac{\partial u_r}{\partial r} = 0 \tag{6.6}$$

where u_r is the radial component of velocity and $u_c(r,t)$ is the convection velocity at the outflow at time t, obtained from the radial component of velocity at the previous time step, i.e., $u_c(r,t) = u_r(r, t - \Delta t)$. The initial condition is given by the potential flow, for the impulsive start of the cylinder from rest.

The nonlinear convection terms of Eq. (6.2) are discretized using a high accuracy DRP scheme which provides near-spectral accuracy. The second order central differencing scheme is used for the Laplace operator and the four-stage Runge–Kutta integration scheme is used for time-marching. A sixth order non-periodic filter is applied on the vorticity field to remove spurious disturbances due to aliasing and other nonlinear numerical instability problems.

A detailed discussion on various numerical approaches in the literature is given in [120]; the authors used the immersed boundary method (IBM) in a Cartesian grid. The Cartesian grid helps avoid aliasing error committed in discretizing dissipation terms in the transformed plane. At the same time, a Cartesian grid in the IBM introduces error due to interpolation at the solid boundary, suppression of which requires forcing terms in mass and momentum conservation equations. Interpolation errors have been discussed in detail in [360]. Also, a higher accuracy numerical solution is needed to reconstruct POD modes.

Discretized Poisson equations are solved using the Bi-CGSTAB variant of the conjugate gradient method of [419]. Solving Eq. (6.3) requires the Neumann boundary condition on the physical surface and at the far field; these are obtained from the normal (η) momentum equation given by

$$\frac{h_1}{h_2}\frac{\partial P_t}{\partial \eta} = -h_1 u\omega + \frac{1}{Re}\frac{\partial \omega}{\partial \xi} - h_1\frac{\partial v}{\partial t} \tag{6.7}$$

Computational results for the flow past a cylinder are shown in Fig. 6.4a for $Re = 60$.

Time variation of vorticity at a near-wake point R ($x_R = 0.5044$, $y_R = 0.0$) with the center of the cylinder at the origin is shown in the top frame. In the bottom frame, we have shown the logarithm of the envelope of the amplitude. In Fig. 6.4b, the experimental result from [393] for $Re = 49$ is shown for the streamwise fluctuating velocity component. For the experimental conditions, the time in seconds in Fig. 6.4b has to be scaled up by a factor of 378.43 to obtain the corresponding non-dimensional time. For both these Reynolds numbers, effects of transient during the process of linear instability play a major role. The slope of the envelope provides the linear temporal growth rate of a normal mode, only if

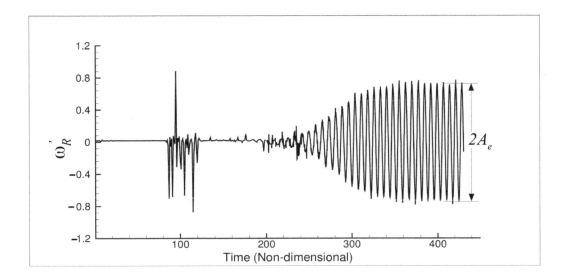

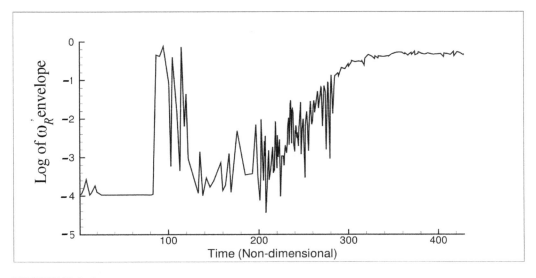

FIGURE 6.4a
Computed time variation of vorticity at the near-wake point R ($x_R = 0.5.44$, $y_R = 0.0$) shown for $Re = 60$ along with its envelope in the bottom frame.

the amplitude of the disturbance is small enough to neglect contributions by other modes and nonlinear intermodal interactions. If there were only a single dominant mode present, then the time variation would appear initially as an exponentially growing curve (when the amplitude is small), which would culminate in time-independent amplitude after the nonlinear "saturation" stage was reached. However, variation of "saturation" amplitude from cycle to cycle indicates the presence of more than one dominant mode. This variation of "equilibrium amplitude" is seen unmistakably in both the experimental and computational

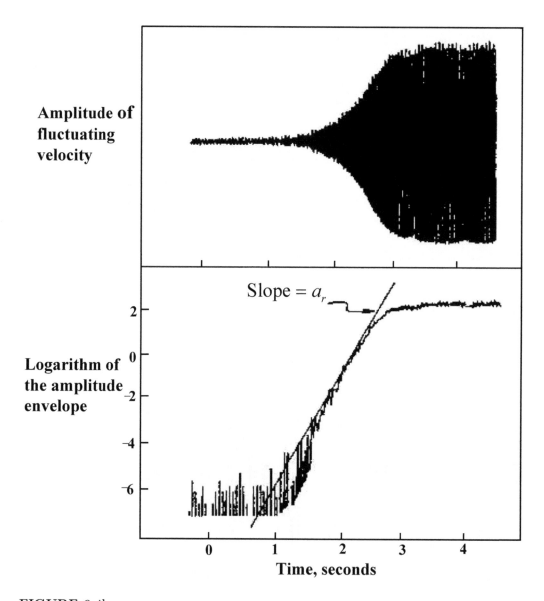

FIGURE 6.4b
Experimental data from [393] for fluctuating velocity and the logarithm of the amplitude envelope for $Re = 49$. Note that the dimensional time of 1 sec is equivalent to 378.435 of the non-dimensional computational time.

results. Some experimental results on the variation of equilibrium amplitude of disturbance velocity in the wake with Re for moderate levels of FST are available in [387]. For a particular case of FST, experimental results from [393] and corresponding computed results, based on solution of Eqs. (3.202) and (3.203) with the appropriate model of FST from [329], are shown together in Figs. 6.5a and 6.5b.

Accuracy of present computations is noted by comparing the numerical results with the available experimental results for fluctuating quantities in Fig. 6.5. Computed equilibrium amplitude of lift coefficient without any FST model is plotted against Reynolds number as discrete symbols in Fig. 6.5a. Plotted equilibrium value is an average of amplitude calculated over a time interval for each Reynolds number case, after the nonlinear "equilibrium stage" in the evolution of lift coefficient is reached. In the figure, the correlation obtained analytically by [254], based on experimental data is shown by the continuous line. Computed results show that the discrete points do not fall on a single smooth curve, despite this, the match between the analytical fit of Norberg and present computations is very good. Data in Norberg was obtained by a single analytical function for a tunnel with turbulent intensity $Tu = 0.06\%$, while the computational data are affected by very low levels of numerical noise of the high accuracy method. Present computations also produce strikingly similar results with the experimental results taken from [393] in Fig. 6.5b, where the mean square streamwise disturbance velocity data is shown plotted as a function of Re.

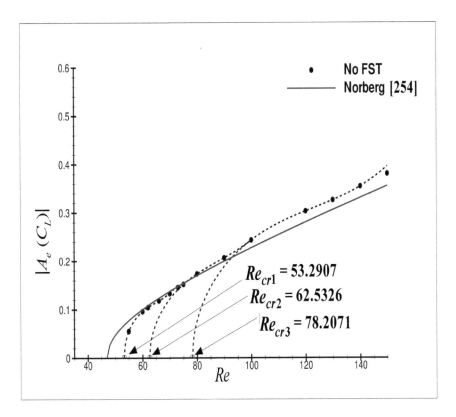

FIGURE 6.5a
Equilibrium amplitude of fluctuating lift variation plotted against Reynolds number, using peak-to-peak computed data, and the experimental correlation shown is from [254].

6.3   Multiple Hopf Bifurcations, Landau Equation and Flow Instability

The perturbation field of a flow undergoing spatio-temporal growth due to physical instability can be expressed by Galerkin-type expansion, in terms of various instability modes present. This formalism has been used variously in [91, 94, 353, 360, 395] to describe fluid flow instability. This Galerkin-type expansion is very common in computing, as well as in POD. The present interest is to use it during instability, which will be later related to POD. We also note that this eigenfunction-type expansion should not be confused with separation of variable used for PDEs. Here we begin by considering an equilibrium flow which is steady, as the case is for flow past a circular cylinder. The theory of stability of a steady basic flow generally gives a spectrum of independent modes with perturbation of the form given for vorticity field by

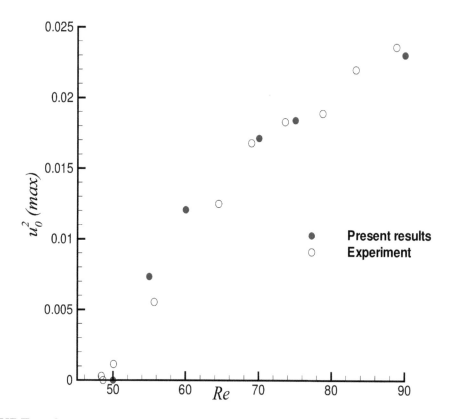

FIGURE 6.5b
Experimental data (blank circle) from [393] and the present computation (filled circle) showing the variation of the amplitude of the fluctuating streamwise velocity component with Reynolds number. The data were obtained at 8 diameters behind the cylinder in the wake. The present computed data have been normalized with the experimental value for $Re = 70$.

$$\omega'(\vec{X}, t) = \sum_{j=1}^{\infty} [A_j(t) f_j(\vec{X}) + A_j^*(t) f_j^*(\vec{X})] \tag{6.8}$$

where A_j's represent time-dependent amplitude functions of the disturbance field, and f_j's belong to a complete set of space-dependent functions satisfying boundary conditions. In Eq. (6.8), quantities with asterisks denote complex conjugates. If the complex amplitude is given by, $A_j(t) = $ constant $e^{s_j t}$ during the linear growth stage, then the individual modes are governed by $\frac{dA_j}{dt} = s_j A_j$. Beyond this linear stage, the disturbance field saturates due to nonlinearity expressed by the following Landau-Stuart-Eckhaus equation, obtained by the eigenfunction expansion process given in [91, 94] as

$$\frac{dA_j}{dt} = s_j A_j + N_j(A_k) \tag{6.9}$$

where $N_j(A_k)$ accounts for all nonlinear interactions among the modes (including self-interaction). This nomenclature for this equation has been proposed in [353, 360] to emphasize that this is more general than the celebrated Stuart–Landau equation developed in [190] and formalized in [395]. Drazin & Reid [91] used it in the context of explaining the role of nonlinearity for subcritical instability and supercritical stability. In [353, 360], the role of multi-mode interactions was highlighted following the proposal in [94], which makes Eq. (6.9) more general, as compared to the Stuart–Landau equation given below in Eqs. (6.10) and (6.11), which only incorporates nonlinear self-interaction, ignoring multi-modal interactions.

Irrespective of the number of modes present, many researchers following Stuart–Landau equation have considered only a single dominant mode with $s_j = \sigma_r + i\omega_1$ representing its linear complex temporal growth exponent. The proposed model in [190] was only for the amplitude of the leading mode, whereas Stuart [395] developed a satisfactory differential equation for a plane parallel flow following the mathematical foundation in [434] from the governing Navier–Stokes equation for both the amplitude and the phase of the most dominant mode. With a single mode in consideration, Landau and Stuart considered the self-interaction term only in Eq. (6.9) as $N_j = -\frac{l}{2} A |A|^2$, where $l = l_r + i l_i$ is the Landau coefficient. Writing $A = |A| e^{i\theta}$, one can rewrite Eq. (6.9) for the amplitude and phase of the leading mode as

$$\frac{d|A|^2}{dt} = 2\sigma_r |A|^2 - l_r |A|^4 \tag{6.10}$$

$$\frac{d\theta}{dt} = \omega_1 - \frac{l_i}{2} |A|^2 \tag{6.11}$$

In linear stability theory, one can superpose modal contribution — a good reason to resort to normal mode analysis. However, for the nonlinear stage of disturbance growth, one cannot superpose contributions calculated from the Stuart–Landau equation by individual modes.

For different flows, different signs of real and imaginary parts of Landau coefficient l lead to different nonlinear effects through Eqs. (6.10) and (6.11). Here we will keep our attention focused on flows for which instability begins as linear temporal instability, i.e., $\sigma_r > 0$ for $Re > Re_{cr}$. One important aspect of this linear instability is the subsequent non-linear saturation, adequately explained by Eq. (6.10) if only l_r is positive. We focus on this type of flow only.

Despite the nonlinearity of Eq. (6.10), it is directly integrable to

$$|A|^2 = \frac{A_0^2}{(A_0/A_e)^2 + [1 - (A_0/A_e)^2] \, e^{-2\sigma_r t}} \tag{6.12}$$

where A_0 is the value of A at $t = 0$. Here $A_e = \sqrt{2\sigma_r/l_r}$ represents the asymptotic value of the solution for $t \to \infty$, for which the left-hand side of Eq. (6.10) is zero. The approach of A to A_e indicates the independence of A_e on A_0 − one reason for which this flow is termed an oscillator. Such a solution is due to a particular combination of $l_r > 0$ and $Re > Re_{cr}$ which takes the temporally growing flow to a strictly time periodic neutral state of *supercritical stability*. For supercritical Re, $\sigma_r \sim (Re - Re_{cr})$ and thus a plot of A_e versus Re will represent a parabolic variation between the two. It is expected that the Stuart–Landau model has better applicability near Re_{cr}.

Similarly, one can also obtain the equilibrium Strouhal number after the nonlinear saturation from Eqs. (6.10) and (6.11) as

$$|A_e|^2 = 2\sigma_r/l_r \quad \text{and} \quad \omega_e = \omega_1 - \sigma_r l_i/l_r \tag{6.13}$$

Hopf bifurcation describes the passage of a dynamical system from a steady state to a periodic state as a typical bifurcation parameter is varied, which in this case is the Reynolds number [129].

6.4 Instability of Flow Past a Cylinder

For steady equilibrium flow past a circular cylinder, we identify the critical Reynolds number corresponding to linear temporal instability as Re_{cr_1}. Similarly, we identify the critical Reynolds number value indicated in Homann's experiment [152] as Re_{cr_2}. Results of the numerical investigations mentioned above relate to study of the flow system unimpeded by noise or perturbations.

Figures 6.5a and 6.5b show discontinuous variation of A_e with Re for the experimental and computational data, which indicates more than one dominant mode during linear instability and subsequent nonlinear saturation. It is generally believed that all these displayed variations originate via a single Hopf bifurcation following Eq. (6.12). However, discontinuities in these figures are either due to the presence of multiple modes active for different Reynolds numbers or the continuous presence of these modes at all Reynolds numbers, with different interactions for different Reynolds number ranges. Variation of A_e with Re seen in experiments and computations can be due to a combination of both these effects. Discontinuous variation is thus indicative of more than one leading mode and one bifurcation − as is commonly modeled through the Stuart–Landau equation. One can interpret the observed variation by multiple Hopf bifurcations originating at different Reynolds numbers, with different segments merging at the discontinuities. The computational data in Fig. 6.5a, shown by discrete symbols, indicate the presence of two such discontinuities to originate from three bifurcations − a result shown first in [353].

An analysis was performed in [353] for the bifurcation sequence noted for the flow in Fig. 6.5a. For the experimental data in Fig. 6.5b, a discontinuity is clearly noted near $Re = 80$ and the computed results in Fig. 6.5a reveal a similar discontinuity for Re above 75. In the analysis, a quartic polynomial for the computed data was fitted in different ranges of Reynolds numbers according to

$$|A_e|^2 = k_{1i}Re + k_{2i}Re^2 + k_{3i}Re^3 + k_{4i}Re^4 \tag{6.14}$$

where Re_{cri} is the critical Reynolds number in the appropriate ranges, as defined in Table 6.1. Values of equilibrium amplitude (A_e) obtained from DNS for different Reynolds numbers are used in Eq. (6.14) for four points in each range of Reynolds numbers indicated in the first column of Table 6.1. Having obtained the constants k_{ij}s in different range used in Eq. (6.14) to draw the dotted lines, as shown in Fig. 6.5a. In the figure, wherever the dotted line in each segment intersects the Reynolds number axis, that point defines the corresponding Re_{cri} of that range and are listed in the last column of Table 6.1. The primary intention here is to show the proximity of the value of Re_{cr2} for the second bifurcation with the Reynolds number for vortex shedding in [152] and the kink noted in (A_e versus Re) curve in [393], as reproduced here in Fig. 6.5b. We also note that though Eq. (6.14) represents a quartic, the cubic and the quartic terms are of lesser importance. Thus, comparisons in Figs. 6.4 and 6.5 not only establish the accuracy of the computed results here, but also support the higher critical Reynolds number observed in Homann [152], a direct attribute of multiple Hopf bifurcations for this flow field.

We note that in [168], the authors have studied the effects of noise on pitchfork and Hopf bifurcations of dynamical systems governed by ordinary differential equations. They reported marginally changed critical parameters for Hopf bifurcation. It is only 0.2% towards subcriticality (which is within the limits of experimental and fitting uncertainties) when measured with respect to the Stuart–Landau model and by only 0.32% towards subcriticality when measured with respect to the power spectrum by the least squares fit. When the effects of the added noise are estimated from the amplitude probability distribution, a shift of 0.43% towards supercriticality was noted for the critical parameter. These observations show only a very small change in critical parameter, as compared to that obtained in different experiments for flow past a cylinder. The overall difference between the range of experimentally obtained Re_{cr} with the first Hopf bifurcation value is more than 50% for this flow. We also note that the observations in [168] are with respect to added white noise. Unfortunately, in many computations (as in [31, 163, 207, 292]), FST effects have been modeled with the help of computer generated uniformly distributed pseudo random numbers. This is possibly with the exception in [342], where a model for FST was developed which is described below, along with the later improvements added to the model. The necessity to model background disturbances appropriately gains in importance following the observation in [442] that near the onset of dynamic instability, time periodic systems act as amplifiers of small periodic perturbations, whose *details depend solely on the type of bifurcation involved.* Thus, it is apparent that in an attempt to develop FST models, one must include all periodic components of background disturbances — especially the ones in the neighborhood of Strouhal frequency, for flow past bluff bodies.

TABLE 6.1
Coefficients used in the saturation amplitude equation given by Eq. (6.14).

Re range	$k_{1i} \times 10^2$	$k_{2i} \times 10^4$	$k_{3i} \times 10^5$	$k_{4i} \times 10^8$	Re_{cri}
53.29-75	-2.0256	9.34156	-1.4345	7.4092	53.2907
62.53- 100	-0.9750	3.3070	-0.3667	1.3940	62.5326
78.20- 150	-1.1630	2.9389	-2.3580	0.6414	78.2071

6.5   Role of FST on Critical Reynolds Number for a Cylinder

In actual flow past a bluff body, omnipresent FST with low amplitude of excitation would be effective during the receptivity stage and would affect transition. This would be different from the computed uniform flow over a smooth geometry. FST may appear as random and is often modeled stochastically. In [329], a model of FST combining deterministic and stochastic approaches has been used in computing flow past a cylinder. In the proposed model, it is noted that the variance or second order statistics represents the magnitude of turbulence fluctuations; skewness or third order statistics represents deviation from symmetric distribution and kurtosis or fourth order statistics describes the flatness of the tail of the distribution.

The FST model described here is based on empty wind tunnel data, whose histogram remains invariant with respect to the origin of the time series, despite the apparent highly disorganized nature of the disturbance component of the velocity signal. The implicit statistical invariance of FST was exploited in [342] where the second and fourth order statistics of the measured disturbance field of a tunnel reported in [102] was matched by a synthetic time series for the oncoming flow through the inflow of Fig. 6.3a. In this original model, third order statistics was neglected. Furthermore, it was conjectured that the disturbance field is isotropic, which allows construction of the vorticity field entering the flow domain via the standard moving-average model of order one, as given in [106]. This type of assumption is commonly employed in LES subgrid scale models to parametrize small scales in the energy spectrum.

First, we describe that part of the model which mimics the small scale isotropic component of FST – representing the tail of the spectrum shown in Figs. 6.1 and 6.2. The time series representing the oncoming small scale fluctuations through the inflow of the domain is given by

$$x_t = e_t + \alpha_f \, e_{t-1}$$

where α_f is a model constant to be fixed and the subscripts indicate the time at which the quantities are evaluated. The variable e_ts are given by normal distribution of zero mean and a standard deviation, σ. Various moments of this time series are given by

$$\mu_1 = E(x_t) = 0, \ \mu_2 = E(x_t^2) = (1 + \alpha_f^2)\sigma^2$$

$$\mu_3 = E(x_t^3) = 0, \ \mu_4 = E(x_t^4) = (1 + \alpha_f^4)3\sigma^4$$

If one has the histogram for the fluctuating velocity $f(x)$, then the above moments can be obtained as

$$\mu_n = \frac{\sum_i [f(x_i - \mu_1)]^n}{\sum_i f(x_i)}$$

where $f(x_i)$ is the discrete i^{th} data point in the histogram. Thus in [342], two parameters, α_f and σ, are obtained from μ_2 and μ_4 evaluated from the histogram.

In experimental facilities, low frequency deterministic sources are mostly due to fans, other rotary components and flow turning which create a flow with vorticity. The FST is parametrized using a moving average model of order one, where a time series is created for the streamwise disturbance component of the velocity field represented by

$$(u_x')_\infty = e_t + \alpha_f \, e_{t-1} + \sum_{j=1}^{N} b_j e^{ik(x-ct)} \qquad (6.15)$$

The last term represents the low frequency component of the FST, which is facility and speed dependent; c is the phase speed of low wavenumber coherent structures. Such a model is used at inflow of the computational domain to mimic background disturbance for solving the Navier–Stokes equation, as shown in Fig. 6.1. It is noted in [329] that the exact value of c is not very important. This is due to the fact that once the disturbances enter the computational domain at any velocity, subsequently the dynamics of the eddies associated with eddies are dictated by the Navier–Stokes equation. Thus, at every time step of integrating Eq. (6.2), the input disturbance field is obtained by using Eq. (6.15) for every point of the inflow. Note that b_j depends on flow speed for a given facility, carrying the spectrum information of the FST for frequencies around the Strouhal number.

The synthetic model of [325] is compared in Fig. 6.6a with an empty tunnel velocity spectrum from a wind tunnel and the match is quite good. The computed flow past a cylinder for $Re = 55$ reported in [329] using this FST model displayed a very good match with Homann's results [152].

Having obtained the time series for the Cartesian components of disturbance velocity at the inflow as $(u'_x)_\infty$ and $(v'_y)_\infty$, one transforms these to components in an orthogonal coordinate (ξ, η) as

$$(u')_\infty = (u'_x)_\infty \, (y_\eta - x_\eta) \, h_1 h_2$$

$$(v')_\infty = (v'_y)_\infty \, (y_\xi - x_\xi) \, h_1 h_2$$

The above FST model has been used to compute a flow field for a turbulence intensity (Tu) of 0.06%, as used in [254]. In Fig. 6.6b, computed A_e versus Re curves between the cases of zero FST and $Tu = 0.06\%$ are reported to note the difference on the first Hopf bifurcation. The same procedure is used, by fitting a quartic relation between A_e^2 and Re, as given in Eq. (6.14). The introduction of a low level of FST brings down Re_{cr1} from

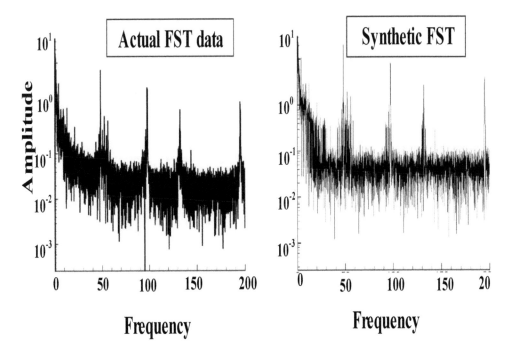

FIGURE 6.6a
FFT of actual and modeled FST data obtained using Eq. (6.15).

53.29 to 49.87, which explains why different experimental facilities report different critical Reynolds numbers. This also helps one understand the role of numerical noise in reporting Re_{cr} by different numerical methods or by using the same method with different numerical parameters. This can explain that the qualitatively different value of Re_{cr} in [152] is due to the totally different working medium used in the experiment. During the receptivity stage, dependence on input disturbance is obvious – which controls primary instability. In [152], highly viscous oil was used as the working medium, which contributed to dissipating background disturbances effectively, bypassing the first Hopf bifurcation.

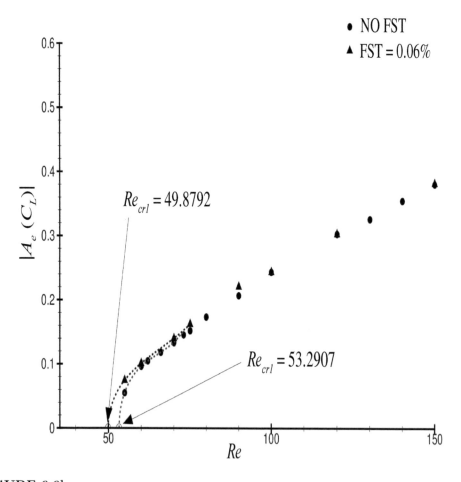

FIGURE 6.6b
Comparison of equilibrium amplitudes of fluctuating lift variation plotted against Reynolds number for cases without and with FST for the level of $Tu = 0.006\%$, corresponding to the case in [254].

6.6 POD Modes and Nonlinear Stability

POD expansion was originally developed in [181] to project complex stochastic spatio-temporal dynamics on to a reduced number of deterministic eigenfunctions while representing almost the full energy or enstrophy. POD related to turbulent flows has been described in [151]. The POD technique that is used most often is the method of snapshots in [371]. Another alternative is given in [356] in the context of reduced order modeling by locating eigenvalues and eigenvectors of the covariance matrix using Lanczos iteration. The reported results here are based on the method of snapshots.

Seemingly chaotic turbulent flows exhibit underlying coherent structures with the help of POD using the method of snapshots. For vortex dominated laminar or transitional flows, the spatio-temporal dynamics are easily interpreted with the help of POD modes — as established in [353, 360] for internal and external flows. In the absence of stochasticity, the POD modes become not only deterministic, but also unique. This observation also allows relating POD modes with the instability modes. Such an approach has been advanced in [353, 360] in representing a physical model for such dynamical systems. In this method, the number of snapshots M used in the analysis is significantly smaller than the number of grid points used. Data obtained by DNS provides the ensemble of snapshots taken at M instants of time. Taking a reduced number of basis functions (from the complete set in a separable Hilbert space) causes the mathematical modes to be projected as POD modes.

In POD, a space-time dependent field ω' is represented by eigenvectors, ϕ's, such that the projection of the latter to the former is maximized, i.e., the quantity $< \omega'\phi >^2$ is maximized subject to given initial and boundary conditions. The angular brackets represent a suitable averaging operation. Although, the POD technique was developed for stochastic dynamical systems, its extension to study flow instabilities was advanced in [353, 360]. Without going into details, one can show that the vector ϕ_i's are the solution of an optimization problem in calculus of variation obtained from [119] as

$$\int_{\text{region}} R_{i,j}(r, r')\, \phi_j(r')\, dr' = \lambda_i \phi_i$$

The kernel of this integral equation is the two-point correlation function given by $R_{ij} = < \omega_i(x)\omega_j(x') >$ of the space-time dependent field. For a system with finite energy content, the Hilbert–Schmidt theory applies to give rise to denumerably infinite POD modes which are orthogonal to each other. If one is studying disturbance quantities as in instability theories, then the disturbance remains finite, even if one is investigating inhomogeneous flow of an infinite extent. This is due to the fact that the disturbance field receives its energy from the equilibrium flow, which has finite energy. Thus, many of the restrictions of using POD for turbulent flows related to the applicability of the Hilbert–Schmidt theory are not present for flows suffering instability.

In interpreting instability, a time horizon is chosen over which a time mean is obtained, which is subtracted from instantaneous realization to get the disturbance component of the field. The velocity field is often used for POD analysis, so that the cumulative sum of the eigenvalues provides a measure of the kinetic energy. But in many work, vorticity field is also used, so that the eigenvalues provide a measure of enstrophy of the investigated flow field. Vorticity being a primary flow quantity of interest, this approach is preferred and moreover, the solution obtained by stream function-vorticity is more accurate. If one defines the disturbance vorticity field as

$$\omega'(\vec{X}, t) = \sum_{m=1}^{M} a_m(t)\, \phi_m(\vec{X}) \tag{6.16}$$

then ϕ_ms are obtained as the eigenvectors of the covariance matrix whose elements are defined as $R_{ij} = (1/M) \int \int \omega'(\vec{X}, t_i)\, \omega'(\vec{X}, t_j)\, d^2\vec{X}$, with $i, j = 1, 2, ..., M$ defined over all the collocation points in the domain. Corresponding eigenvalues give the probability of occurrence and the sum giving the total enstrophy of the system. Their partial sums give the cumulative enstrophy of the system

$$En = \sum_{i=1}^{N} \omega'^2(\vec{X}, t_m)$$

In reporting the multi-modal nature of the flow evolution by POD, the flow past the cylinder for $Re = 60$ was considered in [353] on purpose, for the following reason. In Fig. 6.7, time histories of the lift coefficient for Reynolds numbers of 60, 100 and 250 are depicted. It is evident from the figure that for $Re = 250$, there is only a single dominant mode determining equilibrium variation of lift. In comparison, for $Re = 100$, the presence of more than one mode is evident from the displayed time variation. However, for $Re = 60$, the time history makes it abundantly clear that there are many active modes, not only during the transient phase, but also after the nonlinear "saturation" of the lift coefficient.

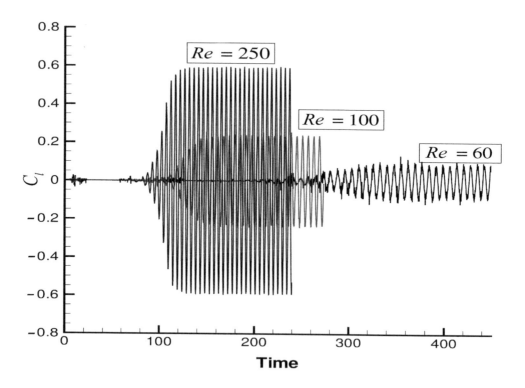

FIGURE 6.7
Computed lift coefficient variation with time for $Re = 100$ and $Re = 250$. Note the increasing presence of multiple modes with decreasing Reynolds number. Subsequently, only the $Re = 60$ case is analyzed for multiple modes in POD analysis and studying the LSE equation.

This flow field for $Re = 60$ was analyzed in [353] by taking the same time series for three different overlapping time ranges to account for effects of transients on the equilibrium flow. First, eigenvalues of the dynamical system are obtained for these three cases presented in Table 6.2. Cases A ($200 \le t \le 430$) and B ($350 \le t \le 430$) were compared mostly to highlight the roles of transients in terms of various POD eigenmodes. Case C included comprehensively the full time range from $t = 0$ to $t = 430$ to highlight the impulsive start of the flow.

According to [402], "many practical engineering problems can be solved approximately, by assuming that the external disturbance is impulsive .... changes of boundary conditions may therefore be considered, with a reasonable accuracy, as impulsive. This is specially true of unsteady separation." Also, it was noted that "the early stages of impulsive viscous flows contain valuable information on flow properties that control the subsequent development of the phenomenon and a thorough understanding of its properties is essential in the study of unsteady flows." Different start-ups have been studied by some researchers which include the work on accelerated flow past a symmetric airfoil in [343] where three prototypical start-ups including impulsive start were considered. We emphasize that impulsive start provides

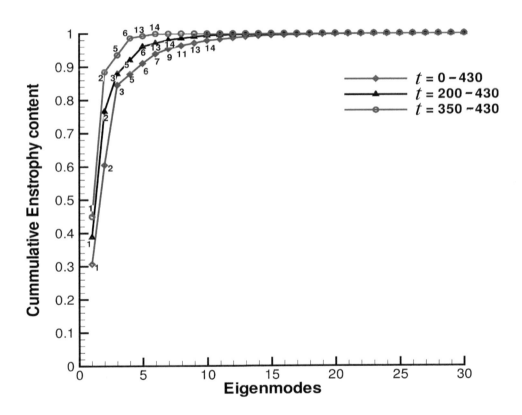

FIGURE 6.8
Cumulative enstrophy distribution among the eigenmodes for the same data set analyzed for different time ranges. The complete range for the data taken from $t = 0$ to 430 was used for the curve shown with rhombus symbols, data taken from $t = 200$ to 430 is shown by the curve with triangles and data taken from $t = 350$ to 430 with circles. Note the data segments with transients require more numbers of modes for an accurate representation by POD.

a consistent initial condition in studying the receptivity aspect of flow fields. This is due to the fact that impulsive start by its very nature excites the widest possible bandwidth of circular frequencies and hence would be able to latch onto the unstable frequency naturally. This is also the reason for which one undertakes study of impulse response and this is adopted for the flows studied here. It is also well known that impulsively started flows are initially inviscid.

It is to be noted that a different start-up can lead a flow to settle down to a different equilibrium solution, as has been shown in [343] for flow past an airfoil and in [309] for flow past a circular cylinder. Sarpakaya [309] studied the effects of uniform acceleration for flow past a circular cylinder by defining a non-dimensional acceleration parameter A_p given by

$$A_p = D\frac{dU_\infty}{dt}/\bar{U}_\infty$$

where D is the diameter of the cylinder and $\bar{U}_\infty$ is the final velocity and U_∞ is the instantaneous velocity. It was noted that for $A_p > 0.27$, the drag coefficient beyond the period of initial acceleration does not measurably depend upon A_p and the flow can be considered as impulsively started. The authors in [343] noted that acceleration has the most pronounced effect when the first bubble forms over the surface. Thus, the decided dependence of the start-up condition is not trivial and the impulsive start is considered a standard benchmark condition and in many experimental set-ups this is usually the case, until and unless it is not practiced on purpose.

In Fig. 6.8, the cumulative enstrophy content of the flow past a circular cylinder for $Re = 60$ is shown for the three cases of Table 6.2, with five snapshots taken in a unit time interval for POD. For Case C, the time range begins from impulsive start to a state where the fluctuating component of flow variables has reached an equilibrium time periodic state. Case B encompasses the range $200 \leq t \leq 430$, which partly includes the late-transient and the saturation stage of flow evolution, while the time interval $351 \leq t \leq 430$ contains only the equilibrium stage. Total enstrophy content during the equilibrium stage requires fewer modes for its correct description, as compared to any case that includes the transient stage. However, for all the cases, the first fifteen modes account for nearly the total amount of enstrophy, with higher modes contributing insignificantly. This is the reason for including only the first fourteen POD modes in Table 6.2. The numbering of POD modes signifies the contribution to enstrophy of a mode, with lower numbered ones contributing larger values. We have ordered the modes in Fig. 6.8 in a numerical sequence, with blanks to indicate missing modes. This convention follows the idea in [81, 209, 253] which stated that the POD modes primarily form pairs and vortex shedding in the wake as a result of interaction of these phase-shifted modes of the pairs. For example, for the flow past a cylinder, first and second mode will always form a pair with the modes roughly 90° phase apart, with almost equal magnitude of the eigenvalues. Deane et al. [81] proposed that vortex shedding is due to interactions between the leading pair of eigenmodes $a_1(t)\,\phi_1(\vec{X})$ and $a_2(t)\,\phi_2(\vec{X})$, which also is responsible for the traveling characteristics of the vortex street. Interestingly, it was pointed out in [253] that the dynamics of cylinder wake are also influenced by the presence of a mode — termed the shift mode — which is a solitary mode without forming a pair.

POD modes appearing in pairs and satisfying the Stuart–Landau equation have been identified as regular POD modes or R-modes. These POD modes have amplitude functions whose time variation resembles the time variation of the vorticity in Fig. 6.4. This is displayed by the instability modes following the Stuart–Landau Eq. (6.10). When modes appear alone and/or do not follow this type of time variation, we term them anomalous modes. Thus, anomalous modes appearing alone are always followed by a missing mode, as shown by ϕ_3 in Fig. 6.9 for Case C. These POD modes' amplitude shows a non-periodic shift with time and hence, are called the shift modes [253] or anomalous T_1-modes. This rationale was used for the nomenclature and numbering of modes in [353, 360]. One encounters another type of anomalous modes which appear in pairs, but whose time variation for the corresponding amplitude function is different from that given by the Stuart–Landau equation; are called T_2-modes. Eigenvalues in various cases of Table 6.2 are arranged and as numbered in Fig. 6.8, become evident by looking at the eigenvectors shown in Figs. 6.9 and 6.10. The numbering sequence was decided upon by looking at the POD results for Case C. It was noted that modes 1 and 2, 5 and 6 and 13 and 14 formed pairs, out of which the first two are the regular ones and the last one is a T_2-mode. The rest of the modes are T_1-modes or anomalous modes of the first kind. Modes are said to form a pair when the eigenvalues are close to each other and the eigenvectors show some resemblance with a phase shift of 90°. One can also note pairing easily by looking at the time variation of POD amplitude functions $a_j(t)$, as shown later.

In Fig. 6.9, eigenvectors for Case C are shown plotted with (ϕ_1, ϕ_2) and (ϕ_5, ϕ_6) representing regular modes and the alternating vortical structures in the wake indicating the above mentioned phase shift noted in the equilibrium flow. But modes (ϕ_{13}, ϕ_{14}) represent the T_2-mode, while the rest of the eigenvectors, including ϕ_3, constitute T_1-modes. For Case C, the presence of four such anomalous modes of the first kind was noted.

The third eigenmode (ϕ_3) shown in Figs. 6.9 and 6.10 has been called here T_1-mode, which has a similar counterpart in [253] called the shift mode. It has been stated that the shift mode is an attribute of the variation of the flow in a slow time scale, with these modes considered as slowly varying waves traveling downstream, with two symmetrically placed vortices about the wake center-line. The T_1-modes were obtained directly from the POD of DNS data, while the shift mode was obtained differently, with 100 snapshots taken uniformly over a single time period defined by the Strouhal frequency. Thus, in analyzing the data for the shift mode, data were sampled that was only a fraction of the full time period of the shift mode. In [353], data were sampled over a large number of time periods corresponding to the Strouhal frequency. The shift mode in [253] has been interpreted to be present at all times. However, Figs. 6.9 and 6.10 show the T_1-modes to be transient in nature.

The shift mode in [253] was constructed as a *mean-field correction*, equal to $(u_0 - u_s)$, where u_0 is the time averaged solution of the unsteady Navier–Stokes equation and u_s is the solution of the steady Navier–Stokes equation. When POD modes were calculated using 20 snapshots in a unit time interval for the $Re = 60$ case in [353], not more than four T_1-modes were noted for Case A.

In Fig. 6.10, the eigenvectors for Case A are shown. This case is characterized by the presence of a transient in part, along with the equilibrium phase of the disturbance growth. One notes the presence of a T_1-mode (ϕ_3) and a T_2-mode represented by (ϕ_{13}, ϕ_{14}). The R-modes (ϕ_1, ϕ_2) and (ϕ_5, ϕ_6) define the periodic wake. The absence of early transient effects in this shorter time interval from $200 \le t \le 430$ is characterized by a lesser number of T_1-modes, which represent essentially the transient effects.

Case B of Table 6.2 is not shown, as the flow is solely in the equilibrium state without transients, and this shows only three pairs of regular modes. The utility of POD analysis with accurate DNS results was clearly noted in the exercise by the total absence of anomalous modes. From Fig. 6.8, one notes that these six nontrivial POD modes account for more than 99% of total enstrophy.

Having obtained the snapshots and the eigenfunctions in the chosen time range, as in Figs. 6.9 and 6.10, one can calculate the POD amplitude functions from Eq. (6.16) as

$$a_i = \frac{\int \int_S \omega' \phi_i \, dS}{\int \int_S \phi_i^2 \, dS} \tag{6.17}$$

by using the orthogonal property of eigenfunctions. It should be noted that to obtain POD amplitudes correctly, one needs to take a large number of snapshots, as in [353, 360]. Otherwise, amplitude functions would be obtained inaccurately. Pairing of modes is taken into account while plotting consecutive amplitudes either together or separately. Such identification becomes even clearer upon viewing the POD amplitude functions depicted in Fig. 6.11 for Case C.

Almost equal magnitudes of successive eigenvalues indicate formation of regular pairs and/or pairs of T_2-modes, while isolated modes are the T_1-modes (as seen, e.g., from the

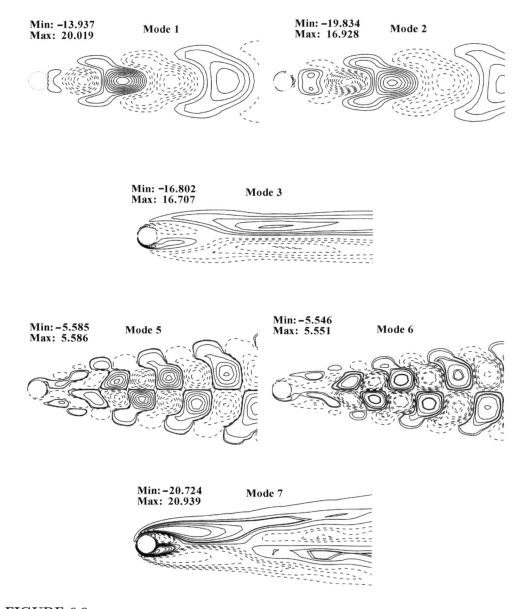

FIGURE 6.9a

Leading eigenfunctions for the data set from $t = 0$ to 430 for $Re = 60$ following an impulsive start. The first two modes, and also the fifth and sixth modes form regular pairs. The third and seventh are anomalous modes of the first kind which appear without forming a pair.

third eigenvalue for Cases A and C in Table 6.2). However, one cannot distinguish between R- and T_2-modes from the eigenvalues only. For this, one looks at the amplitude functions, as evaluated from the DNS data. The amplitude functions of POD modes for Case C are plotted in Fig. 6.11. This figure helps in easy identification of both types of anomalous modes. While it is easy to understand the trend of the time variation of paired modes at early times, for T_1-modes the amplitude varies rapidly in the transient stage, as seen clearly for a_3, a_7, a_9 and a_{11}. Unlike the present approach, researchers have attempted to generate POD amplitude functions from the Navier–Stokes equation. This is difficult, as it amounts to solving a set of stiff differential equations (even when pressure terms are neglected as in [81, 253].

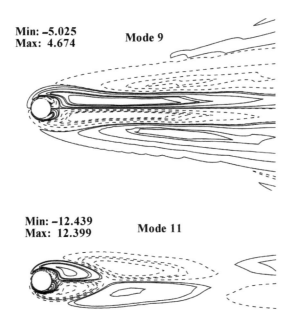

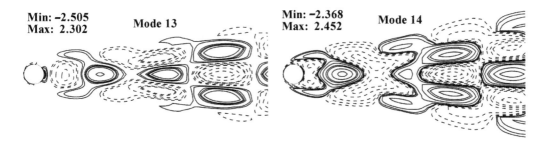

FIGURE 6.9b

Leading eigenfunctions for the data set from $t = 0$ to 430 for $Re = 60$ following an impulsive start. The thirteenth and fourteenth modes constitute an anomalous mode of the second kind; the ninth and eleventh are anomalous modes of the first kind.

The stiffness of the governing differential equations for amplitude functions involving multi-modal interactions has been investigated in [353] with respect to the developed Landau–Stuart–Eckhaus equation. In Fig. 6.11, T_1-modes show rapid excursion during the

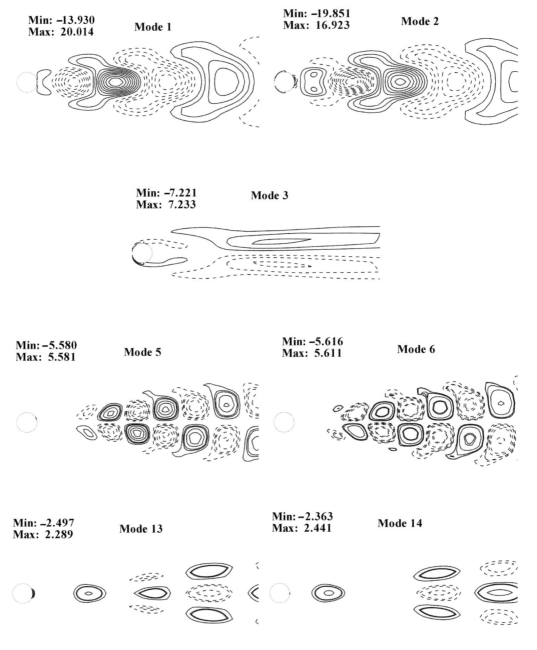

FIGURE 6.10

Eigenfunctions for the data set from $t = 200$ to 430 for $Re = 60$ following an impulsive start. The first two modes and also the fifth and sixth modes form regular pairs. The third mode is the anomalous mode of the first kind. The thirteenth and fourteenth modes constitute an anomalous mode of the second kind.

transient stage. Amplitudes of the higher modes are higher as compared to a_3, yet the third eigenmode dominates the transients − as the contribution of individual modes can be noted from Fig. 6.8.

Evidently for R-modes, eigenfunctions and amplitude functions show a phase shift of $90°$. Thus, it is natural to relate POD modes with instability modes governed by Stuart–Landau equations, in Eqs. (6.10) and (6.11). One of the major contributions in [353, 360] was to relate these two types of descriptions for the disturbance field. This was performed by representing these phase shifted complementary POD modes as the real and the imaginary parts of an instability mode along with the help of a normalization factor given by $\epsilon_j = (\lambda_{2j-1} + \lambda_{2j})/\sum_{l=1}^{M} \lambda_l$ via

$$A_j(t) = \epsilon_j \left[a_{2j-1}(t) + i \ a_{2j}(t)\right] \tag{6.18}$$

$$f_j(\vec{X}) = \frac{1}{2\epsilon_j}\left[\phi_{2j-1}(\vec{X}) - i \ \phi_{2j}(\vec{X})\right] \tag{6.19}$$

This normalization appears feasible, as the eigenvalues of the covariance matrix determine the corresponding contribution to enstrophy of the flow of the overall disturbance field. This procedure to investigate flow instability as a spatio-temporal event for global nonlinear analysis is a powerful tool, provided one can obtain POD modes as accurately as possible. Thus, (a_1, a_2) constitute real and imaginary parts of A_1 and (a_5, a_6) constitute A_3 and so on, for the eigenmodes appearing as pairs. But A_2 has only one eigenvector, as shown in Fig. 6.9, and so will be related to a_3 alone. Furthermore, a_7 forms A_4 and a_9 constitutes A_5 and so on. The POD amplitude modes in lower case are related to instability amplitudes indicated by A_j and are called the LSE-modes in [353].

In Figs. 6.9 and 6.10, while A_1 and A_3 contribute to overall dynamics in an unequal measure, these modes do not act in the same spatial region of the wake. The first instability mode, A_1, is present globally with a dynamic range (as indicated by the minimum and maximum values of vorticity in the figures) that is almost three times that indicated for A_3. Also, the A_3-mode becomes significant only after a few diameters distance in the wake, as shown in Fig. 6.10. The role of the first anomalous mode can explain the success of flow

TABLE 6.2
Eigenvalues in different time ranges for $Re = 60$, for flow past a cylinder.

POD mode number	Eigenvalues for Case A: $200 \leq t \leq 430$	Eigenvalues for Case B: $350 \leq t \leq 430$	Eigenvalues for Case C: $0 \leq t \leq 430$
1	1.662	2.497	0.8918
2	1.617	2.414	0.8674
3	0.484	—	0.7001
4	—	—	—
5	0.175	0.283	0.0941
6	0.174	0.280	0.0938
7	—	—	0.0830
8	—	—	—
9	—	—	0.0441
10	—	—	—
11	—	—	0.0260
12	—	—	—
13	0.042	0.037	0.0227
14	0.040	0.036	0.0217

control in [87, 392, 393]. The problem of controlling the wake of a cylinder by a smaller control cylinder in the near-wake was investigated in these references. It was noted that control was effective when the smaller cylinder was placed in a region that resembles the near-wake contours of ϕ_3 in Fig. 6.9. Apart from symmetry about the wake center-line, the exact shapes of the shift mode of [253] and ϕ_3 of Fig. 6.9 are distinctly different. The location, size and shape of ϕ_3 in Fig. 6.9 define the active region for the control cylinders in [392]. Physically, when the very small control cylinder is placed in this patch, it creates a separation bubble of its own that corresponds to a very low Reynolds number flow. Within this bubble, diffusion is significantly active and damps disturbances. In effect, this weakens A_1 and delays the first Hopf bifurcation. Such control is effective only up to a certain Reynolds number due to the change in mode shape with Re for ϕ_3 and the fact that even if one controls the near-wake mode A_1, A_3 can still destabilize the flow.

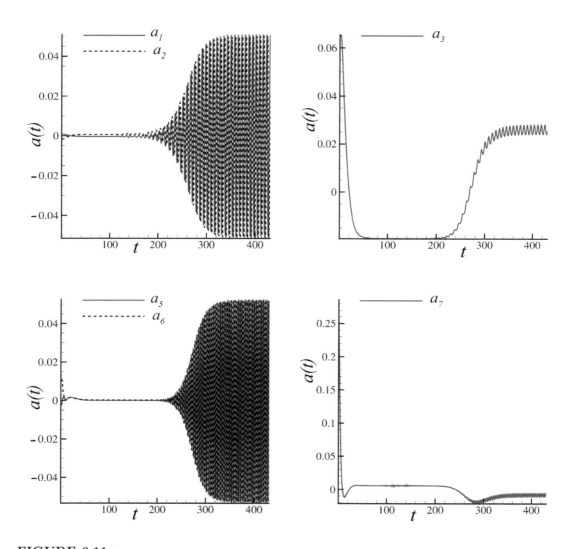

FIGURE 6.11a

Time-dependent amplitude functions of the first seven POD modes of Fig. 6.9 shown for $t = 0$ to 430.

Th role of ϕ_3 in the wake dynamics at post-critical Reynolds number can be understood by looking at the disturbance velocity components given by

$$u_d(x,y,t) = \int \int [\hat{u}\, e^{i(\alpha x - \beta t)} + \hat{u}^*\, e^{-i(\alpha x - \beta t)}]\, d\alpha\, d\beta \tag{6.20}$$

$$v_d(x,y,t) = \int \int [\hat{v}\, e^{i(\alpha x - \beta t)} + \hat{v}^*\, e^{-i(\alpha x - \beta t)}]\, d\alpha\, d\beta \tag{6.21}$$

where quantities with asterisks denote complex conjugates. If one represents variables as a sum of the mean field plus a disturbance component and substitutes these in the Navier–Stokes equation, then one obtains the governing equation in an averaged sense (indicated

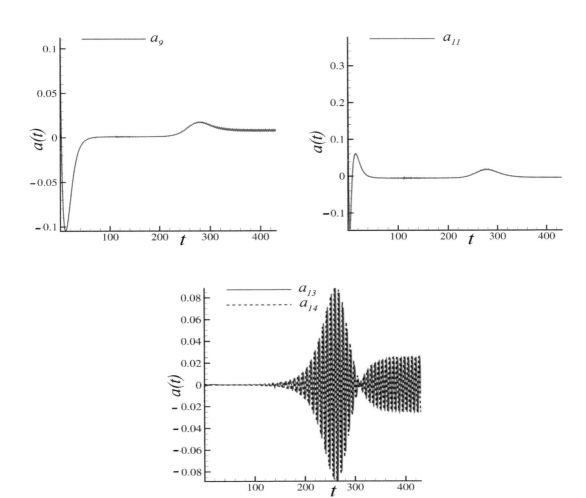

FIGURE 6.11b
Time-dependent amplitude functions of higher POD modes of Fig. 6.9 shown for $t = 0$ to 430. Note that $a_1 - a_2$; $a_5 - a_6$ and $a_{13} - a_{14}$ form pairs; a_3, a_7, a_9 and a_{11} do not form pairs, corresponding to an anomalous mode of the first kind. The pair $a_{13} - a_{14}$ does not follow the time variation given by either the Stuart–Landau or the LSE equation and is an anomalous mode of the second kind.

by angular brackets) for the mean. The resultant equation will have gradient transport contribution terms, similar to Reynolds stress-like terms given by

$$< u_d(x, y, t)v_d(x, y, t) >=< \int \int \left[(\hat{u}\,\hat{v}^* + \hat{u}^*\,\hat{v}) \right.$$

$$\left. + \hat{u}\,\hat{v}\,e^{2i(\alpha x - \beta t)} + \hat{u}^*\,\hat{v}^*\,e^{-2i(\alpha x - \beta t)} \right] d\alpha\,d\beta > \qquad (6.22)$$

Note that in Eq. (6.22), the first term on the right-hand side is phase independent and should be included in the mean-field equations, irrespective of whether one is talking about time or ensemble averaged equations. In [341], this was called the wave induced stress term, studying flow past a compliant wall. This steady streaming is always present in the Navier–Stokes equation for any flow field in the presence of disturbances. This explains why one can have many such terms arising from different modes, with α and β related via the dispersion relation. These types of terms will not be present when the modes appear in pairs. There are absolutely no restrictions on the number of isolated modes and in the present case, the third, seventh, ninth and eleventh are isolated modes, as shown in Fig. 6.11. Note that the mean-field correction term given by Eq. (6.22) is essentially a spatio-temporal contribution and the presented results for the cylinder case show this as a transient term.

6.7   Landau–Stuart–Eckhaus Equation

The Stuart–Landau equation is incapable of explaining the discontinuity in the A_e versus Re curve in Fig. 6.5, implying the inadequacy of considering only the nonlinear self-interaction. In [346], discontinuous variation of A_e was explained with the help of a new equation based on the eigenfunction expansion process of [94] replacing Eq. (6.10). In this approach, the instability amplitude equation is given by Eq. (6.9). Eckhaus [94] introduced the eigenfunction expansion in a Galerkin method to yield the complex amplitude equation including multi-modal nonlinear interactions which also incorporates the self-interaction term of Eq. (6.9).

Since this equation is more generic and includes the Stuart–Landau equation as a special case, we refer to Eq. (6.9) as the Landau–Stuart–Eckhaus (LSE) equation — mindful of the basic idea of incorporating multi-modal interactions. In Eq. (6.9), the term $N_j(A_k)$ includes nonlinear actions of all modes on the jth mode, including the self-interaction term. Eckhaus [94] suggested the nonlinear interaction term is given by $N_j(A_k) = A_j|A_k|^2$.

The role of multiple POD modes in contributing to the actual signal is seen in Fig. 6.12, where the computed data at R ($x_R = 0.5044, y_R = 0$) is plotted and compared with the reconstructed results from the POD analysis for Case A. The top frame compares the DNS data with the reconstructed signal using the POD modes 1, 2, 5 and 6. Reconstruction using these first two regular pairs of POD modes reveals poor agreement during the disturbance growth stage. In contrast, when the anomalous mode (POD modes 13 and 14) is added for reconstruction, significant improvement is noted during the growth stage as well, shown in the bottom frame of Fig. 6.12. It is also noted in the top frame that retaining only the first two regular pairs actually overestimates the solution in comparison to the DNS data during the growth phase, while underestimating it in the nonlinear saturation stage. Thus, it is seen that there are multiple modes that are active during the growth and nonlinear saturation stages in varying degrees. In studying flow instability by the Stuart–Landau equation, one makes the assumption that there is only one single dominant mode dictating the instability and its nonlinear saturation. Even if one makes the allowance that the Stuart–Landau equation can be applied to different modes (as in normal mode analysis), this places a strong restriction on these modes to act independently during the nonlinear saturation stage. An alternative to this was theoretically proposed in [94] via an eigenfunction expansion procedure, but has not been studied before for this flow.

In performing multi-modal analysis in [353], instability mode amplitudes in the LSE equation were linked to POD eigenmodes. For the flow at $Re = 60$, the first two regular pairs' interactions were studied following the suggestions provided in [94, 190], with nonlinear

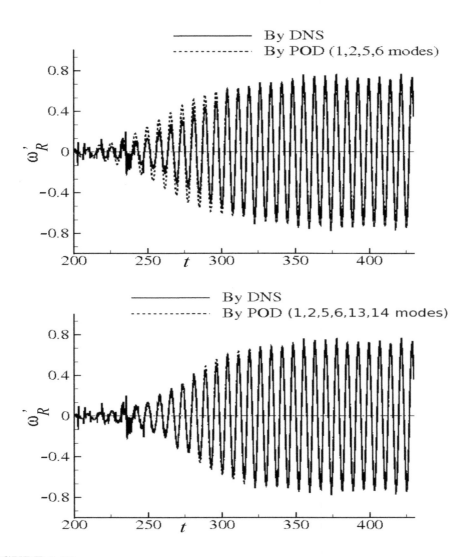

FIGURE 6.12
Comparison of reconstructed POD data with the direct simulation (DNS) results for $Re = 60$ stored at a location ($x_R = 0.5044$, $y_R = 0$). The top frame uses 1, 2, 5 and 6 POD modes for the reconstruction that shows overestimation in the growth phase and underestimation in equilibrium stage. The bottom frame also includes the anomalous 13th and 14th POD modes and this shows a very good agreement with the DNS data at the growth as well as at the equilibrium stage.

interactions (including self-interaction) as given in the following equations. In the LSE formalism, the first two R-modes are given by

$$\frac{dA_1}{dt} = \alpha_1 A_1 + \beta_{11} A_1 |A_1|^2 + \beta_{13} A_1 |A_3|^2 \tag{6.23}$$

$$\frac{dA_3}{dt} = \alpha_3 A_3 + \beta_{31} A_3 |A_1|^2 + \beta_{33} A_3 |A_3|^2 \tag{6.24}$$

These equations do not admit closed form expression for the instability amplitudes. Even obtaining numerical solutions of Eqs. (6.23) and (6.24) is not straightforward, as these constitute a set of stiff differential equations due to different orders of magnitudes for the growth rates of A_1 and A_3. That was evident from the values of β_{ij} obtained in these equations, in [353]. It was pointed out that the POD amplitude functions of Fig. 6.11 are normalized and do not show their relative importance, while the actual normalization is used in describing instability modes. Direct simulation data were used to obtain various coefficients in Eqs. (6.23) and (6.24).

The problem of stiffness arises due to different growth rates of the fundamental solutions for A_1 and A_3 and in [353] this was avoided by using the compound matrix method, described in [4, 91] and in Chap. 3. In the above coupled set of equations, we have two fundamental solutions with growth/decay rates differing by orders of magnitudes making the differential equations stiff. The compound matrix method avoids this by reformulating the problem in terms of new dependent variables. Equations (6.23) and (6.24) were transformed by introducing a new variable, $Y = A_1 A_3$, to remove the stiffness and to obtain a single evolution equation for it as

$$Y^{-1}\frac{dY}{dt} = \alpha + \beta_1 |A_1|^2 + \beta_3 |A_3|^2 \tag{6.25}$$

where $\alpha = \alpha_1 + \alpha_3$, $\beta_1 = \beta_{11} + \beta_{31}$ and $\beta_3 = \beta_{13} + \beta_{33}$. Equation (6.25) can be used along with the direct simulation data for $Re = 60$, to obtain the following complex coefficients: $\alpha = (0.12856, -0.34242)$, $\beta_1 = (-95.722, -1026.225)$ and $\beta_3 = (2039.764, 61376.693)$. One notes that the nonlinear interaction coefficients β_1 and β_3 are of different orders of magnitude. This actually helps one realize that for the overall amplitude, the third mode (A_3) plays only a perturbative role, with the leading order mode satisfying $\frac{dA_1}{dt} = \alpha_1 A_1 + \beta_{11} A_1 |A_1|^2$ − as in the Stuart–Landau equation. The use of direct simulation data allows one to obtain $\alpha_1 = (0.03706, 0.75938)$, which in turn gives $\alpha_3 = (0.09149, -1.10180)$. One also obtains $\beta_{11} = (-20.330, 52.901)$ and $\beta_{31} = (-75.391, -1079.126)$. Having obtained β_{11} and α_1, one can use Eq. (6.23) to obtain the time-averaged value for $\bar{\beta}_{13} = (-3484.41, 8821.94)$. Similarly, one can obtain $\bar{\beta}_{33}$, which also takes a large value. Such large values of nonlinear coefficients indicate that A_3 is very loosely coupled in this equation with A_1.

6.8 Universality of POD Modes

We have seen so far that there is more than one dominant instability mode present in flow past a cylinder for $Re = 60$. These modes have been related to POD modes as a consequence of the proposal in [353, 360], as given by Eqs. (6.18) and (6.19). In these references, using DNS results for two-dimensional flows past a circular cylinder and inside a lid-driven cavity (LDC), the universality of POD modes has been proposed by the similarities of POD modes of these seemingly dissimilar flows. The original goal of POD was to project a stochastic dynamical system onto an optimal reduced order deterministic basis model. In fluid mechanics, the role of POD was thought to assist in analyzing turbulent flows. In [81, 209], flow past a circular cylinder was studied using POD to elucidate vortex shedding. POD of direct simulation of bypass transition data was performed in [356]. Flows past a circular cylinder have also been studied in [253, 369] in a time-averaged sense for the underlying dynamical system. The study in [353] is also on flow instability, which is described in this chapter. While instability modes are considered to be deterministic, can one say the same thing about POD modes, despite relating these two types of modes in Eqs. (6.18) and (6.19)? In early applications of POD, it was performed with the assumption that the dynamical system is characterized by limit cycle oscillation, and hence the resultant spatial modes are appropriate only for the limit cycle. In order to enlarge the range of application to express the transient state as well, the authors in [253, 369] used "the steady solution of the flow as an additional data point in order to achieve a model with correct transient dynamics." This was the reason for proposing the shift mode or mean flow mode in these references. It was also acknowledged that even these additional modeling efforts did not represent fluctuations and there are mismatches between the transient dynamics of the model and the Navier–Stokes solution in terms of the initial growth rates. Siegel et al. [369] concluded that "POD in its original form is not well suited to describing transient data sets." In calculating POD modes for a cavity flow in [164], the authors used a method called the sequential POD to process data from different flow states, by ensuring the modes to be orthogonal. The success of this and the results from [353, 360] show that POD modes are perfectly capable of capturing the transient dynamics and hence Eqs. (6.18) and (6.19) are perfectly valid. However, in [360], the POD modes for flow past a cylinder and inside an LDC have shown the universality of POD modes and the instability modes for vortex dominated flows, despite the fact that one is an external and the other is an internal flow.

To test the universality of POD modes, we first record the variability of these for a circular cylinder at a higher value of $Re = 100$. The eigenvectors are shown in Fig. 6.13 for the data set from $t = 80$ to 221. Altogether, 19 POD modes are required to represent 99% of total enstrophy. This readily shows that with increasing Reynolds numbers, more modes are required to reconstruct the dynamical system. In this case, the third and the eleventh modes are the T_1-modes, as shown in Fig. 6.14. For the pair consisting of the fifth/sixth modes, POD amplitudes show slight overshoot during the late growth stage; still this is termed an R-mode. The seventh/eighth and the ninth/tenth mode pairs' time variation is not given by the Stuart–Landau or LSE equations during the growth phase. For the seventh/eighth mode pair, the amplitude of a_7 displays a wave packet like time variation during the growth phase, while a_8 does not show such behavior. Thus, the instability mode created by a seventh/eighth POD modes depicts an interaction between the R- and T_2-modes. In contrast, the ninth/tenth modes display wave packet like behavior at early times for both the members of the pair. This is similar to the behavior noted in Fig. 6.11 for the thirteenth/fourteenth mode pair of the $Re = 60$ case, where both the members of T_2-modes display wave packet like variations at early times.

Next, we show the generic nature of POD modes of vortex-dominated flows by considering the LDC flow case shown in [360]. The essential point to note here is that this flow is also characterized by linear temporal instability, which is followed by supercritical stabilization by nonlinear action. Thus, one would be interested to track the critical Reynolds number for the flow. In [46], temporal instability of LDC flow was investigated and Re_{cr} for the onset of periodicity was noted to be in the range of $8000 \leq Re_{cr} \leq 8050$. In [305], Re_{cr} was extrapolated to be 8031.92, while in [266], the Hopf bifurcation was predicted to begin at $Re_{cr} = 7402 \pm 4$. In [100] the critical Reynolds number was found to be approximately equal to 8000; in [280] a value of $Re_{cr} = 7763$ was predicted, while in [53] this was reported as $Re_{cr} = 7819$. The exact value for Re_{cr} depends upon the choice of numerical methods, the accuracy with which the boundary conditions are applied (specifically the top lid condition) and added numerical dissipation along with effectiveness of dissipation discretization playing a major role. These numerical aspects have been taken up in detail while developing

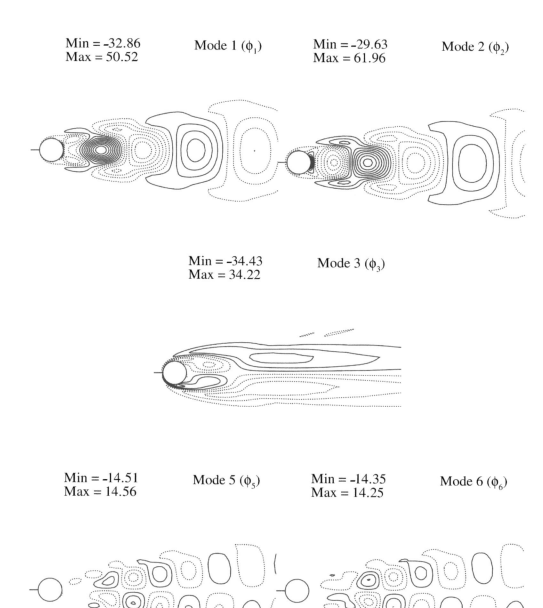

FIGURE 6.13

The first six leading eigenfunctions for the flow past a circular cylinder for $Re = 100$ shown for the data set from $t = 80$ to 221. The first two modes form a pair as do the fifth and sixth modes. The third mode is a T_1-mode. Maximum and minimum values are of the same order for the paired modes.

an extremely accurate combined compact scheme (NCCD) in [340, 361]. For the flow at $Re = 8500$ and 10000, the multi-periodic nature of the vorticity field with respect to time

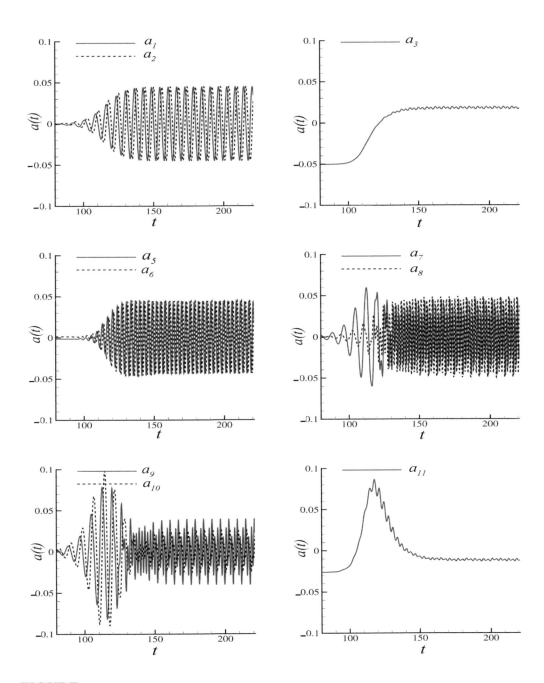

FIGURE 6.14
Time-dependent POD amplitudes for flow past a circular cylinder for $Re = 100$ during $t = 200$ to 350. The modes (a_1, a_2), (a_5, a_6), (a_7, a_8) and (a_9, a_{10}) form pairs; a_3 and a_{11} are T_1-modes. Modes a_5 to a_{10} are anomalous T_2-modes which do not strictly follow the time variation given by the Stuart–Landau and LSE equations.

was noted in these references. It is interesting to note that the solution obtained by using the NCCD scheme reported a new equilibrium solution. In this configuration, one notices a triangular vortex at the core that is surrounded by three secondary vortices as satellites. While, triangular vortices were noted experimentally in [26, 51], perhaps it was the first instance where such triangular vortices were noted computationally in [340, 361] via the solution of the Navier–Stokes equation.

For the flow in LDC, a uniform structured grid of 257 points along each of the x- and y-directions was used. Eqs. (6.1) and (6.2) were solved subject to the following boundary conditions to obtain the solution. On all four walls of the LDC, $\psi = $ constant is prescribed, which arises due to the no-slip and no-penetration boundary conditions; $\omega_b = -\frac{\partial^2 \psi}{\partial n^2}$ is applied on the walls to obtain the boundary vorticity (with n as the wall-normal coordinate on individual segments of the wall) and this is calculated using a Taylor series expansion at the wall boundaries incorporating appropriate velocity conditions at the wall. For this flow, the top lid moves horizontally with a unit nondimensional velocity, while all other walls are stationary. To solve Eq. (6.1), the Bi-CGSTAB method has been used here. The convection and diffusion terms are discretized using the NCCD scheme, which obtains both first and second derivatives simultaneously, whereas for the time derivative, the four-stage, fourth-order Runge-Kutta (RK_4) method is used. These methods have been rigorously analyzed for their spectral resolution and effectiveness in discretizing the dissipation terms along with the dispersion relation preservation (DRP) properties before being implemented to flows. It was noted that the NCCD method is very efficient and provides high resolution along with effective dissipation discretization. Additionally, it has inherent ability to control aliasing error and spurious numerical modes arising out of convection terms [340, 360]. However, this method can be used only with uniform structured grids.

Flow inside an LDC is a canonical example of internal flows that display linear temporal instability and Hopf bifurcation in the parameter space. Starting from a quiescent initial condition inside the cavity, flow evolves with the continual creation of vorticity at the wall segments. During this early stage, the flow is highly transient. Thereafter, small scale disturbances are created near the top right corner of the cavity, as the lid moves from left to right, and distinct vortices are formed due to linear temporal instability. Subsequently, the flow saturates due to nonlinearity to a multi-periodic state. In Fig. 6.15a, an instantaneous snapshot of the flow is shown for $Re = 8500$ at $t = 550$, with three points marked in the frame. In frames 6.15b to 6.15d, vorticity time series are shown for these points, which have structural similarities with the variations shown for the case of a circular cylinder. However, the LDC flow starts from a strongly transient state to a multi-periodic equilibrium state about a non-zero mean. This is in contrast to flow past a cylinder where the lift experienced starts from zero to a time-periodic state with zero mean. Subtle aspects of computing the LDC flow by accurate methods are given in [340, 361], which also discuss the topology and dynamics of various vortical structures. The top lid, as driven from left to right, sets in an orbital motion of two sets of gyrating vortices around the weak triangular vortex at the center, as seen in Fig. 6.15a. Additionally, there are three corner vortices which display initial temporal instability followed by nonlinear saturation. In contrast, at the core of the cavity, vorticity values are orders of magnitude lower. The point P ($x = 0.95, y = 0.95$) also displays a nonzero mean vorticity whose time variation about this mean is symmetric, similar to the time variation of vorticity on the center line of the cylinder wake (with zero mean). Vorticity variations for points at Q ($x = 0.95, y = 0.05$) and R ($x = 0.07, y = 0.17$) are different, with strong monotonic variation for the former. For both these points, time-periodic variation is not symmetric about the mean.

We note that the time series of disturbance vorticity also indicates the time variation of the instability amplitudes $A_j(t)$. Thus, the saturated vorticity disturbance amplitude also indicates the equilibrium amplitude of the most dominant instability mode. In Fig. 6.16, the equilibrium amplitude (A_e) of disturbance vorticity for the flow in the LDC at point P is shown plotted as a function of Re. If the flow were completely dominated by a single mode (that varied continuously with Re), then this variation would simply be parabolic. However, discontinuous variation of A_e indicates multiple Hopf bifurcations, as discussed in [360] for flow past a circular cylinder. In Fig. 6.16, we have drawn three equivalent parabolic segments, meeting the Re axis at three critical values, indicating three possible Hopf bifurcations for the flow in the LDC. Although the figure also indicates other possible bifurcations, the drawn sequences are meant to illustrate multiple bifurcations possible for the type of flows investigated in the present work. Earlier, three definitive Hopf bifurcations

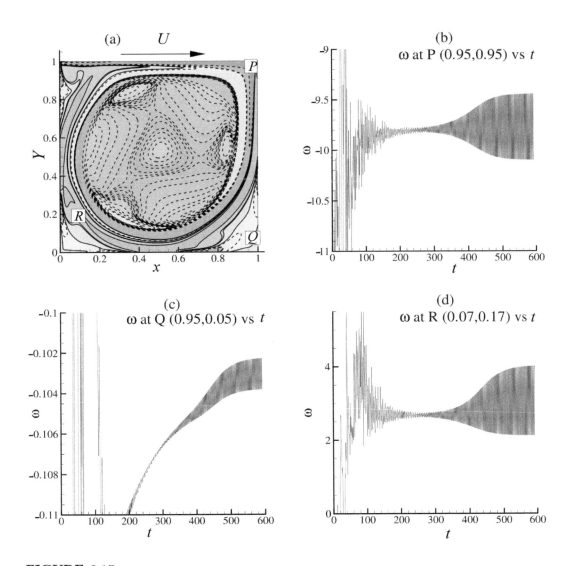

FIGURE 6.15
(a) Computed vorticity flow field of flow inside a square lid-driven cavity for $Re = 8500$ shown at $t = 550.0$. (b)–(d) The time variation of vorticity at P, Q and R (as in frame (a)) plotted.

were indicated for flow past a circular cylinder, based on computational and experimental evidence in Fig. 6.5.

Next we revisit the classification of POD modes used for a circular cylinder in Fig. 6.9, for the LDC flow. The existence of R-, T_1- and T_2-modes would testify to the universal nature of the classification. This flow inside the LDC was investigated for $Re = 8500$ in [360], which is between the second and the third Hopf bifurcations shown in Fig. 6.16. The time interval considered for POD spans $t = 201$ to 592 and the covariance matrix is formed using M snapshots from $t = 201$ to 592 with an interval of 0.2. In Fig. 6.17, the first ten eigenfunctions are shown, following the numbering sequence of POD modes explained for the circular cylinder. For this flow, the first few eigenvalues are obtained as [$\lambda_1 = 0.02917$; $\lambda_2 = 0.02435$;, $\lambda_3 = 0.00301$; $\lambda_5 = 0.00059$; $\lambda_6 = 0.00049$; $\lambda_7 = 3.651 \times 10^{-4}$; $\lambda_9 = 9.82 \times 10^{-5}$; $\lambda_{10} = 7.51 \times 10^{-5}$; $\lambda_{11} = 3.52 \times 10^{-5}$; $\lambda_{12} = 2.83 \times 10^{-5}$]. The numbering of eigenvalues and the reason for missing ones are apparent from the corresponding eigenfunctions in Fig. 6.17. These eigenfunctions indicate that modes (1,2) and (5,6) constitute R-modes.

Having obtained the snapshots and the eigenfunctions in the chosen time range, one can calculate the POD amplitude functions from Eq. (6.17), using the orthogonal property of the

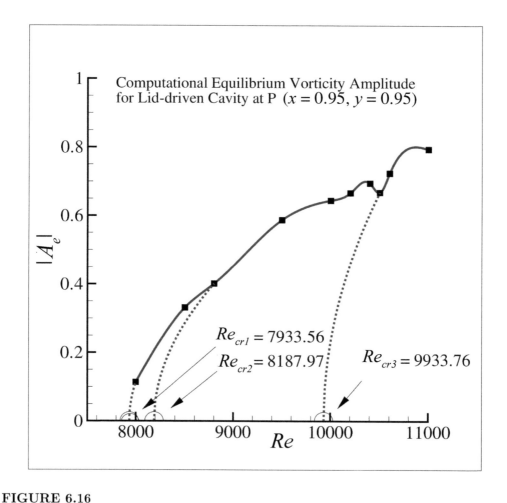

FIGURE 6.16
Equilibrium amplitude of the computed vorticity variation plotted against Re for the flow in the LDC with critical Reynolds numbers for point P ($x_P = 0.95, y_P = 0.95$).

eigenfunctions. It should be noted that to obtain the POD amplitudes correctly, one needs to take the number of snapshots sufficiently large to obtain the amplitude functions accurately. For the flow in the LDC, these amplitudes are shown in Fig. 6.18. Pairing of modes has been taken into consideration in plotting consecutive amplitudes either together or separately.

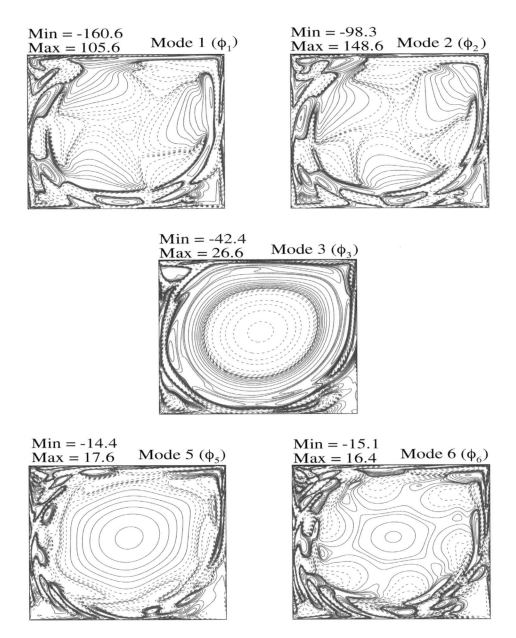

FIGURE 6.17a
The first few eigenfunctions for LDC flow for $Re = 8500$, shown from the data set $t = 201$ to 592. The first two modes form a pair, as do the fifth and sixth modes. The third mode is a T_1-mode. Maximum and minimum values are indicated, which are of the same order for the paired modes.

Such identification becomes clearer upon viewing the POD amplitude functions depicted in Fig. 6.18.

The third and seventh POD modes are isolated, without forming a pair and these are the anomalous T_1-modes. These modes were noted during the transient stage for flow past a cylinder. These T_1-modes for the LDC are characterized by circular vorticity contours in the

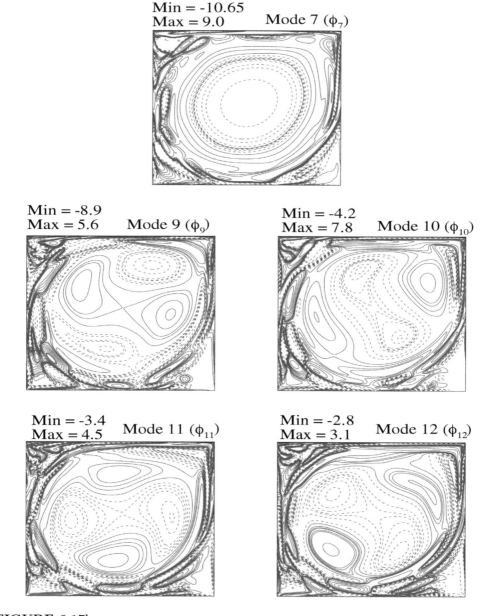

FIGURE 6.17b
Higher eigenfunctions for the LDC flow for $Re = 8500$ shown for the data set from $t = 201$ to 592. The seventh mode is a T_1-mode, while modes (9,10) and (11,12) are T_2-modes. Maximum and minimum values are of the same order for the paired modes.

center of the cavity shown in Fig. 6.17, along with the corner vortices. The isolated ϕ_3-mode implies absence of the fourth mode. From Fig. 6.18 for LDC flow, T_1-modes are present always with influence increasing with time. This indicates a larger shift which saturates

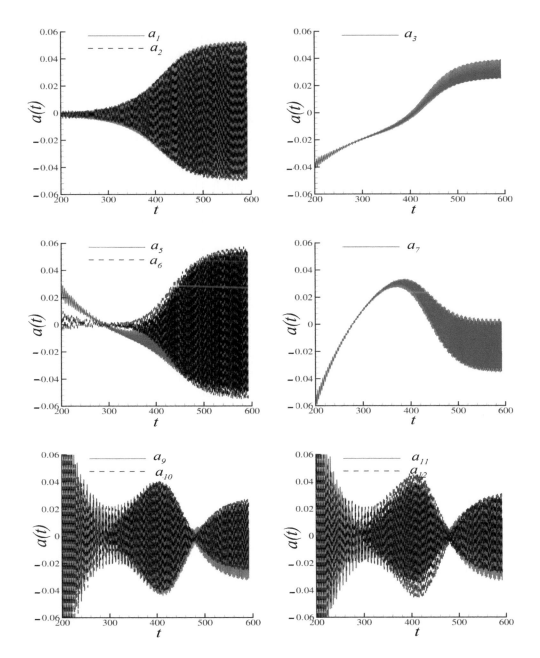

FIGURE 6.18
Time-dependent POD amplitude functions for the LDC flow for $Re = 8500$ during $t = 201$ to 592. The modes (a_1, a_2), (a_5, a_6), (a_9, a_{10}) and (a_{11}, a_{12}) form pairs, with a_3 and a_7 as the T_1-modes. Modes a_9 to a_{12} are anomalous type T_2-modes which do not follow the time variation given by the Stuart–Landau and LSE equations.

after $t = 500$, while there are higher frequency fluctuations with T_1-modes which also create larger r.m.s. contributions.

The mode pairs (1,2) and (5,6) show time variations typical of R-modes in Fig. 6.18. It is noted that ϕ_5 and ϕ_6 have distinct hexagonal core vortical structures, as shown in Fig. 6.17a. The modes (ϕ_9, ϕ_{10}) and (ϕ_{11}, ϕ_{12}) form pairs, but the time variation of amplitude functions corresponds to anomalous modes of the second kind (T_2-mode). No model of instability exists which displays this type of time variation. The depicted 12 eigenmodes (including the blank ones) account for 99.93% of the total enstrophy during the time interval $200 \leq t \leq 592$.

To show that the obtained POD modes have adequate accuracy, in Fig. 6.19 signals for the flow in the LDC is reconstructed. Apart from validating the POD procedure, this also brings into focus roles of specific modes, in particular, roles of the anomalous modes. In Fig. 6,19, reconstructed vorticity values at points P and R are compared with the computed data for the flow in the LDC. For point P, when only the leading pair is used there is mismatch at early times, as well as during the equilibrium stage. It is interesting to note that the addition of the anomalous third mode improves the match with DNS data significantly at early times, without which one notes a large shift. Thus, the third mode plays the role of providing correct shift to the solution. The match improves further when the fifth to seventh and ninth to twelfth modes are added. In the same way, for point R, instability is associated with asymmetric growth; the leading pair shows significant difference between the DNS data and the reconstruction. Just with the addition of the first anomalous mode, one notices significant improvement in the match. This match improves significantly when additional modes are included, as shown in the bottom right frame of Fig. 6.19.

In Fig. 6.20, we have shown the POD amplitudes for the flow in the LDC for $Re = 8500$ during the time interval $200 \leq t \leq 500$. The purpose for doing so is to emphasize the role of transients in determining ensuing instability. Note that (a_5,a_6) and a_7 of Fig. 6.18 exchange numbering of the modes with that shown in Fig. 6.20. This is essentially due to a different quantum of enstrophy contributed by leading instability modes calculated by considering different time intervals. This also shows that taking longer time interval reduces the amplitude of the T_1-mode (a_7). In Fig. 6.20, mean variation for the first two T_1-modes, the third and fifth modes, can be expressed in terms of an eighth order polynomials, $\bar{A}_j(t) = \sum_{k=0}^{8} p_{kj}t^k$, which can be expressed in terms of the following coefficients for $\bar{A}_3$ are given by $p_{03} = -15.52397$, $p_{13} = 0.362503$, $p_{23} = -0.0036543$, $p_{33} = 0.002065$, $p_{43} = -7.142 \times 10^{-8}$, $p_{53} = 1.5467 \times 10^{-10}$, $p_{63} = -2.048 \times 10^{-13}$, $p_{73} = 1.5184 \times 10^{-16}$, $p_{83} = -4.828 \times 10^{-20}$ and for $\bar{A}_5$ these are given by, $p_{05} = -62.4552$, $p_{15} = 1.596186$, $p_{25} = -0.017689$, $p_{35} = 1.10776 \times 10^{-4}$, $p_{45} = -4.283 \times 10^{-7}$, $p_{55} = 1.046 \times 10^{-9}$, $p_{65} = -1.577 \times 10^{-12}$, $p_{75} = 1.34014 \times 10^{-15}$, $p_{85} = -4.915 \times 10^{-19}$. These data fits for $\bar{A}_3$ and $\bar{A}_5$ are indicated by white lines in the second and the third frames of Fig. 6.20. High frequency fluctuations about this mean do not follow LSE-type equations. Even when piecewise cubic B-splines were fitted, one required more than 1000 nodes during $300 \leq t \leq 500$ only.

To test the universality of the POD modes discussed so far for some canonical problems, we also look at another flow where two cylinders are kept side by side and the problem is solved by the same high accuracy schemes used for flows past a single cylinder and the LDC flow. However, because of the multiply connected domain, the problem is solved by the chimera grid technique of [355] and details of this case are available in [365]. The computational geometry and dimensions are shown in Fig. 6.21, where two polar grids around each cylinder are overset onto the background Cartesian grid. The initial condition is given by the inviscid irrotational flow past the cylinders obtained by the method of images, as given in [365]. Here a case of non-dimensional (with respect to diameter of the cylinders) spacing (with respect to diameter of the cylinders) of $s = 2.5$ is discussed. Because of the close proximity of the two cylinders, vortex shedding obtained for $Re = 100$

shows interesting coupling. However, one notices vortex shedding behind each cylinder which exhibits initial temporal instability followed by nonlinear saturation. In this case, passage from the linear to the nonlinear stage of flow development is different from what is noted for flow past a single cylinder at the same Reynolds number. Despite this, the amplitude of

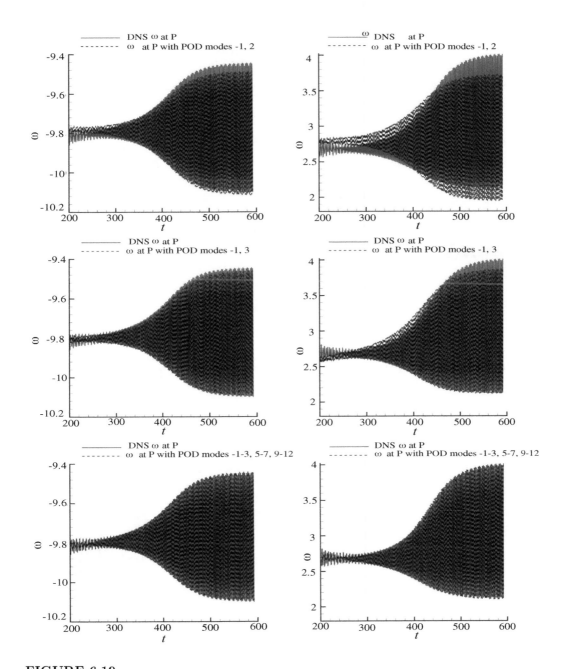

FIGURE 6.19
Comparison of reconstructed POD data with DNS results for LDC flow for $Re = 8500$ at two points P (left) and R (right). Note the proper shift caused by the inclusion of the third mode for both the points and produced the excellent match when 12 modes are included.

the disturbance field reaches a steady state. This steady state amplitude A_e of computed lift experienced by each cylinder is plotted as a function of Re in Fig. 6.22.

The equilibrium amplitudes of the computed lift coefficient are obtained by taking averages of values evaluated over a long time interval for each of the Reynolds number cases, after it has reached the non-linear equilibrium stage. Once again the computed values do not fall on a single parabolic curve, indicating multiple Hopf bifurcations. It is noted that the first bifurcation occurring for $Re = 49.82$ is followed by the second Hopf bifurcation for $Re = 57.36$. When the gap between the cylinders was increased to $s = 3.4$, the corresponding critical Reynolds numbers were 45.72 and 70.12, respectively.

These results show an interesting observation regarding the presence of the second cylin-

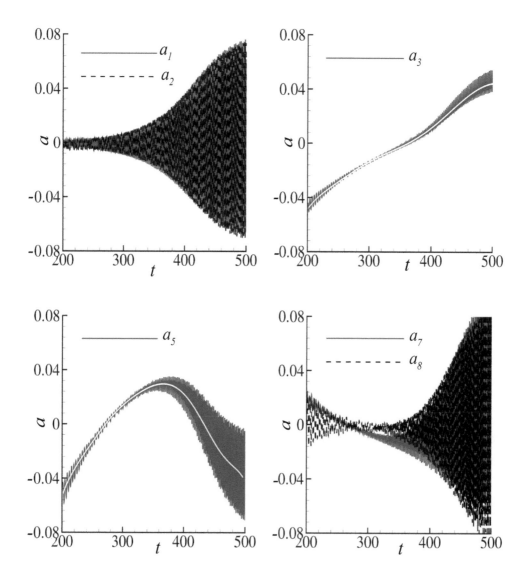

FIGURE 6.20
POD amplitude functions for the flow in an LDC for $Re = 8500$ during $200 \le t \le 500$.

der. When the cylinders are close by, as in the case of $s = 2.5$, the first bifurcation is delayed, while the second bifurcation is advanced. This is consistent with the flow control results in [87, 393], where it was noted that a second smaller cylinder — called the control cylinder — placed in the wake of the first cylinder delays the vortex shedding. However, we note that the interaction between the two cylinders also brings the second bifurcation Reynolds number lower.

The similarity of the present case with that of LDC flow is clearly evident in Fig. 6.23, where the time variation of amplitude function is shown. Time variation of all the modes does not start from the quiescent condition, as is observed for single cylinder R-modes. Instead it starts with a finite amplitude disturbance, resembling LDC flow's transient variation. However, in LDC flow these transients are at very high frequency in Fig. 6.15, as compared to the lower frequency variation seen in Fig. 6.23. With time progressing, the first three modes follow R-modes and saturate to a value indicating a limit cycle. The dissimilarity of these modes with the single cylinder case is solely due to significant interference effects between the two cylinders. It is also noted that the second and third modes form pairs to give rise to LSE instability modes. The first mode indicates that both the cylinders experience identical lift in the equilibrium stage and implies a heaving motion of the two

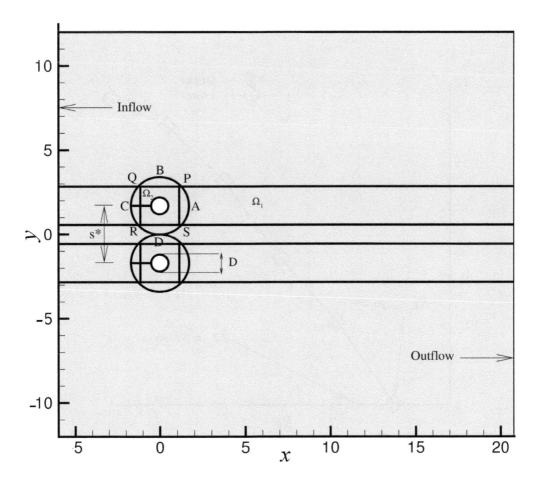

FIGURE 6.21
Sketch of the computational domain, with two cylinders separated by a distance of $s = s^*/D$.

cylinders together. The fourth mode resembles an isolated T_1-mode with a time average shift along with multi-periodic oscillation. For the larger gap cases of 3.4 and 4 in [365], one does not note any mode resembling the first mode of Fig. 6.23, while R-modes and T_1-modes are seen to be present.

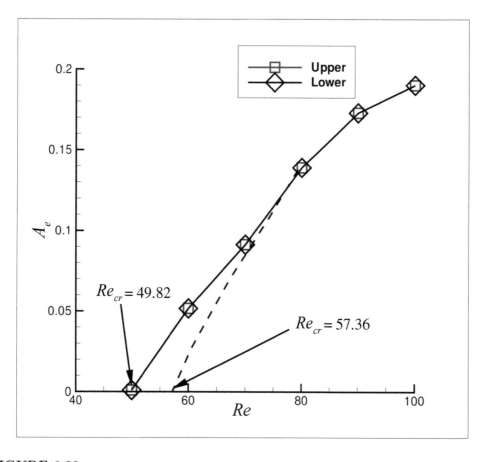

FIGURE 6.22
Equilibrium amplitude of computed lift plotted for both the cylinders, for $s = 2.5$, as a function of Re.

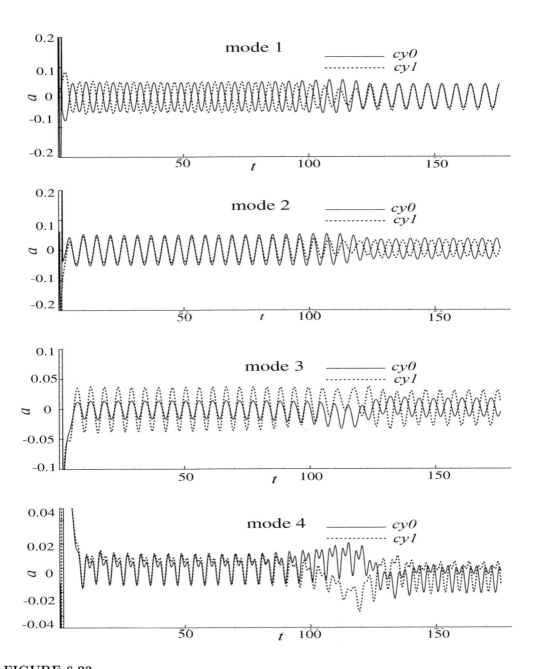

FIGURE 6.23
Time variation of the first four amplitude functions of POD modes for both the cylinders, for $s = 2.5$. Here $cy0$ and $cy1$ refer to the bottom and top cylinders, respectively.

7

Stability and Transition of Mixed Convection Flows

7.1 Introduction

So far we have discussed various aspects of instability and receptivity of pure hydrodynamic flows whose linear dynamics can be followed by solving the fourth order Orr–Sommerfeld equation (OSE). If one includes any one of the effects due to compressibility, heat transfer or surface compliance, then the linear dynamics become more involved, as one needs to solve instead the sixth order OSE. This added complexity also enhances the flow instability, whether one is interested in linear analysis for small excitation amplitudes or for large amplitude excitations for which one would be required to solve the full Navier–Stokes equation. In the present chapter, the effect of heat transfer is studied from both linear and nonlinear perspectives, on instability. Also, focus is on mixed convection flow at low speed, which can be modeled by the Boussinesq approximation. One of the major difficulties of studying mixed convection flows is the lack of available canonical equilibrium flows, even when one is interested in flows past a flat plate. For a constant external flow past a horizontal flat plate, a similarity solution is available due to [315]. We will confine our instability and receptivity studies of this flow only, while reference is made to flow past a vertical plate whose mean flow is obtained by non-similar approaches. In the concluding remarks for this chapter, we will discuss another class of mean flows obtained in [240] which is a generalization of the similarity solution originally introduced in [315].

The spatial stability properties of a mixed convection boundary layer, developing due to a constant external flow over a heated horizontal plate, are studied here first under the linear and quasi-parallel flow assumptions. The main aim of the present work is to find out if there is a critical buoyancy parameter that would indicate the importance of heat transfer in destabilizing mixed convection boundary layers, when the buoyancy effect is given by the Boussinesq approximation. The undisturbed flow used here is the one given by the similarity solution of [315], which requires that the wall temperature vary as the inverse square root of the distance from the leading edge of the plate for the similarity to hold when the boundary layer edge velocity is held constant. The stability of this flow has been investigated by using the compound matrix method (CMM), which allows finding all modes in a chosen range of the complex wavenumber plane for spatial stability analysis by the grid search technique.

Mixed convection flows are important, as these are ubiquitous in nature and in engineering devices. At the global scale in our atmosphere, geophysical fluid dynamics depend critically on mixed convection flow instability properties. Similarly, heating and cooling in electronic devices at the micro scale depend on mixed convection flow transition. For the practical ranges of parameters involved in the governing equations and boundary conditions, mixed convection flows undergo transition from the laminar to the turbulent state due to flow instabilities. Thus, study of instability and receptivity of such flows to different types of disturbance environments is of utmost importance. Here attention is focused only on a vortical disturbance introduced at the wall for the receptivity study, as this is also often studied using linear stability approaches. One can also perform studies involving free stream excitation. Moreover, one can think of exciting the dynamical system by entropic disturbances. The present approach of using vortical excitation is interesting, as it shows that weak vortical excitation can create thermal fluctuation (entropic disturbances).

Linear analysis has been employed with success in studying the onset of instabilities of flows without heat transfer, as narrated in Chap. 3. The subject has a rich history, with early works reported in [143, 408, 310] from German schools of thought led by Sommerfeld and Prandtl, which found a class of wavy disturbances, named after Tollmien and Schlichting. Following this early success in obtaining viscous instability, it has been proposed since then that transition from laminar to turbulent flows occurs via amplification of TS waves, whose origin is traced to viscous instability of the equilibrium flow, as given by the eigen solution of the Orr–Sommerfeld equation. The predicted onset of theoretical instability over a flat plate was experimentally demonstrated subsequently by the classical vibrating ribbon experiment of [316]. This experiment validated the onset of linear instability only for flows without heat transfer. Conversion of a linear disturbance field to a nonlinear field, which ultimately leads to flow transition has never been established conclusively, until very recently. This aspect is often overlooked in relating the complete process of transition as a consequence of the growth of TS waves. We have seen in Chap. 5 that transition is often led by the bypass route through the amplification of the spatio-temporal front and not the TS wave packets. These recent developments are summarized at the end of the book in an epilogue.

Heat transfer effects at very low speed are often modeled by buoyancy effects in the conservation equations. Due to such effects, instabilities in mixed convection flows at low speed differ considerably from instabilities of flows without heat transfer. In isothermal flows, instabilities due to hydrodynamic effects are discussed in Chaps. 3 to 5. In mixed convection flows, instabilities additionally depend on energy transfer due to buoyancy effects as modeled in the momentum equation. The consequence of these effects is to induce an additional pressure gradient, altering the equilibrium flow as compared to flows without heat transfer. In [41], it is noted that for natural and mixed convection flows, instability arises due to the growth of small disturbances. Temporal linear stability analysis was performed in [238] for mixed convection flow along an isothermal vertical flat plate for which the equilibrium flow was obtained by a local non-similarity method. Most early linear stability studies used non-similar mean flows, which are essentially local in nature, with non-similar terms treated as new variables. Boundary layer equations are augmented by differentiating the transformed boundary layer equation with respect to new variables. This leads to accuracy and uniqueness problems of the solution obtained by this approach, as the authors in [41] have noted that *this measure makes the accuracy of local nonsimilarity method difficult to assess.* It is in this context, the similarity profile in [315] for flow past a horizontal plate becomes important. This is perhaps the only known similarity profile for mixed convection flows when the edge velocity is held constant. When the edge velocity varies as x^m, a generalization is possible for the mean flow, as given in [240].

For natural convection flows, instability is a narrow-band phenomenon, while for forced convection boundary layers, the unstable frequency band is wider. Buoyancy forces cannot be neglected, even when these are small, if the convection velocities are small or the temperature difference between the surface and the ambiance is large. A comprehensive review of heat transfer aspects of mixed convection can be found in [118]. For some flows, induced body forces are either parallel or anti-parallel to the mean convection direction, as in the case of mixed convection flow past vertical plates.

Flow and heat transfer properties are more complex for mixed convection flows past inclined or horizontal plates. This is due to the fact that the buoyancy forces induce a longitudinal pressure gradient which will directly alter flow and heat transfer rates. Eckert & Soehngen [93] (as reported in [41]) have shown flow visualization pictures indicating transition to turbulence for the natural convection problem, originating as an instability of small disturbances. For natural convection flow past inclined flat plates in [385], the authors have reported generation of an array of longitudinal vortices. In [206, 450], the authors have experimentally investigated instability of flow past an inclined plate and reported the presence of two modes depending on the inclination angle of the plate. For inclination angles less than 14° with respect to the vertical, the authors reported wave-like instability. Such wave-like instability has also been studied in [64]. For inclination angles greater than 17°, for the natural convection problem, it is noted experimentally that the disturbance field is dominated by vortices and is often termed vortex instability. This vortex mode of instability is present for both the horizontal and inclined plates. This has been variously studied in [135, 445, 366, 122, 237, 433], among many other such studies. An alternate viewpoint presented in [139] classifies instability of forced-convection boundary layers over horizontal heated plates in terms of the two prototypical instabilities: the Rayleigh–Benard type, which is usually described for a closed convection system heated from below, and the Tollmien–Schlichting type that is typical of isothermal open flows, as in wall-bounded shear layers triggered by viscous instability.

As the instability in mixed convection flows starts off as growing small disturbances, linear stability is often used to analyze such flows. Analysis has been traditionally performed using temporal theory, as in [238] for mixed convection flow along an isothermal vertical flat plate. The effect of buoyancy was studied by the temporal theory of a flow perturbed by a small buoyancy-induced body force. Primary mean flow was obtained by the local non-similarity method and the authors reported that for assisting flows, the effect of buoyancy is to stabilize the flow. One should, however, note the critique of various non-similar flow descriptions used in instability studies given in [41]. For an inclined plate, the earliest instability study was in [64] and for the horizontal isothermal plate [65]. The results of these studies indicated that the flow along vertical and inclined plates is more stable when the buoyancy force aids external convection and that the stability is decreased as the inclination angle approaches the horizontal. Also noted for horizontal plates is the tendency of flow to become more unstable when buoyancy force is directed away from the surface.

Despite the distinction between temporal and spatial methods, the neutral curve is identical and is of vital interest for instability and transition. In [155], results were reported using linear spatial theory with a parallel flow approximation for free-convection flow past heated, inclined plates. The spatial stability results of natural convection flow over inclined plates were reported in [415], providing the eigen spectrum.

When the Boussinesq approximation is adopted in the Navier–Stokes equation, the effect of the buoyancy force appears in terms of Gr/Re^2, where Gr is the Grashof number and Re is the Reynolds number defined in terms of appropriate length, velocity and temperature scales. However, in [193, 386], the authors have shown that for boundary layers, the equivalent buoyancy parameter changes to $K = Gr/Re^{5/2}$. Experimental investigations in [433, 122] have also demonstrated that the onset of instability always occurs at the same value of K, showing its importance as the relevant buoyancy parameter. Similarity profiles derived in [315] and [240] are also given in terms of K alone. One of the main aims here is to identify a K_{cr}, beyond which the transport property changes qualitatively for a mixed convection flow past horizontal plate. In [66], the authors have noted significant buoyancy effects for $K_x \geq 0.05$ and $K_x \leq -0.03$, for aiding and opposing flows past a horizontal plate, where $K_x = Gr_x/Re_x^{5/2}$.

The addition of heat transfer effects necessitates incorporating the energy equation, which requires solving the sixth order stability equation instead of the fourth order equation for isothermal flows. Tumin [415] has used both Chebyshev collocation and Runge–Kutta integration schemes to solve the ninth order stability equation. In the collocation method used, 70 modes were taken and the Runge–Kutta method required the Gram–Schmidt orthonormalization procedure to maintain linear independence of the components of fundamental solutions. Spatial stability properties of a mixed convection boundary layer developing over a heated horizontal flat plate were investigated under the linear and quasi-parallel assumptions by solving the Orr–Sommerfeld equation in [358] with the equilibrium flow of [315] by using the CMM. Recent and comprehensive results of spatial linear stability and nonlinear receptivity calculations are given in [416].

One of the major obstacles in studying instability is that the fundamental solutions of the composite system display variation of functions at totally dissimilar rates when the independent variable changes. This problem of *stiffness* has already been discussed in Chap. 3. Three main approaches in solving hydrodynamic stability problems are based on (see [91, 314] for the description) (i) the matrix method using spectral or finite difference discretization, (ii) shooting methods using orthogonalization or orthonormalization and (iii) shooting methods using the CMM. The Chebyshev collocation method belongs to the first category, while the Runge–Kutta method with orthonormalization or the Kaplan filter belongs to the second category. Bridges & Morris [42] have noted that *the matrix method leads to the appearance of "spurious" eigenvalues.* For plane Poisseuille flow, the authors in [42] reported both stable and unstable spurious eigenvalues with large distinguishable magnitudes. According to [5], such spurious eigenvalues are created due to fracturing of the continuous spectrum for problems in the infinite domain. A QR algorithm based on Chebyshev polynomials for the spatio-temporal study of mixed convection boundary layers developing over a vertical plate was used in [230]. It has also been noted [5] that an orthogonalization/orthonormalization procedure causes the numerical solution to be non-analytic and on infinite domains there is the added problem that asymptotically correct boundary conditions may not preserve analyticity. Kaplan filter-based methods also suffer from this source of error. In contrast, the CMM does not suffer from any of these problems. The CMM has been re-developed using exterior algebra in a coordinate free context in [4]. The CMM has been used in [322] for receptivity analysis and reporting the eigenspectrum of the Blasius boundary layer. Allen & Bridges [5] solved the stability problem of Ekman boundary layer interacting with a compliant surface by solving a sixth order system. Ng & Reid [249] developed the CMM for general fourth and sixth order systems. For the mixed convection linear instability problem over a heated horizontal flat plate reported here, the CMM is extensively used.

Linear stability theory is incapable of describing the genesis of a complete wave system. For example, in [314] a sketch of a TS wave is shown − away from a monochromatic exciter − omitting the immediate neighborhood of the exciter. It has been noted before that eigenvalue formulation implicitly views the disturbance source to be placed at the wall and the created disturbances decay with height. In a noisy experimental device, there are distributed disturbance sources which are not monochromatic in constitution.

We consider laminar two-dimensional motion of fluid past a hot semi-infinite plate, with the free stream velocity and temperature denoted by $U_\infty = $ const. and T_∞. We focus our attention on the top of the plate, for which the temperature is T_w that is, greater than T_∞, while assuming the leading edge of the plate as the stagnation point. Governing equations are written in dimensional form (indicated by the quantities with asterisks), along with the Boussinesq approximation to represent the buoyancy effects, for the velocity and temperature fields in [118]. Earlier experimental works faced difficulty in trying to achieve the ideal condition required in theoretical analysis. Background disturbances form an omnipresent source of disturbance which is difficult to quantify. This also violates the assumption inherent in the linear instability formulation that disturbances decay with height as one approaches the free stream. All these can be avoided by adopting the receptivity approach with the appropriate noise model.

For flows without heat transfer, in [16] and references contained therein, attempts have been made to provide a mathematical framework of receptivity analysis. In [319, 322], the linear receptivity of a zero pressure gradient boundary layer to wall excitation was studied using the parallel flow approximation for a flow without heat transfer. Despite the absence of nonlinear and nonparallel effects in these studies, significant advances were reported. The local solution in the immediate neighborhood of the exciter is clearly brought out by receptivity analysis, in addition to the effects of all the modes present. Further advances in linear spatio-temporal receptivity analysis by the Bromwich contour integral method have been made in [349, 350], with the finding that the spatially stable Blasius boundary layer for a fixed frequency excitation displays a spatio-temporal growing wave front. The difference in speed between signal speed with group velocity of eigenmodes creates a detached wave front from an asymptotic TS wave. Without the spatio-temporal approach, it was not possible to predict the existence of a spatio-temporal front.

7.2   Governing Equations

We consider the laminar two-dimensional motion of fluid past a hot semi-infinite plate, with the free stream velocity and temperature denoted by U_∞ and T_∞, respectively. We focus our attention on the top of the plate, whose temperature distribution is $T_w(x)$, which is greater than T_∞, while assuming the leading edge of the plate as the stagnation point. The governing equations are written in dimensional form (indicated by the quantities with asterisks), along with the Boussinesq approximation to represent the buoyancy effect, for the velocity and temperature fields as [118]

$$\nabla^* \cdot \vec{V}^* = 0 \tag{7.1}$$

$$\frac{D\vec{V}^*}{Dt^*} = \vec{g}_j \, \beta_t \, (T^* - T_\infty) - \frac{1}{\rho} \nabla^* p^* + \nu \, \nabla^{*2} \vec{V}^* \tag{7.2}$$

$$\frac{DT^*}{Dt^*} = \alpha \, \nabla^{*2} T^* + \frac{\nu}{C_p} \, \Phi_v + \frac{q}{\rho C_v} \tag{7.3}$$

which give the evolution of velocity and temperature fields with time. Here $\frac{D}{Dt^*}$ represents the substantial derivative, $g_j = [0, \, g, \, 0]^T$ represents the gravity vector, β_t is the volumetric thermal expansion coefficient, ν is the kinematic viscosity and α is the thermal diffusivity. In this analysis, both the viscous dissipation (Φ_v) and heat source terms (q) will not be considered in the energy equation.

In order to non-dimensionalize the above system of equations, a length scale (L), a velocity scale (U_∞), a temperature scale $(\Delta T_L = T_w(L) - T_\infty)$ and a pressure scale (ρU_∞^2) are adopted. Instead of prescribing L, we have used a Reynolds number based on L, given by 10^5. The non-dimensionalized form of the above system of equations is obtained as

$$\nabla \cdot \vec{V} = 0 \tag{7.4}$$

$$\frac{D\vec{V}}{Dt} = \frac{Gr}{Re^2} \, \theta \hat{j} - \nabla p + \frac{1}{Re} \nabla^2 \vec{V} \tag{7.5}$$

$$\frac{D\theta}{Dt} = \frac{1}{RePr} \nabla^2 \theta \tag{7.6}$$

where $\vec{V} = \frac{\vec{V}^*}{U_\infty}$, $\theta = \frac{T^* - T_\infty}{\Delta T_L}$, $Gr = \frac{g\beta \Delta T_L L^3}{\nu^2}$, $Re = \frac{U_\infty L}{\nu}$, $Pr = \frac{\nu}{\alpha}$ and $\hat{j}$ is the unit vector in the wall-normal direction. The Grashof number (Gr) gives the ratio of buoyancy force to viscous force present in the fluid and the Richardson number (Ri) or Archimedes number, given by $\frac{Gr}{Re^2}$, shows the relative dominance of natural to forced convection. Positive and negative signs of the Richardson number refer to assisting and opposing flows. In the mixed convection regime, the Richardson number is of order one. The Prandtl number (Pr) used in the present study is 0.71, a value for air as the working medium.

It was already noted that the buoyancy parameter (K) is more relevant for boundary layers. Hence, the instability of this boundary layer is also dependent upon this parameter defined as

$$K = \frac{Gr}{Re^{\frac{5}{2}}} = g\,\beta_t\,[T_w(L) - T_\infty]\,(L\nu)^{\frac{1}{2}}\,U_\infty^{-\frac{5}{2}} \tag{7.7}$$

The buoyancy parameter can be defined alternatively in terms of Ri as $K = \frac{Ri}{\sqrt{Re}}$. The value of $K = 0$ indicates flow over a flat plate without any heat transfer, which is the same as the Blasius flow over a flat plate without any buoyancy effects. $K > 0$ corresponds to an assisting flow in which the flow occurs above a heated flat plate and $K < 0$ corresponds to an opposing flow where a fluid flows over a cooled flat plate. K switches sign for flow below the corresponding plates.

7.3   Equilibrium Boundary Layer Flow Equations

When Reynolds numbers are not too low, momentum and energy transport through gradients is limited to a narrow region inside the boundary layer. The application of the boundary layer approximation to the Navier–Stokes equations yields the following set of equations for two-dimensional incompressible flow expressed in Cartesian coordinates by

$$\frac{\partial U}{\partial X} + \frac{\partial V}{\partial Y} = 0 \tag{7.8}$$

$$U\frac{\partial U}{\partial X} + V\frac{\partial U}{\partial Y} = -\frac{\partial P}{\partial X} + \frac{\partial^2 U}{\partial Y^2} \tag{7.9}$$

$$0 = K\theta - \frac{\partial P}{\partial Y} \tag{7.10}$$

$$U\frac{\partial \theta}{\partial X} + V\frac{\partial \theta}{\partial Y} = \frac{1}{Pr}\frac{\partial^2 \theta}{\partial Y^2} \tag{7.11}$$

where $X = x^*/L$, $Y = y^*\sqrt{Re}/L$, $U = u^*/U_\infty$, $V = v^*\sqrt{Re}/U_\infty$, $\theta = \frac{T^* - T_\infty}{T_w(L) - T_\infty}$ and $P = p^*/(\rho_\infty U_\infty^2)$. The above equations are solved subject to following boundary conditions: i) at the wall, $(Y = 0$ and $X > 0)$: $U = V = 0$ and ii) at the free stream, $(Y \to \infty)$: $U = 1$, $p^* = p_\infty$ and $\theta = 0$.

It is noted that the buoyancy parameter K appears in the Y-component of the momentum equation, fixing the wall-normal pressure gradient, incorporating buoyancy induced effects. Integrating the Y-momentum equation with respect to Y and using the boundary condition $Y \to \infty$: $P = 0$, one gets the streamwise pressure gradient as $-K\int_Y^\infty \theta_X dY$, where the subscript in θ indicates a partial derivative. Hence, depending upon the sign of K, one can create either an adverse or a favorable streamwise pressure gradient in the boundary layer. It is inappropriate to even call these adverse or favorable pressure gradients — as is customary for flows without heat transfer. This will be readily evident from the computed solutions shortly. For the present investigation of flow over a heated flat plate, as K is positive, buoyancy results in a pressure gradient accelerating the flow along the streamwise direction. This term, therefore, usually induces non-similarity in the governing equations, unless something special is performed.

7.3.1 Schneider's Similarity Solution

The similarity solution derived here for Eqs. (7.8) to (7.11) is as given in [315]. Introducing the stream function (Ψ) automatically satisfies the continuity equation, Eq. (7.8). Substituting the above discussed expression for streamwise pressure gradient in the X-momentum equation and the energy equation, we get

$$\Psi_Y \Psi_{XY} - \Psi_X \Psi_{YY} - K \int_Y^\infty \theta_X dY = \Psi_{YYY} \tag{7.12}$$

$$\Psi_Y \theta_X - \Psi_X \theta_Y = \frac{1}{Pr}\theta_{YY} \tag{7.13}$$

where subscripts X and Y denote partial derivatives with respect to X and Y, respectively.

For the above system of equations to admit a similarity solution, the wall temperature distribution given by $\theta_w \propto X^{-1/2}$ has to be adopted, where $\theta_w = \frac{T_w(x^*)-T_\infty}{T_w(L)-T_\infty}$. The similarity transform which converts the above system of PDEs in X and Y into an ODE in η — the similarity variable — is defined by $\eta = YX^{-1/2}$. By defining a similarity variable as here, the above system of PDE's in X and Y is converted into an ODE with respect to the independent variable η. Transformations for the dependent variables used are $\Psi = X^{1/2}f(\eta)$ and $\theta = \theta_w \Theta(\eta)$, yielding the following system of equations

$$2f''' + ff'' + K\eta\Theta = 0 \tag{7.14}$$

$$\frac{2}{Pr}\Theta'' + f\Theta' + f'\Theta = 0 \tag{7.15}$$

In these equations, a prime indicates a derivative with respect to the independent variable, η, and these have to be solved subject to the following boundary conditions at $\eta = 0$: $f = f' = 0$ and $\Theta = 1$ and as $\eta \to \infty$: $f' = 1$ and $\Theta \to 0$. Equation (7.14) is integrated analytically once to obtain the following equation

$$\frac{2}{Pr}\Theta' + f\Theta = 0 \tag{7.16}$$

Thus, the similarity profile depends only on K, with dependence on Re implicit through the definition of the Y coordinate and η. It can be seen from Eq. (7.16) and the boundary condition at $\eta = 0$: $f = 0$ that irrespective of the buoyancy parameter, adiabatic conditions exist all over the plate surface. All the heat transfer occurs singularly from the leading edge only.

The similarity solution for the problem under consideration is given for the quantities f', f''', Θ and Θ' as a function of η in Figs. 7.1 to 7.3, in frames (a), (b), (c) and (d), respectively, for different values of K. Here f' represents the streamwise velocity, f''' its second derivative, Θ represents the temperature and Θ' its wall-normal derivative. Figure 7.1 shows the result for a very low value of buoyancy parameter ($K = 1 \times 10^{-6}$). In Fig. 7.1(a), the velocity profile monotonically grows following the boundary condition at $\eta = 0$ to $\eta = \eta_{max}$, where it tends to unity. Similarly from Fig. 7.1c, the temperature profile is seen to reach the free stream conditions, where it is zero according to the non-dimensionalization adopted. In Fig. 7.1b, the inset provides a detailed view of the portion enclosed by the rectangle. It helps to detect an inflection point, if there exists one. For this K, the result indicates no inflection point to be present. Fig. 7.2 shows the mean flow quantities plotted for $K = 3 \times 10^{-3}$. The general trend for the variation of mean flow quantities remains the same. But Fig. 7.2b indicates an inflection point for $\eta \simeq 9$. The same trend is observed in Fig. 7.3, for $K = 9 \times 10^{-2}$, which implies a relatively greater heat transfer due to buoyancy effects. In addition, we see an overshoot in the velocity profile in Fig. 7.3a. It has been noted in Fig. 7.2 of [315] that high values of K result in a velocity overshoot within the boundary layer. These observations tell us that for low values of K, although the second derivative of streamwise velocity remains zero at the wall and at the free stream, enhanced heating of the plate surface causes an inflection point at an intermediate height. From Figs. 7.1 to

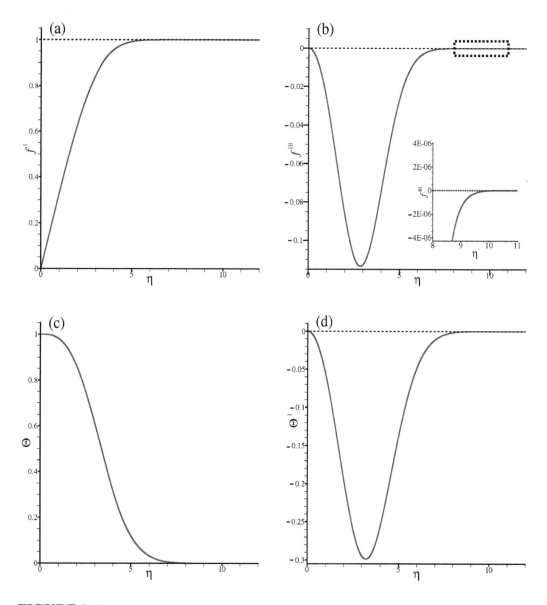

FIGURE 7.1
Variation of mean flow quantities obtained for $K = 1 \times 10^{-6}$.

7.3, it is noted that as K is increased, variation of mean parameters progressively occurs within a smaller range of η, which means that the mean flow quantities and their gradients achieve free stream conditions more rapidly, implying higher gradients in the boundary layer. Increasing K, i.e., increasing the temperature difference between the plate surface and the free stream, the boundary layer shrinks and the point of inflection moves closer to the wall. A similar overshoot of velocity within the boundary layer has been reported in [240].

From Eq. (7.10), it is understood that the effect of heat transfer is to introduce a wall normal pressure variation within the boundary layer. This is contrary to flow without heat

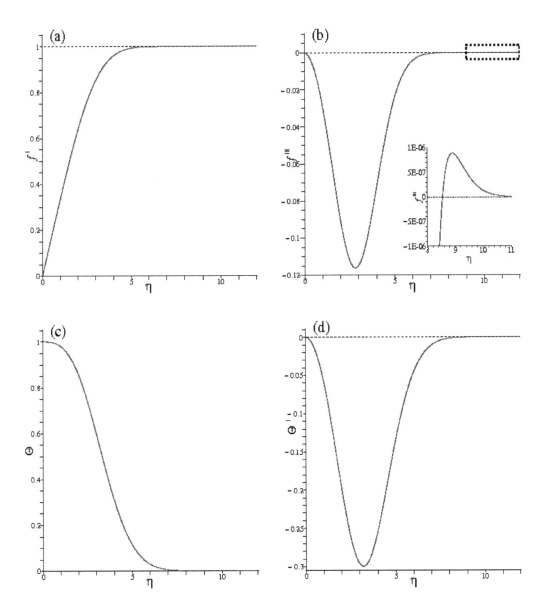

FIGURE 7.2
Variation of mean flow quantities obtained for $K = 3 \times 10^{-3}$.

transfer, where pressure is impressed upon the boundary layer by the outer inviscid flow, which remains invariant with height within the boundary layer, i.e., pressure within the boundary layer is a constant. This is the major aspect which sets apart mixed convection boundary layers from isothermal flows. The wall normal pressure variation within the boundary layer modifies the streamwise pressure gradient in the X-momentum equation which is given by

$$-K \int_Y^\infty \theta_X \, dY$$

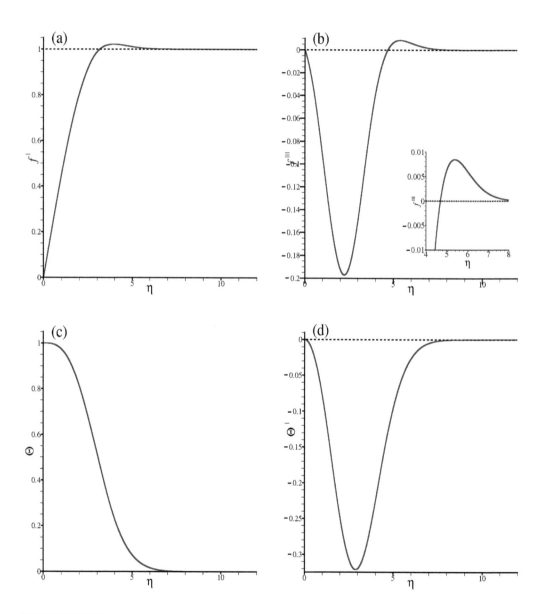

FIGURE 7.3

Variation of mean flow quantities obtained for $K = 9 \times 10^{-2}$.

We have

$$\theta = \theta_w \Theta = X^{-1/2}\Theta, \quad \eta = YX^{-1/2}$$

Note that the temperature gradient is with respect to X, while keeping Y fixed, i.e., $\frac{\partial\theta}{\partial X}|_Y$ and with wall temperature varying as $X^{-1/2}$, this would be

$$\frac{\partial}{\partial X}\left[X^{-1/2}\Theta(\eta)\right] = -\frac{1}{2}X^{-3/2}\Theta + X^{-1/2}\frac{\partial\Theta}{\partial\eta}\frac{\partial\eta}{\partial X}$$

As

$$\frac{\partial\eta}{\partial X} = -\frac{\eta}{2X}$$

thus

$$\frac{\partial\theta}{\partial X}|_Y = -\frac{1}{2X^{3/2}}[\Theta + \eta\,\Theta']$$

Therefore, the induced streamwise pressure gradient is given by

$$-K\int_Y^\infty \theta_X dY = -\frac{K}{2X^{3/2}}\left[\int_\eta^\infty (\Theta + \eta\,\Theta')X^{1/2}d\eta\right]$$

$$= -\frac{K}{2X}\left[\int\Theta d\eta + \eta\Theta - \int\Theta d\eta\right]$$

Using the above relations and integrating the pressure gradient term, we get the following form for the streamwise pressure gradient

$$-K\int_Y^\infty \theta_X\, dY = -\frac{K}{2X}\left[\eta\Theta\right]_\eta^\infty = -\frac{K}{2X}\eta\Theta \tag{7.17}$$

The boundary condition $\Theta \to 0$ for $\eta \to \infty$ and Eq. (7.17) show that buoyancy effects do not impose a pressure gradient at the wall and at the far stream. But at all intermediate heights within the boundary layer, there is a height dependent streamwise pressure gradient. For heated plates, K has a positive value and hence the pressure gradient within the mixed convection boundary layer is negative, which accelerates the flow. The value $K\eta\Theta$ determines the magnitude of this pressure gradient at any given streamwise location, which is plotted as a function of η for various buoyancy parameter (K) values in Fig. 7.4, certain aspects of which are noteworthy. The location of the maximum pressure gradient within the boundary layer moves towards the plate as K is increased. Also, its magnitude is of the same order as that of the corresponding K. The maximum value of favorable pressure gradient along with the location at which it occurs (η_{fpg}) are given in Table 7.1.

In flows without heat transfer, it is well known that a favorable pressure gradient stabilizes the flow, which is impressed upon the boundary layer by the outer inviscid flow. Pressure remains invariant inside the shear layer in the wall-normal direction. But for flows with heat transfer, we note the presence of a differential streamwise pressure gradient within the boundary layer. The effect of this favorable pressure gradient is to accelerate the flow differentially within the shear layer, which results in a velocity overshoot, responsible for the presence of an inflection point. This is clearly dependent on the value of K. The higher the value of K, the higher will be the magnitude of the velocity overshoot. Also, Fig. 7.4 and Table 7.1 show that at higher values of K, the maximum of the pressure gradient moves closer to the plate surface. Fig. 7.5 shows the contours of pressure gradient expression as given in Eq. (7.17) for values of K under discussion in the (x, y)-plane ($x = x^*/L$ and $y = y^*/L$). The η level at which the maximum pressure gradient occurs is also marked by a dotted line in Fig. 7.5.

The existence of a K-dependent inflection point for the mean flow provides the necessary condition for inviscid instability, according to the Rayleigh–Fjørtoft theorem, as explained in Chap.1. Temporal growth of instabilities occurs within the shear layer once the sufficient conditions are met, as noted in Rayleigh's theorem. Therefore, we are dealing with an equilibrium flow which has an inherent tendency towards inviscid temporal instability. This is in stark contrast to flows without heat transfer, where phenomenologically one observed spatial growth of disturbances. Thus, one notes the propensity of the flows with heat transfer to display tendencies of spatial and temporal instabilities. When dealing with three-dimensional instabilities in Chap. 8, we will note the presence of similar instabilities for the cross-flow component of the flow displaying temporal instability, while the streamwise flow shows spatial instability for small background disturbances.

Figure 7.6 shows variation of the inflection point with respect to K, which indicates that the inflection point moves closer to the wall as K is increased. It helps us to conclude that the mean flow becomes more and more unstable as the wall is heated, due to the presence of an inflection point nearer to the wall. Fig. 7.7 shows the quantity $f'''(f' - f'_{ip})$ plotted as a function of η for the buoyancy parameters discussed here. f'_{ip} is the streamwise velocity at the inflection point. A necessary condition for inviscid temporal instability according to the Fjørtoft theorem is that $f'''(f' - f'_{ip}) \leq 0$ anywhere within the shear layer, with equality only at $\eta = \eta_{ip}$.

From Fig. 7.7, one notices that the necessary condition of the Fjørtoft theorem is satisfied for $K = 3 \times 10^{-2}$ and 9×10^{-2} only, as shown in the insets of the figure. Cases with smaller values of K do not satisfy the necessary condition and do not ensure temporal instability, as these two higher values of K show the mixed convection flow to experience inviscid temporal instability. Thus, we note that the mean flow over a heated plate indicates the possibility of the existence of an inflection point, which in turn is indicative of temporal instability of flow. It is natural to inquire if the basic equilibrium flow also has the tendency of spatial instability, as it is common for flow over a flat plate in the absence of heat transfer. Whether such spatial instability is augmented or not due to the presence of buoyancy effects is discussed in the next section.

TABLE 7.1
Streamwise pressure gradient induced by buoyancy effects for a heated flat plate for different buoyancy parameters.

Case	K	η_{fpg}	Maximum Pressure Gradient Parameter ($K\eta\Theta$)
1	1×10^{-6}	2.62250	1.86443×10^{-6}
2	1×10^{-3}	2.61816	1.86144×10^{-3}
3	3×10^{-3}	2.61000	5.55667×10^{-3}
4	3×10^{-2}	2.52144	5.36419×10^{-2}
5	9×10^{-2}	2.38344	1.51597×10^{-2}

7.4 Linear Spatial Stability Analysis of the Boundary Layer over a Heated Plate

Here the stability equations for two-dimensional plane flows have been derived, starting from the nondimensional form of the Navier–Stokes equation with the Boussinesq approximation for heat transfer effects given by

$$\frac{\partial \tilde{u}}{\partial \tilde{x}} + \frac{\partial \tilde{v}}{\partial \tilde{y}} = 0 \tag{7.18}$$

$$\frac{\partial \tilde{u}}{\partial \tilde{t}} + \tilde{u}\frac{\partial \tilde{u}}{\partial \tilde{x}} + \tilde{v}\frac{\partial \tilde{u}}{\partial \tilde{y}} = -\frac{\partial \tilde{p}}{\partial \tilde{x}} + \frac{1}{\tilde{Re}}\left[\frac{\partial^2 \tilde{u}}{\partial \tilde{x}^2} + \frac{\partial^2 \tilde{u}}{\partial \tilde{y}^2}\right] \tag{7.19}$$

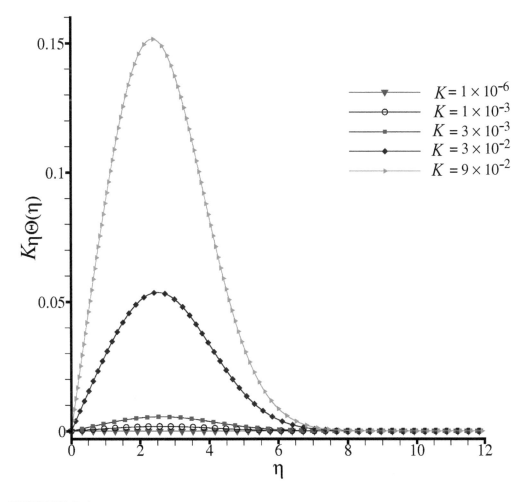

FIGURE 7.4
Buoyancy induced pressure gradient term $K\eta\Theta$ plotted as a function of similarity variable η.

$$\frac{\partial \tilde{v}}{\partial \tilde{t}} + \tilde{u}\frac{\partial \tilde{v}}{\partial \tilde{x}} + \tilde{v}\frac{\partial \tilde{v}}{\partial \tilde{y}} = \frac{\tilde{Gr}}{\tilde{Re}^2}\tilde{\theta} - \frac{\partial \tilde{p}}{\partial \tilde{y}} + \frac{1}{\tilde{Re}}\left[\frac{\partial^2 \tilde{v}}{\partial \tilde{x}^2} + \frac{\partial^2 \tilde{v}}{\partial \tilde{y}^2}\right] \tag{7.20}$$

$$\frac{\partial \tilde{\theta}}{\partial \tilde{t}} + \tilde{u}\frac{\partial \tilde{\theta}}{\partial \tilde{x}} + \tilde{v}\frac{\partial \tilde{\theta}}{\partial \tilde{y}} = \frac{1}{\tilde{Re}Pr}\left[\frac{\partial^2 \tilde{\theta}}{\partial \tilde{x}^2} + \frac{\partial^2 \tilde{\theta}}{\partial \tilde{y}^2}\right] \tag{7.21}$$

For stability analysis, dependent variables like velocity, pressure and temperature are

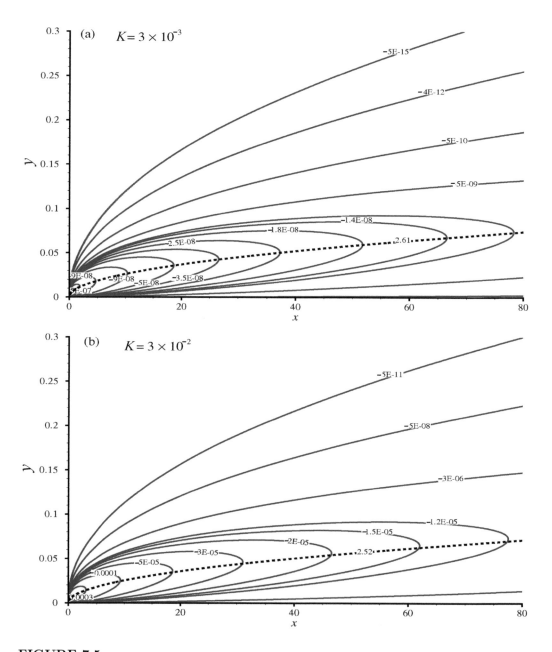

FIGURE 7.5
$-K\eta\Theta/2X$ contours plotted in the (x, y)-plane for the indicated values of K. η_{fpg} is marked by a dotted line.

split into two components, a time independent equilibrium quantity and a fluctuating disturbance quantity. Thus, flow variables are represented as

$$\tilde{u}(\tilde{x}, \tilde{y}, \tilde{t}) = \bar{U}(\tilde{x}, \tilde{y}) + \epsilon\, \hat{u}(\tilde{x}, \tilde{y}, \tilde{t}) \tag{7.22}$$

$$\tilde{v}(\tilde{x}, \tilde{y}, \tilde{t}) = \bar{V}(\tilde{x}, \tilde{y}) + \epsilon\, \hat{v}(\tilde{x}, \tilde{y}, \tilde{t}) \tag{7.23}$$

$$\tilde{p}(\tilde{x}, \tilde{y}, \tilde{t}) = \bar{P}(\tilde{x}, \tilde{y}) + \epsilon\, \hat{p}(\tilde{x}, \tilde{y}, \tilde{t}) \tag{7.24}$$

$$\tilde{\theta}(\tilde{x}, \tilde{y}, \tilde{t}) = \bar{T}(\tilde{x}, \tilde{y}) + \epsilon\, \hat{\theta}(\tilde{x}, \tilde{y}, \tilde{t}) \tag{7.25}$$

In the above, left-hand side quantities represent total flow variables. On the right-hand side, quantities with an overbar indicate the steady mean components, whereas quantities multiplied with ϵ are the unsteady perturbation quantities. These perturbation components are considered one order of magnitude lower than the mean components, for linear instability

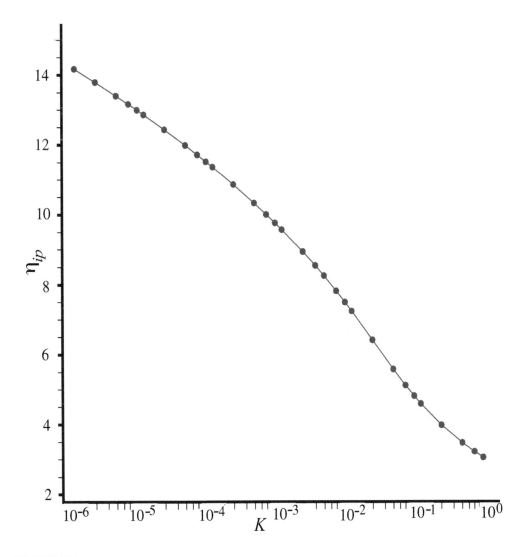

FIGURE 7.6
Location of inflection point (η_{ip}) plotted as a function of K.

studies. To obtain the governing stability equations, the parallel flow assumption is invoked which is equivalent to using $\bar{U} = \bar{U}(\tilde{y})$, $\bar{V} = 0$ and $\bar{T} = \bar{T}(\tilde{y})$. For viscous instability of a parallel flow over a semi-infinite flat plate, a relevant length scale is the local displacement thickness (δ^*). The reference velocity scale is U_∞ and $\Delta T(x^*)$ is the reference temperature scale. The temperature scale is defined as the difference between the local plate surface temperature $T_w(x^*)$ and the free stream temperature T_∞. Therefore, the temperature scale is given by $\Delta T(x^*) = (T_w(x^*) - T_\infty)$. It is noted that in Eqs. (7.18) to (7.20), the Reynolds number and the Grashof number are defined based on these reference quantities:

$$\tilde{Re} = \frac{U_\infty \delta^*}{\nu} \quad \text{and} \quad \tilde{Gr} = \frac{g\beta_t \, \Delta T(x^*) \, \delta^{*3}}{\nu^2}$$

Equations (7.22) to (7.25) are substituted in Eqs. (7.18) to (7.21) and by retaining only $O(\epsilon)$ terms, the governing linearized equations can be obtained. In a formal normal mode analysis

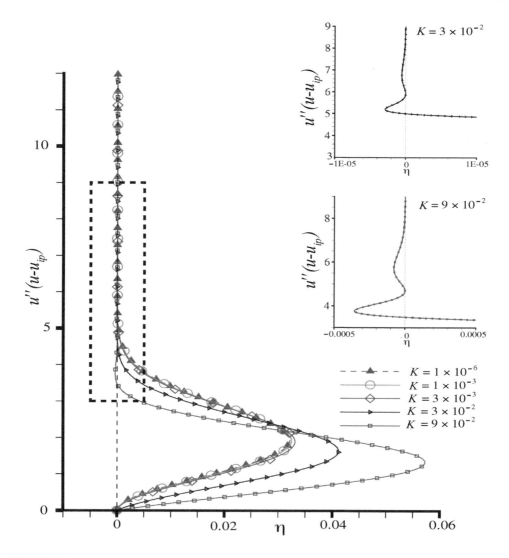

FIGURE 7.7
The quantity $f'''(f' - f'_{ip})$ plotted as a function of η for the indicated values of K.

carried out with linearized governing equations, disturbance quantities are expressed in the Fourier space, irrespective of whether one is studying temporal and/or spatial instability

$$[\hat{u}, \ \hat{v}, \ \hat{p}, \ \hat{\theta}] = [\gamma(\tilde{y}), \ \phi(\tilde{y}), \ \pi(\tilde{y}), \ h(\tilde{y})] \ e^{i(k\tilde{x}-\beta_0\tilde{t})} \qquad (7.26)$$

where $\gamma(\tilde{y})$, $\phi(\tilde{y})$, $\pi(\tilde{y})$ and $h(\tilde{y})$ are the disturbance complex amplitude functions for the corresponding physical perturbation quantity. The non-dimensional wavenumber is defined as $k = k^*\delta^*$ and β_0 is the non-dimensional circular frequency defined as $\beta_0 = \beta_0^*\delta^*/U_\infty$. On substituting Eq. (7.26) into the linearized governing equations, one gets the following system of ODEs

$$ik\gamma + \phi' = 0 \qquad (7.27)$$

$$i(k\bar{U} - \beta_0)\gamma + \bar{U}'\phi = -ik\pi + \frac{1}{\tilde{Re}}(\gamma'' - k^2\gamma) \qquad (7.28)$$

$$i(k\bar{U} - \beta_0)\phi = \frac{\tilde{Gr}}{\tilde{Re}^2}h - \pi' + \frac{1}{\tilde{Re}}(\phi'' - k^2\phi) \qquad (7.29)$$

$$i(k\bar{U} - \beta_0)h + \bar{T}'\phi = \frac{1}{\tilde{Re}Pr}(h'' - k^2h) \qquad (7.30)$$

Primes in these equations refer to differentiation with respect to $\tilde{y}$. Here the mean quantities used are obtained from Schneider's similarity solution. For the similarity solution wall temperature distribution is given by

$$\theta_w = x^{-1/2}$$

Also, the following relations are needed to use the mean flow quantities in the disturbance equations, $\tilde{y} = \eta/C_\delta$, so that

$$C_\delta = \int_0^\infty (1 - f') \, d\eta$$

The displacement thickness is

$$\delta^*(x^*) = \sqrt{\frac{\nu x^*}{U_\infty}} \int_0^\infty (1 - f') \, d\eta$$

Thus

$$\delta^* = \frac{C_\delta x^*}{\sqrt{Re_{x^*}}}$$

Using the above relations, it is seen that

$$\frac{\tilde{Gr}}{\tilde{Re}^{5/2}} = K \, C_\delta$$

The above transformations are required to incorporate the buoyancy parameter (K) in the stability equations.

One can eliminate π and f by combining Eqs. (7.28) and (7.29) and substituting in Eq. (7.27). This results in a system of ODEs, where one uses the notation $\hat{K} = KC_\delta$ to obtain the final form of the Orr–Sommerfeld equations for mixed convection flows as

$$i(k\bar{U} - \beta_0)(k^2\phi - \phi'') + ik\bar{U}''\phi = \hat{K}k^2h - \frac{1}{\tilde{Re}}(\phi^{iv} - 2k^2\phi'' + k^4\phi) \qquad (7.31)$$

$$i(k\bar{U} - \beta_0)h + \bar{T}'\phi = \frac{1}{\tilde{Re}Pr}(h'' - k^2h) \qquad (7.32)$$

Since the differentiation of mean flow quantities in the OSE are with respect to $\tilde{y}$ and the mean flow quantities f and Θ obtained from the similarity solution are functions of η, derivatives of mean quantities are to be incorporated into the OSE as

$$\bar{U} = f'$$

$$\bar{U}'' = f''' C_\delta^2$$

$$\bar{T}' = \Theta' C_\delta$$

Equations (7.31) and (7.32) represent a coupled sixth order system of ODEs in the independent variable $\tilde{y}$, which shows that the temperature and velocity fields are coupled for mixed convection flow. In order to solve it, six boundary conditions are to be specified. At the wall, for the disturbance field one incorporates the homogeneous condition. Additionally, for the instability of the mean flow which has a finite source of energy, as $\tilde{y}$ approaches the free stream, disturbance quantities must decay to zero. These conditions are stated in the physical plane for stability analysis as at $\tilde{y} = 0$:

$$\hat{u}, \quad \hat{v} \text{ and } \quad \hat{\theta} = 0 \tag{7.33}$$

And as $\tilde{y} \to \infty$:

$$\hat{u}, \quad \hat{v}, \quad \hat{\theta} \to 0 \tag{7.34}$$

7.4.1 Fundamental Solutions of the OSE

Equations (7.31) and (7.32) along with the boundary conditions of Eqs. (7.33) and (7.34) constitute a homogeneous system of equations with homogeneous boundary conditions – an eigenvalue problem.

Boundary conditions corresponding to Eqs. (7.33) and (7.34) in the spectral plane are given at $\tilde{y} = 0$ by

$$\phi, \quad \phi' = 0; \quad h = 0 \tag{7.35}$$

Also as $\tilde{y} \to \infty$:

$$\phi, \quad \phi', \quad h \to 0 \tag{7.36}$$

The general solution to the system of coupled ODEs given by Eqs. (7.31) and (7.32) can be written as

$$\phi = a_1\phi_1 + a_2\phi_2 + a_3\phi_3 + a_4\phi_4 + a_5\phi_5 + a_6\phi_6 \tag{7.37}$$

$$h = a_1h_1 + a_2h_2 + a_3h_3 + a_4h_4 + a_5h_5 + a_6h_6 \tag{7.38}$$

with the disturbance quantities decaying as $(\tilde{y} \to \infty)$.

To understand the analytical structure of the fundamental solutions of the OSE, we look at Eqs. (7.31) and (7.32) at the free stream, where these become constant coefficient ODEs, as at the free stream we have $\bar{U} = 1$, $\bar{U}'' \approx 0$ and $\bar{T}' \approx 0$. Immediately, one notices that the disturbance energy equation, Eq. (7.32), modifies as follows

$$h'' - [i\tilde{R}ePr(k - \beta_0) + k^2]h = 0 \tag{7.39}$$

This indicates that the energy equation decouples from the momentum equation at the free stream, implying that entropic disturbance can exist in the free stream in isolation. However, the assumed nature of the homogeneous far field condition does not apply to such a disturbance field. When entropic disturbances are applied at the free stream, these have properties like those described for free stream modes in Chap. 3. The fundamental solution of (7.39) is obtained in terms of the characteristic modes at the free stream as

$$h_\infty = a_5 e^{-S\tilde{y}} + a_6 e^{S\tilde{y}} \tag{7.40}$$

where $S = \sqrt{k^2 + i\tilde{R}e Pr(k - \beta_0)}$. To satisfy free stream homogeneous boundary conditions, one must have $a_6 = 0$ for real $(S) > 0$. Since Eq. (7.39) is a second order ODE, it is clear that as $\tilde{y} \to y_\infty$, $h_{1\infty}, h_{3\infty} \to 0$. At the free stream, the momentum equation is given by

$$\phi^{iv} - [2k^2 + i\tilde{R}e(k - \beta_0)]\phi'' + [k^4 + i\tilde{R}e(k - \beta_0)k^2]\phi = \hat{K}\tilde{R}ek^2 h \tag{7.41}$$

It is evident from Eqs. (7.39) and (7.41) that at the free stream, although the temperature field is decoupled from the velocity field, there is coupling between the velocity and temperature fields due to convection. The temperature field can be considered as forcing for the momentum equation at the free stream. Thus, the solution of Eq. (7.41) consists of a homogeneous part and a particular integral. The homogeneous part of solution of Eq. (7.41) is given by

$$(\phi_H)_\infty = a_1 e^{-k\tilde{y}} + a_2 e^{k\tilde{y}} + a_3 e^{-Q\tilde{y}} + a_4 e^{Q\tilde{y}} \tag{7.42}$$

where $Q = \sqrt{k^2 + i\tilde{R}e(k - \beta_0)}$. Rewriting the momentum equation at the free stream in terms of S and Q and substituting for h with h_∞, we get

$$\phi^{iv} - [k^2 + Q^2]\phi'' + [k^2 Q^2]\phi = \hat{K}\tilde{R}ek^2\, a_5 e^{-S\tilde{y}} \tag{7.43}$$

This gives the particular integral for the solution of Eq. (7.41) as

$$(\phi_{PI})_\infty = a_5\Gamma e^{-S\tilde{y}} \tag{7.44}$$

where

$$\Gamma = \frac{\tilde{R}e\, \hat{K}\, k^2}{S^4 - (k^2 + Q^2)S^2 + k^2 Q^2}$$

for positive real parts of k, Q and S. In order to have a decaying solution at the free stream, Eq. (7.42) demands that $a_2 = a_4 = 0$. Therefore

$$\phi_\infty = a_1 e^{-k\tilde{y}} + a_3 e^{-Q\tilde{y}} + a_5\Gamma e^{-S\tilde{y}} \tag{7.45}$$

Thus, we obtain the solution to the coupled system of Eqs. (7.31) and (7.32) subject to homogeneous boundary conditions at the wall (Eq. (7.35)) and the free stream (Eq. (7.36)) as

$$\phi = a_1\phi_1 + a_3\phi_3 + a_5\phi_5 \tag{7.46}$$

$$h = a_1 h_1 + a_3 h_3 + a_5 h_5 \tag{7.47}$$

with the asymptotically decaying solution components having the following form at the free stream

$$h_{1\infty} = h_{3\infty} \to 0$$

$$h_{5\infty} \sim e^{-S\tilde{y}}$$

$$\phi_{1\infty} \sim e^{-k\tilde{y}}$$

$$\phi_{3\infty} \sim e^{-Q\tilde{y}}$$

$$\phi_{5\infty} \sim \Gamma e^{-S\tilde{y}}$$

where real parts of k, Q and S are all positive.

In flows where boundary layer assumptions are used, Re is high to make $|Q|$ and $|S| >> |k|$. This causes the fundamental solutions of the OSE to vary at dissimilar rates with respect to $\tilde{y}$. The fundamental solutions vary at different orders of magnitude, near and far away from the wall. This makes the sixth order OSE a *stiffer* system of equations so that direct integration of Eqs. (7.31) and (7.32) is not possible. On starting the integration directly from the free stream with corresponding initial conditions, as $\tilde{y}$ decreases, ϕ_3 and ϕ_5 will grow more rapidly compared to ϕ_1 and so does the error associated with these. The round off error (inherent in all numerical methods) in ϕ_3 and ϕ_5 would dominate ϕ_1. Thereby, the fundamental solutions lose their linear independence. This is the cause of exploding *parasitic error*. The control of parasitic error and stiffness is avoided here by using the CMM, as introduced in Chap. 3 for the fourth order system and extended here for the sixth order OSE, as discussed below.

7.4.2   Compound Matrix Method for the Sixth Order OSE

Here, the CMM is applied to the sixth order system given by Eqs. (7.31) and (7.32). Instead of solving for ϕ and h directly, the CMM solves an auxiliary system of equations for the compound variables. These variables are obtained from combinations of ϕ_j's, h_j's and their higher derivatives, which are governed by the analytical structure of the solution at the free stream. All these new variables have the same rate of variation with respect to $\tilde{y}$, which removes the stiffness problem of the initial ODE. For the system of ODEs defined by Eqs. (7.31) and (7.32), the original boundary value problem is again converted into an initial value problem, as was noted in Chap. 3, for flows without heat transfer. With the initial condition at the free stream known, this new set of equations is to be solved by marching from the free stream to the wall.

The OSE given by Eqs. (7.31) and (7.32) can be expressed as a system of six first order ODEs given by

$$\{\mathbf{r'}\} = [A]\,\{\mathbf{r}\} \tag{7.48}$$

where $\{\mathbf{r}\} = [\phi, \phi', \phi'', \phi''', h, h']^T$, and matrix $[A]$ is given as

$$[A] = \begin{bmatrix} 0 & 1 & 0 & 0 & 0 & 0 \\ 0 & 0 & 1 & 0 & 0 & 0 \\ 0 & 0 & 0 & 1 & 0 & 0 \\ -a & 0 & b & 0 & c & 0 \\ 0 & 0 & 0 & 0 & 0 & 1 \\ e & 0 & 0 & 0 & d & 0 \end{bmatrix}$$

where $a = k^4 + i\tilde{R}e\ k\ \bar{U}'' + i\tilde{R}e\ k^2(k\bar{U} - \beta_0)$, $b = 2k^2 + i\tilde{R}e(k\bar{U} - \beta_0)$, $c = \tilde{R}e\ \hat{K}\ k^2$, $d = k^2 + i\tilde{R}e\ Pr\ (k\bar{U} - \beta_0)$ and $e = \tilde{R}e\ Pr\ \bar{T}'$.

Following the notations and methodology in [5], a new set of variables is created, all of which vary with $\tilde{y}$ at the same rate. For the sixth order system, this is equivalent to projecting the solution on a subspace of C^6, with the help of three decaying boundary conditions for $\tilde{y} \to y_\infty$, into $\Lambda^3(C^6)$. Similarly, the boundary conditions at $y = 0$ define a second three-dimensional subspace of C^6. The problem is thus reduced to linking these two three-dimensional subspaces, satisfying Eqs. (7.31) and (7.32). Any subspace spanned by three linearly independent vectors, $\{\phi_1, \phi_3, \phi_5\}$, is an admissible modes of solution for the OSE, for $y \to \infty$. This is represented notationally as a point [5], $\phi_1 \wedge \phi_3 \wedge \phi_5$, in the vector space $\Lambda^3(C^6)$, where Λ is the wedge product.

Now we introduce six basis variables, defined as $e_1 = [\phi_1\ \phi_3\ \phi_5]$, $e_2 = [\phi_1'\ \phi_3'\ \phi_5']$, $e_3 = [\phi_1''\ \phi_3''\ \phi_5'']$, $e_4 = [\phi_1'''\ \phi_3'''\ \phi_5''']$, $e_5 = [h_1\ h_3\ h_5]$, $e_6 = [h_1'\ h_3'\ h_5']$ in C^6. All the elements of $e_i \wedge e_j \wedge e_k$ form the basis for $\Lambda^3(C^6)$, which has a dimension of $^6C_3 = 20$. With the help of the above mentioned basis vectors, we define the complete solution matrix in the following form

$$\begin{bmatrix} \phi_1 & \phi_3 & \phi_5 \\ \phi_1' & \phi_3' & \phi_5' \\ \phi_1'' & \phi_3'' & \phi_5'' \\ \phi_1''' & \phi_3''' & \phi_5''' \\ h_1 & h_3 & h_5 \\ h_1' & h_3' & h_5' \end{bmatrix} \tag{7.49}$$

The new variables for which numerical integration is carried out are obtained as the 20 (3×3) minors of the solution matrix given by Eq. (7.49), which are the second compounds here. The compound variables can be represented as

$$y_{ijk} = [e_i e_j e_k]^T$$

where $i = 1, 2, 3, 4; j = i+1, ..., 5;$ and $k = j+1, ...6$. Writing the compound matrix variables in a lexicographic fashion, i.e., $y_1 \equiv y_{123}; y_2 \equiv y_{124} \ ... \ y_{20} \equiv y_{456}$ we get

$$\mathbf{y_1} = \phi_1(\phi_3'\phi_5'' - \phi_5'\phi_3'') - \phi_3(\phi_1'\phi_5'' - \phi_5'\phi_1'') + \phi_5(\phi_1'\phi_3'' - \phi_3'\phi_1'') \tag{7.50}$$

$$\mathbf{y_2} = \phi_1(\phi_3'\phi_5''' - \phi_5'\phi_3''') - \phi_3(\phi_1'\phi_5''' - \phi_5'\phi_1''') + \phi_5(\phi_1'\phi_3''' - \phi_3'\phi_1''') \tag{7.51}$$

$$\mathbf{y_3} = \phi_1(\phi_3'h_5 - \phi_5'h_3) - \phi_3(\phi_1'h_5 - \phi_5'h_1) + \phi_5(\phi_1'h_3 - \phi_3'h_1) \tag{7.52}$$

$$\mathbf{y_4} = \phi_1(\phi_3'h_5' - \phi_5'h_3') - \phi_3(\phi_1'h_5' - \phi_5'h_1') + \phi_5(\phi_1'h_3' - \phi_3'h_1') \tag{7.53}$$

$$\mathbf{y_5} = \phi_1(\phi_3''\phi_5''' - \phi_5''\phi_3''') - \phi_3(\phi_1''\phi_5''' - \phi_5''\phi_1''') + \phi_5(\phi_1''\phi_3''' - \phi_3''\phi_1''') \tag{7.54}$$

$$\mathbf{y_6} = \phi_1(\phi_3''h_5 - \phi_5''h_3) - \phi_3(\phi_1''h_5 - \phi_5''h_1) + \phi_5(\phi_1''h_3 - \phi_3''h_1) \tag{7.55}$$

$$\mathbf{y_7} = \phi_1(\phi_3''h_5' - \phi_5''h_3') - \phi_3(\phi_1''h_5' - \phi_5''h_1') + \phi_5(\phi_1''h_3' - \phi_3''h_1') \tag{7.56}$$

$$\mathbf{y_8} = \phi_1(\phi_3'''h_5 - \phi_5'''h_3) - \phi_3(\phi_1'''h_5 - \phi_5'''h_1) + \phi_5(\phi_1'''h_3 - \phi_3'''h_1) \tag{7.57}$$

$$\mathbf{y_9} = \phi_1(\phi_3'''h_5' - \phi_5'''h_3') - \phi_3(\phi_1'''h_5' - \phi_5'''h_1') + \phi_5(\phi_1'''h_3' - \phi_3'''h_1') \tag{7.58}$$

$$\mathbf{y_{10}} = \phi_1(h_3h_5' - h_5h_3') - \phi_3(h_1h_5' - h_5h_1') + \phi_5(h_1h_3' - h_3h_1') \tag{7.59}$$

$$\mathbf{y_{11}} = \phi_1'(\phi_3''\phi_5''' - \phi_5''\phi_3''') - \phi_3'(\phi_1''\phi_5''' - \phi_5''\phi_1''') + \phi_5'(\phi_1''\phi_3''' - \phi_3''\phi_1''') \tag{7.60}$$

$$\mathbf{y_{12}} = \phi_1'(\phi_3''h_5 - \phi_5''h_3) - \phi_3'(\phi_1''h_5 - \phi_5''h_1) + \phi_5'(\phi_1''h_3 - \phi_3''h_1) \tag{7.61}$$

$$\mathbf{y_{13}} = \phi_1'(\phi_3''h_5' - \phi_5''h_3') - \phi_3'(\phi_1''h_5' - \phi_5''h_1') + \phi_5'(\phi_1''h_3' - \phi_3''h_1') \tag{7.62}$$

$$\mathbf{y_{14}} = \phi_1'(\phi_3'''h_5 - \phi_5'''h_3) - \phi_3'(\phi_1'''h_5 - \phi_5'''h_1) + \phi_5'(\phi_1'''h_3 - \phi_3'''h_1) \tag{7.63}$$

$$\mathbf{y_{15}} = \phi_1'(\phi_3'''h_5' - \phi_5'''h_3') - \phi_3'(\phi_1'''h_5' - \phi_5'''h_1') + \phi_5'(\phi_1'''h_3' - \phi_3'''h_1') \tag{7.64}$$

$$\mathbf{y_{16}} = \phi_1'(h_3h_5' - h_5h_3') - \phi_3'(h_1h_5' - h_5h_1') + \phi_5'(h_1h_3' - h_3h_1') \tag{7.65}$$

$$\mathbf{y_{17}} = \phi_1''(\phi_3''h_5 - \phi_5'''h_3) - \phi_3''(\phi_1'''h_5 - \phi_5'''h_1) + \phi_5''(\phi_1'''h_3 - \phi_3'''h_1) \tag{7.66}$$

$$\mathbf{y_{18}} = \phi_1''(\phi_3'''h_5' - \phi_5'''h_3') - \phi_3''(\phi_1'''h_5' - \phi_5'''h_1') + \phi_5''(\phi_1'''h_3' - \phi_3'''h_1') \tag{7.67}$$

$$\mathbf{y_{19}} = \phi_1''(h_3h_5' - h_5h_3') - \phi_3''(h_1h_5' - h_5h_1') + \phi_5''(h_1h_3' - h_3h_1') \tag{7.68}$$

$$\mathbf{y_{20}} = \phi_1'''(h_3h_5' - h_5h_3') - \phi_3'''(h_1h_5' - h_5h_1') + \phi_5'''(h_1h_3' - h_3h_1') \tag{7.69}$$

Now if we represent

$$\mathbf{y} = [\mathbf{y_1}, \ \mathbf{y_2}, \ \mathbf{y_3}, \ ..., \ \mathbf{y_{20}}]^{\mathbf{T}}$$

it can be shown that **y** satisfies the following system of differential equations

$$\{y'\} = [\mathbf{B(y)}] \{y\} \tag{7.70}$$

where the matrix $[B]$ has been obtained in [249, 5] for general fourth and sixth order systems. The elements of $[B]$ are obtained in terms of the elements of matrix $[A]$ in Eq. (7.48). On expanding Eq. (7.70), the auxiliary system of equations in terms of the compound variables is obtained as

$$y'_1 = y_2 \tag{7.71}$$

$$y'_2 = by_1 + cy_3 + y_5 \tag{7.72}$$

$$y'_3 = y_4 + y_6 \tag{7.73}$$

$$y'_4 = dy_3 + y_7 \tag{7.74}$$

$$y'_5 = cy_6 + y_{11} \tag{7.75}$$

$$y'_6 = y_7 + y_8 + y_{12} \tag{7.76}$$

$$y'_7 = dy_6 + y_9 + y_{13} \tag{7.77}$$

$$y'_8 = by_6 + y_9 + y_{14} \tag{7.78}$$

$$y'_9 = by_7 + dy_8 + cy_{10} + y_{15} \tag{7.79}$$

$$y'_{10} = y_{16} \tag{7.80}$$

$$y'_{11} = -ay_1 + cy_{12} \tag{7.81}$$

$$y'_{12} = y_{13} + y_{14} \tag{7.82}$$

$$y'_{13} = ey_1 + dy_{12} + y_{15} \tag{7.83}$$

$$y'_{14} = ay_3 + by_{12} + y_{15} + y_{17} \tag{7.84}$$

$$y'_{15} = ey_2 + ay_4 + by_{13} + dy_{14} + cy_{16} + y_{18} \tag{7.85}$$

$$y'_{16} = ey_3 + y_{19} \tag{7.86}$$

$$y'_{17} = ay_6 + y_{18} \tag{7.87}$$

$$y'_{18} = ey_5 + ay_7 + dy_{17} + cy_{19} \tag{7.88}$$

$$y'_{19} = ey_6 + y_{20} \tag{7.89}$$

$$y'_{20} = ey_8 - ay_{10} + by_{19} \tag{7.90}$$

where a, b, c, d and e are as defined in Eq. (7.48). The independent variable in Eqs. (7.71) to (7.90) is $\tilde{y}$. Since all the compound variables (y_1 to y_{20}) are defined in terms of ϕ_1, ϕ_3, ϕ_5, h_1, h_3, h_5 and their higher derivatives, and since all of these have an analytical expression for their free stream conditions, it is possible to obtain the conditions for the compound variables as $\tilde{y} \to \infty$. This has already been pointed out as the conversion of the original boundary value problem with wall and far stream boundary conditions into an initial value problem for the compound variables starting from the free stream. Due to the definition of second compounds, all these variables have an identical rate of exponential variation with respect to $\tilde{y}$, given by the exponent $-(Q + S + k)$. This effectively removes the stiffness problem of the OSE. Thus, it no longer remains a stiff system and Eqs. (7.71) to (7.90) could be integrated in a straightforward way as an initial value problem by any standard ODE solving techniques, provided the conditions of y_1 to y_{20} as $\tilde{y} \to \infty$ are known.

7.4.3   Initial Conditions for an Auxiliary System of Equations

The asymptotic conditions of fundamental solutions of Eqs. (7.31) and (7.32) are given for $\tilde{y} \to \infty$ when real $[k, Q, S] > 0$ as

$$\phi_1 \sim e^{-k\tilde{y}}$$

$$h_1 \sim 0$$

$$\phi_3 \sim e^{-Q\tilde{y}}$$

$$h_3 \sim 0$$

$$\phi_5 \sim \Gamma e^{-S\tilde{y}}$$

$$h_5 \sim e^{-S\tilde{y}}$$

Substituting the above given asymptotic conditions into Eqs. (7.50) to (7.69) and scaling by $e^{(k+Q+S)\tilde{y}_\infty}$, we get the following set of initial conditions for the system of ODEs given by Eqs. (7.71) to (7.90)

$$y_{1\infty} = \Gamma[S^2(k-Q) + Q^2(S-k) + k^2(Q-S)] \tag{7.91}$$

$$y_{2\infty} = \Gamma[-S^3(k-Q) - Q^3(S-k) - k^3(Q-S)] \tag{7.92}$$

$$y_{3\infty} = (k-Q) \tag{7.93}$$

$$y_{4\infty} = S(Q-k) \tag{7.94}$$

$$y_{5\infty} = \Gamma[S^3(k^2-Q^2) + Q^3(S^2-k^2) + k^3(Q^2-S^2)] \tag{7.95}$$

$$y_{6\infty} = (Q^2 - k^2) \tag{7.96}$$

$$y_{7\infty} = S(k^2 - Q^2) \tag{7.97}$$

$$y_{8\infty} = (k^3 - Q^3) \tag{7.98}$$

$$y_{9\infty} = S(Q^3 - k^3) \tag{7.99}$$

$$y_{10\infty} = 0 \tag{7.100}$$

$$y_{11\infty} = \Gamma[S^3(kQ^2 - k^2Q) + Q^3(Sk^2 - S^2k) + k^3(QS^2 - Q^2S)] \tag{7.101}$$

$$y_{12\infty} = Qk(k-Q) \tag{7.102}$$

$$y_{13\infty} = QSk(Q-k) \tag{7.103}$$

$$y_{14\infty} = Qk(Q^2 - k^2) \tag{7.104}$$

$$y_{15\infty} = -QSk(Q^2 - k^2) \tag{7.105}$$

$$y_{16\infty} = 0 \tag{7.106}$$

$$y_{17\infty} = Q^2 k^2 (k-Q) \tag{7.107}$$

$$y_{18\infty} = Q^2 Sk^2 (Q-k) \tag{7.108}$$

$$y_{19\infty} = 0 \tag{7.109}$$

$$y_{20\infty} = 0 \tag{7.110}$$

One can numerically solve the OSE by integrating Eqs. (7.71) to (7.90) from $\tilde{y}_\infty$ to the wall ($\tilde{y} = 0$), using the four-stage Runge–Kutta integration technique for ODEs, with the initial conditions given by Eqs. (7.91) to (7.110).

7.4.4   Dispersion Relation

The original problem defined by Eqs. (7.31) and (7.32) along with the boundary conditions of Eqs. (7.33) and (7.34), represent a homogeneous system of ODEs with homogeneous boundary conditions at the wall and at the free stream, which makes it an eigenvalue problem. Thus, for a fixed set of parameters like $\tilde{Re}$, β_0 and $\hat{K}$, there are non-trivial values of k which satisfy the governing equations and boundary conditions as the eigenvalues of the problem. Any such eigenvalue represents a complex wavenumber, with the imaginary part indicating the growth or decay rate of disturbances and the real part indicating the wave length of the corresponding disturbances in the flow field (in terms of the reference length δ^* of the parallel boundary layer). When CMM equations are marched from the free stream towards the wall, it is seen that the wall boundary conditions are not necessarily satisfied, unless the complex wavenumber chosen is an eigenvalue of the problem. Thus, for a particular mean flow, there are combinations of k and β_0 which satisfy the homogeneous boundary conditions. The relation between k and β_0 is known as the dispersion relation of the physical problem which interrelates spatial and temporal scales of the flow in the spectral plane.

For CMM formulation, the dispersion relation is obtained by satisfying the wall boundary conditions for the disturbance components of velocity and temperature given by Eq. (7.33). These relations, when written in terms of fundamental solution components, give the following

$$a_1\phi_1 + a_3\phi_3 + a_5\phi_5 = 0 \tag{7.111}$$

$$a_1\phi_1' + a_3\phi_3' + a_5\phi_5' = 0 \tag{7.112}$$

$$a_1h_1 + a_3h_3 + a_5h_5 = 0 \tag{7.113}$$

The characteristic determinant of the system of Eqs. (7.111) to (7.113) turns out to be the third member of the second compound set, as given in Eq. (7.52), which is

$$D_r + iD_i = y_3 = 0 \text{ at } \tilde{y} = 0 \tag{7.114}$$

7.4.5   The Grid Search Method and the Newton–Raphson Technique for Obtaining Eigenspectrum

For spatial stability analysis, the non-dimensional circular frequency of excitation, β_0, is a fixed real quantity. The objective is to find suitable values of complex k which satisfy the dispersion relation. These become the eigenvalues of the OSE for a particular set of parameters ($\tilde{Re}$, $\hat{K}$, Pr and β_0). In the following, the methodology is explained with reference to the typical buoyancy parameter, $K = 3 \times 10^{-3}$.

For the grid search method employed here (as also done using the CMM in [322]), the following values of $Pr = 0.70$, $\tilde{Re} = Re_{\delta^*} = 3000$ and $\beta_0 = 0.05$ have been used. We have to find k's which satisfy Eq. (7.114). Equations (7.71) to (7.90) are solved using the initial conditions specified by Eqs. (7.91) to (7.110), starting from $\tilde{y} = \tilde{y}_{max}$, a user specified value. The maximum ordinate $\tilde{y}_{max}$ is defined as $\tilde{y}_{max} = \eta_{max}/C_\delta$. The reported results are for $\eta_{max} = 12$. As defined earlier, C_δ is a function of K indirectly. For $K = 3 \times 10^{-3}$, it is found that $C_\delta = 1.69596$. The numerical method used for marching the solution is the four-stage, fourth order Runge–Kutta method, with 2000 points for the chosen $\tilde{y}_{max}$. In the grid search method, auxiliary equations are solved for a predefined range of k_{real} and k_{imag}. Here the range of complex wavenumber is given as $0 \le k_{real} \le 4$ and $-0.5 \le k_{imag} \le 3.5$. A total of 301 points are taken in each direction. Thus, we obtain the second compound corresponding to the dispersion relation given by $D = D_r + D_i = y_3$ for each (k_{real}, k_{imag}) combination in the above mentioned range. Since the dispersion relation $D = 0$ is a complex quantity, both D_r and D_i have to be simultaneously zero for a given (k_{real}, k_{imag}) combination for that particular k to qualify as an eigenvalue. In order to locate these eigenvalues, $D_r = 0$ and $D_i = 0$ contours in the chosen range of the (k_{real}, k_{imag})-plane are plotted together. Such a plot for the above mentioned parameters is shown in Fig. 7.8a for $K = 3 \times 10^{-3}$. The solid lines in the figure represent $D_r = 0$ contours and the dotted lines indicate $D_i = 0$ contours. It is seen in Fig. 7.8a that there are 13 points at which the grid search method predicts potential eigenvalues. All of these need not be true eigenvalues, as this is a pictorial representation dependent upon plotting procedure which also involves data interpolation. However, the grid search method results provide initial guesses of the eigenspectrum of the system in the chosen part of the complex wavenumber plane. Such eigenvalues can be tested by the Newton–Raphson technique for authenticity and this technique also helps to *polish* the eigenvalues. Note that all calculations performed here are restricted to the positive k_{real} axis.

The eigenspectrum obtained by the grid search method for the case of Fig. 7.8a is shown in Table 7.2. The Newton–Raphson method is basically used to polish off approximate roots of an equation to a required degree of accuracy in an iterative manner, as described next. In Table 7.2, there are 13 possible eigenvalues within the chosen range. The Newton–Raphson technique would distinguish between the physical and spurious modes. The following procedure is adopted to check whether a particular k indicated by the grid search method represents an actual eigenvalue of the system and to obtain it to the required degree of accuracy. The value of k given by the grid search is taken as the initial guess. This starting value is denoted as k_n. Let D_n be the expression for D at the wall, as defined by Eq. (7.114), corresponding to k_n. Next, k_n is perturbed by a small amount, 0.00001, to get k_{n+1} and the corresponding D is calculated as D_{n+1}. The value of k in the next iteration, k_{n+2}, is obtained through the relation

$$k_{n+2} = k_n - \frac{D_n (k_{n+1} - k_n)}{D_{n+1} - D_n} \tag{7.115}$$

The expression for D corresponding to k_{n+2} is denoted D_{n+2}. This iteration is repeated until the latest value of D_{n+2} satisfies the dispersion relation at the wall. Ideally, computations are preformed until D_{n+2} reaches a computational tolerance ($\tilde{\epsilon}$), i.e., when $|D| \leq \tilde{\epsilon}$ at $\tilde{y} = 0$. The choice of $\tilde{\epsilon}$ depends on the precision of computations and for the present double precision calculations, $\tilde{\epsilon}$ is chosen to be 10^{-12}. If the value of $|D|$ does not satisfy the convergence criterion, then the corresponding initial guess is considered a spurious value and removed. There may be cases in which the convergence criterion is met but the value of k obtained by the Newton–Raphson search is far removed from the initial guess. Such initial guesses are also considered spurious ones. Sometimes it also happens that two or more different approximate eigenvalues predicted by the grid search can converge to a single true eigenvalue after polishing is done by the Newton–Raphson technique, as noted in a few cases of Table 7.2.

We now discuss the above mentioned procedure as applied to the grid search result shown in Fig. 7.8(a). Initial guesses of eigenvalues for the previously mentioned parameters are given in Table 7.2, along with the corresponding polished values of these modes by the Newton–Raphson technique.

In Table 7.2, SM in the remarks column represents a spurious mode. Here, second and third columns show k_{real} and k_{imag}, respectively, as obtained from the grid search method. The fourth and fifth columns show the corresponding polished values obtained from the Newton–Raphson technique. Consider the first entry in the table, which is referred to as Mode 1. The polished value is in close proximity to the grid search result, implying this is a genuine eigenvalue. As the value of k_{imag} for this eigenvalue is positive, it is a stable mode.

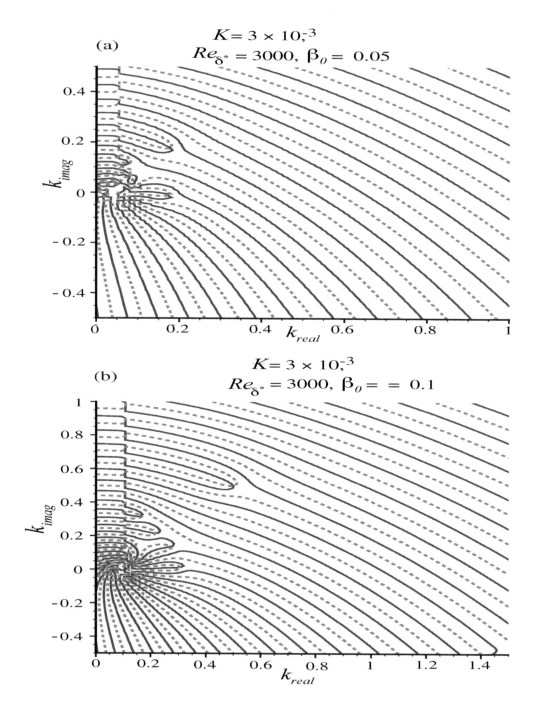

FIGURE 7.8
Eigenspectrum of the OSE obtained through the grid search method at the seed points for $K = 3 \times 10^{-3}$ at $Re_{\delta^*} = 3000$. (a) For $\beta_0 = 0.05$. (b) $\beta_0 = 0.1$. In both these frames, solid lines represent $D_r = 0$ and dotted lines represent $D_i = 0$.

Similarly, the second entry is also a genuine eigenvalue and its polished value shows that this (Mode 2) is also a stable mode. Similarly, the third entry (Mode 3) is also found to be a genuine eigenvalue which is stable. It is noted that although the grid search method predicts four different eigenvalues represented by entries 4 to 7, the Newton–Raphson technique shows all of these to converge to a single eigenvalue (Mode 4). This shows that the grid search method indicates spurious or fractured modes, contaminated by plotting technique. Entries 8 to 11 show that polishing the grid search results causes the final value to converge to a value far removed from the initial guess. Also from the grid search result, it could be clearly observed that around $k_{real} = 0.3$, the contours show no tendency to intersect. This leads us to conclude that these entries are also spurious ones. The twelfth and thirteenth entries converge to Mode 5, with k_{imag} as negative. This is a genuine eigenvalue, indicating it is an unstable mode.

7.4.6 Neutral Curve and Wavenumber Contours

Spatial amplification contours are obtained by plotting k_{imag} contours in the $(Re_{\delta*}, \beta_0)$-plane, with the $k_{imag} = 0$ contour in this plane known as the neutral curve. The region

TABLE 7.2
Grid search and Newton–Raphson polishing results for $K = 3.0 \times 10^{-3}$, $Pr = 0.7$, $Re_{\delta*} = 3000$ and $\beta_0 = 0.05$.

Case	k_{real} (GS*)	k_{imag} (GS*)	k_{real} (NR**)	k_{imag} (NR**)	Remarks
1	0.176886	0.177115	0.176927	0.176814	MODE 1
2	0.0769187	0.114972	0.075867	0.113723	MODE 2
3	0.0757979	0.0761263	0.0767555	0.0751849	MODE 3
4	0.0793309	0.0793309	0.0827282	0.0660410	MODE 4
5	0.0824809	0.0673818	0.0827282	0.0660410	MODE 4
6	0.0756609	0.0628665	0.0827282	0.0660410	MODE 4
7	0.0776794	0.033027	0.0827282	0.0660410	MODE 4
8	0.0918642	0.0180722	0.3495202	−0.116062	SM
9	0.0563055	−0.00279656	0.327198	−0.130203	SM
10	0.0836735	−0.00165521	0.315595	−0.135621	SM
11	0.0791002	0.0203019	0.380901	−0.0804865	SM
12	0.172713	−0.0124779	0.175352	−0.0117661	MODE 5
13	0.0929694	0.000959116	0.175352	−0.0117671	MODE 5

* GS - Grid Search and ** NR - Newton–Raphson

within the neutral curve has negative values of k_{imag} and outside, k_{imag} is positive. Hence, all disturbances within the neutral curve are unstable, i.e., these experience amplification as the disturbance convects downstream. In order to obtain the neutral curve, the above procedure, including grid search and Newton–Raphson techniques, can be employed for each and every combination of $Re_{\delta*}$ and β_0 within the region chosen for the neutral curve. However, such a procedure is a tedious task to perform. Instead, we adopt an efficient method of obtaining the real and imaginary parts of the least stable eigenvalue, as explained below.

This procedure also employs the CMM along with the Newton–Raphson technique. Having obtained a genuine eigenvalue for a particular $(Re_{\delta*}, \beta_0)$ combination, one marches along constant $Re_{\delta*}$ or constant β_0 to find neighboring eigenvalues, using this eigenvalue as the seed point. In Table 7.2, Mode 5 is the lone unstable mode for $K = 3 \times 10^{-3}$ at $Re_{\delta*} = 3000$ and $\beta_0 = 0.05$ and this is taken as the seed value from which calculations

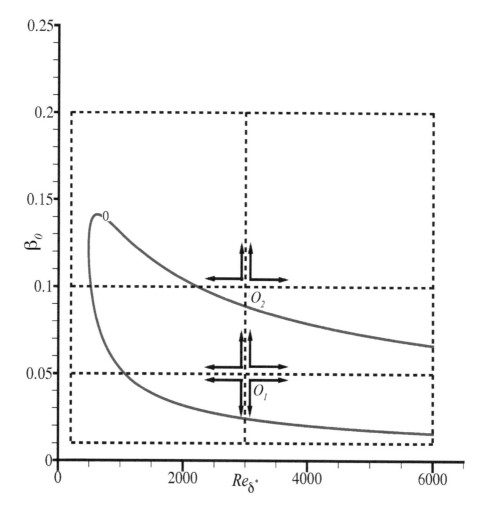

FIGURE 7.9
Methodology to generate stability properties for $K = 3 \times 10^{-3}$. Points O_1 and O_2 are the seed points from which calculations are started in the $(Re_{\delta*}, \beta_0)$-plane. From O_1, the lower four rectangular regions marked by dotted lines were swept out. From O_2, the upper two rectangular regions were covered. The neutral curve is shown as a reference.

begin. We concentrate on a domain in the (Re_{δ^*}, β_0)-plane spanned by the boundaries $Re_{\delta^*} = 6000$ on the right, $Re_{\delta^*} = 3000$ on the left, $\beta_0 = 0.05$ on the top and $\beta_0 = 0.01$ at the bottom.

The seed eigenvalue ($k = k_{real} + k_{imag}$) at $Re_{\delta^*} = 3000$ and $\beta_0 = 0.05$, is $(0.175352716 - i1.176611831 \times 10^{-2})$. This is marked as O_1 in Fig. 7.9 and one marches along $\beta_0 = 0.05$ to the right by changing Re_{δ^*} to $Re_{\delta^*} = 3001$. In order to find the correct eigenvalue at the new (Re_{δ^*}, β_0) coordinates at (3001, 0.051), the eigenvalue at (3000, 0.05) is taken as the initial guess for the Newton–Raphson technique. This procedure is continued successively to obtain eigenvalues along this $\beta_0 = 0.05$ line, until the right boundary of the (Re_{δ^*}, β_0) plane [(6000, 0.05) in this calculation] is reached. In the same way, one starts from the same seed point and obtains eigenvalues for the Re_{δ^*} line, shifting to (3000, 0.049) by decreasing β_0. This variation in β_0, given by $d\beta_0 = 0.001$, is found to be adequate. Having obtained the eigenvalue for (3000, 0.049) by using the eigenvalue at (3000, 0.05) as the initial guess in the Newton–Raphson technique, one again marches by increasing the Reynolds number until the point at (6000, 0.049) is reached along the $\beta_0 = 0.049$ line. This procedure is continued until the lowest β_0 level is reached.

From the seed point (3000, 0.05), a similar procedure can be adopted to find the eigenvalues in four adjacent patches by suitably changing the (Re_{δ^*}, β_0) coordinates as indicated in Fig. 7.9. This method was used to obtain the spatial amplification contours within the limits $200 \leq Re_{\delta^*} \leq 6000$ and $0.01 \leq \beta_0 \leq 0.1$. Any other mode from Table 7.2 chosen as the seed value would not have resulted in the neutral curve, as these remain stable within the (Re_{δ^*}, β_0) range studied in the calculation.

Another grid search was carried out at the point (3000, 0.1) in the (Re_{δ^*}, β_0)-plane, as shown in Fig. 7.8b. This is marked as point O_2 in Fig. 7.9. Zero contours of D_r are denoted by solid lines, whereas those of D_i are indicated by dotted lines in Fig. 7.8b. From this point, calculations were extended to $\beta_0 = 0.2$ for the above mentioned range of Re_{δ^*}. Table 7.3 shows the eigenspectrum for $Re_{\delta^*} = 3000$ and $\beta_0 = 0.1$ (with other parameters remaining the same). Only the genuine eigenvalues obtained through the Newton–Raphson technique following the grid search procedure are shown, which show six stable modes. The first five modes (Mode 6 to Mode 10) are highly stable compared to Mode 11. But Mode 11 is closer to neutral stability due to its relatively low value of k_{imag} and is chosen as the seed value, with $k = 0.307233105 + i1.0500144 \times 10^{-2}$. As calculations proceed in the (Re_{δ^*}, β_0)-plane, this eigenvalue becomes unstable at lower values of Re_{δ^*} and provides the neutral curve.

Carrying out the above procedure, we obtain the neutral curve for a given mean flow defined by the value of buoyancy parameter K. A similar procedure was adopted to generate neutral curves for all the buoyancy parameters discussed in the following section.

7.4.7 Precision in Computing

We have already mentioned that the convergence criterion for the numerical value of D at the wall is set to $\tilde{\epsilon} = 10^{-12}$ for double precision calculations. If one tries for higher degree of accuracy by using $\tilde{\epsilon} \leq 10^{-12}$, then one faces convergence problems for computations using double precision, as can be understood from Eq. (7.115), which has difference terms $(k_{n+1} - k_n)$ in the numerator and $(D_{n+1} - D_n)$ in the denominator. Due to round off errors, even before the convergence criterion $|D| \leq \tilde{\epsilon}$ (with $\tilde{\epsilon} \leq 10^{-12}$) is satisfied, the value of k_n and k_{n+1} becomes almost identical. Due to lower precision calculations, their difference is rounded off to zero. So the latest value of k, given by k_{n+2}, becomes identical to k_n or k_{n+1}. This means, in the following iteration, since the values of k_n and k_{n+1} are equal, corresponding D_n and D_{n+1} become the same, which results in division by zero, terminating the computations.

This precision problem is not a drawback of the CMM or the spurious nature of eigenvalues. This is solely an issue of precision in computing. If we use quadruple precision, the convergence criteria $\tilde{\epsilon}$ can be set to a much lower value of 10^{-30}. Thus, the difference in k at successive iterations will be evaluated to greater precision and the round off error will creep in only at a considerably later stage. In the present calculations, the lowest value of β_0 up to which eigenvalues have been calculated is $\beta_0 = 0.01$. This is for relatively lower values of K having an order of magnitude between 10^{-6} and 10^{-3}. It is noted that as K increases, calculations at lower β_0 become more difficult. For example, in the spatial amplification contours for $K = 7 \times 10^{-2}$, calculations have been restricted to a minimum value of $Re_{\delta*} = 400$ for β_0 values less than 0.03 in the $(Re_{\delta*}, \beta_0)$-plane. Beyond this range, which is closer to the origin, a double precision Newton–Raphson procedure has convergence problems. Here the successive iterations keep on diverging from each other. Either convergence is not achieved or the calculations culminate in division by zero (in Eq. (7.115)). By using quadruple precision computations, one should be able to take the calculations to a larger region in the $(Re_{\delta*}, \beta_0)$-plane, especially towards the origin and closer to the axes.

7.4.8 Results of the Linear Spatial Stability Theory

Results obtained from the linear theory are shown with the contours of k_{imag} and k_{real} in the $(Re_{\delta*}, \beta_0)$-plane. Once the neutral curve is obtained, one can predict whether the combination of $Re_{\delta*}$ and β_0 makes the flow stable or unstable. In Fig. 7.10, contours for imaginary and real parts of the complex wavenumber are shown in frames (a) and (b), respectively, for $K = 1 \times 10^{-6}$. In Fig. 7.10a, the neutral curve is the zero contour of k_{imag}, with the region inside the neutral curve having negative values for k_{imag}. Any wave disturbance within the shear layer having a corresponding $(Re_{\delta*}, \beta_0)$ combination coming inside the neutral curve will show amplification. If the same disturbance comes outside the neutral curve, the wave will be attenuated. In Fig. 7.10a two rays are marked OA_1 and OA_2 which correspond to constant non-dimensional physical frequency (F), defined as follows:

$$F = \beta_0 / Re_{\delta*} = 2\pi\nu\tilde{f}/U_\infty^2$$

where $\tilde{f}$ is the excitation frequency in Hz. When the flow is excited by a fixed frequency source, the response can be tracked by following such rays corresponding to the constant exciter frequency. In Fig. 7.10, OA_1 corresponds to $F = 4 \times 10^{-5}$ and OA_2 corresponds

TABLE 7.3
Grid search and Newton–Raphson polishing results for
$K = 3.0 \times 10^{-3}$, $Pr = 0.7$, $Re_{\delta*} = 3000$ and $\beta_0 = 0.1$.

Sl.No	k_{real} (GS*)	k_{imag} (GS*)	k_{real} (NR**)	k_{imag} (NR**)	Remarks
1	0.500317	0.49984	0.499095	0.499009	MODE 6
2	0.161561	0.315957	0.163418	0.314808	MODE 7
3	0.215604	0.214703	0.216794	0.215080	MODE 8
4	0.138329	0.135435	0.138014	0.135314	MODE 9
5	0.269636	0.135197	0.270929	0.134605	MODE 10
6	0.305247	0.0108736	0.307233105	0.0105001	MODE 11

*GS - Grid Search and **NR - Newton–Raphson

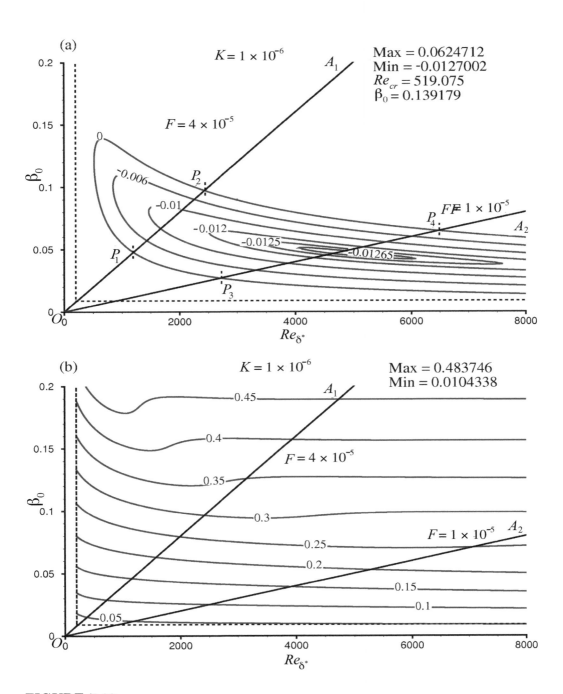

FIGURE 7.10
(a) Growth rate (k_{imag}) contours along with the neutral curve shown plotted in the (Re_{δ^*}, β_0)-plane for $K = 1 \times 10^{-6}$; (b) the real part of wavenumber (k_{real}) contours shown plotted in the (Re_{δ^*}, β_0)-plane for the same K. Both frames show two constant non-dimensional frequency rays: $F = 4 \times 10^{-5}$ represented by OA_1 and $F = 1 \times 10^{-5}$ represented by OA_2.

to $F = 1 \times 10^{-5}$. The point P_1 indicates where the ray OA_1 enters the neutral curve and P_2 represents the point where it exits the neutral curve. P_1 corresponds to $Re_{\delta*} = 1195.71$

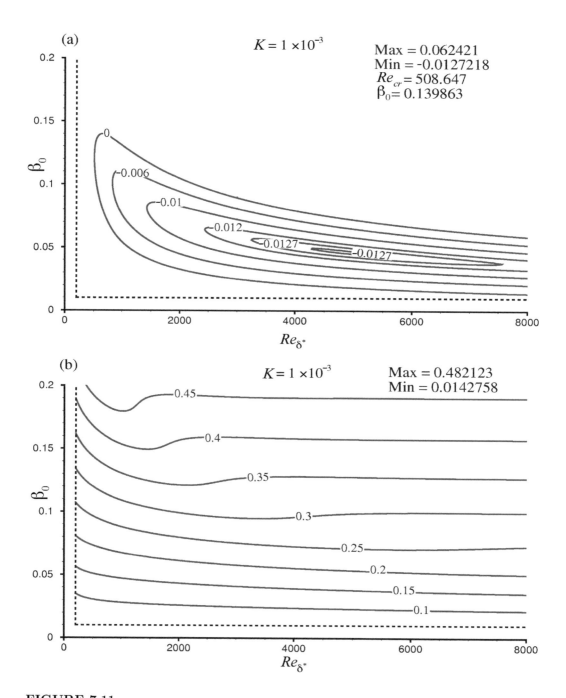

FIGURE 7.11
(a) Growth rate (k_{imag}) contours along with the neutral curve shown plotted in the $(Re_{\delta*}, \beta_0)$-plane for $K = 1 \times 10^{-3}$; (b) the real part of wavenumber (k_{real}) contours shown plotted in the $(Re_{\delta*}, \beta_0)$-plane for the same K.

and P_2 corresponds to $Re_{\delta*} = 2431.9$. A disturbance created at $F = 4 \times 10^{-5}$ at a location outside and below the neutral curve will be attenuated until it reaches P_1 corresponding to $Re_{\delta*} = 1195.71$. Once the disturbance enters the neutral curve, its amplitude will grow. This amplitude growth first increases to a maximum as it moves towards the center of the neutral curve, where we have maximum negative values for k_{imag}. Then the growth rate falls off as it moves towards the exit point P_2 where $Re_{\delta*} = 2431.9$. Beyond this location, the disturbance is again in a stable region where its amplitude decays. A similar behavior is noted for $F = 1 \times 10^{-5}$ with corresponding entry and exit points denoted by P_3 and P_4. For $F = 1 \times 10^{-5}$ being a relatively lower frequency, it remains within the neutral curve for a longer range of $Re_{\delta*}$ and hence experiences prolonged amplification compared to $F = 4 \times 10^{-5}$. Figure 7.10b shows the corresponding k_{real} contours, which give us an idea of the length scale associated with the disturbance as it convects through the shear layer. Under the parallel flow assumption, it is believed that the flow adjusts itself to local conditions, dictating growth or damping, as well as fixing the wavelength of the associated disturbances. The k_{real} contours also indicate that the unstable TS waves have a maximum wavenumber around $k_{real} = 0.35$, to indicate the smallest unstable wavelength which the shear layer can sustain. Using the relation $k_{real\ max} = 2\pi\delta*/\tilde{\lambda}_{min}$, where $\tilde{\lambda}_{min}$ represents the smallest unstable TS waves, which for this flow is $\tilde{\lambda}_{min} \approx 18\delta*$. In Fig. 7.10a, one notices that the upper and the lower branch of the neutral curve come closer to each other at higher $Re_{\delta*}$, which implies that this is indeed a viscous instability, as with $Re_{\delta*} \to \infty$, the flow does not have any spatial instability.

Certain critical parameters are noted from Fig. 7.10a. The lowest $Re_{\delta*}$ at which a disturbance of any frequency becomes unstable is the critical Reynolds number (Re_{cr}) and is given by the left most point of the neutral curve. The highest non-dimensional circular frequency above which all disturbances remain stable is the critical circular frequency. Figure 7.10a shows these critical parameters along with the range of k_{imag}. The minimum value of k_{imag} determines the maximum spatial growth rate.

Next, we note how heat transfer modifies spatial stability characteristics by obtaining neutral curves for a range of K values. Frames (a) and (b) in Figs. 7.10 to 7.15 show the contours for real and imaginary parts of the complex wavenumber, respectively, for different K values. K varies from a minimum value of $K = 1 \times 10^{-6}$ to the high value of $K = 9 \times 10^{-2}$. One distinct feature in the neutral curves in frame (a) of Figs. 7.10 to 7.15 is the value of Re_{cr} for $K = 1 \times 10^{-6}$, 519.075, which falls off monotonically to 118.983 for $K = 9 \times 10^{-2}$. In Fig. 7.16a, this variation of critical Reynolds number as a function of K is plotted. It is noted that up to $K = 1 \times 10^{-3}$, Re_{cr} varies less from its value at $K = 1 \times 10^{-6}$. But above $K = 1 \times 10^{-3}$, a small increase in K results in a relatively larger drop in Re_{cr}. This trend becomes more acute as we move towards $K = 1 \times 10^{-2}$ and above. This implies that for a hotter plate, disturbances become unstable at a shorter distance from the singular source at the leading edge of the plate, if the linear theory is to hold. The maximum unstable circular frequency is also seen to grow as K increases, as can be observed from its variation with K plotted in Fig. 7.16b. The maximum unstable β_0 for $K = 1 \times 10^{-6}$ is 0.139179, which for $K = 9 \times 10^{-2}$ is 0.170975. Another parameter of interest is the maximum relative spatial growth rate $[(-k_{imag})_{max}]$. For $K = 1 \times 10^{-6}$, this is $(-k_{imag})_{max} = 0.01276615$ and for $K = 3 \times 10^{-2}$, this changes to $(-k_{imag})_{max} = 0.013019$. The value of $(-k_{imag})_{max}$ determines the maximum growth rate of the disturbances within the neutral curve and is an indicator of linear instability. But we note that, beyond $K = 3 \times 10^{-2}$, $(-k_{imag})_{max}$ shows a slight decrease and it ends up at $(-k_{imag})_{max} = 0.012184$ for $K = 3 \times 10^{-2}$. The variation of $(-k_{imag})_{max}$ with respect to K is shown is Fig. 7.16c. At higher values of K, maximum growth rate $(-k_{imag})_{max}$ also shifts towards lower $Re_{\delta*}$. It also gives us a reference solution to infer how much heat transfer has destabilized the mixed convection boundary layer as compared to the Blasius solution by comparing Re_{cr}, maximum unstable

β_0 and $(-k_{imag})_{max}$ for different values of K with corresponding quantities for the Blasius solution.

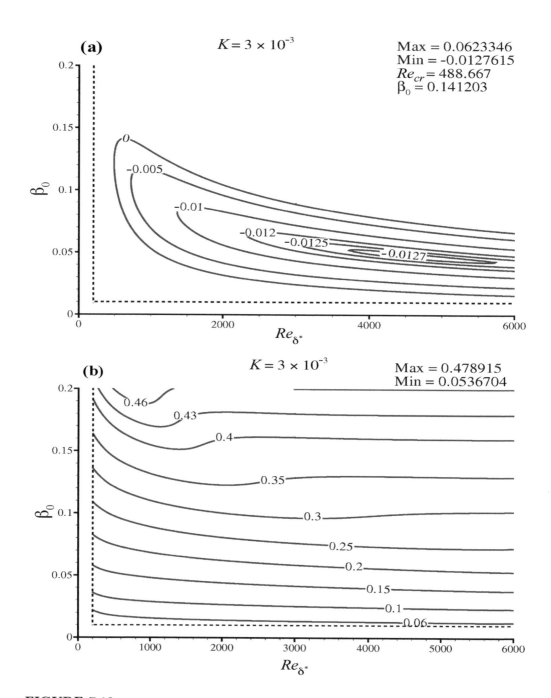

FIGURE 7.12
(a) Growth rate (k_{imag}) contours along with the neutral curve shown plotted in the (Re_{δ^*}, β_0)-plane for $K = 3 \times 10^{-3}$; (b) the real part of wavenumber (k_{real}) contours shown plotted in the (Re_{δ^*}, β_0)-plane for the same K.

The shape of the neutral curve in frame (a) of Figs. 7.10 to 7.15 suggests that it has a tendency to become narrower at higher $Re_{\delta*}$ as K is increased. Up to $K = 1 \times 10^{-2}$, the

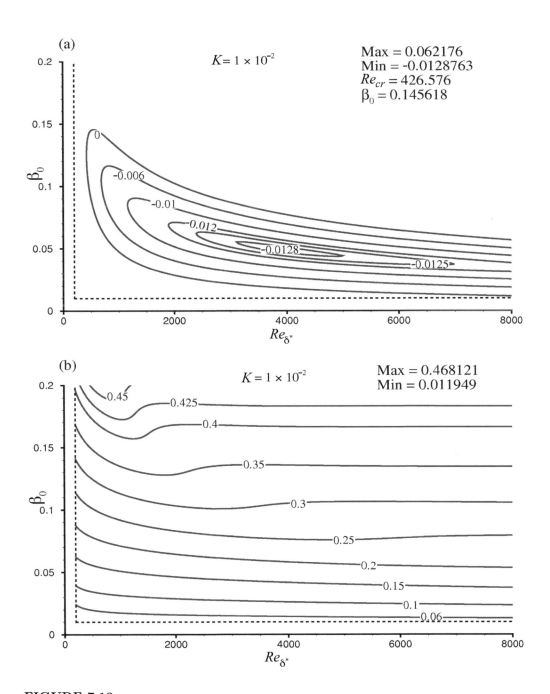

FIGURE 7.13
(a) Growth rate (k_{imag}) contours along with neutral curve shown plotted in the ($Re_{\delta*}, \beta_0$)-plane for $K = 1 \times 10^{-2}$; (b) the real part of wavenumber (k_{real}) contours shown plotted in the ($Re_{\delta*}, \beta_0$)-plane for the same K.

general shape of the neutral curve does not show considerable change except for the trends discussed above. But it is seen that any increment in K beyond $K = 1 \times 10^{-2}$ causes a

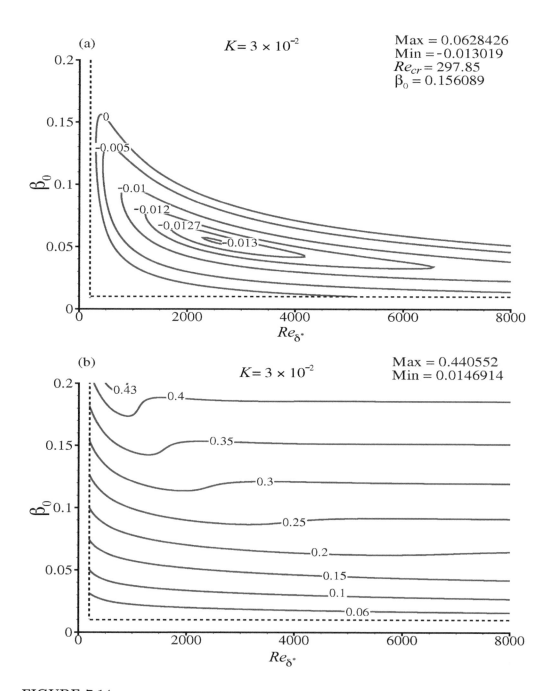

FIGURE 7.14
(a) Growth rate (k_{imag}) contours along with the neutral curve shown plotted in the (Re_{δ^*}, β_0)-plane for $K = 3 \times 10^{-2}$; (b) the real part of wavenumber (k_{real}) contours shown plotted in the (Re_{δ^*}, β_0)-plane for the same K.

considerable elongation of the upper part of the neutral curve and it is also pushed down

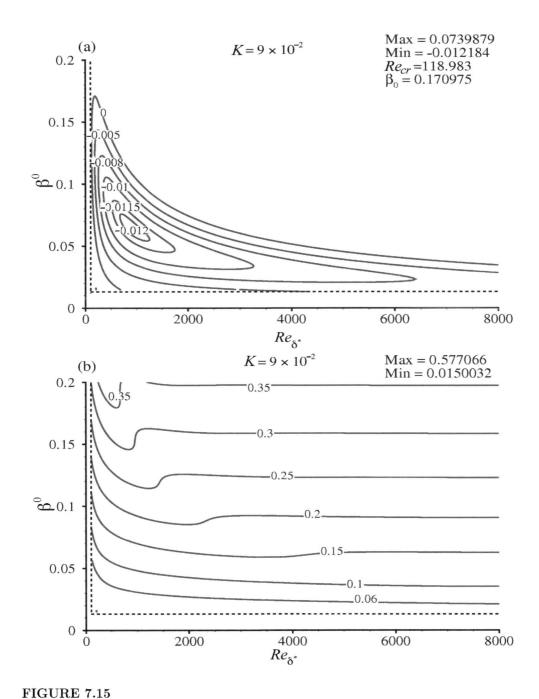

FIGURE 7.15
(a) Growth rate (k_{imag}) contours along with the neutral curve shown plotted in the (Re_{δ^*}, β_0)-plane for $K = 9 \times 10^{-2}$; (b) the real part of wavenumber (k_{real}) contours shown plotted in the (Re_{δ^*}, β_0)-plane for the same K.

towards lower ranges of β_0. Hence, at these K values, the flow becomes susceptible to a wider range of frequencies nearer to the leading edge.

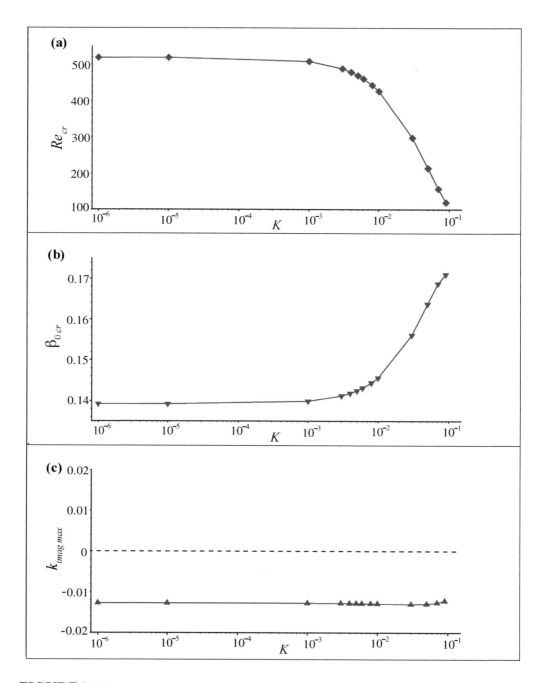

FIGURE 7.16
Variation of critical parameters as obtained from the linear spatial stability theory plotted as a function of K: (a) Re_{cr} variation with K; (b) maximum unstable frequency variation with K; (c) $k_{imag\ min}$ variation with K.

Frames (b) in Figs. 7.10 to 7.15 show the k_{real} contours for the corresponding buoyancy parameters. It shows that at lower values of β_0, k_{real} has very little variation with respect to $Re_{\delta*}$. Also, at higher values of $Re_{\delta*}$, k_{real} contours remain parallel for all circular frequencies. But one notes that there exists sharper variation of k_{real} at higher β_0 and lower $Re_{\delta*}$. These variations become more severe and get distributed to lower ranges of β_0 as K is increased. In the contour plots presented for the real and imaginary parts of the complex wavenumber for various buoyancy parameters, dotted lines parallel to the $(Re_{\delta*}, \beta_0)$-axes show the region to which calculation of the eigenvalues has been restricted. As mentioned before, double precision calculations become difficult closer to the origin in the $(Re_{\delta*}, \beta_0)$-plane.

Figures 7.17, 7.18 and 7.19 show the group velocity contours for $K = 3 \times 10^{-3}$, $K = 3 \times 10^{-2}$ and $K = 9 \times 10^{-2}$, respectively, in the $(Re_{\delta*}, \beta_0)$-plane. Here also one follows a constant physical frequency ray to understand what happens to disturbances created by a

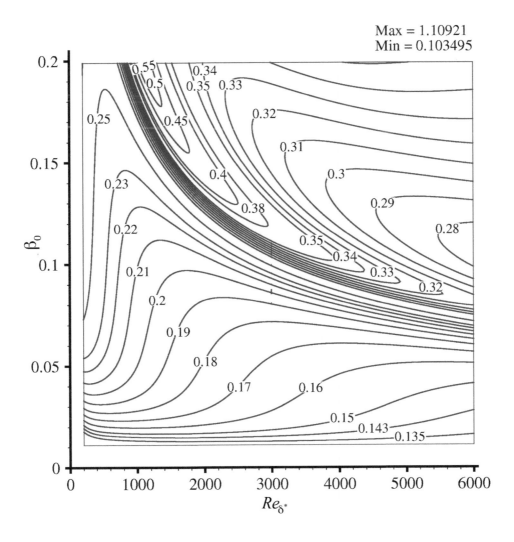

FIGURE 7.17
Group velocity (V_g) contours for $K = 3 \times 10^{-3}$.

fixed frequency exciter. The primary inference one draws from Figs. 7.17 to 7.19 is that the group velocity is positive over the calculated domain, which implies that the mode is downstream propagating for all buoyancy parameters. From these contours it is clear that, if the exciter is near the leading edge of the plate, disturbances will have a group velocity of the order of 0.2 nearer to the exciter and 0.3 further away from it. It can be seen from the group velocity contours that these two regions are separated by a zone of sharp variation common to Figs. 7.17 to 7.19. Therefore the downstream propagating disturbances will have maximum group velocity at an intermediate $Re_{\delta*}$ determined by this zone of sharp variation. Such a zone is also noted for the Blasius profile. It is also seen that as K increases, this zone becomes narrower and the maximum value of group velocity for a constant frequency disturbance increases.

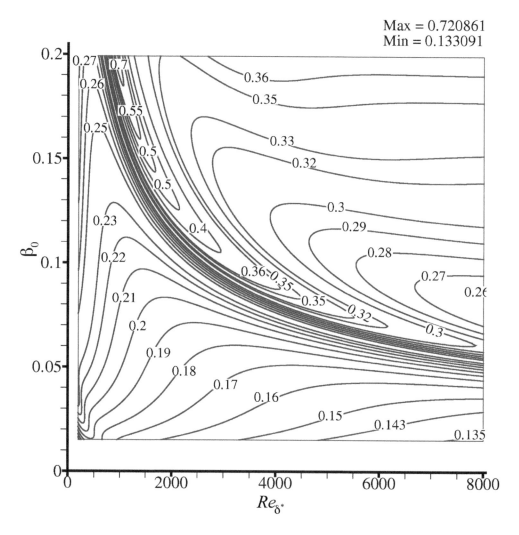

FIGURE 7.18
Group velocity (V_g) contours for $K = 3 \times 10^{-2}$.

7.5 Nonlinear Receptivity of Mixed Convection Flow over a Heated Plate

The results presented in Secs. 7.3 and 7.4 depend on the existence of the similarity profile in [315]. This similarity solution provides typical velocity and temperature profiles for which instability studies have been performed, similar to that being performed in pure hydrodynamics for a zero pressure gradient Blasius boundary layer. However, the formulation for the similarity profile requires a horizontal semi-infinite adiabatic plate with a singular heat source and sink at the leading edge. Such conditions lead to a wall temperature distribution according to $(T_w - T_\infty) \propto \pm x^{-1/2}$. For such a wall temperature variation, it is noted that the buoyancy parameter K defined in Eq. (7.7), exists which does not change along the plate. While $K = 0$ corresponds to the Blasius solution, for $K > 0$ one obtains a unique

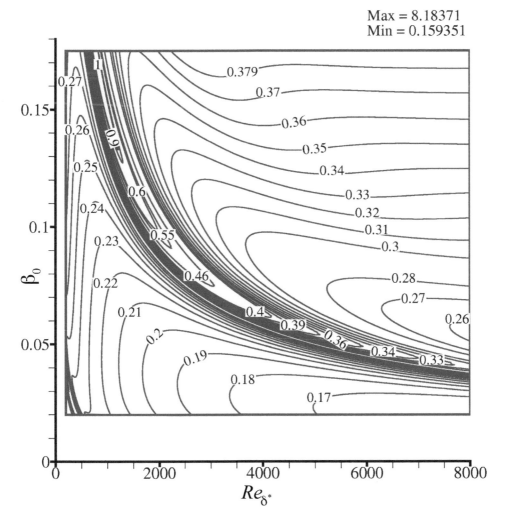

Max = 8.18371
Min = 0.159351

FIGURE 7.19
Group velocity (V_g) contours for $K = 9 \times 10^{-2}$.

similarity solution. Interestingly, in [315, 2, 149, 223], it has been pointed out that two self-similar solutions exist for $K_c \leq K < 0$.

Noshadi & Schneider [255] have discussed the importance of non-uniqueness of equilibrium flow and have shown that this can be simulated using commercial software. When two branches of solution exist for a cooled plate for $K_c \leq K < 0$, the upper branch similarity solution was obtained approximately in this reference. However, in other regime of the parameter space for a cooled plate, a similarity solution does not exist. As the main emphasis of the present work is to study physical instability of mixed convection flow, we focus upon the heated plate case with $K > 0$ for the equivalent similarity profile in [315].

Due to a very special requirement of the boundary condition for the wall temperature for the existence of a similarity solution, it is natural to inquire whether such a similar velocity profile can be obtained directly from the solution of the Navier–Stokes equation. In this context, the contribution in [255] is noteworthy. At the same time, one would like to explore whether one can obtain results which match more closely with the similarity solution. Having obtained such a similarity solution from the Navier–Stokes equation, one can also study its receptivity to vortical or entropic excitation. Based on the observations in Chap. 5, it is equally interesting to check whether spatio-temporal wave fronts are also created in mixed convection flow by a simultaneous blowing-suction (SBS) strip.

The SBS strip exciter used here is the same that was used in [98], as shown in Fig. 7.20.

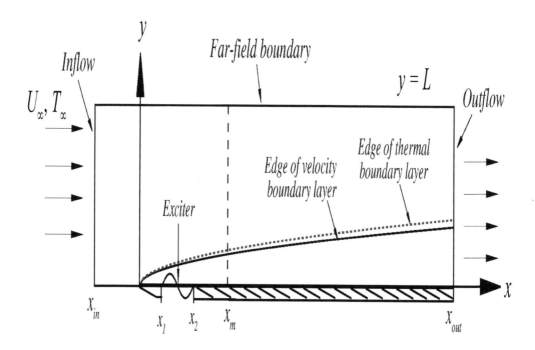

FIGURE 7.20

Schematic diagram showing the computational domain for simulation of flow over a semi-infinite heated plate with a simultaneous blowing and suction harmonic exciter placed at the wall. A stretched tangent hyperbolic grid is taken from x_{in} to x_m and a uniform grid thereafter. A similar stretched grid is used in wall-normal direction, as defined in the text.

This type of receptivity study was repeated in [325], the results of which have been presented in Chaps. 3 and 4, employing a very high accuracy method to investigate nonlinear, nonparallel aspects of boundary layer receptivity by solving the Navier–Stokes equation in a (ψ, ω)-formulation, with attention focused on local and asymptotic parts of the receptivity solution. The inclusion of the growth of the shear layer for monochromatic excitation exhibits the limited extent of the region where the TS wave packet is found. This suggests a need to perform a receptivity calculation by solving the full time-dependent Navier–Stokes equation. In this section, this is reported while tracking all components of the receptivity solution for flow with heat transfer, created by the SBS strip exciter.

Fasel & Konzelmann [98] provided numerical simulation of the Navier–Stokes equation in a primitive variable formulation to study the effects of wall disturbances on flow over a flat plate to simulate the experiments of [316]. In [325], the same flow was solved using significantly lower levels of excitation, as compared to the excitation used in [98]. For mixed convection flow over a vertical plate, in [423], DNS was performed to validate the linear spatial stability analysis and the results matched in most qualitative details. The linear theory results in [423] showed the existence of double loops in the neutral curve for opposing flow past a vertical plate when $Ri_x \geq 0.02$. For the same set of parameters, DNS showed that disturbances grow in two streamwise stretches, in accordance with the dual lobes predicted by the linear analysis.

For the study of nonlinear receptivity of a heated plate, an equilibrium state is first obtained by solving the conservation equations in a derived variable formulation. The physical plane is defined in terms of non-dimensional Cartesian coordinates (x, y), for which the same reference length has been used that makes X and x same, while Y and y are related as $Y = y\sqrt{Re}$. The flow is computed in the transformed orthogonal (ξ, ζ)-plane used for grid refinement wherever necessary. In the transformed (ξ, ζ)-plane, the vorticity-transport equation (VTE), the stream function equation (SFE) and the energy equation are obtained as in [423]

$$h_1 h_2 \frac{\partial \omega}{\partial t} + h_2 u \frac{\partial \omega}{\partial \xi} + h_1 v \frac{\partial \omega}{\partial \zeta} = \frac{1}{Re}\left[\frac{\partial}{\partial \xi}\left(\frac{h_2}{h_1} \frac{\partial \omega}{\partial \xi} \right) + \frac{\partial}{\partial \zeta}\left(\frac{h_1}{h_2} \frac{\partial \omega}{\partial \zeta} \right) \right] + K\sqrt{Re}\frac{\partial}{\partial \xi}(h_2 \theta) \quad (7.116)$$

$$\frac{\partial}{\partial \xi}\left(\frac{h_2}{h_1} \frac{\partial \psi}{\partial \xi} \right) + \frac{\partial}{\partial \zeta}\left(\frac{h_1}{h_2} \frac{\partial \psi}{\partial \zeta} \right) = -h_1 h_2 \omega \quad (7.117)$$

$$h_1 h_2 \frac{\partial \theta}{\partial t} + h_2 u \frac{\partial \theta}{\partial \xi} + h_1 v \frac{\partial \theta}{\partial \zeta} = \frac{1}{RePr}\left[\frac{\partial}{\partial \xi}\left(\frac{h_2}{h_1} \frac{\partial \theta}{\partial \xi} \right) + \frac{\partial}{\partial \zeta}\left(\frac{h_1}{h_2} \frac{\partial \theta}{\partial \zeta} \right) \right] \quad (7.118)$$

with the contra-variant components of the velocity vector given by

$$u = \frac{1}{h_2} \frac{\partial \psi}{\partial \zeta}$$

$$v = -\frac{1}{h_1} \frac{\partial \psi}{\partial \xi}$$

where h_1 and h_2 are the scale factors of the transformation given by $h_1 = (x_\xi^2 + y_\xi^2)^{\frac{1}{2}}$ and $h_2 = (x_\zeta^2 + y_\zeta^2)^{\frac{1}{2}}$. In the transformed grid, ξ is in a direction parallel to the wall and ζ is in the wall-normal direction. Hence, the transformation is given by $x = x(\xi)$ and $y = y(\zeta)$, so that $h_1 = x_\xi$ and $h_2 = y_\zeta$. The parameters Re and Pr have been used as 10^5 and 0.71, respectively, as mentioned before.

7.5.1   Boundary and Initial Conditions

As described above, Eqs. (7.116) to (7.118) are solved to obtain equilibrium flow for receptivity calculations. Once it is obtained, vortical excitation of the flow is imposed through appropriate wall boundary conditions. Here we discuss the necessary boundary conditions enforced for both equilibrium flow and for the excitation field used in receptivity studies. It is noted that for receptivity calculations, dependent variables in Eqs. (7.116) to (7.118) are composed of a primary quantity (denoted by an overbar) and a disturbance quantity (e.g., $\omega = \bar{\omega} + \omega_d$), and the boundary conditions used here are applicable for total quantities.

For calculating the equilibrium flow, free stream conditions are prescribed at the inlet and the top boundary of the domain, which yield

$$\frac{\partial \bar{\psi}}{\partial \zeta} = h_2 \tag{7.119}$$

$$\bar{\omega} = 0 \tag{7.120}$$

$$\bar{\theta} = 0 \tag{7.121}$$

At the wall, appropriate boundary conditions for primary quantities are given by

$$\bar{\psi}_w = \bar{\psi}_o = \text{constant} \tag{7.122}$$

$$\bar{\omega}_w = -\frac{1}{h_2^2} \frac{\partial^2 \bar{\psi}}{\partial \zeta^2} \tag{7.123}$$

$$\theta_w = \frac{1}{\sqrt{x}} \tag{7.124}$$

Receptivity analysis by vortical wall excitation is carried out by an SBS strip at the wall. Inlet and top boundary conditions remain the same, but wall boundary conditions for stream function and vorticity change as follows:

$$\psi_w = \bar{\psi}_o + \psi_{wp} \tag{7.125}$$

$$\omega_w = -\frac{1}{h_1 h_2} \frac{\partial}{\partial \xi} \left(\frac{h_2}{h_1} \frac{\partial \psi}{\partial \xi} \right) - \frac{1}{h_2^2} \frac{\partial^2 \psi}{\partial \zeta^2} \tag{7.126}$$

In writing Eq. (7.126), one notices that the wall is no longer a $\psi =$ constant line, due to mass transfer through the SBS strip. Consider the SBS strip to be placed between x_1 and x_2, as shown in the schematic diagram in Fig. 7.20. The wall perturbed stream function value (ψ_{wp}) in Eq. (7.125) can be obtained from the no-slip condition and the expression for wall-normal velocity given by

$$u_d = 0, \quad v_d = A(x)\,\sin(\beta_L t) \tag{7.127}$$

where β_L is the non-dimensional disturbance frequency based on length scale (L) and $A(x)$ is the amplitude of the disturbance velocity. The amplitude function $A(x)$ is related to $A_m(x)$ as $A(x) = \alpha_1 A_m(x)$, where α_1 is the amplitude control parameter and $A_m(x)$ is defined in [98, 423] as follows.
For $x_1 \leq x \leq x_{st}$:

$$A_m = 15.1875 \left(\frac{x - x_1}{x_{st} - x_1} \right)^5 - 35.4375 \left(\frac{x - x_1}{x_{st} - x_1} \right)^4 + 20.25 \left(\frac{x - x_1}{x_{st} - x1} \right)^3 \tag{7.128}$$

And for $x_{st} \leq x \leq x_2$:

$$A_m = -15.1875 \left(\frac{x_2 - x}{x_2 - x_{st}} \right)^5 + 35.4375 \left(\frac{x_2 - x}{x_2 - x_{st}} \right)^4 - 20.25 \left(\frac{x_2 - x}{x_2 - x_{st}} \right)^3 \qquad (7.129)$$

where $x_{st} = (x_1 + x_2)/2$. This excitation produces vorticity disturbances which create experimentally observed downstream propagating waves in flows without heat transfer.

At the outflow, boundary conditions enforced for both mean and perturbed flows are as shown below.

$$\frac{\partial v}{\partial x} = 0 \qquad (7.130)$$

$$\frac{\partial \omega}{\partial t} + U_c \frac{\partial \omega}{\partial x} = 0 \qquad (7.131)$$

$$\frac{\partial \theta}{\partial t} + U_c \frac{\partial \theta}{\partial x} = 0 \qquad (7.132)$$

The convective speed of disturbances through the outflow, U_c, in the Sommerfeld radiative outflow conditions of Eqs. (7.131) and (7.132) is taken as the free stream speed. These types of boundary conditions are well suited to better approximate the flow over a semi-infinite plate. The success of the present methodology in reproducing the similarity profile solution is related to the use of the high accuracy method with this outflow boundary condition, as compared to that used in [255], where fully developed conditions were used at the outflow boundary.

The initial condition for starting the equilibrium flow is given by the inviscid solution as

$$\frac{\partial \bar{\psi}}{\partial \zeta} = h_2 \qquad (7.133)$$

$$\bar{\omega} = 0 \qquad (7.134)$$

$$\bar{\theta} = 0 \qquad (7.135)$$

The computational domain is defined by $-0.05 \leq x \leq 80$, $0 \leq y \leq 2$. In the wall-normal direction, a stretched tangent hyperbolic grid is used which gives the y-coordinate for the jth point as

$$y_j = y_{max} \left[1 - \frac{\tanh[b_y(1 - \zeta_j)]}{\tanh b_y} \right]$$

Here $y_{max} = 2$ and $b_y = 2$ have been used for appropriate grid clustering with a total of 501 points taken in the wall-normal direction. In the streamwise direction, the domain is divided in two segments: the first segment extends from $x_{in} = -0.05$ to $x_m = 10$, which contains points once again distributed by a tangent hyperbolic function, and for the ith point the coordinate is obtained as

$$x_i = x_{in} + (x_m - x_{in}) \left[1 - \frac{\tanh[b_x(1 - \xi_i)]}{\tanh b_x} \right]$$

In the second segment from $x_m = 10$ to $x_{out} = 80$, we have used uniform distribution of points. For the present calculations, we have used $b_x = b_y$ for the stretched grid. Altogether, 4501 points have been used in the streamwise direction.

In Fig. 7.21, similarity solutions obtained by solving Eqs. (7.14) and (7.15) are compared with the direct simulation results of Eqs. (7.116) to (7.118) for $K = 3 \times 10^{-3}$. In the top frame, the wall-normal temperature gradient is plotted as a function of the streamwise coordinate at the indicated time instants. According to the condition given in Eq. (7.16) for the similarity profile of [315], this quantity should be equal to zero over the whole plate, except at the leading edge. Such discontinuous heat transfer is difficult to simulate

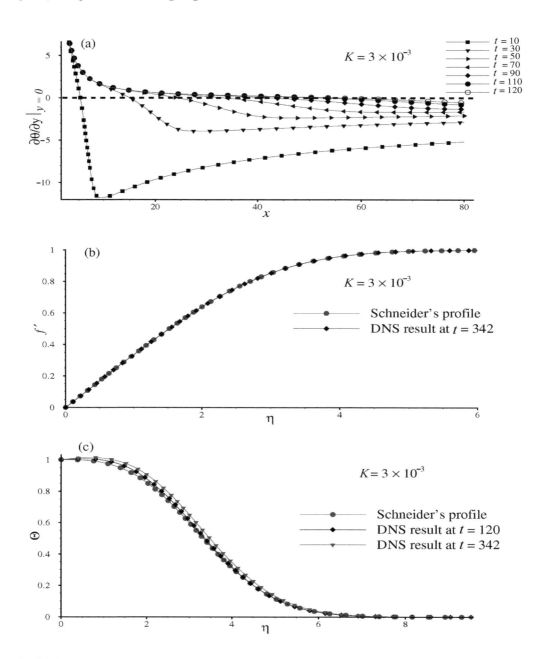

FIGURE 7.21
Mean flow profiles for velocity and temperature fields. (a) Wall-normal temperature gradient at the plate obtained by direct simulation at indicated times. (b) Velocity profile obtained by direct simulation at $t = 342$ compared with the similarity profile of [315]. (c) Temperature profile obtained by direct simulation at $t = 120$ and 342 compared with the same similarity profile. Profiles are obtained at a location where the Reynolds number based on local displacement thickness is 3758.15.

and implemented through the wall temperature distribution, $\theta_w \propto x^{-1/2}$. In the code, this condition is simulated by prescribing a large temperature value near the leading edge of the plate. The first point nearest to the leading edge is at $x = 8.759 \times 10^{-5}$ and the temperature has been taken as $\theta = 106.849$. The temperature gradient displays progressive improvement of the adiabatic condition over a longer segment of the plate with time. This improvement is also noted in the velocity profile shown in Fig. 7.21b at a location where the Reynolds number based on displacement thickness is given by $Re_{\delta*} = 3758.151$. This comparison shows a perfect match at $t = 342$ of the direct simulation results with similarity profiles. Interestingly, the temperature profile in Fig. 7.21c shows a slight degradation between the results at $t = 120$ and 342. This is a direct consequence of computational inability to exactly follow the wall temperature distribution for the similarity condition of [315].

The presented results in Fig. 7.21b, show the velocity profile to match excellently, while the temperature profile shows an acceptable match. For receptivity calculations, solution of the Navier–Stokes equation at $t = 120$ has been taken as the equilibrium solution, whose receptivity to vortical excitation is studied here for $K = 3 \times 10^{-3}$.

The results in Fig. 7.22 correspond to a low value of the buoyancy parameter $K = 3 \times 10^{-3}$. In this figure, three rays OA_1, OA_2 and OA_3 are marked which correspond to fixed physical frequency cases for which numerical simulations are performed. These rays indicate constant non-dimensional physical frequency F. For fixed frequency disturbances, one should track disturbances along such rays, as indicated by OA_1, OA_2 and OA_3 in Fig. 7.22a. All these rays cut the neutral curve (i.e., $k_{imag} = 0$ curve) at two points. These two points indicate the extent where the growth of disturbances is indicated by the linear theory. If a particular constant frequency ray enters and exits the neutral loop at x_a and x_b, with corresponding amplitudes A_i and A_o upon entry and exit, then as per the linear theory

$$A_o = A_i \, \exp\left[\int_{x_a}^{x_b} -k_{imag} \, dx\right]$$

This defines a cumulative growth rate as $\hat{n} = \int_{x_a}^{x_b} -k_{imag} \, dx$.

In Table 7.4, we have listed details of relevant parameters as derived from linear stability analysis obtained by the CMM for these three different frequencies. All these cases are run for $K = 3 \times 10^{-3}$. In the table we have listed (i) x_{ex} and $Re_{\delta*}(x_{ex})$ are x and $Re_{\delta*}$, respectively, at the exciter location; (ii) x_a and $Re_{\delta*}(x_a)$ evaluated at the entry point of the $F = $ const. ray into the neutral curve; (iii) x_b and $Re_{\delta*}(x_b)$ evaluated at the exit point of the $F = $ const. ray from the neutral curve; (iv) the maximum value of growth rate $(-k_{imag})_{max}$ along the $F = $ const. ray and (v) cumulative growth rate $\hat{n}$, as defined above.

In all subsequent discussions and references in figures, the indicated times refer to the time following the onset of excitation. In Eq. (7.127), the amplitude of excitation for the SBS strip is taken as a small fraction (α_1) of the value taken in [98]. A typical response field for a low amplitude excitation case ($\alpha_1 = 0.0002$) is shown in Fig. 7.23a at $t = 150$ to display various components of the solution. In the figure, the location of the exciter is marked by a vertical arrowhead, which begins at $x_1 = 0.99$ and ends at $x_2 = 1.22$. Corresponding Reynolds numbers based on displacement thickness are 541 and 600.65, respectively. The result is shown for the streamwise disturbance velocity plotted for the tenth ζ-line from

TABLE 7.4
Linear stability properties for various cases with $K = 3 \times 10^{-3}$.

F	x_{ex}	$Re_{\delta*}$ at x_{ex}	x_a	$Re_{\delta*}$ at x_a	x_b	$Re_{\delta*}$ at x_b	Maxm. $-k_{imag}$	$\hat{n}$
0.000084835	1.0994	562.3	2.077	773.0	6.768	1395.2	0.00861	2.4736
0.000045240	1.9985	758.1	4.120	1088.6	17.049	2214.4	0.01099	5.6970
0.000039119	2.3227	817.3	4.854	1181.5	21.015	2458.5	0.01137	6.6789

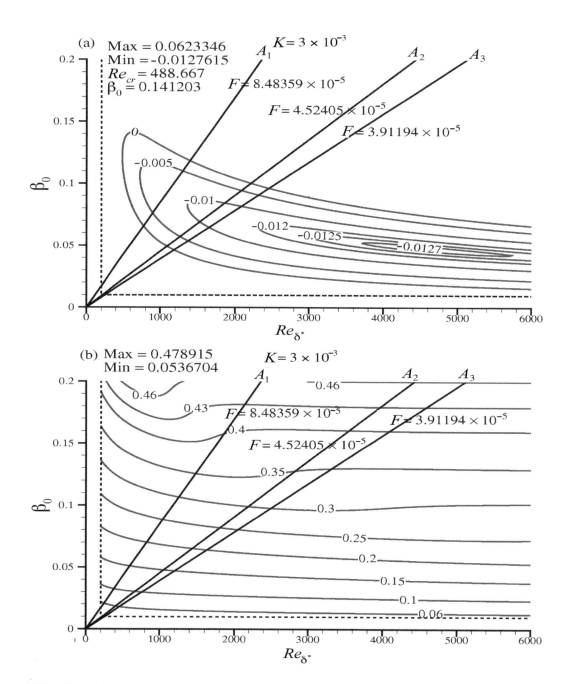

FIGURE 7.22
Imaginary and real parts of the wavenumber obtained by the linear spatial stability theory
obtained by the compound matrix method for the buoyancy parameter, $K = 3 \times 10^{-3}$,
using the similarity profiles of [315]. (a) Growth rate contours along with the neutral curve
shown plotted in the $(Re_{\delta*}, \beta_0)$-plane along with three rays with indicated non-dimensional
frequencies: $F = 8.48359 \times 10^{-5}$ for OA_1; $F = 4.52405 \times 10^{-5}$ for OA_2 and $F = 3.91194 \times 10^{-5}$ for OA_3. (b) The real part of the wavenumber shown plotted in the $(Re_{\delta*}, \beta_0)$-plane.

the plate, at a non-dimensional height of $y = 0.005464$. The disturbance component of the solution is obtained by subtracting the equilibrium solution from the instantaneous

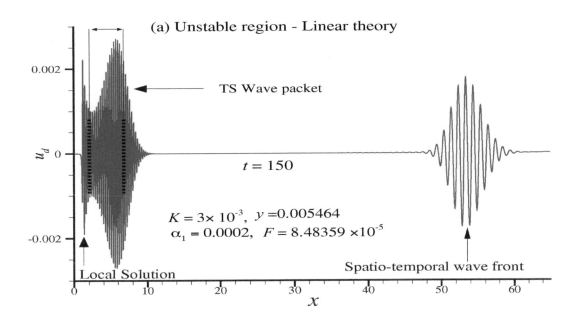

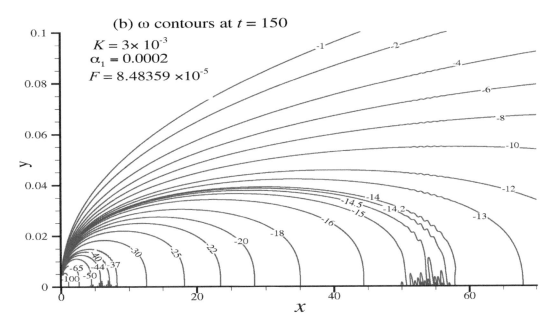

FIGURE 7.23
(a) Navier–Stokes solution for $K = 3 \times 10^{-3}$ and $F = 8.48359 \times 10^{-5}$ shown at $t = 150$ after the onset of excitation at $y = 0.005463$ for the disturbance field. The exciter is located between $x = 0.99$ and $x = 1.22$ and $\alpha_1 = 0.0002$. The linear unstable region is marked by vertical lines. (b) Vorticity contours shown in the computational domain at the same time and one notes a TS wave packet and spatio-temporal wave front.

solution of Eqs. (7.116) to (7.118). In Fig. 7.23a, the extent over which the linear spatial theory indicates instability is marked by two vertical dotted lines. Due to growth of the shear layer and changing stability properties, one notices a wave packet consisting of TS waves. It is noted that the disturbance velocity computed by nonlinear calculation shows the delayed growth phase, as compared to the linear theory. Also, the nonlinear response starts showing the decay of solution earlier, as compared to the linear theory prediction. Such differences between linear and nonlinear approaches have been earlier shown in [325] for flow without heat transfer and in [423] for flow past a heated vertical plate. In Fig. 7.23a, one also notes the local component of the solution in the immediate neighborhood of the exciter. Such a local component is noted in all receptivity solutions reported following linear and nonlinear approaches by the authors. However, the spatio-temporal front shown in this figure on the right, centered around $x \approx 54$, has not been reported earlier for mixed convection flows.

Differences between linear and nonlinear approaches were noted in [325] as being very significant at higher excitation frequencies for flows without heat transfer. Furthermore, at lower heights a spatially growing vortical field was noted, while the velocity field indicated growth at higher heights, where vorticity displayed spatial decay. To see if such behavior prevails here, too, vorticity contours are plotted in the indicated domain in Fig. 7.23b at $t = 150$. In this figure, the total vorticity field is shown and the spatio-temporal wave front is identified as undulations in various contours. There are also convecting separation bubble-like structures seen in Fig. 7.23b, which are the spatio-temporal wave front shown in Fig. 7.23a.

The evolution and transport of the spatio-temporal wave front is tracked in Fig. 7.24, where the streamwise disturbance velocity component (u_d) is shown at the indicated times as a function of x in different frames. Results are shown for the twentieth ζ-line from the wall, at a non-dimensional height of $y = 0.0119995$. One notes the TS wave packet centered around almost the same location. In all frames of the figure, vertical dotted lines indicate the onset of growth and decay of TS waves obtained by the spatial linear theory. The onset of instability obtained by the Navier–Stokes solution matches excellently with the linear stability result. However, the growth phase of the linear theory is over-predicted. Thus, the mismatch of the onset of instability in Fig. 7.23a is due more to the local component of the solution in close proximity to the wall. It is interesting to note that the local component of the solution is absent in Fig. 7.24 at this higher height. This confirms the analytical explanation provided for the local solution in Chap. 3 using Tauber's theorem. We have noted that the TS wave packet remains stationary with low amplitude pulsation with time. In contrast, the spatio-temporal front shows propagation with constant speed, as is evident from frames shown in this figure at an identical time interval.

In Fig. 7.25, the disturbance temperature field (θ_d) for the case of Figs. 7.23 and 7.24 is shown. It is noted that a vortical excitation can create a disturbance temperature field and vice versa. However, in Fig. 7.25a, we note the created θ_d display a decaying packet for $y = 0.005464$, except for a narrow stretch in the middle where the disturbance remains neutral. In Fig. 7.25b, total temperature field (θ) contours are shown in the computational domain. Here one notes a very weak wave front in the outer part of the boundary layer. The fact that such a front only exists in the outer part of the shear layer is also noted when θ_d is plotted as a function of x for $y = 0.0119995$ at different time instants in Fig. 7.26. For this height, growth of θ_d correlates better with the linear theory. However, from this figure one can barely see the spatio-temporal wave front at $t = 144$ and 180.

Thus, one notices a distinct wave front associated with the streamwise disturbance velocity field when the boundary layer is excited by vortical excitation, as shown in Figs. 7.23 and 7.24. It is also noted in Fig. 7.24 that the peak-to-peak amplitude of the wave front continuously increases in space and time, while the amplitude of the TS wave packet

remains the same with mild pulsation. Thus, even though the eigenmodes have a group
velocity with which energy propagates, ensuing disturbances created by the exciter grow

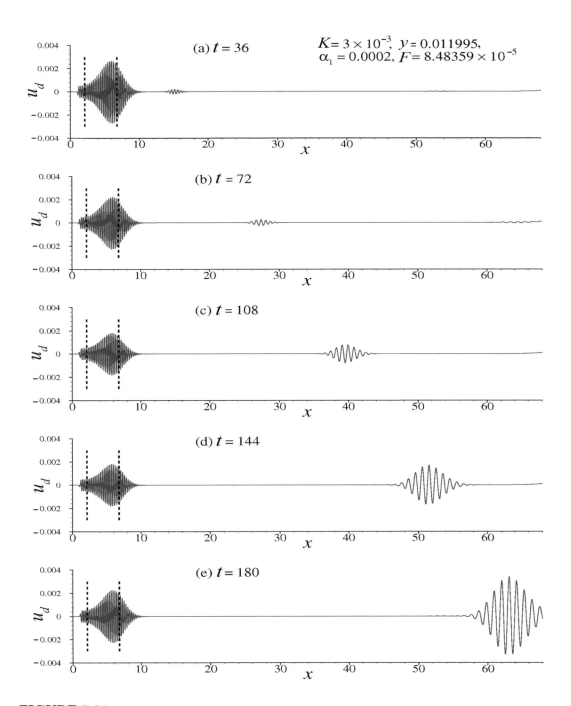

FIGURE 7.24
Variation of u_d as a function of x shown at the indicated times for $K = 3 \times 10^{-3}$, $F = 8.48359 \times 10^{-5}$ and $\alpha_1 = 0.0002$ at $y = 0.011995$. Entry and exit points of $F = 8.48359 \times 10^{-5}$ ray into and from the neutral curve are indicated by vertical dotted lines in each frame.

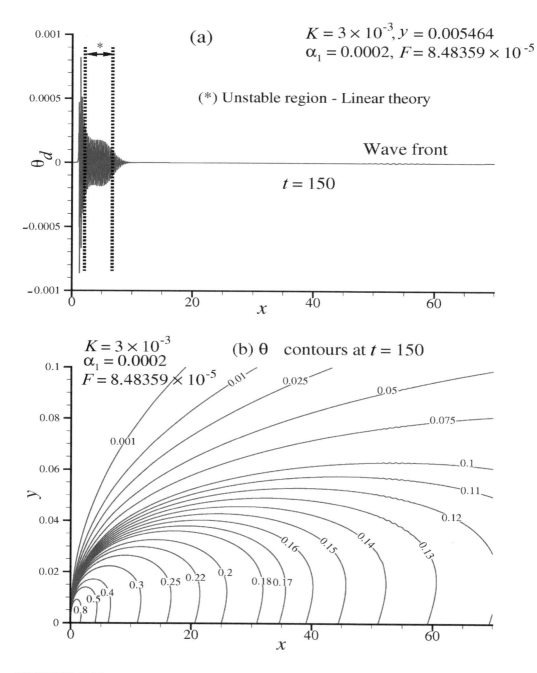

FIGURE 7.25

(a) Variation of disturbance temperature θ_d as a function of streamwise distance shown at $t = 150$ after the onset of excitation for the case of figure 7.23. At this height, θ_d is seen to be spatially stable, while the velocity field was noted as unstable. (b) Temperature contours shown in the computational domain at $t = 150$ after the onset of excitation and one can note only the spatio-temporal wave front, which has lower amplitude near and far from the plate.

and decay according to the k_{imag} variation in Fig. 7.22a, as these propagate downstream. Spatio-temporal wave front properties are not predicted by either the spatial or temporal

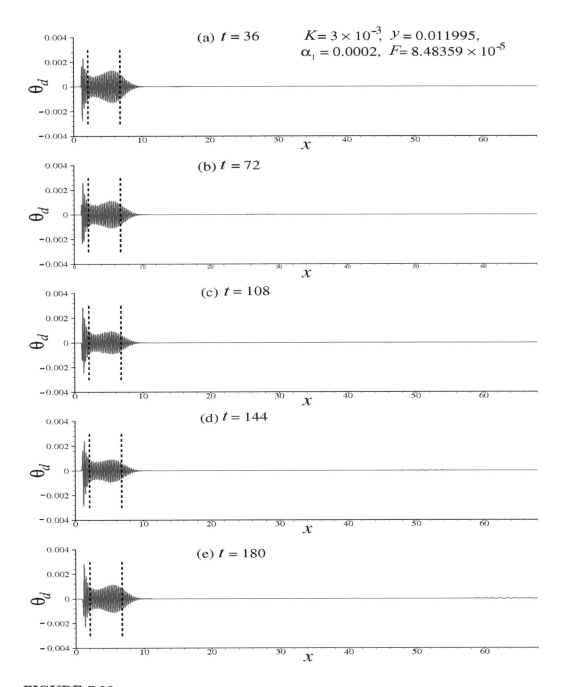

FIGURE 7.26
Variation of θ_d as a function of x shown at the indicated times for $K = 3 \times 10^{-3}$, $F = 8.48359 \times 10^{-5}$ and $\alpha_1 = 0.0002$ for $y = 0.011995$. Entry and exit points of the $F = 8.48359 \times 10^{-5}$ ray into and from the neutral curve are indicated by vertical dotted lines in each frame.

theories. This was captured by spatio-temporal approach of the Bromwich contour integral method in [349], using a linear receptivity theory, for flow without heat transfer, as discussed in Chap. 5.

Variation of maximum peak to peak amplitude of the wave front with space and time is represented in Fig. 7.27. Following the usual convention, variation with space is shown here in terms of $Re_{\delta*}$. For the case of lower input amplitude ($\alpha_1 = 0.0002$), initially the wave front amplitude decays up to a certain distance and thereafter it increases exponentially. This initial decay up to $Re_{\delta*} = 2401.32$ is essentially due to dispersion, as noted from the increase in wavelength of the wave front, as it starts from the exciter and passes through the TS wave packet. As implied in Fig. 7.24, after $t \simeq 150$ the wave front amplitude surpasses the TS wave packet amplitude. Thus, the transition phenomenon interpreted as the growth

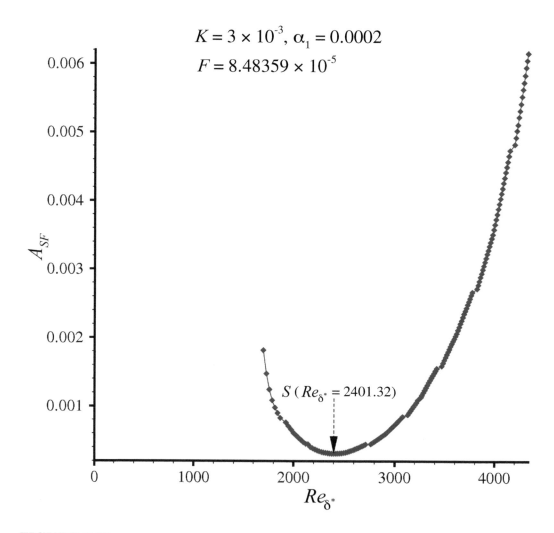

FIGURE 7.27
Variation of the amplitude of the spatio-temporal wave front plotted against streamwise distance represented by $Re_{\delta*}$ for $\alpha_1 = 0.0002$ and $F = 8.48359 \times 10^{-5}$. Note that the amplitude of the wave front first decreases up to a location marked S and thereafter increases exponentially until it emerges from the computational domain.

of disturbances is likely to occur due to the wave front rather than due to the TS wave packet for these low amplitude cases. This is contrary to the classical viewpoint, which considers the growth of TS waves to be the sole determinant of transition. The wave front is tracked here by solving the Navier–Stokes equation, which includes both nonlinear and nonparallel effects. Existence of the wave front is shown here for a two-dimensional disturbance field for two-dimensional mean flow and this is unlike the transient growth reported in [49, 291], who attributed subcritical transition to non-orthogonal or non-normal modes for flows without heat transfer. However, it has been noted [411] that the so-called transient growth by a non-normal mode is valid only for three-dimensional flows. Thus, the dominant growth of the disturbance wave front in two-dimensional flows, as seen here and in [349], is more generic.

This approach of nonlinear receptivity study helps to identify the effects of amplitude and frequency of input excitation imposed through the SBS strip. In Fig. 7.28, we have plotted the amplitude of the peak-to-peak variation of the wave front when frequency of excitation is kept the same ($F = 4.52405 \times 10^{-5}$) for the cases with $\alpha_1 = 0.002$, 0.01 and 0.05. Variation of amplitude of the spatio-temporal wave front for the $\alpha_1 = 0.0002$ and $F = 8.48359 \times 10^{-5}$ case is included here for reference. The highest amplitude considered here is still only 5% of the amplitude considered in [98]. Response of the flow appears qualitatively different for the case of $\alpha_1 = 0.0002$ from the other two higher amplitude cases in Fig. 7.28. The amplitude of the wave front starts off from a significantly higher level and remains almost the same for the higher amplitude cases, as compared to the lowest amplitude case shown in this figure. The lowest amplitude case displays an exponential growth, starting from a level which is three orders of magnitude lower. It is shown below that higher amplitude cases are strongly affected by unsteady separation on the plate. Such unsteady separation has been shown in vortex-induced instability studies in [199, 331] and is identified as bypass transition in Chap. 4. However, in proposing the nomenclature of bypass transition, Morkovin [232] predicted it as a phenomenon which appears without the dominant presence of TS waves. In the present high amplitude cases, we see the presence of both a TS wave packet and a spatio-temporal growing wave front simultaneously. However, the TS wave packet plays a subdominant role, as compared to the wave front, which is different from the wave packet noted in Figs. 7.23 and 7.24. A higher amplitude of excitation causes unsteady flow separation on the plate. Thus, unsteady separation is one hallmark of bypass transition.

In Fig. 7.29, we have compared the peak-to-peak amplitude of the wave front for the cases with two lower frequencies of $F = 3.91194 \times 10^{-5}$ and 4.52405×10^{-5}, for $\alpha_1 = 0.01$, which correspond to 1% of the amplitude considered in [98]. These higher amplitude cases display bypass transition. For reference, we have also shown the lower amplitude, higher frequency case of Fig. 7.27 in this frame. This again shows the dominant role of bypass events for higher excitation amplitudes. In the following, the morphology of events is shown with the help of u_d and the disturbance stream function (ψ_d) for these two cases of higher amplitude and lower frequencies.

The case of $F = 3.91194 \times 10^{-5}$ and $\alpha_1 = 0.01$ is shown in Fig. 7.30(a) where u_d is plotted against x for $y = 0.011995$ at the indicated time instants. This case is interesting, as the linear theory results for this frequency in Fig. 7.22a predict a TS wave packet which is significantly under-predicted by this nonlinear receptivity solution. Here, the disturbance upon entering the neutral loop shows the TS wave packet for a shorter distance, beyond which unsteady separation is noted with a very localized wave front, always staying ahead of the following separation bubbles. The wave front itself induces unsteady separation whose scales are significantly larger. In Fig. 7.30(a), the TS wave packet remains stationary, while the bypass transition-induced separation bubbles continue to elongate, making the extent of the transition zone longer with time. Also, the wave front evolves with time for this case. At $t = 10$, one can only find the local solution and the evolving TS wave packet. However, at $t = 30$, one notices the wave front induces unsteady separation immediately behind the wave front. These unsteady separations introduce smaller length scales, as compared to the leading wave front. Thus, we have three distinct length scales noted in this figure beyond $t = 30$: (i) the TS wave packet represents the smallest length scale (related to k_r);

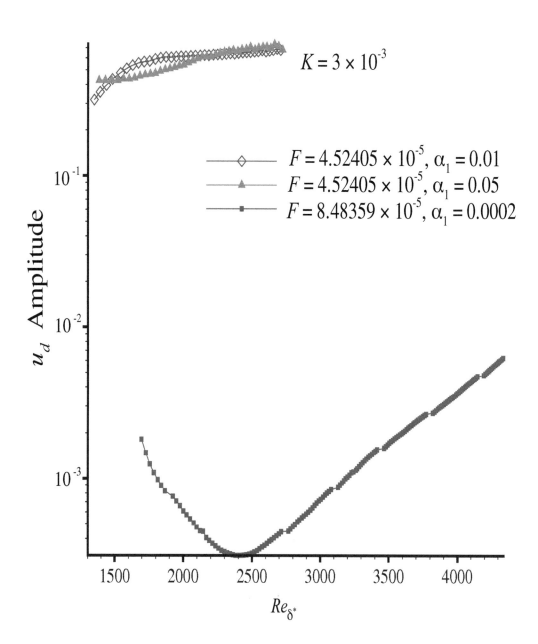

FIGURE 7.28
Variation of the amplitude of dominant disturbance fields for the indicated combinations of excitation parameters and $K = 3 \times 10^{-3}$. The lowest amplitude case is the same one shown in Fig. 7.13, showing the spatio-temporal wave front as the dominant disturbance field, while the higher amplitude cases suffer bypass transition.

(ii) unsteady separation bubbles leading the TS wave packet with an intermediate length scale and (iii) the longest length scale associated with the leading wave front. One can also notice a *transition* from the TS wave packet to the leading disturbance front; the latter

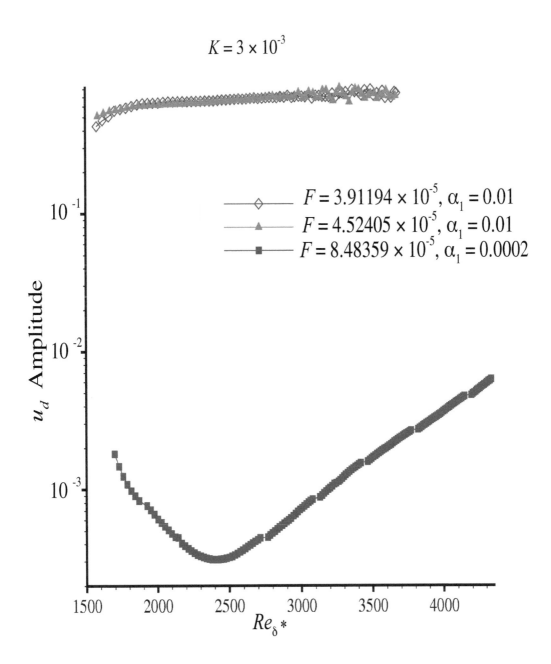

$$K = 3 \times 10^{-3}$$

FIGURE 7.29
Variation of the amplitude of dominant disturbance fields for the indicated combinations of excitation parameters and $K = 3 \times 10^{-3}$. Higher amplitude cases ($\alpha_1 = 0.01$) show bypass transition, and disturbance amplitudes for both the frequencies are three orders of magnitude higher.

has a non-zero mean, as compared to the TS wave packet, which displays a zero mean, in the region it is present. One notes the leading wave front as a single spatially decaying packet, while its maximum peak-to-peak amplitude increases with time. The role of this wave front in entraining the following unsteady separation bubbles is shown subsequently. Of prime importance during bypass transition is the presence of two dominant length scales near the exciter. However, farther downstream, the entraining action of the spatio-temporal front creates intermittent packets of disturbances, which makes the disturbance spectrum wide-band.

In Fig. 7.30(b), the case of $F = 4.52405 \times 10^{-5}$ and $\alpha_1 = 0.01$ is shown by plotting u_d against x for $y = 0.011995$ at the indicated time frames. For this frequency, the linear stability theory predicts a lesser extent of the TS wave packet, as compared to the case of Fig. 7.30(a). However, the displayed u_d obtained from the Navier–Stokes equation does not follow this limit. Once again, flow is dominated by unsteady separation bubbles forming on the plate, as explained above for the lower frequency case. The displayed solution up to $t = 120$ clearly indicates the intermittent nature of the leading wave front. Also, the TS wave packet is seen to coexist with bypass events of unsteady separation in the near-field of the exciter.

In Fig. 7.31(a), ψ_d contours are plotted at the indicated times for the case of Fig. 7.30(a), which shows formation of separation bubbles near the wall. For the purpose of elucidation, we have only plotted the wall and near-wall streamlines to indicate the nature of events which take place during bypass transition. As seen in the figure, a bypass event is characterized by wall-normal eruptions, which are present at subsequent times. To indicate scales of the eruptions, the shear layer is marked by its displacement thickness distribution. Evidently, bypass transition causes unsteady separations, indicated by a continuous stream of bubbles originating from the exciter. The left most vortical structure is due to the local solution and the subsequent small scale vortical structures are due to the TS wave packet. This is followed by unsteady separations, which indicate the bypass event. The leading front is associated with the tallest structure at the right. One also notes the strength of eruption keeps increasing with time, which is associated with the leading front.

In Fig. 7.31(b), ψ_d contours are shown plotted for the case of Fig. 7.30(b). Once again, one notes a bypass event following the description provided above. However, the sequence is shown for a longer time interval and one can see further evolution of the spatio-temporal wave front and the following events in which the ensuing bubbles become wider at later times, indicating enrichment of lower wavenumbers in the spectrum with time.

For spatial instability, the amplitude of disturbances at a station x given by $A(x)$ is related to initial amplitude A_0 at $x = 0$ to be given by (here $x = 0$ refers to the location of the onset of instability for that particular frequency) the linear instability theory as

$$A(x) = A_0 e^{-\int_0^x k_{imag}\, dx}$$

In classical stability analysis, the exponent in Eq. (7.135) is termed the n factor. For flows without heat transfer, it is empirically correlated with transition, where the n factors are modeled to take different values in the presence of different strain rates. The n factor can also be obtained as

$$n = \ln\left[\frac{A(x)}{A_0}\right] \tag{7.136}$$

One can compare the equivalent growth rate factor (n) for the spatio-temporal wave front for the case of $\alpha_1 = 0.0002$ with the bypass transition cases of $\alpha_1 = 0.01$ and 0.05 for different non-dimensional frequencies, F. In these cases, the dominant growth of disturbances is by these two routes, i.e., by the spatio-temporal growing wave front for low

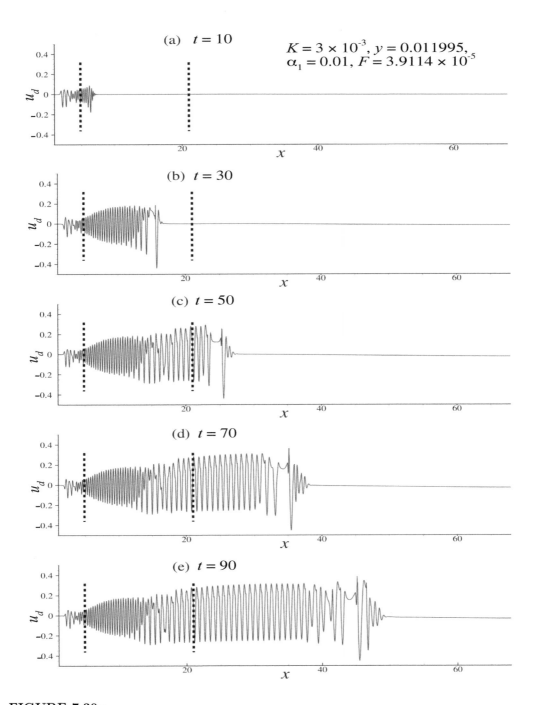

FIGURE 7.30a
Variation of u_d with x at $y = 0.011995$ for $K = 3 \times 10^{-3}$, $F = 3.9114 \times 10^{-5}$ and $\alpha_1 = 0.01$; at the indicated time instants after the onset of excitation. One notices effects of unsteady separation in exciting larger length scale, as compared to the TS wave packet.

amplitude excitation and by the bypass mechanism of unsteady separation induced at the surface. Growth rate factors of these routes have been shown and compared in Fig. 7.32.

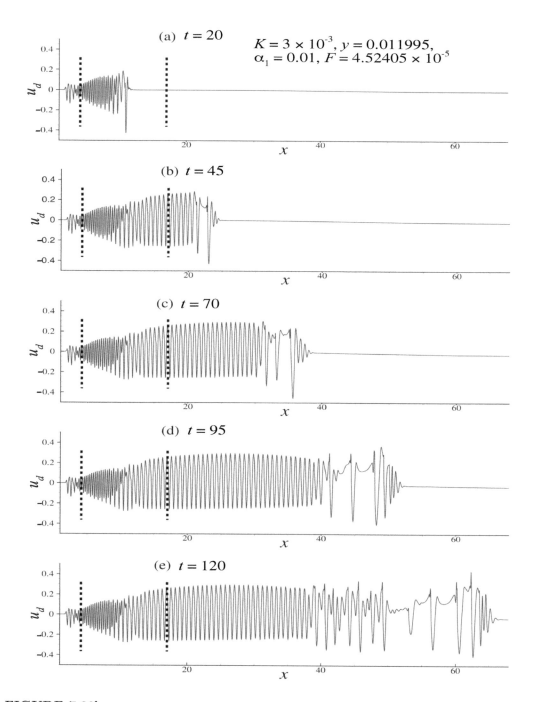

FIGURE 7.30b
Variation of u_d with x at $y = 0.011995$ for $K = 3 \times 10^{-3}$, $F = 4.524 \times 10^{-5}$ and $\alpha_1 = 0.01$ at the indicated time instants after the onset of excitation.

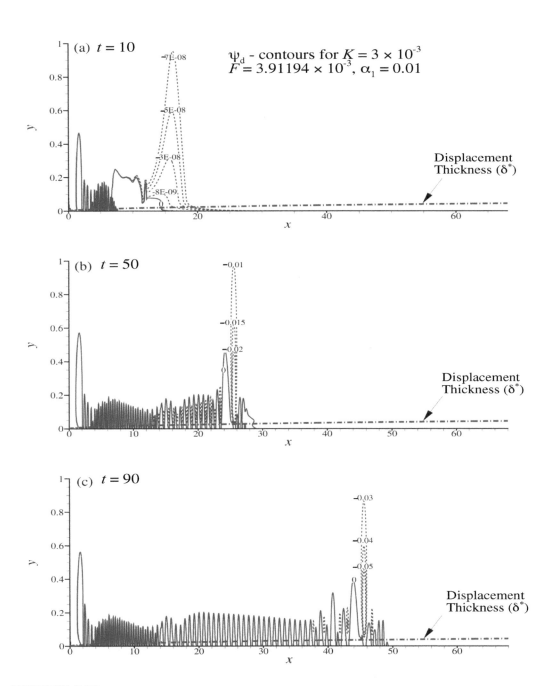

FIGURE 7.31a
Disturbance stream function (ψ_d) contours for $K = 3 \times 10^{-3}$, $F = 3.91194 \times 10^{-5}$ and $\alpha_1 = 0.01$, with the wall streamline (solid line) showing propagating unsteady separation bubbles right from the exciter. The displacement thickness of the equilibrium flow is marked by a dash dotted line. Also shown is the strong vortical eruption ahead of the leading front at all times.

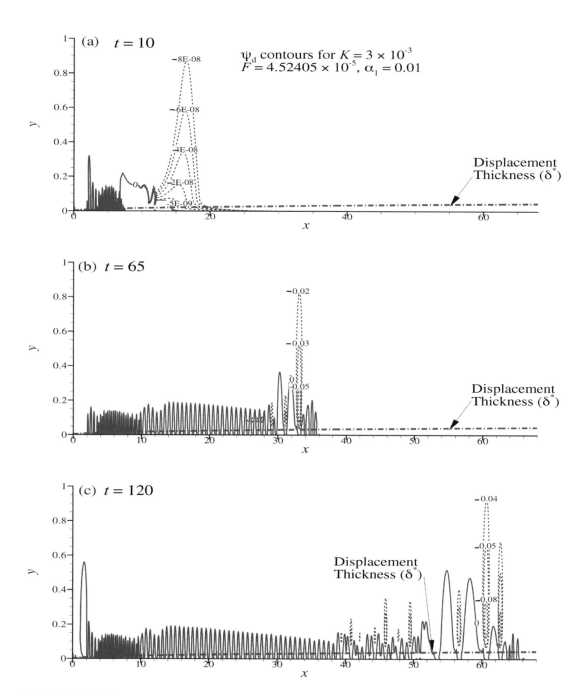

FIGURE 7.31b

Disturbance stream function (ψ_d) contours for $K = 3 \times 10^{-3}$, $F = 4.52405 \times 10^{-5}$ and $\alpha_1 = 0.01$, with the wall streamline (solid line) showing propagating unsteady separation bubbles right from the exciter, corresponding to the case of Fig. 7.30b. With time strength and frequency of vortical eruptions increase due to more secondary instabilities.

For the lowest amplitude excitation, the spatio-temporal front grows to a higher amplitude as compared to the unstable TS wave packet. This is despite the fact that its amplitude is seen to decrease initially (as noted in Fig. 7.27) and thereafter it grows continually. It is generally believed that for low amplitude excitations, the TS wave packet is responsible for eventual transition where linear growth is followed by some qualitatively noted secondary [24] and nonlinear mechanisms [39, 28]. The present receptivity calculations clearly reveal that this is not the case. Instead, the spatio-temporal wave front suffers maximum growth leading to its nonlinear breakdown for the lowest amplitude case studied here. In Fig. 7.32, the n factor for this case is calculated as a function of space from the lowest amplitude onwards and is seen to increase up to 3 by the time it leaves the outflow of the computational domain. Here the computational domain is of limited extent and the nonlinear growth of the wave front is not seen. In comparison, for other higher amplitude cases displaying bypass transition, this growth factor reaches a value of around 0.5 only. This might give an illusory impression that the spatio-temporal front is a more powerful mechanism of transition. In

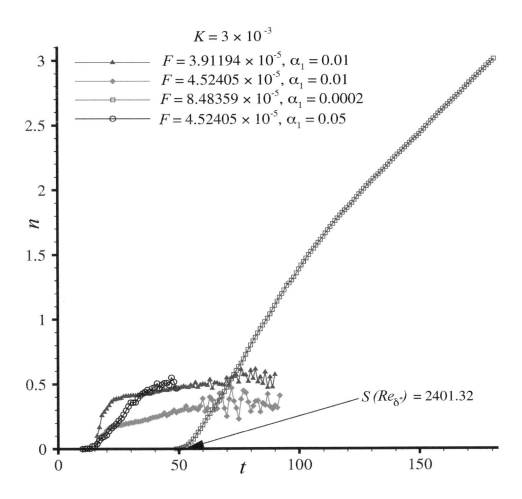

FIGURE 7.32
Equivalent growth rate (n factor, as defined by Eq. (7.136)) of the spatio-temporal wave front (for $\alpha_1 = 0.0002$) compared with three other higher amplitude cases which display bypass transition.

the actual case of high amplitude excitation, it is just the opposite. When bypass events take place, levels of disturbance at the onset are three orders of magnitude higher − as shown in Fig. 7.28. Also, this points to the fact that developing semi-empirical methods of transition based on stability theories for the ratio of A/A_0 are misleading. Instead one must perform receptivity calculations to predict transition for flows with and without heat transfer.

7.5.2   Eigenfunction Structure and DNS of the Mixed Convection Problem

Having obtained the eigenvalues from the linear instability theory, it is possible to obtain the corresponding eigenfunctions using the CMM. The eigenfunction can be written in terms of the decaying fundamental modes as

$$\phi = a_1\phi_1 + a_3\phi_3 + a_5\phi_5 \tag{7.137}$$

For the sixth order system, one can form the auxiliary system of equations by eliminating a_1, a_3 and a_5 from the definition of the second compounds and

$$\phi' = a_1\phi_1' + a_3\phi_3' + a_5\phi_5' \tag{7.138}$$

$$\phi'' = a_1\phi_1'' + a_3\phi_3'' + a_5\phi_5'' \tag{7.139}$$

$$\phi''' = a_1\phi_1''' + a_3\phi_3''' + a_5\phi_5''' \tag{7.140}$$

$$\phi^{iv} = a_1\phi_1^{iv} + a_3\phi_3^{iv} + a_5\phi_5^{iv} \tag{7.141}$$

$$\phi^{v} = a_1\phi_1^{v} + a_3\phi_3^{v} + a_5\phi_5^{v} \tag{7.142}$$

For example, using Eqs. (7.138) to (7.142) and using the definition of the second compounds, one obtains the following eigenfunction equation

$$y_1\phi''' - y_2\phi'' + y_5\phi' - y_{11}\phi = 0 \tag{7.143}$$

Alternatively, one can also obtain the following auxiliary equation for evaluating the eigenfunction

$$y_1\phi^{iv} - y_3\phi'' + y_6\phi' - y_{12}\phi = 0 \tag{7.144}$$

There are many other possible auxiliary equations which can be derived and used in principle to obtain the eigenfunctions. In the context of fourth order systems, the multiplicity of such equations was considered a source of confusion [91], which was resolved later in [320] by looking at the eigenfunction equations for their correct asymptotic behavior when $y \rightarrow Y_\infty$. As the problem allows only three decaying modes for the present case, it is necessary to ensure that the chosen eigenfunction equation also displays the same asymptotic decay, without any spuriously growing mode(s). For example, the solution of Eq. (7.143) displays correct decay rates of $-|k|$, $-|Q|$ and $-|S|$, while Eq. (7.144) has the asymptotic variation given by the characteristic exponents as $-|k|$, $-|Q|$, $-|S|$ and $(|k| + |Q| + |S|)$. Thus, Eq. (7.144) has a violently growing spurious mode and surely this equation cannot be used. As Eq. (7.143) already displays correct solution behavior, we will not look for any other auxiliary equations.

However, instead of showing the eigenfunction from linear analysis, here we have shown the streamwise disturbance velocity profile inside the shear layer in Fig. 7.33, as obtained for the case of $K = 3 \times 10^{-3}$ and $\alpha_1 = 0.002$ by solving the nonlinear receptivity problem. The profile resembles strongly the eigenfunction obtained for the streamwise disturbance velocity profile for the pure hydrodynamic case, without heat transfer by linear stability analysis. One notices an inner maximum and a flatter outer maximum, for which the amplitude varies by a factor of almost four. Also, note that the plot is shown with $y = y^*/L$ as the independent variable.

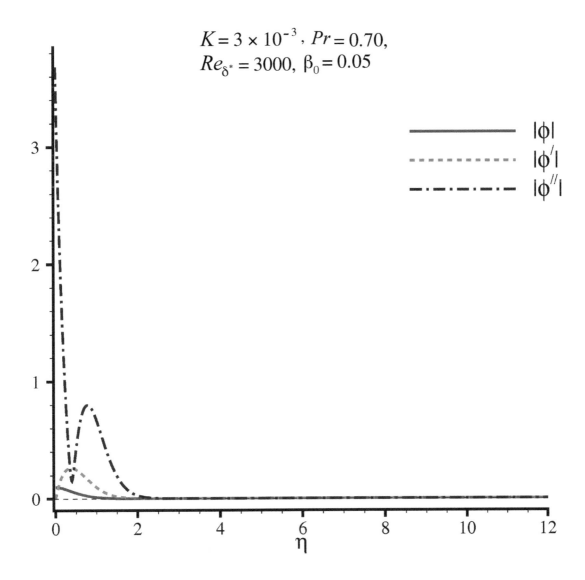

FIGURE 7.33
Streamwise disturbance velocity variation with height for the case of $K = 3 \times 10^{-3}$ and $\alpha_1 = 0.002$.

7.6   Concluding Remarks

In the introductory remarks for this chapter, we have noted that the linear and nonlinear stability and receptivity calculations reported are for an equilibrium flow for which the shear layer edge velocity is constant. To ensure the existence of a similarity profile for this external flow, Schneider [315] considered a special type of wall temperature distribution, where heat transfer takes place only at the leading edge. Thus, the heat transfer takes place like a Dirac delta function at the leading edge. For any general dynamical system, this type of input constitutes an impulse response. This was reported in [319, 322] to excite instability waves in the Blasius boundary layer via localized wall excitation. The solution method rested on modeling it as the signal problem and the complete spatio-temporal analysis by using the Bromwich contour integral for the disturbance amplitude given by the Orr–Sommerfeld equation. Creation of TS waves in these references was the first such attempt for this canonical equilibrium flow.

In comparison for mixed convection flows there is no linearized receptivity analysis available in the literature. In this chapter, we have provided linear instability results which are supplemented by nonlinear receptivity calculations for an equilibrium flow described in Sec. 7.3. However, we have noted that such an equilibrium similarity flow imposes a singular requirement of heat transfer from the plate. It is heartening to note that the Navier–Stokes equation can be solved to obtain such an equilibrium flow, as seen in Fig. 7.21. The resultant receptivity study performed by solving the Navier–Stokes equation brings out certain features of linear instability results obtained by solving the Orr–Sommerfeld equation. More importantly, nonlinear aspects of the solution show similarity between instabilities of flows with and without heat transfer, i.e., the instability is dominated by the spatio-temporal front for a low amplitude of excitation and not the TS wave packet. Such similarities also extend to a higher amplitude of excitation for these two classes of flow. Thus, a valuable synthesis is achieved between the material in Chap. 5 and what is presented in this chapter.

One of the limitations of the instability and receptivity studied here is that the boundary condition imposes an adiabatic condition at the wall, as noted from Eq. (7.16). This is due to the requirement of similarity for a uniform external flow. If instead one is interested in an isothermal wall boundary condition, then the corresponding equilibrium flow has to be found. In a recent paper, Mureithi & Denier [240] have extended the similarity concept in [315] for external flows of the kind given by $U_e(X) = X^m$ and the similarity coordinate by $\eta = Y X^{(m-1)/2}$. Self-similar solutions can be obtained using the transformation

$$U = X^m f'(\eta); \quad \theta = X^{(5m-1)/2}\Theta(\eta); \quad P = -U_e^2/2 + X^{2m}q(\eta)$$

to obtain the governing equations

$$f''' = n(f'^2 - 1) - \frac{1}{2}(n+1)ff'' + 2nG_0 q + \frac{1}{2}(n-1)\,G_0\eta q'$$

$$q' = \Theta$$

$$\Theta'' = \frac{Pr}{2}\left[(5n-1)\Theta f' - (n+1)f\Theta'\right]$$

These have to be solved subject to the boundary conditions

$$f(0) = f'(0) = 0, \quad \Theta(0) = 1, \quad f'(\infty) = 1; \quad q(\infty) = \Theta(\infty) = 0$$

Note that these boundary conditions imply that the plate is kept at a constant temperature, given by $\Theta(0) = 1$. It is also interesting to note that the authors of [240] obtained a velocity overshoot, exactly for the same reason that has been ascribed for a similar overshoot in Figs. 7.2 and 7.3. However, the authors in [240] performed only inviscid linear instability studies and made the claim that the complex dispersion relation would indicate absolute instability via satisfaction of the Briggs–Bers criterion [29, 153]. At this point in time, there are no reported receptivity studies for flows − with or without heat transfer − which show one to one correspondence between absolute instability and computed response field. It is possible now to investigate the similarity profile in [240] for linear and nonlinear receptivity analysis and show this link.

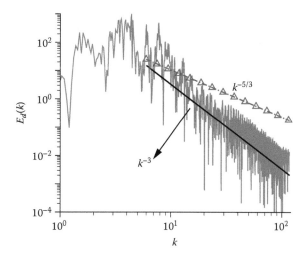

Figure 2.12
A typical energy spectrum of a two-dimensional flow over a flat plate. Note that the spectrum varies as k^{-3} for an intermediate wavenumber range [323].

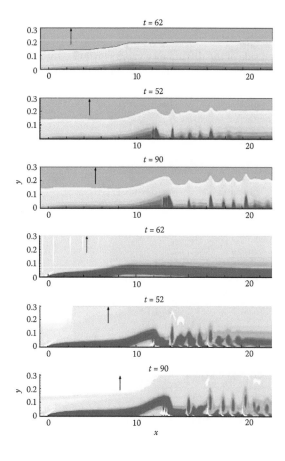

Figure 4.12
Stream function (top three panels) and vorticity contours plotted at the indicated times. The same contour values are plotted for each quantity. Arrowheads at the top of each frame indicate the instantaneous streamwise location of the free stream vortex.

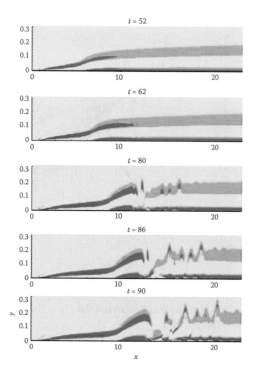

Figure 4.13
Contours of the right-hand side of the disturbance energy Eq. (4.5). The negative contours are shown by darker blue color in the figure, indicating the energy source for disturbance quantities.

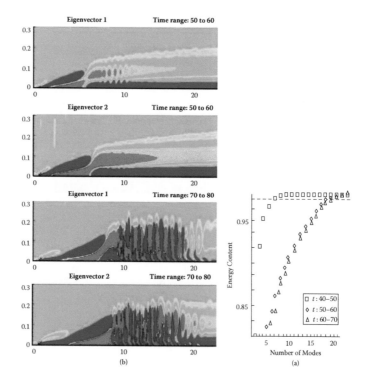

Figure 4.14
(a) The sum of a specific number of eigenvalues divided by their total sum, indicating disturbance enstrophy content. The dotted line indicates the 99% level. (b) The first two eigenvectors of disturbance vorticity during the indicated time ranges are shown.

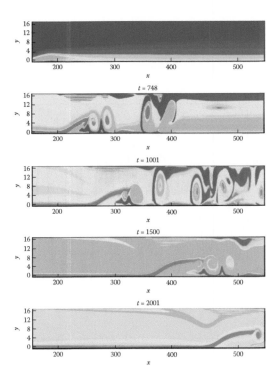

Figure 4.16
Vorticity contours for the case of LEC by a counter-clockwise rotating vortex.

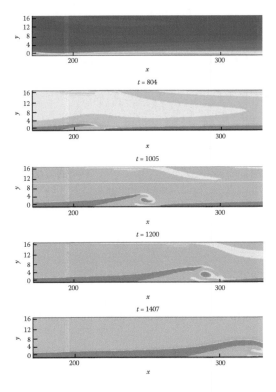

Figure 4.17
Vorticity contours for the case of a convecting negative vortex with strength given by $|\Gamma|/\nu = 126.638$ and the other parameters same as in the previous case. Details of this are available in [334] and [356]

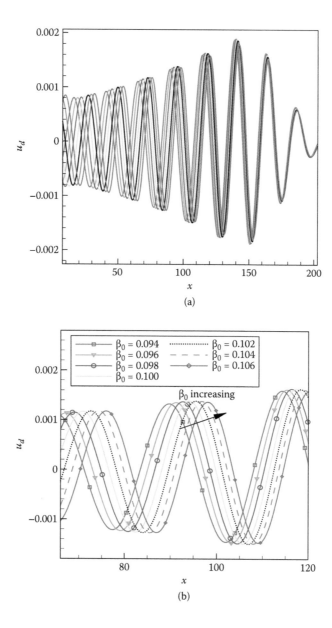

Figure 5.6

(a) Streamwise disturbance velocity plotted as a function of x at $t = 411.5$ for different β_0 with $Re_{\delta^*} = 1000$ and results shown at $y = 0.278$. (b) Enlarged view of (a) shown for selected frequencies.

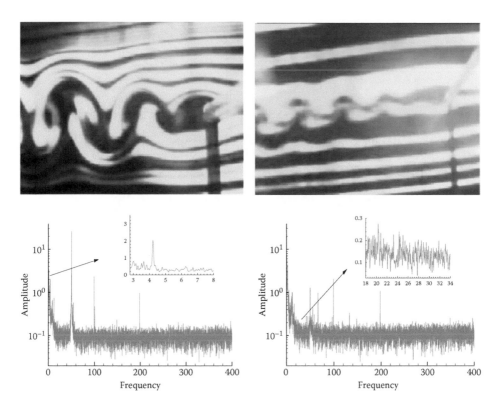

Figure 6.1
Experimental observation of flow past a cylinder for $Re = 53$ with $D = 5$ mm; $U\infty = 17$ cm/s; $Tu = 0.007$ (left) and (b) $D = 1.8$ mm; $U\infty = 46.9$ cm/s; $Tu = 0.01$ (right). FFT of the empty tunnel disturbances at the same speed is shown below the visualization pictures. Insets show disturbance amplitudes near the Strouhal frequency.

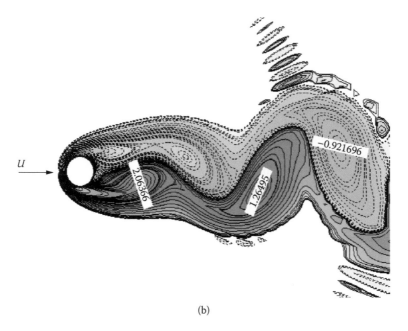

(b)

Figure 6.3b
Computed vorticity contours for the flow at $Re = 75$, when no FST is introduced through the inflow.

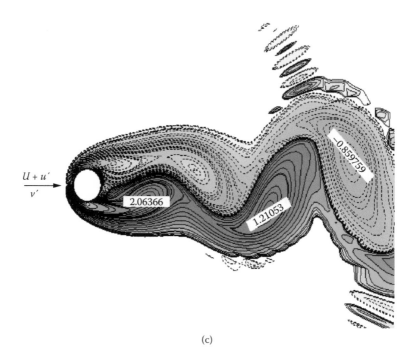

(c)

Figure 6.3c
Computed vorticity contours for the flow at $Re = 75$ with an FST level given by 0.06% corresponding to the data of [254].

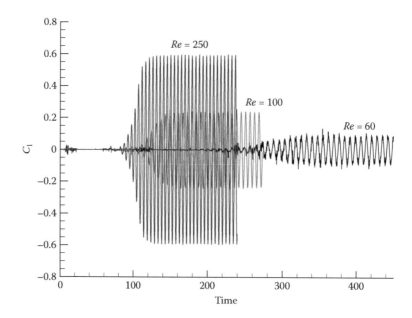

Figure 6.7
Computed lift coefficient variation with time for $Re = 100$ and $Re = 250$. Note the increasing presence of multiple modes with decreasing Reynolds number. Subsequently, only the $Re = 60$ case is analyzed for multiple modes in POD analysis and studying the LSE equation.

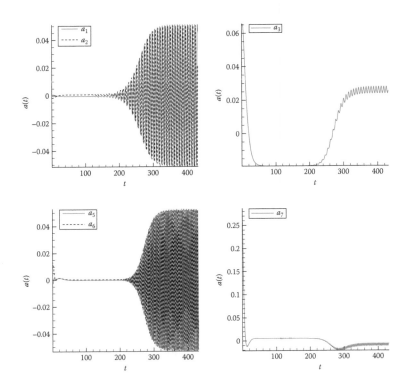

Figure 6.11a
Time-dependent amplitude functions of first seven POD modes of Fig. 6.9 shown for $t = 0$ to 430.

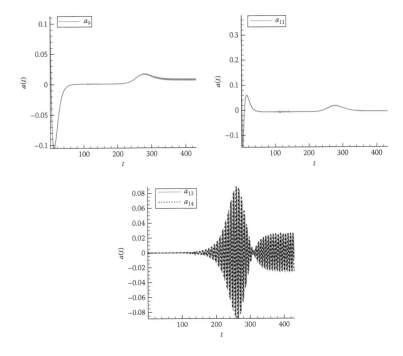

Figure 6.11b
Time-dependent amplitude functions of higher POD modes of Fig. 6.9 shown for $t = 0$ to 430. Note that a_1–a_2; a_5–a_6 and a_{13}–a_{14} form pairs; a_3, a_7, a_9 and a11 do not form pairs corresponding to an anomalous mode of the first kind. The pair a_{13}–a_{14} does not follow the time variation given by either the Stuart–Landau or the LSE equation, and is an anomalous mode of the second kind.

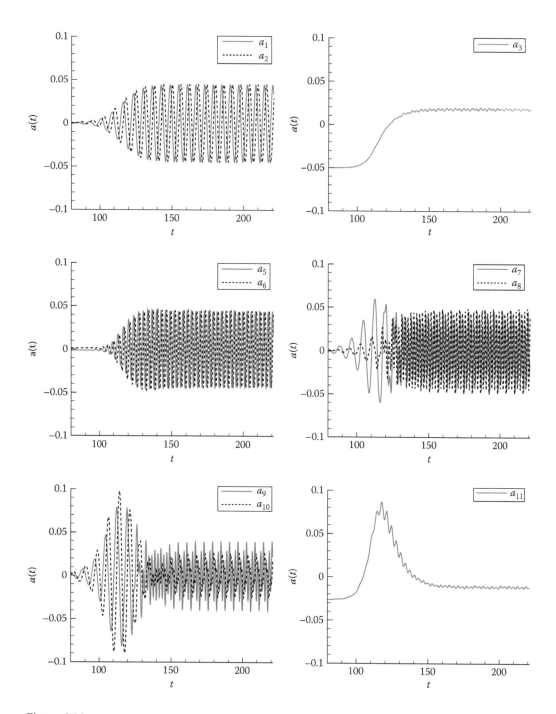

Figure 6.14

Time-dependent POD amplitudes for flow past a circular cylinder for $Re = 100$ during $t = 200$ to 350. The modes (a_1, a_2), (a_5, a_6), (a_7, a_8) and (a_9, a_{10}) form pairs, a_3 and a_{11} are T_1-modes. Modes a_5 to a_{10} are anomalous T_2-modes which do not strictly follow the time variation given by Stuart–Landau and LSE equations.

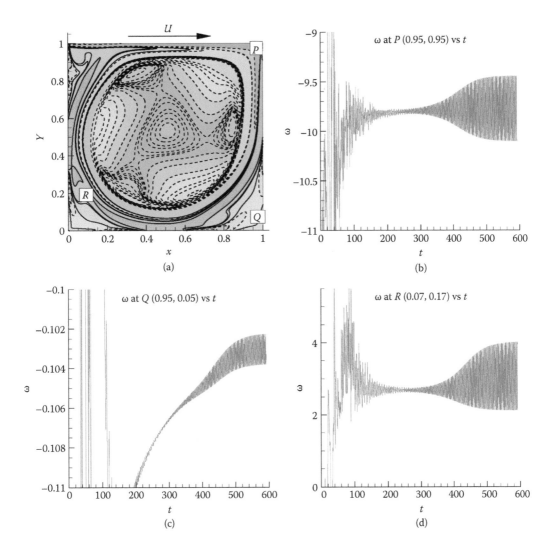

Figure 6.15

(a) Computed vorticity flow field of flow inside a square lid-driven cavity for $Re = 8500$ shown at $t = 550.0$. (b)–(d) The time variation of vorticity at P, Q and R (as in frame (a)) are shown plotted.

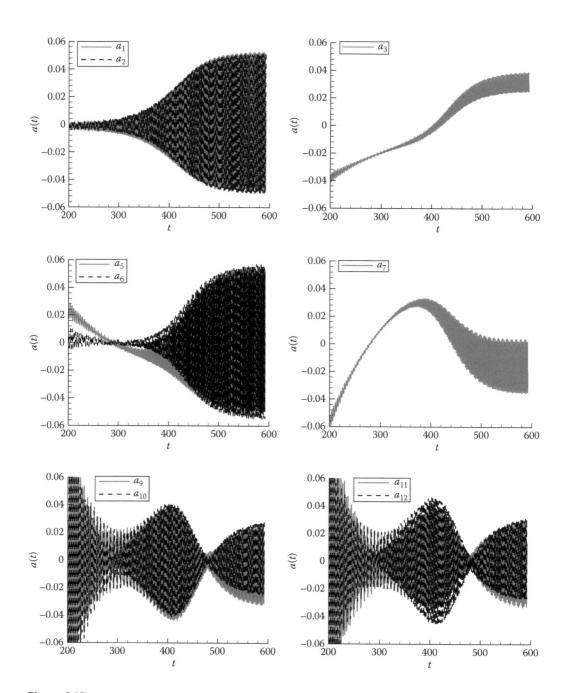

Figure 6.18

Time-dependent POD amplitude functions for the LDC flow for $Re = 8500$ during $t = 201$ to 592. The modes (a_1, a_2), (a_5, a_6), (a_9, a_{10}) and (a_{11}, a_{12}) form pairs, with a_3 and a_7 as the T_1-modes. Modes a_9 to a_{12} are anomalous type T_2-modes which do not follow the time variation given by Stuart–Landau and LSE equations.

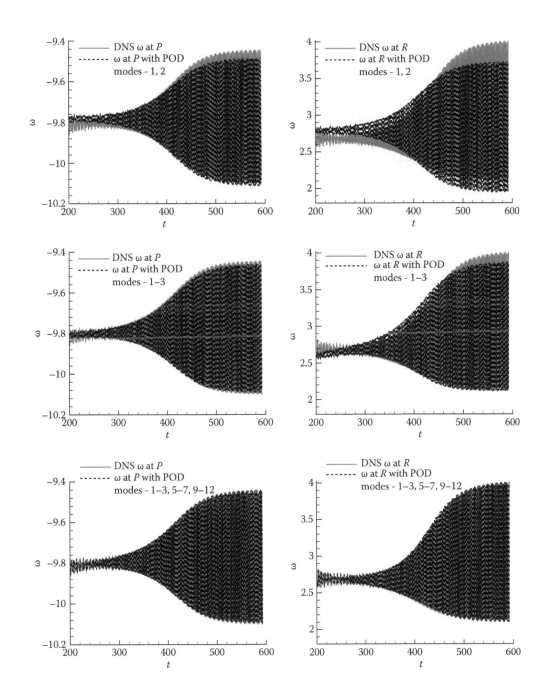

Figure 6.19
Comparison of reconstructed POD data with DNS results for LDC flow for $Re = 8500$ at two points P (left) and R (right). Note the proper shift caused by the inclusion of the third mode for both the points and produced an excellent match when twelve modes are included.

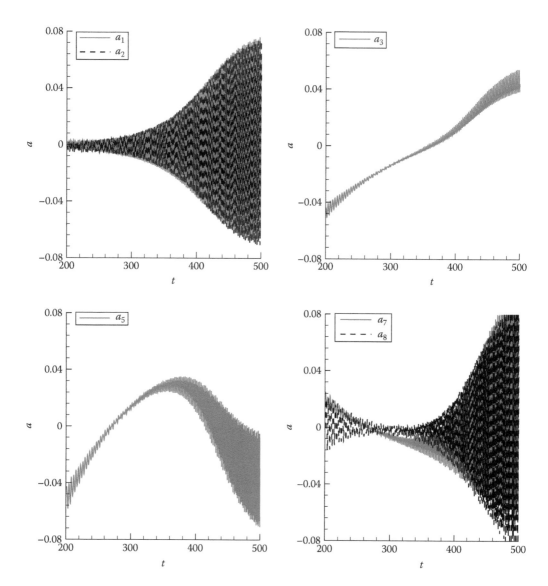

Figure 6.20
POD amplitude functions for the flow in an LDC for $Re = 8500$ during $200 \le t \le 500$.

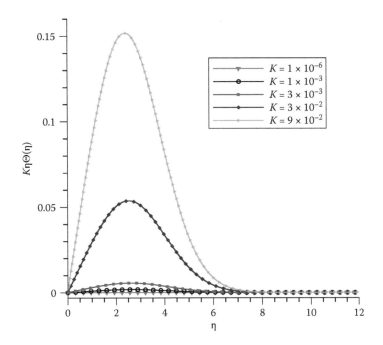

Figure 7.4
Buoyancy induced pressure gradient term $K\eta\Theta$ plotted as a function of similarity variable, η.

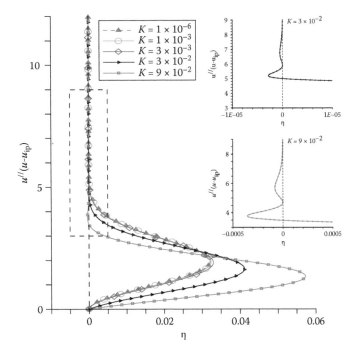

Figure 7.7
The quantity $f'''(f' - f'_{ip})$ plotted as a function of η for indicated values of K.

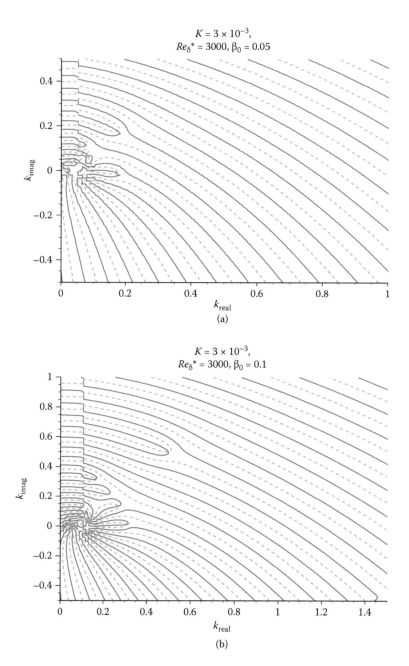

Figure 7.8

Eigenspectrum of the OSE obtained through the grid search method at the seed points for $K = 3 \times 10^{-3}$ at $Re_{\delta*} = 3000$. (a) For $\beta_0 = 0.05$. (b) $\beta_0 = 0.1$. In both these frames solid lines represent $D_r = 0$ and dotted lines represent $D_i = 0$.

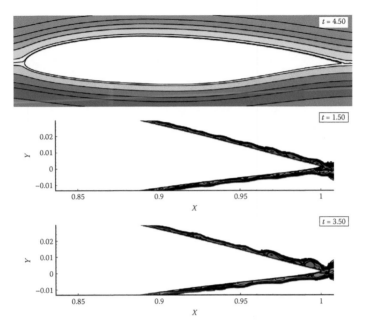

Figure 9.32
Stream function contours for flow past SHM1 airfoil are shown at the indicated time instants for $Re = 10.3×10^6$ when the airfoil is kept at zero AOA. The top frame shows the flow field around the full airfoil, while the bottom frames show the zoomed view of the flow field near the trailing edge, showing the presence of small bubbles indicative of bypass transition.

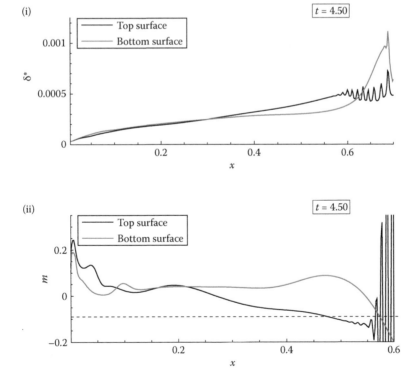

Figure 9.33
Variations of the displacement thickness (δ^*) on top and bottom surfaces of SHM1 airfoil are shown in the top frame, at $t = 4.5$. The corresponding variation of steady flow separation parameter $m = \dfrac{x}{U_e}\dfrac{\partial U_e}{\partial x}$ on the top and bottom surfaces of the airfoil is shown in bottom frame.

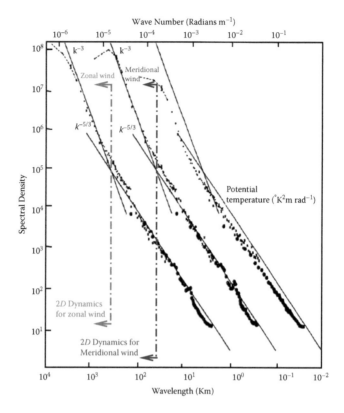

Figure 10.1
Kinetic energy and temperature spectrum near the tropopause from Nastrom and Gage [245].

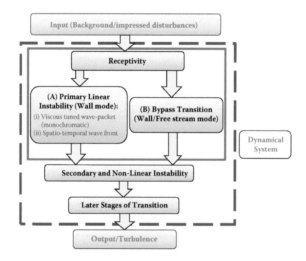

Figure 10.2
Block diagram showing different routes of turbulence generation.

8

Instabilities of Three-Dimensional Flows

8.1 Introduction

Here we review and explain instabilities of external three-dimensional flows which is complicated and seemingly contradictory to instabilities noted for two-dimensional flows. Instabilities for three-dimensional flows are more complex, as these can occur not only for the streamwise component of the flow, but additional instabilities of the cross flow component of the velocity profile need to be considered. Additionally, for aerospace applications, one has also to study the problem of leading edge contamination, i.e., instability of the flow along the attachment line, which was discussed in Chap. 4.

In Chap. 7, we have noted that a favorable pressure gradient created due to buoyancy effects can also give rise to an inflectional velocity profile, which is instrumental in triggering additional temporal instability over and above the spatial instability. For flow over aircraft wings, similar mechanisms are present. Due to these complications, specifically for cross flow instabilities in a favorable pressure gradient region for a swept wing configuration, a detailed understanding of these spatial and temporal mechanisms in isolation and in conjunction is necessary. For flight at high subsonic and transonic speeds, a conflicting requirement for aircraft design forces us to keep the sweep-back angle as small as possible, without compromising the laminar flow requirement over the wing to keep skin friction drag low, while not being adversely affected by wave drag. Wing sweep back is needed at higher Mach numbers to keep wave drag under control. In this chapter, we discuss various aspects of instabilities which occur in three-dimensional flows. To begin with, we note the reason why three-dimensional flows are more complicated.

8.2 Three-Dimensional Flows

Two different classes of three-dimensional boundary layers are found in many external or internal flows of engineering interest. These are the *boundary sheets* and *boundary regions*, with the former exemplified by a boundary layer forming over a swept wing (excluding the root and tip region of the wing), as shown in Fig. 8.1, where the boundary layer is thin and parallel to the underlying surface. The free shear layer in the wake is a continuation of the boundary layer over the wing and is also classified as the *boundary sheet*. The *Boundary region* forms at the wing-fuselage junction near the wing tips in the trailing vortex and is characterized by a confluence of the *boundary sheets* coming from the top and bottom surface of the wing. Such a classification also holds for internal flows in ducts (see [55] for details).

In *boundary sheets*, length scales in the x- (streamwise) and z-directions (spanwise) are both of order one and so the Navier–Stokes equation in the z-direction can be simplified similar to that applied for the equation in the x-direction. The *boundary regions* or the slender shear layers (as they are alternately known) are seen to occur in a duct or in the wing-body junction and have a scale of order δ in the y- and z-directions. The governing equations for different types of three-dimensional boundary layers are given in [55, 57, 71, 283]. In the following, we will mostly keep our attention focused on the *boundary sheet* type of flow and will refer to it by the generic name of boundary layer. When such a boundary layer forms over the wing, it is instructive to find out how the cross flow originates.

For three-dimensional flows, as the external velocity vector is not in the direction of the pressure gradient, the external streamlines on the surface are curved, which gives rise to the cross flow. Consider the flow shown in Fig. 8.2, where a negative pressure gradient ($\frac{\partial p}{\partial z} < 0$) is applied in the spanwise direction. This pressure gradient is related to the curvature of the external streamline and is dynamically balanced by a centrifugal force, i.e.,

$$\frac{\partial p}{\partial z} = -\frac{\rho u_e^2}{r_e}$$

where r_e is the radius of curvature of the external streamline. Closer to the wall, but still outside the shear layer where the velocity is u, we must have $\frac{\partial p}{\partial z} = -\frac{\rho u^2}{r}$, with r as the radius of curvature of the local streamline.

The pressure and hence $\frac{\partial p}{\partial z}$ do not vary in the wall-normal direction, and hence r is less than r_e, as u is less than u_e. Thus, the crosswise pressure gradient $\frac{\partial p}{\partial z}$ skews the streamlines differently at different heights, causing a component of the flow shown in Fig. 8.2a, with the no-slip condition ensuring zero cross flow at the wall. Prandtl [283] termed this the secondary flow of the first kind or skew-induced secondary flow. When $\frac{\partial p}{\partial z}$ changes sign, the external flow displays an inflection point, as shown in Fig. 8.2b. When Eq. (8.4) given below

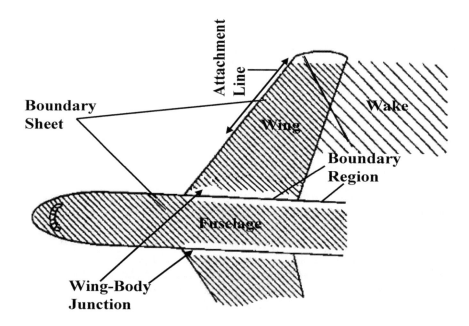

FIGURE 8.1
Typical three-dimensional boundary layers over a transport aircraft.

is viewed at the wall, one has $u = v = w = 0$ and thus

$$\frac{\partial p}{\partial z} = \frac{\partial \tau_{zy}}{\partial y}$$

As a consequence, the cross flow displays reverse flow near the wall in the cross flow direction. Thus, the cross flow profile is S-shaped, because near the wall flow reacts more rapidly, with inertia playing a subdominant role. This is the mechanism of a pressure driven three-dimensional boundary layer, as one would find in flow over a swept wing.

In the following, three-dimensional boundary layer equations are stated in the local Cartesian coordinate system. For more details on three-dimensional boundary layer equations in a streamline, body-oriented coordinate system one can refer to [57]. Here the equations are written for laminar flow, as would be required for the equilibrium flow whose stability or receptivity is studied.

$$\frac{\partial u}{\partial x} + \frac{\partial v}{\partial y} + \frac{\partial w}{\partial z} = 0 \tag{8.1}$$

$$u\frac{\partial u}{\partial x} + v\frac{\partial u}{\partial y} + w\frac{\partial u}{\partial z} = -\frac{1}{\rho}\frac{\partial p}{\partial x} + v\frac{\partial^2 u}{\partial y^2} \tag{8.2}$$

$$\frac{\partial p}{\partial y} = 0 \tag{8.3}$$

$$u\frac{\partial w}{\partial x} + v\frac{\partial w}{\partial y} + w\frac{\partial w}{\partial z} = -\frac{1}{\rho}\frac{\partial p}{\partial z} + v\frac{\partial^2 w}{\partial y^2} \tag{8.4}$$

These equations are solved subject to the wall ($y = 0$) boundary conditions

$$u = w = 0 \quad \text{and} \quad v = v_w(x, z) \tag{8.5a}$$

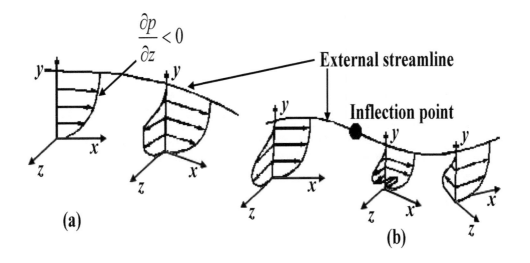

FIGURE 8.2
Secondary flow generated by different spanwise pressure gradients. (a) Flow with $\frac{\partial p}{\partial z} < 0$ (b) Flow where $\frac{\partial p}{\partial z}$ changes sign.

and at $y = \delta$:

$$u = u_e \quad \text{and} \quad w = w_e \tag{8.5b}$$

Boundary conditions at the wall show the possibility of mass transfer, which is often needed to obtain equilibrium flow in the presence of suction. Receptivity studies of such an equilibrium flow can be pursued either in a linear or nonlinear framework, similar to that studied for two-dimensional flows discussed in Chaps. 3 to 7. This may be required for laminar flow control applications via suction. At the edge of the boundary layer, the edge velocity components are related to the pressure gradients by

$$u_e \frac{\partial u_e}{\partial x} + w_e \frac{\partial u_e}{\partial z} = -\frac{1}{\rho} \frac{\partial p}{\partial x} \tag{8.6a}$$

$$u_e \frac{\partial w_e}{\partial x} + w_e \frac{\partial w_e}{\partial z} = -\frac{1}{\rho} \frac{\partial p}{\partial z} \tag{8.6b}$$

The above equations for a three-dimensional boundary layer can be further simplified under certain conditions. Problems of practical engineering interests arise where the flow becomes independent of one independent variable. Such degenerate cases of three-dimensional flows having three nonzero velocity components dependent on two independent variables are found, for example, in flow near the leading edge of a swept body with constant cross section and *infinite* span in the attachment line plane of Fig. 8.1, a problem discussed in Chap. 4. Similar simplification is possible for flow over a radial diffuser [437] or flow over a straight tapered swept wing [37]. Of specific interest is flow past an infinite swept wing [11] or yawed cylinder [278], which is discussed next.

8.3   Infinite Swept Wing Flow

Here we define two coordinate systems with respect to the sketch of a wing shown in Fig. 8.3. The wing fixed coordinate system (WFCS), (X, Y, Z), is fixed to the wing with the Z-axis along the swept leading edge; X is in the plane of the wing orthogonal to the Z-direction and Y is perpendicular to the plane of the wing in the wall-normal direction. Additionally, we introduce a coordinate system which is the external streamline fixed coordinate system (ESFCS) indicated by the (x, y, z)-system. One notices that Y and y are essentially the same coordinate.

For flows over a yawed cylinder or infinite swept wings of constant cross section, the flow is independent of the spanwise coordinate, Z. For such a flow, there will be a constant spanwise component of velocity, but the derivatives of all physical variables with respect to Z are to be set equal to zero. The co-ordinate system is as shown in Fig. 8.3, where φ is the sweep angle.

The governing boundary layer equations given in Eqs. (8.1) to (8.4) simplify for an infinite swept wing to

$$\frac{\partial u}{\partial X} + \frac{\partial v}{\partial Y} = 0 \tag{8.7}$$

$$u \frac{\partial u}{\partial X} + v \frac{\partial u}{\partial Y} = -\frac{1}{\rho} \frac{\partial p}{\partial X} + v \frac{\partial^2 u}{\partial Y^2} \tag{8.8}$$

$$u \frac{\partial w}{\partial X} + v \frac{\partial w}{\partial Y} = v \frac{\partial^2 w}{\partial Y^2} \tag{8.9}$$

These are supplemented by the equations

$$\frac{\partial P}{\partial Y} = \frac{\partial P}{\partial Z} = 0$$

These equations show that u and v are independent of w and the resultant equations are just like that for a two-dimensional boundary layer. At the edge of the boundary layer, u and w attain values u_e and w_e. From Eq. (8.9), it is noted that at the free stream $\frac{\partial w}{\partial x} = 0$ and thus w_e is a pure constant, fixed as w_∞. There is, however, a difference between the solution procedures of general three-dimensional and two-dimensional boundary layers. For the solution of three-dimensional boundary layers, one needs initial conditions on two intersecting planes. For some external flows, one can take advantage of available symmetry conditions from physical considerations. For example, one can use the conditions near the attachment line of bodies, as explained next.

8.4   Attachment Line Flow

The line joining stagnation points on each section of the body constitutes the symmetry plane. On this plane of symmetry, the flow is two dimensional except for the cross flow derivative. The flow in this plane is called the attachment line flow. The attachment line is a streamline on the body on which both the cross flow velocity and the cross flow pressure gradient are identically zero. If we represent the flow in the vicinity of the stagnation region, as in Fig. 8.4, then w and $\frac{\partial p}{\partial z}$ are identically zero, making the z-momentum equation singular along the attachment line.

However, the degeneracy can be removed by writing another equation obtained by differentiating the z-momentum equation with respect to z and treating $w_z = \frac{\partial w}{\partial z}$ as the new dependent variable with additional symmetry conditions on the attachment line, i.e.,

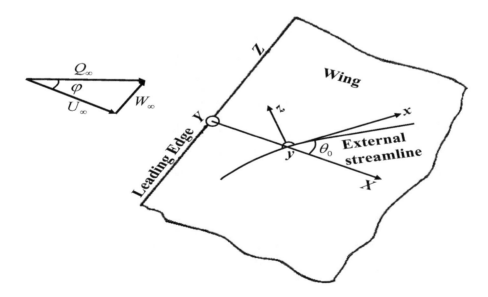

FIGURE 8.3
Coordinate systems over an infinite swept wing.

$$\frac{\partial u}{\partial z} = \frac{\partial v}{\partial z} = \frac{\partial^2 w}{\partial z^2} = 0$$

The resultant equations are given by

$$\frac{\partial u}{\partial x} + \frac{\partial v}{\partial y} + w_z = 0 \tag{8.10}$$

$$u\frac{\partial u}{\partial x} + v\frac{\partial u}{\partial y} = u_e\frac{\partial u_e}{\partial x} + \frac{1}{\rho}\frac{\partial}{\partial y}\left(\mu\frac{\partial u}{\partial y}\right) \tag{8.11}$$

$$u\frac{\partial w_z}{\partial x} + v\frac{\partial w_z}{\partial y} + w_z^2 = u_e\frac{\partial (w_z)_e}{\partial x} + (w_z)_e^2 + \frac{1}{\rho}\frac{\partial}{\partial y}\left(\mu\frac{\partial w_z}{\partial y}\right) \tag{8.12}$$

The boundary conditions for these equations at $y = 0$:

$$u = v = 0 = w_z \tag{8.13a}$$

and at $y = \delta$:

$$u = u_e(x, z) \quad \text{and} \quad w_z = w_{ze} \tag{8.13b}$$

The boundary layer equations are not solved in the form given above. Instead these are transformed first and solved in the transformed plane, which allows taking larger steps in the x- and z-directions. The scaled variables in the transformed plane vary slower as compared to their variations in the physical un-transformed plane. Next, we obtain the boundary layer equations in the transformed plane.

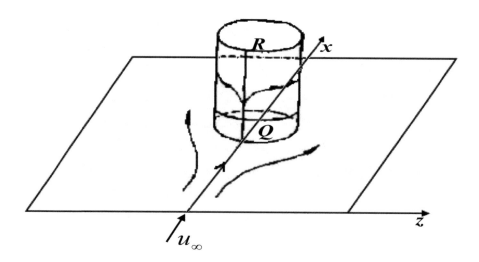

FIGURE 8.4
Limiting streamlines for a cylinder attached to a plate to indicate flow in the vicinity of the stagnation region indicated by the line element QR.

8.5   Boundary Layer Equations in the Transformed Plane

Here the boundary layer equations, Eqs. (8.1) to (8.4), are expressed using a transformation for the independent variables similar to the Falkner–Skan transformation used for two-dimensional flows

$$x \equiv x, \ z \equiv z \ \text{and} \ \eta = y\sqrt{\frac{u_e}{\nu x}} \tag{8.14}$$

This is a non-orthogonal transformation that stretches the $0(\delta)$ shear layer in the y-variable to a thickness that is $0(1)$ in the η-variable. Apart from stretching the wall-normal coordinate, this transformation also removes the singularity of the governing parabolic partial differential equation at $x = 0$ and $z = 0$.

For the dependent variables, two-component vector potentials ϕ and $\hat{\psi}$ are introduced such that

$$u = \frac{\partial \hat{\psi}}{\partial y} \tag{8.15a}$$

$$w = \frac{\partial \phi}{\partial y} \tag{8.15b}$$

The continuity equation provides

$$v = -\left(\frac{\partial \hat{\psi}}{\partial x} + \frac{\partial \phi}{\partial z}\right) \tag{8.15c}$$

Furthermore, one can define the non-dimensional vector potential components f and g with the help of

$$\hat{\psi} = (u_e \nu x)^{1/2} f(x, z, \eta) \tag{8.16a}$$

$$\phi = w_e \left(\frac{\nu x}{u_e}\right)^{1/2} g(x, z, \eta) \tag{8.16b}$$

so that $u = u_e f'$ and $w = w_e g'$, with a prime indicating a derivative with respect to η. The governing equations for f and g are obtained as [57]

$$f''' + m_1 f f'' + m_2(1 - f'^2) + m_5(1 - f'g') + m_6 g f'' =$$

$$x\left[f'\frac{\partial f'}{\partial x} - f''\frac{\partial f}{\partial x} + m_7\left(g'\frac{\partial f'}{\partial z} - f''\frac{\partial g}{\partial z}\right)\right] \tag{8.17}$$

$$g''' + m_1 f g'' + m_4(1 - f'g') + m_3(1 - g'^2) + m_6 g g'' =$$

$$x\left[f'\frac{\partial g'}{\partial x} - g''\frac{\partial f}{\partial x} + m_7\left(g'\frac{\partial g'}{\partial z} - g''\frac{\partial g}{\partial z}\right)\right] \tag{8.18}$$

where

$$m_2 = \frac{x}{u_e}\frac{\partial u_e}{\partial x}; \ m_1 = \frac{m_2 + 1}{2}; \ m_3 = \frac{x}{u_e}\frac{\partial w_e}{\partial z}; \ m_4 = \frac{x}{w_e}\frac{\partial w_e}{\partial x};$$

$$m_5 = \frac{w_e}{u_e}\frac{x}{u_e}\frac{\partial u_e}{\partial z}; \ m_6 = m_3 - \frac{m_5}{2} \ \text{and} \ m_7 = \frac{w_e}{u_e}$$

Boundary conditions for the solution of Eqs. (8.17) and (8.18) at $\eta = 0$:

$$f, f', g, g' = 0 \tag{8.19a}$$

and at $\eta = \eta_\infty$:

$$f', g' \to 1 \tag{8.19b}$$

Once again, to solve these equations we need initial conditions on two intersecting planes. Thus, we start from the symmetry plane: the flow at the attachment line that has been described in the last section. The attachment line equations can also be transformed by a set of transformations via a different two-component vector potential as

$$u = \frac{\partial \hat{\psi}}{\partial y} \quad \text{and} \quad \frac{\partial w}{\partial z} = \frac{\partial \phi}{\partial y} \tag{8.20}$$

So that $v = -\left(\frac{\partial \hat{\psi}}{\partial x} + \phi \right)$. Once again, one can introduce non-dimensional vector potential f with Eq. (8.16a) and the second non-dimensional vector potential is introduced via

$$\phi = \left(\frac{dw}{dz} \right)_e \left(\frac{\nu x}{u_e} \right)^{1/2} g_1(x, z, \eta) \tag{8.21}$$

Substitution and simplification yield the following equations for the attachment line boundary layer

$$f''' + m_1 f f'' + m_2(1 - f'^2) + m_5(1 - f' g_1') + m_3 g_1 f'' = x \left[f' \frac{\partial f'}{\partial x} - f'' \frac{\partial f}{\partial x} \right] \tag{8.22}$$

$$g_1''' + m_1 f g_1'' + m_4(1 - f' g_1') + m_3(1 - g_1'^2) + m_3 g_1 g_1'' = x \left[f' \frac{\partial g_1'}{\partial x} - g_1'' \frac{\partial f}{\partial x} \right] \tag{8.23}$$

Here the definition of all the terms remains the same except

$$g_1' = \frac{(\partial w / \partial z)}{(\partial w / \partial z)_e} \quad \text{and} \quad m_4 = \frac{x}{(\partial w / \partial z)_e} \frac{\partial}{\partial x} \left(\frac{\partial w}{\partial z} \right)_e$$

And the corresponding boundary conditions at $\eta = 0$ are:

$$f, f', g_1, g_1' = 0 \tag{8.24a}$$

And at $\eta = \eta_\infty$:

$$f', g_1' \to 1 \tag{8.24b}$$

8.6   Simplification of Boundary Layer Equations in the Transformed Plane

All the above flow equations in the transformed plane [55] can be further simplified if external flow takes special forms. Introducing non-dimensional vector potential f, g of Eqs. (8.16) and if external flow is of the form [57]

$$u_e = Ax^m z^n \text{ and } w_e = Bx^r z^s \tag{8.25}$$

with $m_2 = m$; $r = m_2 - 1$; and $s = n + 1$, then the governing equations, Eqs. (8.17) and (8.18), simplify to the following equations, assuming the corresponding laminar to be similar.

$$f''' + \frac{m+1}{2} f f'' + m(1 - f'^2) + \frac{B}{A}\left\{ \frac{n+2}{2} g f'' + n(1 - f' g') \right\} = 0 \tag{8.26}$$

$$g''' + \frac{m+1}{2} f g'' + (m-1)(1 - f' g') + \frac{B}{A}\left\{ \frac{n+2}{2} g g'' + (n+1)(1 - g'^2) \right\} = 0 \tag{8.27}$$

These equations were solved in [141] for different values of B, A, m and n. Cebeci & Cousteix [57] point out that such external flows are rotational and exhibit *overshoots* in u profile due to non-uniformities in total pressure. Further simplifications arise when $r = s = n = 0$ and arbitrary m makes the external flow irrotational, which results in the well-known Falkner–Skan–Cooke equations given by

$$f''' + \frac{m+1}{2} f f'' + m(1 - f'^2) = 0 \tag{8.28}$$

$$g''' + \frac{m+1}{2} f g'' = 0 \tag{8.29}$$

These equations further simplify for $m = 1$, which represents the equations from which the similarity solution of the attachment line flow can be obtained. These equations were used to obtain the equilibrium flow for LEC studied in Chap. 4 [334] to study bypass transition caused by a convecting vortex outside the shear layer, but which remains strictly in the attachment line plane.

Similarly one can obtain a non-similar solution for flow past an infinite swept wing by noting the independence of flow properties with z from Eqs. (8.17) and (8.18) as

$$f''' + m_1 f f'' + m_2(1 - f'^2) = x\left[f' \frac{\partial f'}{\partial x} - f'' \frac{\partial f}{\partial x} \right] \tag{8.30}$$

$$g''' + m_1 f g'' = x\left[f' \frac{\partial g'}{\partial x} - g'' \frac{\partial f}{\partial x} \right] \tag{8.31}$$

Note that Eq. (8.30) is independent of g. This makes Eq. (8.31) a linear equation. In the above, non-similarity arises due to right-hand side terms in Eqs. (8.30) and (8.31). Having obtained various three-dimensional equilibrium solutions, it is possible to study their stability.

8.7 Instability of Three-Dimensional Flows

It is natural to expect that in some respects instabilities of three-dimensional flows would be similar to instabilities of two-dimensional flows due to similarities of the mean flow obtained via the Falkner–Skan–Cooke and Falkner–Skan similarity transforms utilized, respectively. However, during flight tests on swept wing aircraft in RAE between 1951 and 1952, it was found that the flow instability and transition zone moved towards the wing leading edge abruptly, which is not compatible with what one would expect from two-dimensional flow analysis. Details of these experiments and a plausible explanation can be found in [131, 262, 263]. Beyond a certain speed the transition point moved close to the leading edge and it was observed that the leading edge radius and sweep angle are the determining factors. Two aspects stood out. First, in some cases the instability was noted in the accelerated part of the flow, which according to two-dimensional flow theory should have stable TS waves and make the flow overall more stable. Second, in some cases the point of transition moved right at the leading edge, at the attachment line itself, which according to two-dimensional flow instability has a critical Reynolds number that is an order of magnitude higher than the zero pressure gradient boundary layer, as discussed in Chap. 3. These flight experiments performed on experimental aircraft AW52 at RAE led to the conclusion in [131] that *no laminar flow is present on normal wings of any appreciable size and speed if the sweep angle exceeds roughly* 20^0. Even today in designing aircrafts, this statement is used as a rule of thumb.

Furthermore, sublimation patterns on the surface indicated external streamline aligned streaks which can be interpreted as stationary waves. These streaks are also regularly spaced in the spanwise direction − vividly demonstrated experimentally in the laboratory in [13]. All this evidence establishes the destabilizing effects of sweep back and bring to the fore the observation that transition occurs in a region of strong negative pressure gradients occurring at large free stream speeds. Also, note the transition occurs downstream of the streaks.

Overall, instability of flow over a swept wing has similarities with mixed convection flow over a heated horizontal plate, in terms of exhibiting both spatial and temporal instabilities. This is due to the presence of an inflection point in the streamwise velocity component for mixed convection flow, with an inflection point in the cross flow profile of flow over swept wing. Such inflection points give rise to temporal instability, following Rayleigh–Fjørtoft theorem.

8.7.1   Effects of Sweep Back and Cross Flow Instability

Once the destabilizing effect of a swept back wing was recognized, further experiments followed using wing tunnels [8, 133], confirming the flight test results at RAE. These were followed by further flight experiments reported in [6, 47] on a straight swept wing. Subsequently, Boltz et al.[35] carried out extensive wind tunnel experiments focusing on effects of Reynolds number, angle of attack and sweep angle on transition location. Further experiments on this topic were reported in [13, 15, 225], where the authors studied flow over an ONERA D aerofoil profile. Although the model had the usual ONERA D profile over the top surface between $x/c = 0.2$ and $x/c = 1$, it had a special cambered leading edge and lower surface, as shown in Fig. 8.5. Note that ONERA D profile is a symmetric section and this modified section was used to highlight the cross flow instability.

As the cross flow instability was noted in that part of the swept wing where the flow was accelerating, in this ONERA D model the instability was mimicked by appropriately created accelerated flow by choosing negative angles of attack with a specific sweep angle. This was the motivation for the design of the leading edge of the ONERA D wing shown in Fig. 8.5. This is also seen in the velocity profiles shown for the suction side at the indicated angles of attack in Fig. 8.5. Following this interesting idea of simulating swept back wing flow on a straight wing, Saric & Yeates [307] created swept wing flow on a flat plate with a swept leading edge and the accelerated flow was created by a bump on the roof of the tunnel. While this last experiment avoided curvature effects, this also precluded the attachment line problem.

In all these cases for sufficiently large Reynolds numbers, transition was seen to occur in regions of a strong negative pressure gradient. Also, wall visualization techniques (sublimation, china-clay, oil flow, etc.) indicated the presence of streaks upstream of the transition location. The fact that the flow becomes unstable in favorable pressure gradient regions indicates that the instability is not related to the usual instability that arises for the flow in the streamwise direction. Thus, researchers have focused their attention on linking the ensuing instability of the flow component which is created by skewing the profile by the imposed pressure gradient, as discussed in Sec. 8.1. We have already noted the presence of a cross flow (that is in the perpendicular to the external streamlines direction) created by such skewing by a spanwise pressure gradient and it is natural to investigate its instability. Near the leading edge of a swept wing, both the surface and the streamlines are highly curved due to the combined effect of the pressure gradient and sweep angle, turning the flow inboard. A schematic of the flow shown in Fig. 8.6 is taken from [295].

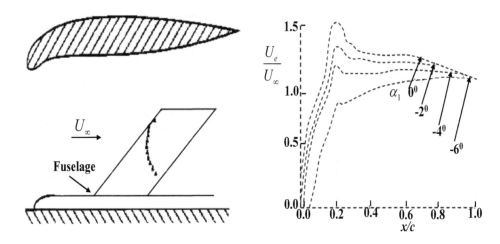

FIGURE 8.5
ONERA D airfoil fitted with a cambered leading edge for the wing. The edge velocity distribution is shown on the right-hand side [13, 15, 54, 225].

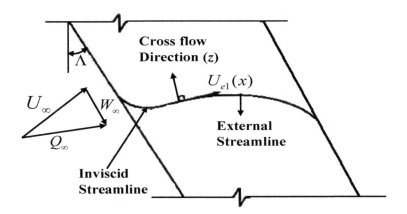

FIGURE 8.6
Schematic of streamline at the edge of a three-dimensional shear layer over a swept wing [295].

Such sharp changes of curvature of flow are also noted near the wing trailing edge, as well. Lower inertia of flow close to the body causes these changes to be more pronounced in creating the cross flow, as shown in Fig. 8.3. Also as shown in Fig. 8.2, the pressure gradient dictates whether this cross flow is S-shaped or not. Irrespective of the spanwise pressure gradient sign, this component of the flow displays a maximum in the interior of the boundary layer and goes to zero at the wall and at the edge of the shear layer. In Fig. 8.7, computed velocity profiles for the experiments in [307] are taken from [294] to give one an idea about the relative magnitudes of the streamwise and cross flow components in the three-dimensional boundary layer.

The velocity components normalized by the edge velocity show that the maximum cross flow velocity is only 3% and this is typical of all reported measured velocities in tunnels and in flight. Despite the fact that this component is significantly small as compared to the streamwise flow component, its effect on the flow field is more profound due to its instability caused by the presence of an inflection point in the velocity profile. The presence of an inflection point has been shown to cause violent inviscid instability, as explained by Rayleigh's inflection point theorem and its extended version given by Fjørtoft's theorem in Chap. 3. It has also been explained in [295] that such instability causes cross flow vortex structures similar to Kelvin's cats eye, with the axes aligned along the streamwise direction. While these vortices are of the dimension of boundary layer thickness, they all rotate in the same direction, and thus they are unlike Goertler vortices. These cross flow vortex structures were computed in [294] for the experimental results reported in [307]. Thus, according to Reed & Saric, cross flow instability was brought into focus in [131] for the first time and was interpreted in [262, 263], by Squire (as an addendum to Owen & Randall's 1952 report) and in [394], independently. Much of the work on instability of three-dimensional flows has been reviewed in [11, 277, 295].

8.8 Linear Stability Theory for Three-Dimensional Flows

In Chap. 3, we developed a general linear stability equation for a three-dimensional disturbance field in a three-dimensional parallel mean flow as

$$\phi^{iv} - 2(\alpha^2 + \beta^2)\phi'' + (\alpha^2 + \beta^2)^2\phi =$$

$$iRe\{(\alpha U_1 + \beta W_1 - \omega)[\phi'' - (\alpha^2 + \beta^2)\phi] - (\alpha U_1'' + \beta W_1'')\phi\} \tag{8.32}$$

where U_1 and W_1 are the streamwise and cross flow profiles; α and β are the wavenumber components in the streamwise and cross flow directions and ω is the circular frequency of the disturbance field.

This equation is written in a special Cartesian coordinate system which is defined in Sec. 8.10. As we have mentioned, the cross flow instability is related to severe turning of streamlines near the leading and trailing edges of the wing; thus it has often been noted that one should write the equilibrium and/or the stability equations retaining the curvature terms. However, neglect of curvature effects in solving the first order boundary layer equation seems justifiable, as the curvature terms are of the same order of magnitude as the neglected higher order terms in the boundary layer equation. Neglecting higher order terms and retaining curvature terms do not appear to be consistent. This is not so for the stability equations, which have been written for orthogonal curvilinear coordinate systems in [11, 54, 56, 69, 218]. According to Cebeci [54], despite earlier confusion about curvature effects originating from the work reported in [218], the work by Cebeci et al. [56] has established the correctness of stability solutions without curvature effects. In discussing the following, we therefore refer to stability equations developed in the Cartesian coordinate system without curvature terms.

8.8.1 Temporal Instability of Three-Dimensional Flows

Stuart, in writing a special section in [133], has shown that the Orr–Sommerfeld equation [Eq. (8.32)] can be reduced to a two-dimensional problem in studying the temporal stability problem by introducing the following quantities

$$\alpha_\psi^2 = \alpha^2 + \beta^2 \ \text{ and } \ U_\psi = U_1 \cos\psi + W_1 \sin\psi \tag{8.33}$$

where $\tan\psi = \beta/\alpha$ relates the real wavenumbers in the streamwise and cross flow directions, as shown in Fig. 8.3, with U_1 and W_1, as defined later in Eqs. (8.45). The Orr–Sommerfeld equation simplifies to

$$\phi^{iv} - 2\alpha_\psi^2\phi'' + \alpha_\psi^4\,\phi =$$

$$iRe\left\{(\alpha_\psi U_\psi - \omega)[\phi'' - \alpha_\psi^2\,\phi] - \alpha_\psi\frac{d^2 U_\psi}{dy^2}\phi\right\} \tag{8.34}$$

Here U_ψ represents the projection of velocity in the direction of the wavenumber vector, $\vec{\alpha}_\psi$ and Eq. (8.34) represents a two-dimensional Orr–Sommerfeld equation in the direction of the wavenumber vector. All benefits of the above transformation disappear in spatial theory, where complex α, β do not allow such manipulations. The above temporal approach is used in the COSAL code developed in [216], where one has five real parameters $(Re, \omega_r, \omega_i, \alpha, \beta)$. For a given Reynolds number Re and real frequency ω_r, if the wavenumber angle ψ is fixed, then the other two remaining parameters can be obtained from the dispersion relation as an

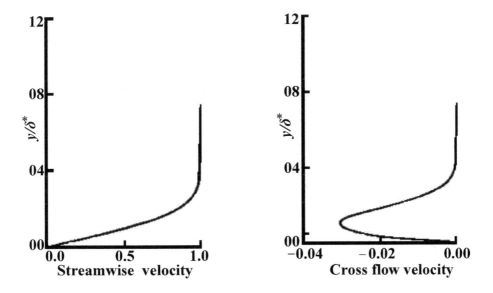

FIGURE 8.7
Computed streamwise and cross flow components of velocity inside a three-dimensional shear layer [294].

eigenvalue. After obtaining the growth rate ω_i for different values of ψ, one can calculate the amplification rate $n = \ln\left(\frac{A}{A_0}\right)$ by using the generalized Cauchy–Riemann transformation from

$$n = \int_{x_0}^{x} \frac{\omega_i}{|V_g|} dx' \tag{8.35}$$

where the group velocity is given by $V_g = \left[\left(\frac{\partial\omega}{\partial\alpha}\right)_{\psi,Re}, \left(\frac{\partial\omega}{\partial\beta}\right)_{\psi,Re}\right]$ and x' is the arc length along the group velocity direction, which is most of the time close to the external streamline direction. In the above envelope method, wave propagation direction ψ is sought along which ω_i is maximum and this maximum value is then integrated according to Eq. (8.35). In order to reduce computer time, Srokowski & Orszag [388] in their SALLY code looked at only stationary waves, $\omega_r = 0$.

8.8.2   Spatial Instability of Three-Dimensional Flows

Note that spatial stability problems are more complicated as we have six parameters now $(Re, \omega_r, \alpha_r, \alpha_i, \beta_r, \beta_i)$, as compared to five parameters in the temporal formulation. In [12, 214], the authors tried to circumvent this by a special form of spatial theory for three-dimensional problems. In these formulations, spanwise wavenumber β is assumed real and along with the real frequency ω_r for a fixed Re, one is left with the task of finding the complex α as the eigenvalue. A similar approach was also used in [345] in explaining the Klebanoff mode, which originates as a three-dimensional disturbance field due to very low circular frequency excitations. Arnal et al. [12] improved the procedure by maximizing the amplification rate α_i, with respect to all possible β, because in the presence of all possible directions of excitations (as given by various values of β), in an actual scenario, only the maximum growth will prevail in an asymptotic sense. However, this approach is not strictly a true spatial instability investigation.

True spatial calculations were later formulated and reported in [58, 246] by deriving a propagation condition. The eigenvalue formulation in these studies is based on complete spatial theory, where the relationship between the two wavenumbers α and β is not assumed as in [12, 214], but is computed making use of group velocity and the saddle point of the dispersion relation, as given in these references. If one assumes the disturbance to be of the normal mode form

$$q'(x,y,z,t) = q(y)\, e^{i(\alpha x + \beta z - \omega t)} \tag{8.36}$$

In the spatial theory developed in [58, 246], propagation of a wave packet is considered instead of normal mode analysis of instability theory, such that the disturbance field is expressed by

$$Q(x,y,z,t) = \int q(y)\, e^{i(\alpha x + \beta z - \omega t)} d\beta \tag{8.37}$$

Note that in the above equation, one is considering a fixed frequency disturbance and the integration range is defined in terms of β only in the complex plane along the Bromwich contour described in Chap. 3, the other wavenumber being an analytical function of β and related to it via

$$\alpha = \alpha(\beta) \tag{8.38}$$

Using this inter-relationship between the two wavenumbers, Eq. (8.37) can be alternately written as

$$Q(x, y, z, t) = \int q(y) \, e^{ix(\alpha(\beta) + \beta\frac{z}{x} - \frac{\omega}{x}t)} d\beta \tag{8.39}$$

For large values of x (away from the excitation source) following a ray given by $z/x =$ constant and $t/x =$ constant, one can use the saddle point method or the method of stationary phase [27] to find the leading contribution to Q. This occurs when the phase of the exponential term is stationary and would occur when the derivative of the phase with respect to β is zero for a fixed frequency disturbance field, i.e.,

$$\left(\frac{\partial\alpha}{\partial\beta}\right)_{\omega, Re} = \left(\frac{\partial\alpha_r}{\partial\beta_r}\right)_{\omega, Re} + i\left(\frac{\partial\alpha_i}{\partial\beta_r}\right)_{\omega, Re} = -\frac{z}{x} \tag{8.40}$$

As α is an analytic function of β, one can use the Cauchy–Riemann relation for the above to get

$$\left(\frac{\partial\alpha_r}{\partial\beta_r}\right)_{\omega, Re} = \frac{\partial\alpha_i}{\partial\beta_i} = -\frac{z}{x} \tag{8.41a}$$

$$\left(\frac{\partial\alpha_i}{\partial\beta_r}\right)_{\omega, Re} = \frac{\partial\alpha_r}{\partial\beta_i} = 0 \tag{8.41b}$$

This immediately implies, according to the saddle point method, that the real part of $(\partial\alpha/\partial\beta)_{\omega, Re}$ is related to wave propagation angle ϕ_1 by

$$\left(\frac{\partial\alpha}{\partial\beta}\right)_{\omega, Re} = -\tan\phi_1 \tag{8.42}$$

Equations (8.41) define the complex quantity $\beta^* = \beta_r^* + i\beta_i^*$ in terms of z/x. Moreover, along a given ray in the (x, z)-plane, β^* is constant.

Angle ϕ_1 also provides the group velocity direction, since

$$\left(\frac{\partial\alpha}{\partial\beta}\right)_{\omega, Re} = -\frac{\partial\omega/\partial\beta|_{\alpha, Re}}{\partial\omega/\partial\alpha|_{\beta, Re}}$$

with α and β related by Eq. (8.41), the amplitude of the disturbance is dominated by the exponential factor $\exp[-(\alpha_i x + \beta_i z)]$ or $\exp[-x(\alpha_i + \beta_i z/x)]$. Using Eq. (8.40), one can see the equivalent streamwise growth is given by $\Gamma = \alpha_i - \beta_i\left(\dfrac{\partial\alpha}{\partial\beta}\right)_{\omega, Re}$. The sign of Γ and the direction of propagation of the disturbances would indicate whether the waves are growing, decaying or staying neutral with streamwise distance. In an operational sequence, the following steps are followed:

(1) For a given frequency and at a streamwise station x one chooses a given ray, z/x, i.e., the disturbance propagation angle is fixed; then we have $(\partial\alpha/\partial\beta)_{\omega, Re}$ along that direction.

(2) Divide the β_r^* axes into intervals and identify these by their mid-points. Using the mid-point value, one iterates on the value of β_i^* which satisfies Eq. (8.41b).

(3) Once α and β are found for a particular disturbance propagation angle, the same procedure can be repeated for other angles to obtain the maximum amplification rate $\Gamma(x)$ for that x. This amplification rate can be integrated with respect to x to obtain the total amplification factor.

(4) As in two-dimensional calculations, here also the same exercise is repeated for different frequencies to find the critical frequency that leads to the highest integrated amplification rate.

The above method is similar to that described for two-dimensional flows in Figs. 3.5 and

3.6. However, we emphasize that the above approaches are all based on the signal problem and do not include the spatio-temporal approach as advocated in Chaps. 4 to 7. This is an area which has not been undertaken for linear studies. In comparison, nonlinear studies based on the Navier–Stokes equation require use of DRP methods, as described in Chap.2. This would require also using massive computational resources.

There is an alternative method called the envelope or the zarf method which was suggested in [58], where a neutral curve is defined and called the zarf. This is to mimic the procedure for two-dimensional flows to track constant frequency disturbances (signal problem), as shown in Fig. 3.2. While Re and ω are real, α and β are in general complex and only on the neutral curve will these be real. Despite this, there is a plethora of possible neutral curves in three-dimensional space. To circumvent this complication of non-uniqueness of such curves, in [58] it was suggested taking a special curve that the authors have termed the *absolute neutral curve*, which has the properties of α, β and $\frac{\partial \alpha}{\partial \beta}$ to be all real along the zarf. Before we close this section, we note that these complications can be easily avoided by resorting to full time dependent receptivity analysis instead, as noted for two-dimensional flows given in Sec. 3.6, using the Bromwich contour integral method developed in [319, 322, 349, 350].

8.9   Experimental Evidence of Instability on Swept Wings

The collected experimental evidence so far has established that there are many mechanisms in action for three-dimensional flows as compared to two-dimensional flows. For a swept wing, the flow can suffer primarily the following instabilities: (a) streamwise instability (SI), (b) cross flow instability (CFI), (c) leading edge contamination (LEC). It is noted that this classification scheme is an idealized one, as the instability mechanisms would not appear in isolation and more importantly the spatio-temporal route has not been addressed even today. Despite this, experiments have been carefully designed to highlight individual mechanisms one at a time, as described next.

The elaborate experiments in [35] using the NASA Ames 12 ft tunnel facility with very low turbulent intensity (0.02%), an *infinite* swept back wing of 1.22 meter chord was tested which had a sweep angle of 20^0 with a basic profile of NACA 64_2A015 aerofoil section. In Fig. 8.8 [11], the experimental locations of transition points are marked as discrete points along with theoretical predictions of the same by various criteria for the above three mechanisms. These criteria are discussed in a later section.

The x-axis in the figure is the Reynolds number based on chord (Re_c) and the nondimensional transition point location (x_{tr}) is shown along the ordinate. At low Reynolds numbers, transition occurs beyond mid-chord in the region of a positive (adverse) pressure gradient corresponding to streamwise instability (SI). However, beginning at $Re_c = 20 \times 10^6$, transition point shifts forward, settling rather closer towards the leading edge in the negative (favorable) pressure gradient zone. As explained before, this corresponds to cross flow instability (CFI). In [35], another set of experiment was performed using a wing with the higher sweep back angle of 50^0. These results are also shown in Fig. 8.8, which indicates no transition due to streamwise instability (SI). Instead, it was noticed that transition point moved more rapidly upstream towards the leading edge, as compared to the 20^0 sweep angle case. At $Re_c < 14 \times 10^6$, the transition point moved at the leading edge, which is attributed to leading edge contamination (LEC).

The implications of these mechanisms of instability were reported effectively in [11], with the help of flow past an infinite swept wing. The reason for undertaking studies with an infinite swept wing is not only for the ease of analysis it affords, but this is due to its canonical status, the same accorded to the Blasius boundary layer or to the Falkner–Skan boundary layer for two-dimensional flows. In Fig. 8.9, it is shown how one can use infinite swept wing results for any real finite swept wing. For a given section, an equivalent infinite swept wing is constructued with the local chord, while the sweep angle of the original wing is maintained by aligning the quarter chord.

In the figure at any spanwise station, an equivalent infinite swept wing is constructed with the sweep angle determined by the quarter chord line. This is akin to using a quasi-parallel approximation for predicting instability and transition for two-dimensional flow over an airfoil. For this reason, in the following we discuss infinite swept wing flow and its stability.

8.10   Infinite Swept Wing Boundary Layer

Referring to Fig. 8.3, for the flow past a swept wing we noted the sweep angle φ and the two axes systems for the flow near the leading edge. The free stream velocity Q_∞ is split into a component normal to the leading edge, U_∞, and a component parallel to the leading edge, W_∞. One associates the WFCS with the XYZ-axes system shown in the figure. Additionally, the coordinate system xyz is attached to the ESFCS. As noted earlier, $Y \simeq y$ and hence no distinction is made between these two. In these two coordinate systems, velocity $Q(y)$ inside the boundary layer is split into $U(y)$ and $W(y)$ in WFCS and $U_1(y)$ and $W_1(y)$ in ESFCS. It is these last components (streamwise and cross flow profiles) that are used in the stability equations, Eqs. (3.34) and (8.34). At the edge of the boundary layer (y_e), these components are given as $U(y_e) = U_e$, $W(y_e) = W_e$ and $U_1(y_e) = U_{1e} = (U_e^2 + W_e^2)^{1/2}$, $W_1(y_e) = 0$. The assumption of an infinite swept wing is expressed by an invariance of flow variables along the span. The mean flow is obtained from the solution of the following equations.

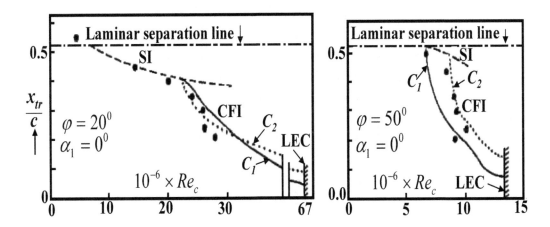

FIGURE 8.8
Experimental and calculated transition locations for flow over a swept wing. Experiments are as reported in [35] and corresponding calculations are as in [10]. SI indicates streamwise instability, CFI refers to cross flow instability and LEC indicates leading edge contamination. Note that there are two CFI empirical criteria following C_1 and C_2.

$$\frac{\partial U}{\partial X} + \frac{\partial V}{\partial Y} = 0 \tag{8.43}$$

$$U\frac{\partial U}{\partial X} + V\frac{\partial U}{\partial Y} = U_e\frac{dU_e}{dX} + \nu\frac{\partial^2 U}{\partial Y^2} \tag{8.44}$$

$$U\frac{\partial W}{\partial X} + V\frac{\partial W}{\partial Y} = \nu\frac{\partial^2 W}{\partial Y^2} \tag{8.45}$$

One notes that the first two equations are decoupled from the third and can be solved in that sequence. Also, these equations are the same as were given in Sec. 8.2, where we proved following Eq. (8.9) that at the free stream $W_e = W_\infty$.

To solve Eqs. (8.43) to (8.45), one needs the velocity profile at $x = 0$, which generally is obtained from the attachment line equations corresponding to an infinite swept wing. At any station, having obtained $U(y)$ and $W(y)$ from the solution of the above equations, one obtains the streamwise and cross flow profiles from

$$U_1(y) = U\cos\theta_0 + W\sin\theta_0 \tag{8.46a}$$

$$W_1(y) = -U\sin\theta_0 + W\cos\theta_0 \tag{8.46b}$$

where the angle between the X and x axes is obtained from $\tan\theta_0 = \frac{W_e}{U_e}$. If ψ represents the angle formed by the wavenumber vector and the external streamline, then one can define U_ψ as in Eq. (8.32) and solve the Orr–Sommerfeld equation given by Eq. (8.34). In [11], a complementary angle ϵ has been defined with respect to the cross flow direction, i.e.,

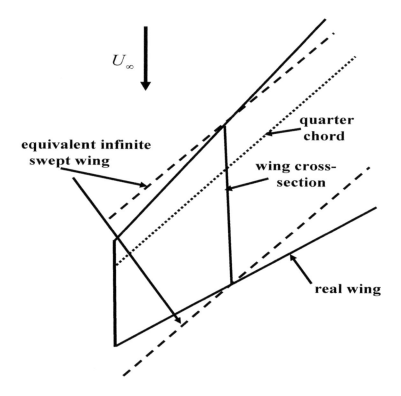

FIGURE 8.9
Equivalent infinite swept wing approximation for an arbitrary finite swept wing.

$\epsilon = \pi/2 - \psi$, and used to study the stability of the velocity vector projected in all directions from $\epsilon = 0^0$ to 90^0, and these profiles are obtained from

$$U_\epsilon = U_1 \sin \epsilon + W_1 \cos \epsilon \qquad (8.47)$$

By varying ϵ from 0^0 to 90^0, one moves continuously from the cross flow to the stream-wise profile. At a given streamwise location, a complete temporal calculation using Eq. (8.34) reduces the three-dimensional instability calculation to a sequence of two-dimensional calculations on the family of profiles U_ϵ. In Fig. 8.10, we show a set of such profiles for different values of ϵ from 0^0 (cross flow profile) to 90^0 (streamwise profile).

In these figures, a cross indicates the location of an inflection point for the profile, with the streamwise profile representing the typical two-dimensional boundary layer profile in the negative pressure gradient and hence its first derivative is monotonic without any inflection point. There is a range of low values of ϵ starting from the cross flow profile for which an inflection point is always present and represents vulnerability to inviscid instability established by the Rayleigh–Fjørtoft theorem. Of all the angles in this range, one would be interested in one particular angle (ϵ_I) for which the inflection point occurs at a height where the local mean velocity is zero. This profile is termed the critical profile in [11].

Figure 8.11 shows representative stability diagrams for $\epsilon = 0^0$, ϵ_I and 90^0. The Reynolds number Re is defined by the same reference length x.

One can draw the following conclusions from the frames in Fig. 8.11:

(i) The streamwise profile is very stable, as one notes a very high critical Reynolds number. Also, as Re tends to large values, the neutral curve collapses, indicating no unstable frequencies, and thus the observed instability is caused by viscous action only at finite Reynolds numbers.

(ii) However, due to the presence of inflection points in the velocity profiles, the other two cases are highly unstable, with very low critical Reynolds number. Also, the amplification rates are significantly higher and there is always a wide range of unstable waves at very large Reynolds numbers, implying the instability is dominated by an inviscid mechanism which does not go away, even in the limit of $Re \to \infty$.

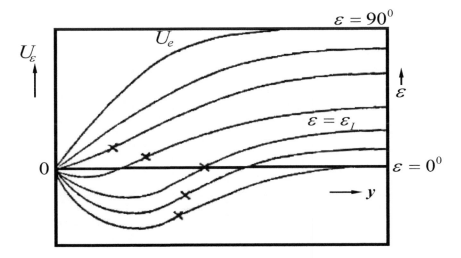

FIGURE 8.10
Projected velocity profiles according to Eq. (8.47) [11].

(iii) The critical profile turns out to be the most unstable among the three profiles shown and more importantly, one can see that zero frequency waves are amplified, which corresponds to standing waves. Such unstable standing waves are seen for a range of ϵ around ϵ_I, say $\epsilon = 2$ to 6 degrees. Note that the experiments in [13] showed the presence of streaks in wall visualization, which are the standing wave patterns amplified by the profiles for values of ϵ in the neighborhood of ϵ_I.

Thus, the results from this infinite swept wing study clearly explain the observed experimental phenomenon at large enough chord Reynolds number for swept wing flow, near the leading edge in the negative pressure gradient region. While the above description tells us about the onset of instability near the leading edge of an infinite swept wing, one may also like to inquire about the downstream development of such initial unstable disturbances.

A description of it can be followed by noting that the equation of the external streamline is given by

$$\frac{dz}{W_e} = \frac{dx}{U_e}$$

Since for the infinite swept wing flow, $W_e = W_\infty$, hence the equation of the streamline can be alternately written as $\frac{dz}{dx} = \frac{W_\infty}{U_e}$, which in turn implies that the streamline curvature $(\frac{d^2 z}{dx^2})$ changes sign at an abscissa location (x_M), where U_e is maximum (i.e., where $\frac{dU_e}{dx} = 0$). Thus, x_M is also the location where the pressure gradient changes sign. That in turn implies that across x_M the pressure gradient changes from a negative pressure gradient to a positive pressure gradient. This is demonstrated in Fig. 8.12, showing the evolution of laminar flow developing over a swept wing.

In this figure, β_0 is the angle between the external streamline and the wall streamline directions. Near the leading edge, the negative pressure gradient displays a full cross flow that increases with height initially and then decreases back to zero at the edge of the shear layer, with nowhere becoming negative. This causes the external streamline to curve inwards in the first quadrant of the xz-plane. However, as x_M is approached, the pressure gradient reduces to zero and that makes the cross flow component also weaker. This causes the wall streamline to deviate towards the streamwise direction, as indicated by the reduction of β_0. When the longitudinal pressure gradient changes sign, somewhere downstream of that location the cross flow velocity component $W_i(y)$ itself can switch sign, close to the wall,

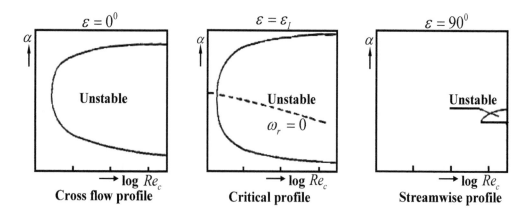

FIGURE 8.11
Calculated neutral curves for some of the representative velocity profiles shown in Fig. 8.10 [11].

giving rise to an S-shaped velocity profile. Thus, the S-shaped cross flow profile occurs towards the end of the negative pressure gradient zone and may not necessarily be the most critical profile. This aspect is highlighted again while discussing the Falkner–Skan–Cooke profile in the next section. When the positive pressure gradient is very strong, the cross flow velocity can become fully negative inside the shear layer. It is in this region, $U_1(y)$ can exhibit an inflection point which can imply strong streamwise instability in that part of the flow. Once again, this region is dominated by streamwise instability and not by cross flow instability.

In Sec. 8.5, we noted the equilibrium flows expressed in the transformed plane, with Eqs. (8.17) and (8.18) providing equations for an infinite swept wing. If one considers the external flow description given by Eq. (8.25), then by the additional choices of $r = s = n = 0$, these equations further reduce to the Falkner–Skan–Cooke equations given by Eqs. (8.28) and (8.29). Fixing the exponent m, the Falkner–Skan–Cooke profile provides a canonical velocity profile that can be viewed as fulfilling the same role as the Blasius profile did for two-dimensional flow instability, and this is studied next.

8.11 Stability of the Falkner–Skan–Cooke Profile

This is a similarity profile describing an infinite swept wing and the attachment line flow and is helpful in explaining the passage from streamwise to cross flow instability. For the infinite swept wing, the outer inviscid velocity components in the normal and parallel to the leading edge are given by $U_e = kx^m$, $W_e = W_\infty$. In this description, we use the following similarity variables

$$\eta = \left(\frac{m+1}{2}\right)^{1/2}\left(\frac{U_e}{\nu x}\right)^{1/2} y$$

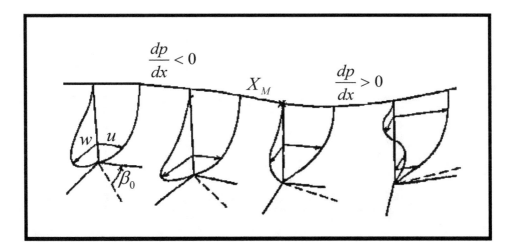

FIGURE 8.12
Sketch of a laminar boundary layer developing over a swept wing. The angle β_0 represents the orientation of external streamline with respect to wall streamline.

$$\frac{U}{U_e} = \bar{F}'(\eta) \text{ and } \frac{W}{W_e} = \bar{G}(\eta) \tag{8.48}$$

With the above transformations, boundary layer equations reduce to the following ordinary differential equations

$$\bar{F}''' + \bar{F}\,\bar{F}'' + \beta_h(1 - \bar{F}'^2) = 0 \tag{8.49}$$

$$\bar{G}'' + \bar{F}\,\bar{G}' = 0 \tag{8.50}$$

where $\beta_h = \frac{2m}{m+1}$ and the attachment line flow can be further obtained corresponding to $m = \beta_h = 1$. The angle between the external streamline and the normal to the leading edge is $\theta_0 = \tan^{-1}\frac{W_e}{U_e}$ and so the streamwise and cross flow profiles are given by

$$U_1 = U\cos\theta_0 + W\sin\theta_0 \text{ and } W_1 = -U\sin\theta_0 + W\cos\theta_0$$

Note that

$$\cos\theta_0 = U_e/(U_e^2 + W_e^2)^{1/2}; \; \sin\theta_0 = W_e/(U_e^2 + W_e^2)^{1/2}$$

and

$$U_{1e} = (U_e^2 + W_e^2)^{1/2}; \; W_{1e} = 0$$

Thus

$$U_1/U_{1e} = \bar{F}'\cos^2\theta_0 + \bar{G}\sin^2\theta_0 \tag{8.51}$$

$$W_1/U_{1e} = (\bar{G} - \bar{F}')\sin\theta_0\cos\theta_0 \tag{8.52}$$

Thus, it is possible to construct streamwise and cross flow profiles depending upon two

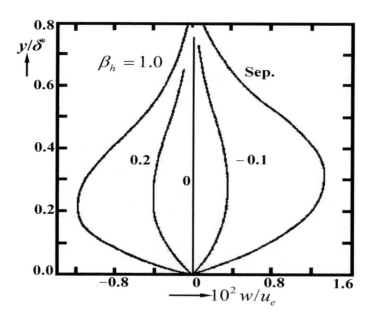

FIGURE 8.13
Cross flow velocity profile obtained from solution of the Falkner–Skan–Cooke equation [11].
The extreme right curve is for the case of β_h for which separation is incipient.

quantities: (i) the pressure gradient parameter β_h and (ii) the cross flow parameter θ_0. Note that W_1 is maximum for $\theta_0 = \pi/4$. Also, for $\beta_h = 0$, from Eqs. (8.48) and (8.49) one notes that $\bar{F}' = \bar{G}$, which leads one to the Blasius profile: $U_1/U_{1e} = \bar{F}'_0$ and $W_1/U_{1e} = 0$. Figure 8.13 shows the cross flow velocity profile for $\theta_0 = \pi/4$ and for different β_h.

For $\beta_h < 0$, the whole cross flow profile changes sign. The Falkner–Skan-Cooke (FSC) profiles cannot represent S-shaped profiles. As discussed in the previous section, the S-shaped profile and fully reversed profiles occur in the downstream location, where cross flow instability is not the most critical event. In [11, 213], stability properties of the FSC profiles have been reported. Of specific interest, are the results shown in Fig. 8.14, where the critical Reynolds number is plotted as a function of β_h for the velocity profile with the cross flow parameter $\theta_0 = \pi/4$, only for zero frequency disturbances.

The dotted line in Fig. 8.14 is for the two-dimensional Falkner–Skan profile ($\theta_0 = 0$) and it is noted that for all adverse pressure gradient flows ($\beta_h < 0$), the FSC profiles are more stable than the two-dimensional Falkner–Skan profile. However, there is a critical $\beta_h = 0.07$, above which three-dimensional flow is more unstable than the corresponding two-dimensional flow. This is another instance, in which Squire's theorem is violated. The Reynolds number is defined with respect to x in these calculations. This also shows that there is a possible existence of stationary waves for three-dimensional flows, as compared to two-dimensional flows, which do not support stationary waves. This aspect is discussed next for three-dimensional flows, mainly from the point of view of unstable stationary waves.

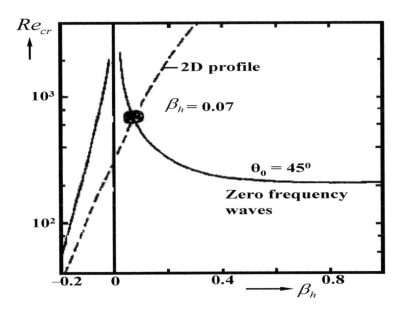

FIGURE 8.14
Critical Reynolds number variation with pressure gradient parameter, β_h, for zero frequency disturbances. The dotted line shows the corresponding two-dimensional critical Reynolds number for the Falkner–Skan profile [11].

8.12 Stationary Waves over Swept Geometries

There is sufficient experimental and theoretical evidence for the existence of unstable stationary waves for three-dimensional flows. This was clearly shown for the Falkner–Skan–Cooke profile in Fig. 8.14. In [13, 225], experimental results have been reported for a swept wing with an ONERA D profile; special features of the model are also discussed and shown in Fig. 8.5. The sweep angle for the model was 40^0, while the angle of attack was -8^0 and the wing chord was 35cm. These angles were decided so that the external velocity normal to the leading edge (U_e) is always accelerating over all possible x/c, as shown in Fig. 8.15, for $Q_\infty = 81$ m / s.

The large acceleration over the top surface is obtained via the large sweep angle, as well as the negative angle of attack. The measured spanwise variation of the external velocity component is shown for different streamwise stations in Fig. 8.16.

When the measured velocity is plotted at a height very close to the wall and in the laminar part of the flow, one notes spanwise waviness with a spanwise wavelength corresponding to the streak spacing noted in the wall visualization (shown in Fig. 12 of [11]). In the transitional part of the flow with an onset at $x/c = 0.30$, the flow evolution in time is noted to be completely chaotic. The external flows displayed in Fig. 8.16 are time averaged values. Also, time averaged flow velocity in the transitional part of the flow is higher than the measured flow in the turbulent part of the flow. Another experiment was reported in [11], for a lower free stream speed of 48 m/s, to make more accurate measurements. At this lower speed, the wavelength of the streaks (λ^*) increased, with a thicker boundary layer and transition location moving back to $x_T/c = 0.85$. In Fig. 8.17, experimental and theoretical estimates of the spanwise streak spacing are compared using the data from [11].

It is noted that λ^*/δ practically remains constant with a value close to 4. An increase in λ^* caused a reduction in the number of streaks, with sublimation technique results showing that some of the streaks broaden and coalesce, while others vanish. Stability calculations explain this behavior, as seen from Fig. 8.18, displaying temporal amplification rates of stationary waves.

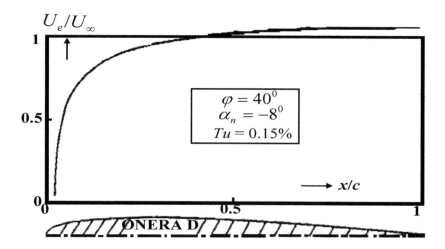

FIGURE 8.15
Edge velocity over a swept wing with an ONERA D profile [11].

The abscissa of this figure is the angle ϵ and the results are for a station, $x/c = 0.69$, with the curve parametrized with λ^* in mm. The results show a narrow range of ϵ (in the

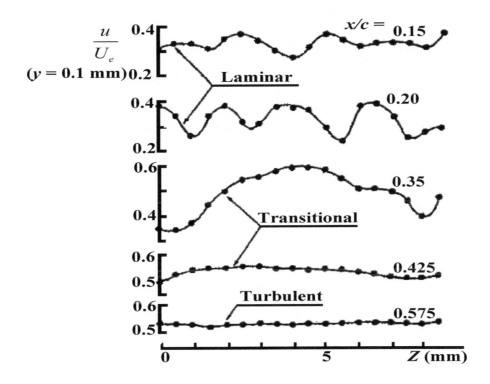

FIGURE 8.16
Streamwise velocity variation with spanwise coordinate for the experimental results reported in [13].

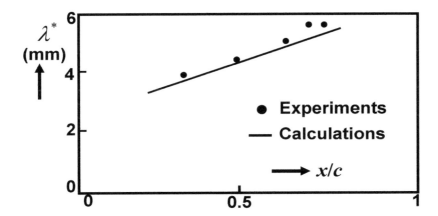

FIGURE 8.17
Streamwise variation of stationary streak spacing in the spanwise direction [11].

range of 1.5^0 and 3.3^0). However, the wavelength range is quite large. In the experiments of [306] on a swept flat plate the flow acceleration was created by a wall bump on the top wall of the tunnel. For a sweep angle of 25^0 and a free stream speed of 10 to 14 m/s, the stationary streaks were rendered visible on the wall by the sublimation technique and the streaks have a wavelength of 1 cm, which is also similar to the experimental observations in [13, 225] and calculated results from linear stability theory. It is interesting to note that in [306], the authors found a dominating structure with a wavelength of 5 mm from hot wire measurements away from the wall. Reed [293] tried to explain this variation arising due to streamwise and cross flow wave interaction via a secondary instability mechanism. Poll [278], experimenting on a swept cylinder, also reported measurements on the characteristics of streaks. He found that the streak direction was close, but not properly aligned with the critical profile direction, ϵ_I. Whether stationary streaks are singularly responsible for flow transition or not can be conclusively found out if measurements on traveling waves are obtained and the relative magnitudes of amplifications compared.

8.13   Traveling Waves over Swept Geometries

While stationary waves are spectacularly visible, traveling waves are found to be more unstable. Linear stability calculation results are shown in Fig. 8.19 and it is seen that the most unstable traveling waves correspond to $f^* = 400$ Hz.

The figure indicates a large band of frequencies to be unstable, while the experiments reveal the presence of only a few dominant frequencies. For example, in the experiment of [10], at $x/c = 0.4$ a dominant frequency of 70 Hz was noted, with a fluctuation amplitude of 0.20 of the external velocity that is about twice the value that one would measure in

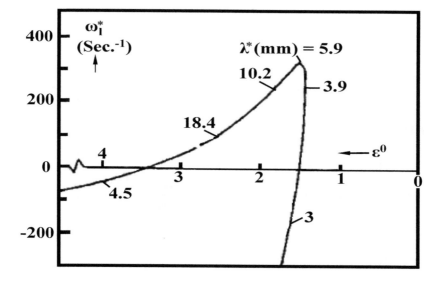

FIGURE 8.18
Temporal growth rates of stationary streaks at a fixed streamwise station for different flow orientations [11].

the turbulent part of the flow. At $x/c = 0.60$, the fluctuations are less coherent and are of smaller amplitude, with high frequency oscillations superimposed on larger amplitude, low frequency oscillations. Similar measurements and patterns were also recorded in [278], but at a different dimensionless frequency. Arnal [11] raised a few questions about (a) the possible absence of those frequencies in the experimental facility which were found to be most unstable by linear theory, (b) the possible effects of pressure gradients and (c) traveling waves created as secondary instability of the primary instability due to stationary waves and these still remain as open questions that can be unequivocally addressed by linear and nonlinear receptivity analyses.

8.14 Attachment Line Problem

Attachment line instability as a problem of bypass transition has been discussed in detail in Chap. 4, with respect to the work in [334]. Here we address the corresponding linear stability problem. The attachment line is a particular streamline that divides the flow into two, with one branch following the upper surface and another branch following the lower surface. The situation is depicted in Fig. 8.20 for flow past a swept cylinder.

In the ESFCS, x coincides with the attachment line and the edge velocity is given by $U_{1e} = Q_\infty \sin \varphi$. The attachment line boundary layer profile can be obtained from the FSC profile by setting $m = \beta_h = 1$ and $\theta_0 = \pi/2$ in Eqs. (8.51) and (8.52). This yields $U_1/U_{1e} = \bar{G}$, $W_1 = 0$. Similarly, one simplifies the governing equations as

$$\bar{F}''' + \bar{F}\,\bar{F}'' + (1 - \bar{F}'^2) = 0 \tag{8.53}$$

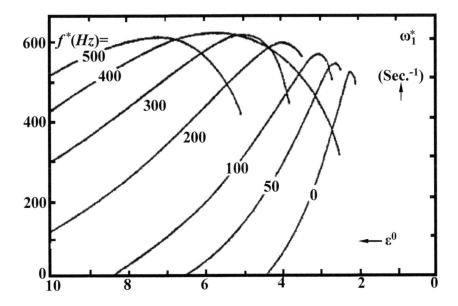

FIGURE 8.19
Temporal growth rates of traveling wave disturbances [11].

$$\bar{G}'' + \bar{F}\,\bar{G}' = 0 \tag{8.54}$$

To solve this equation, one can introduce the independent variable

$$\eta = y\sqrt{\frac{k}{\nu}} \text{ where } k = \left(\frac{dU_{1e}}{dx}\right)_{x=0}$$

From the continuity equation, one gets

$$\frac{V_1}{U_{1e}} = -(k\nu)^{1/2}\bar{F}(\eta)$$

One solves Eq. (8.53) first for $\bar{F}$ and then the second equation for $\bar{G}$. From these solutions, one can obtain the boundary layer integral properties as

$$\delta^* = \int_0^\infty \left(1 - \frac{U_1}{U_{1e}}\right) dy = 1.026\left(\frac{\nu}{k}\right)^{1/2} \tag{8.55}$$

$$\theta = \int_0^\infty \frac{U_1}{U_{1e}}\left(1 - \frac{U_1}{U_{1e}}\right) dy = 0.404\left(\frac{\nu}{k}\right)^{1/2} \tag{8.56}$$

Thus the shape factor is given by $H = \frac{\delta^*}{\theta} = 2.54$. As discussed in Chap. 3, for this lower value of the shape factor the critical Reynolds number is higher at 230 (based on momentum thickness) as compared to 200 for the Blasius boundary layer, for which $H = 2.59$. The linear theory shows the flow to be more stable, as opposed to what is observed often in experiments and in real flight conditions. This aspect has been clearly explained in Chap. 4 following [334].

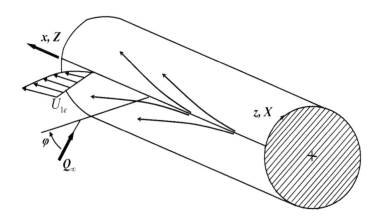

FIGURE 8.20
Schematic of attachment line flow for an infinite swept cylinder [11].

8.15 Empirical Transition Prediction Method for Three-Dimensional Flows

Here one begins with the assumption that transition will be caused by any one of the dominant mechanisms discussed in this chapter. Thus, transition criteria are applied for each of these mechanisms and it is assumed that the boundary layer will cease to be laminar when any one of these is satisfied.

8.15.1 Streamwise Transition Criterion

Streamwise velocity profiles are closer to two-dimensional flow profiles and hence the criterion would be identical when applied along the external streamlines. One can apply any one of the methods described for two-dimensional flows or the modified method due to Granville, as given in [13].

8.15.2 Cross Flow Transition Criteria

An early attempt in providing a transition criterion due to cross flow instability was in [262, 263]. However, it overestimates the instability and is not used anymore. Beasley [25] developed a criterion based on a cross flow profile W_1 and a Reynolds number defined from $\delta_2 = \int_0^\infty \frac{W_1}{U_{1e}} dy$ as $Re_{\delta 2} = \frac{U_{1e}\delta_2}{\nu}$. Beasley [25] proposed that transition would occur whenever $Re_{\delta 2} \geq 150$. While this is a simple criterion, the following separate criteria proposed in [13] are found more satisfactory.

(a) **C1 Criterion:** The above criterion of Beasley ($Re_{\delta 2} \geq 150$) does not involve any parameter depending upon streamwise location, and hence its applicability is limited. It has been reported in [13], by comparing experimental data with a computed laminar boundary layer, that transition does not depend uniquely upon $Re_{\delta 2}$. Better results were obtained by correlating a transition Reynolds number with a parameter linked to the streamwise velocity profile. When shape factor (H) is used to represent the latter, the following empirical relation represents the experimental data quite well.

$$[Re_{\delta 2}]_{tr} = \frac{300}{\pi} \tan^{-1}\left[\frac{0.106}{(H_{tr} - 2.3)^{2.05}}\right] \tag{8.57}$$

This criterion remains valid for $H < 2.7$, beyond which the streamwise criterion is found to be important. Figure 8.21 shows general agreement with the ensemble of experimental data.

(b) **C2 Criterion:** This method uses linear stability theory along with Stuart's theorem. To study the stability along a perturbation direction, making an angle ϵ with the cross flow direction, we define a velocity parameter

$$\frac{U_\epsilon}{U_{\epsilon e}} = \frac{U_1}{U_{1e}} + \frac{W_1}{U_{1e}} \cot \epsilon \tag{8.58}$$

where $U_{\epsilon e} = U_{1e} \sin \epsilon$ with $\epsilon = 0$ and π excluded. Similarly, define a parametric displacement thickness and a Reynolds number by

$$\delta_{1\epsilon} = \int_0^\infty \left[1 - \frac{U_\epsilon}{U_{\epsilon e}}\right] dy \tag{8.59}$$

$$Re_{\delta_{1\epsilon}} = \frac{U_{\epsilon e}\delta_{1\epsilon}}{\nu} = Re_{\delta_1} \sin \epsilon + Re_{\delta_2} \cos \epsilon \tag{8.60}$$

Next, one describes the stability of the U_ϵ profiles and obtains a $(Re_{\delta_{1\epsilon}})_{cr}$ by stability analysis. For any experimental case, one can define another function

$$g(\epsilon) = \frac{(Re_{\delta_{1\epsilon}})_{cr}}{Re_{\delta_{1\epsilon}}} \tag{8.61}$$

Obviously if $g(\epsilon) > 1$, then we have a stable condition. However, if $g(\epsilon) < 1$, then the flow in that direction is unstable. For a three-dimensional flow at any given abscissa, g depends on ϵ and on the Reynolds number based on chord (Re_c). Indeed, $Re_{\delta_{1\epsilon}}$ varies as $\sqrt{Re_c}$, whereas $(Re_{\delta_{1\epsilon}})_{cr}$ is invariant. Thus, if $g_1(\epsilon)$ and $g_2(\epsilon)$ correspond to Re_{c1} and Re_{c2}, then

$$\frac{g_1}{g_2} = \sqrt{\frac{Re_{c2}}{Re_{c1}}} \tag{8.62}$$

That is, when the chord Reynolds number increases, g decreases and the range of unstable directions widens.

8.15.3   Leading Edge Contamination Criterion

In Sec. 8.14, we found the momentum thickness of the attachment line boundary layer to be given by Eq. (8.56):

$$\theta = \int_0^\infty \frac{U_1}{U_{1e}}\left(1 - \frac{U_1}{U_{1e}}\right) dy = 0.404\left(\frac{\nu}{k}\right)^{1/2}$$

where $k = \left(\frac{dU_{1e}}{dx}\right)_{x=0}$. Therefore, the Reynolds number based on the momentum thickness is given by

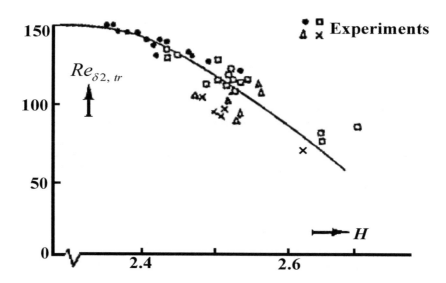

FIGURE 8.21
Correlation of experimental transition location point with the calculation using the C1 criterion [11].

$$Re_\theta = \frac{Q_\infty \theta \sin \varphi}{\nu} \qquad (8.63)$$

Experimentally, it has been noted that transition by leading edge contamination occurs when $Re_\theta > 100$. This can be used as the criterion for leading edge contamination. In Fig. 8.8, various estimates of transition given by the criteria discussed above are shown for the experimental data of [35] for the two sweep back angle cases. The two cross flow criteria give results which are close to experimental values, as well as to themselves. The lower sweep angle case showed a smooth passage from streamwise to cross flow instability with the increase of the chord Reynolds number. Streamwise instability occurred where the pressure gradient was positive, while cross flow instability occurred in a negative pressure gradient region. There are no available experimental data by which the leading edge contamination criterion could be tested. For the higher sweep angle case, the instability is due to cross flow only for Re_c greater than 6 million — as seen for the $\varphi = 50^\circ$ case shown in Fig. 8.8. For the straight wing ($\varphi = 0^\circ$) case in Fig. 8.8, instability is seen to be due to streamwise instability for $Re_c \leq 20 \times 10^6$.

9

Analysis and Design of Natural Laminar Flow Airfoils

9.1 Introduction

To improve aircraft performance one can gainfully reduce drag, which will translate into better range and endurance, as well as economical operation. Drag breakdown of a subsonic transport airplane is shown in Fig. 9.1 [426], highlighting the importance of fuselage drag after the flow over the wing is rendered laminar by some passive means. This can bring down the wing drag from roughly 27% of total viscous drag to 15%. This seems feasible, as [428] has noted that the profile drag of 0.0085 for a fully turbulent airfoil at a Reynolds number of 10^7 comes down to 0.0010 if the flow is kept completely laminar. Thus, it is imperative to understand how such laminar flow can be achieved over an aircraft wing. To begin with, one must understand the airfoil drag contribution and how this can be pegged to a lower level. In this regard, a natural laminar flow (NLF) airfoil maintains extensive laminar flow solely by means of a favorable pressure gradient. This is a passive means to control the flow over an airfoil, as opposed to various possibilities investigated via active control. From the desired performance features of a regional aircraft, one can infer the desired aerodynamic characteristics of an NLF airfoil for such applications. Such an aircraft would have a chord Reynolds number in the range of about 4 to 17 million and a cruise Mach number in the range of 0.4 to 0.6.

Even when only the aerodynamic performance of an NLF airfoil is considered, one must factor in all the different requirements on the airfoil for all segments of flight. That is, one must consider not only the cruise performance, but also the characteristics of the airfoil during landing, take off and climb. The latter may be of significant importance for short haul operations. Such design considerations, in the vocabulary of design and optimization, would be termed multi-objective optimization. In addition, one must also consider various constraints arising from considerations of structural design, manufacturing practices, other geometric restrictions and constraints from aero-elasticity. It is worth noting that such multi-objective, multi-constrained optimization may lead to a conflict for the objective function.

There are two main approaches in airfoil/wing design: direct optimization and inverse design methods (see [186]). In the former, the airfoil/wing design is performed via numerical optimization of an objective function (say, minimization of drag) by linking an analysis tool (say a fluid dynamic solver, which could be as simple as a panel method code to a sophisticated CFD solver solving the Navier–Stokes equation) with an automated design procedure. The design variables are the wing/airfoil geometry parameters they are iterated in the design loop while satisfying the constraints. Some early examples of direct methods are to be found in [50] and [147]. In those formulations, the basic shape was prescribed by a baseline profile and shape functions are superposed linearly over it. However, these methods prove to be expensive when the number of shape functions increased.

In the inverse approach, the design variables are parameters defining physical variables like the pressure distribution. Some of the early successful design methods of this class are reported in [121, 430]. Sometimes this results in shapes which are not physically meaningful or difficult to manufacture. To circumvent these, the authors in [77] have proposed a numerical optimization scheme that specifically provides a method for parameterizing a target pressure distribution.

We also mention that there are other methods of aerodynamic design; notable among them is the use of control theory pioneered in [161]; by genetic algorithm (GA) as in [424] and by the Newton–Krylov algorithm for aerodynamic design using the Navier–Stokes equation for multi-objective optimization, as in [247].

Before one proceeds further, the nomenclature schemes used to identify airfoils in terms of important aerodynamic and geometric design parameters are specified next.

9.2   Airfoil Nomenclature and Basic Aerodynamic Properties

Although the first patented airfoil is due to H. F. Philips in 1884 (see [7]), in 1930s, the National Advisory Committee for Aeronautics (NACA) started systematic airfoil designs based on experiments. This started with the four-digit series airfoils, which are characterized by thickness and camber distribution along the mean chord line. Thereafter, five-digit series airfoils came into existence which additionally included the design criterion of a specific lift coefficient indicated in tenths by the first digit of the nomenclature multiplied by 3/2.

One of the most widely used family of NACA airfoils is the six-series laminar flow airfoils, developed during the Second World War. Laminar flow was obtained by keeping the minimum pressure point as aft as possible, preferably without flow separation during

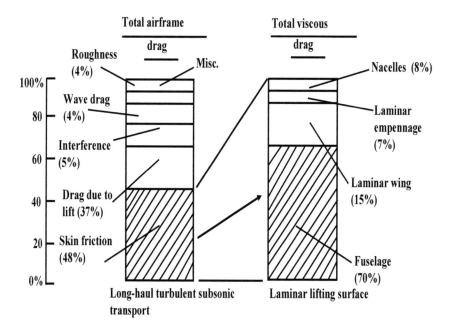

FIGURE 9.1
Drag breakdown of a subsonic transport airplane.

the pressure recovery stage in the rear of the airfoil. In the nomenclature, the first digit identifies the series; the second digit gives the location of minimum pressure in tenth of chord from the leading edge; the third digit is the design lift coefficient in tenths and the last two digits give the maximum thickness in hundredths of chord. For example, in the NACA 67-314 airfoil, the first digit 6 indicates the series, the minimum pressure occurs at 0.7c for the basic symmetric thickness distribution at zero lift, the design lift coefficient is 0.3 and the airfoil is 14% thick. Note that this section served as the basis for the design of NLF(1)-0414F, a specially designed section for NLF applications in the 1980s [429]. We should also note that there are other variations of airfoil nomenclature systems that will be discussed as we define the airfoils. Abbott & Doenhoff [1] is a repository of airfoil data of the classic NACA airfoils up to 1949. It contains coordinates of various sections and their properties based on an extensive experimental data set. In the summary of airfoil data in this reference, there is one NACA-67 series airfoil: NACA 67,1-215. "Even for this section, there are no data above $Re = 6 \times 10^6$, presumably because the higher Reynolds number data produced little laminar flow" [429].

We emphasize here that the NACA series airfoils designed during late 1930s and early 1940s were based exclusively on experimental data. During the 1970s, NASA designed a series of low speed airfoils with superior performance as compared to the NACA airfoils. These were designed computationally by the panel method and boundary layer calculations. Subsequently, wind tunnel tests were conducted to verify the design. These resulted in the first such section for general aviation, called GA (W)-1, which has been redesignated the LS(1)-0417 airfoil. The abbreviation LS stands for low speed, 04 stands for the design cruise lift coefficient $c_l = 0.4$ and it is a 17% thick aerofoil. It is noted that GA (W)-1 is similar to supercritical airfoils designed by R. Whitcomb in 1965 for high speed applications. The earliest general aviation airfoils are derivative of this and the letter W in GA (W) is a recognition of Whitcomb's original contribution to this. In Fig. 9.2, the shape of this 17% thick airfoil is shown; it has a design lift coefficient of 0.4. This airfoil is characterized by a large leading edge radius (0.08c) in order to flatten the usual peak in pressure coefficient near the nose. Also note the large camber near the trailing edge forming a cusp, accounting for the aft loading of the airfoil.

Other airfoils in this series are the LS(1)-0409 and LS(1)-0413 sections [221, 222]. Compared to standard NACA airfoils with the same thickness XX, these new LS(1)-04XX airfoils have (i) approximately 30% higher $c_{l,max}$, (ii) approximately 50% increase in L/D at a lift coefficient of 1.0 (see Anderson [7]), a value of lift coefficient typically used during climb.

We have already noted that NLF airfoils are designed using computers, while the "6-digit" series NACA airfoils were based on wind tunnel experiments. The experiments are costly, time consuming and sometimes restrictive in application for a particular parameter range. The details of the early researches can be found in [1, 158, 269, 399]. We have already noted that NACA 67,1-215 was designed with data available for Reynolds numbers of up to six million only. Thus, this is one of the reasons that for powered aircrafts, NLF airfoils

FIGURE 9.2
The section profile of the LS(1)-0417 airfoil (formerly known as the GA(W)-1 airfoil).

were not based on six digit series airfoils. However, the main reasons the six digit series airfoils were not used are related to the difficulties of manufacturing and operations. The manufacturing techniques of the 1930s and 1940s used rivets and surface waviness of the finished wing surfaces gave rise to a much larger drag value than expected.

In Fig. 9.3, the drag polar and pitching moment curves are shown for NACA $66_2 - 015$ (a symmetric section), taken from [1]. We note that the drag polar shows the well known *drag bucket*, implying a very narrow range of lift coefficient for which drag shows optimum behavior. The reason for this is explained shortly. However, attention is drawn to the fact that the *drag bucket* disappears when the airfoil is tested with standard roughness (as practiced in NACA and explained in [1]) for a Reynolds number of six million. The standard roughness relates to applying a strip with grit near the leading edge of the airfoil.

One also notices the adverse effect of the roughened airfoil by noting that there is severe

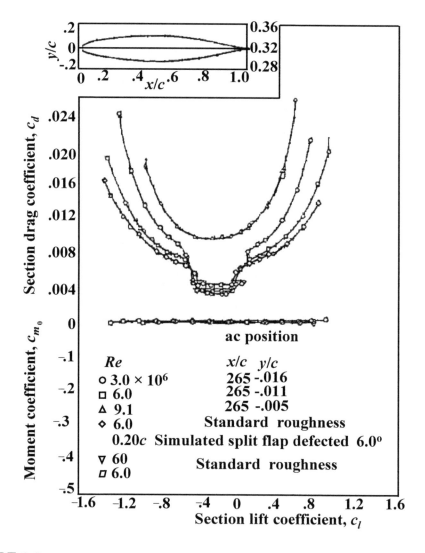

FIGURE 9.3
Drag polar and pitching moment coefficient of the NACA 66_2-015 airfoil.

degradation of $c_{l_{max}}$ with the standard roughness (as defined by NACA and given in [1]) in Fig. 9.3.

The presence of large rivets and surface roughness also cause non-optimum drag behavior with an increase in Reynolds number. This was clearly noted in the series of flight test experiments performed in RAE and reported in [130, 132, 375]. These authors flight tested the King Cobra aircraft with a smoothed and non-wavy wing surface. Viken [428] has explained that *roughness induces streamwise disturbance vortices, which add a crossflow component to the mean boundary layer flow and tends to distort the two-dimensional Tollmien–Schlichting (T-S) disturbances three dimensionally. Once these oscillations are distorted three-dimensionally they amplify much quicker to the critical level which causes transition.* This effect necessitates the minimization of all avoidable surface roughness, whether from poor construction techniques or from contamination during use (insects, dirt etc.). Also, wing surfaces made of metal by conventional techniques tend to buckle, which sets up small scale waviness on that part of a wing surface subjected to compression loads. This aspect was also confirmed in the above flight experiments at RAE, which recorded a large increase of the width of the *drag bucket* when the amplitude to wavelength ratio of the airfoil surface was reduced from 0.00125 to 0.0005. The upper end of the drag rose from 0.25 to 0.48 for the increased lift coefficient from cruise to climb conditions. It is reassuring to note that with advanced composite materials one can fabricate wing surfaces to reduce surface roughness (removing these due to rivets, joints and surface discontinuities) and surface waviness which are inherent with metal wings.

The reason for the *drag bucket* in the six-digit NACA airfoil will be apparent when one looks at the pressure distribution. A typical calculated pressure distribution is shown in Fig. 9.4 for the NACA 67,1-015 airfoil.

As this is a symmetric airfoil, the middle curve corresponds to both the upper and

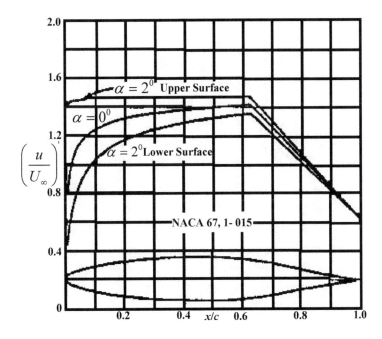

FIGURE 9.4
Pressure distribution over the NACA 67,1-015 airfoil.

lower surfaces for zero angle of attack. The other case corresponds to an angle of attack for which the lift coefficient is 0.2. One notices a relatively flat favorable pressure gradient up to $x = 0.62c$, for the zero angle of attack case. For the lifting case, on the upper surface the pressure gradient flattens further and makes the flow susceptible to earlier transition — explaining the narrowing of the *drag bucket*. The other reason for the abrupt increase in drag coefficient at higher angles of attack is due to the pressure recovery beyond $x/c = 0.62$ being linear. Such linear pressure recovery is not good for preventing flow separation of the corresponding turbulent flow following flow transition.

The NACA six-digit airfoils were partly designed by using Theodorsen's method (see [1] for details). One of the good aspects of the six-digit series airfoils is their very low pitching moment. This near zero pitching moment results from very low aft chord loading of the airfoil. This is at the cost of a bad design pressure recovery feature of the airfoil.

Thus, the six-digit airfoils could be used for a single design point at low Reynolds numbers. The change of philosophy in redesigning NLF airfoils in NASA in the 1970s and 1980s has been well explained in [378, 379] and is explained below with respect to Fig. 9.5 showing a typical drag polar of the NACA six-series airfoil (shown by the dotted line) and its desired improvement (shown by the solid line).

For the shown drag polar in the figure, one can sketch the corresponding desired pressure distribution over the airfoil for different lift coefficients. Point A corresponds to cruise condition ($c_l = 0.4$) for which one can expect the airfoil to experience the lowest profile drag. In NACA designs of that era a constraint was put on the upper surface favorable pressure gradient to be sustained up to $(x/c)_{us} \leq 0.3$, while the lower surface pressure gradient was kept unconstrained. The lowest drag point dictates the choice of camber of the airfoil and if this is pushed downwards, i.e., for a lower c_l, then the maximum lift point would also be lowered proportionately from point C in Fig. 9.5. The change in design philosophy to multi-point optimization for the airfoil is indicated by the choice of point B (at $c_l = 1.0$), as compared to a single design point of older NACA six-series designs. This lift coefficient corresponds

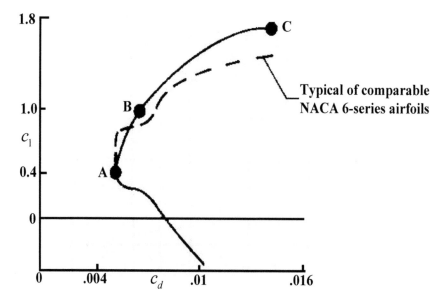

FIGURE 9.5
Sketch of drag polar for an ideal NLF airfoil [378].

to maximum climb rate. This compromise is necessary to have the transition point on the upper surface move slowly and steadily towards the leading edge with increasing c_l. This feature of smoothly varying transition point with increasing angle of attack was brought about by changing the leading edge shape of the airfoil. Typical pressure distributions for operational points A and B are sketched in Fig. 9.6.

There are certain features of the pressure distribution in Fig. 9.6 that are worth highlighting. For point A, the favorable pressure gradient on the upper surface is determined by the applied constraint. Beyond $0.30c$, a short region of adverse pressure gradient is purposely built in to promote efficient transition and further aft a steeper concave pressure recovery is introduced which produces lesser drag, as compared to six-digit series airfoils with linear pressure recovery, shown in Fig. 9.4. Pressure recovery in the aft portion of the airfoil upper surface where the flow is turbulent has been a subject of detailed research in the past [391] and we will come back to this discussion later. It is also necessary to talk about the lower surface pressure distribution as that determines many things — although it is treated as unconstrained. While such a pressure distribution helps in the quest for lower profile drag by incorporating a shallow favorable pressure gradient over the forward portion, this

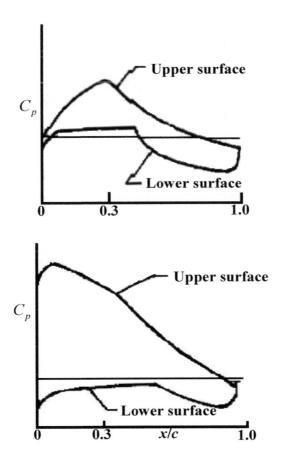

FIGURE 9.6
Typical pressure distribution for the design points of Fig. 9.5 (top figure for point A and bottom figure for point B).

is followed by a rather abrupt and sharper concave pressure recovery (with respect to the top surface) brought about by a fair amount of aft camber. Such camber helps to produce a high maximum lift coefficient. The absolute magnitude of this camber is constrained by the pitching moment requirement $c_m \geq -0.10$, a very significant requirement for operation at relatively higher Mach numbers. Also, notice that at higher lift coefficients the pressure distribution does not display a suction spike (that is typical of thin and sharp leading edges of older airfoils). This is incorporated by designing airfoils with thicker leading edges and detailed tailoring of leading edge thickness distribution. These features of pressure distribution are explained below. But, before that, let us explain the nomenclature of the NASA NLFs designed in 1970s and 1980s. For airfoil NLF(1)-0416, the designation implies natural laminar flow for NLF in the beginning. The number 1 in the parenthesis refers to the airfoil number in the design series; the number 04 refers to the design c_l (i.e., 0.4 in this case) for the cruise condition and the last two digits provide the maximum (t/c) of the airfoil. If an additional letter F appears after the last digit, as in NLF(1)-0416F, that implies the airfoil is designed with a flap. The constraints and corresponding calculations using the method of [96] for the design from [379] are given in Table 9.1.

While the extent of the upper surface favorable pressure gradient at cruise c_l was expected to be less than equal to $0.3c$, the calculated value was found to be $0.26c$, an acceptable violation. The maximum (t/c) value of 0.16 was met. However, the pitching moment coefficient at zero lift was desired to be above -0.10, which was seen to be slightly violated with a calculated value of -0.11.

Viken [428] has reported design of two airfoils as DESB154 and DESB165. It is interesting to note that the design proceeded with hand drawing an airfoil (based on experience!) and then calculating corresponding (i) pressure distribution, (ii) boundary layer and its (iii) stability properties — in that sequence. This procedure was carried through iteratively until

TABLE 9.1
Calculated section properties with prescribed constraints.

	Objective/constraints	Calculated
$(C_l)_{max}$ at $Re = 3.0 \times 10^6$	≥ 1.76	1.64
C_d at $Re = 4.0 \times 10^6$	Similar to six-series airfoil	0.0063

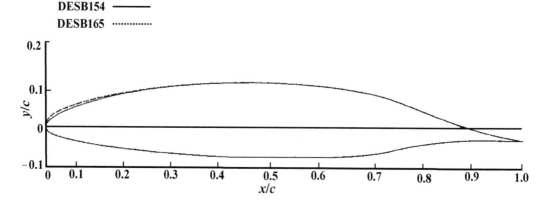

FIGURE 9.7
Comparison of airfoil profiles designed in [428].

the design goals were met for a Mach number $M = 0.4$ and c_l varying in the range from 0.15 to 0.95. Both the airfoils are 14.3% thick and designed to have accelerated flow all the way up to $x/c = 0.70$ on both the surfaces for cruise configuration. Maximum thickness occurs at approximately 45% of the chord and the section pitching moment coefficient about the quarter chord point is -0.0882 at the design point c_l of 0.45. This pitching moment coefficient varied from -0.833 to -0.0934 when c_l was varied from 0.15 to 0.95. All these data are for a Reynolds number of 10 million. The DESB154 airfoil was designed with a much sharper leading edge than is normally seen on airfoils of this type to reduce high negative C_p values at low angles of attack, giving a wider low drag c_l range. Also, to retain lower drag at cruise, the trailing edge has a sharp edge to give a thinner boundary layer at the trailing edge. However, the thin leading edge compromised $(c_{l,max})$ performance. This led to the development of the DESB165 airfoil, which is a modification of the DESB154 airfoil obtained by thickening the leading edge portion (only up to $0.2c$). Thickness was superposed directly onto the leading edge, while retaining the aft shape as much as possible. Both the airfoils are shown in Fig. 9.7 to indicate the changes made at the leading edge in the DESB154 airfoil to obtain the DESB165 airfoil.

In Fig. 9.8, pressure distributions are shown for the same angle of attack for these two airfoils. The inviscid pressure calculations for these two cases are shown in Fig. 9.8 at $\alpha = -0.954$ degrees.

Calculated results show a difference in the third decimal place for c_l and c_m. Thus, this indicates that the lift coefficient does not change much for lower angles of attack, while the relative change in the pitching moment coefficient is higher. Of course the changes at higher angles of attack will be more profound.

Further evidence for the effects of changing the airfoil leading edge shape is provided in [429], which explains the design of the NLF(1)-0414F airfoil starting from the baseline design of the DESB159 airfoil. In Fig. 9.9, the leading edge of the NLF(1)-0414F airfoil is compared with the DESB159 and NACA 67-314 airfoils.

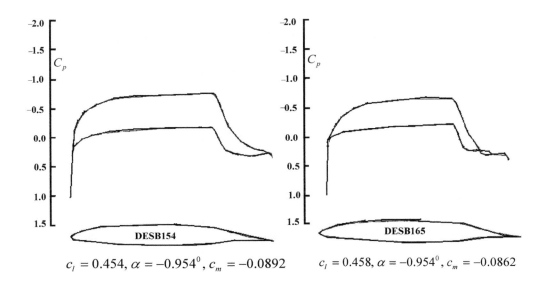

$$c_l = 0.454, \alpha = -0.954^0, c_m = -0.0892 \qquad c_l = 0.458, \alpha = -0.954^0, c_m = -0.0862$$

FIGURE 9.8
Pressure distribution over DESB154 and DESB165 airfoils for the same angle of attack [428].

The leading edge of the DESB159 airfoil was modified to achieve an acceptable low drag c_l range for the zero flap deflection case. For lower angles of attack which provide low drag operation for the DESB159 airfoil, the sharp leading edge helps suppress large negative pressure peaks. However, when the angle of attack is increased, the same sharp leading edge causes large negative pressure peaks as a result of the centripetal forces to turn the flow around the bend. In this case also an additional thickness is superposed on the baseline airfoil profile, merging the two at $x/c = 0.15 - 0.20$, to simply reduce the negative pressure peak. The essential idea is to turn the flow when the speeds are lower, allowing a smaller radius of curvature, which is subsequently increased with increasing speed. It is interesting to note that for the DESB159 and NACA 67-314 airfoils, the smallest radius of curvature is at the leading edge, while for the NLF(1)-0414F airfoil the smallest radius of curvature is placed on the lower surface.

9.3   Pressure Distribution and Pressure Recovery of Some Low Drag Airfoils

Other geometric differences exist between NACA six-series and NLF airfoils apart from the shape and thickness distribution near the leading edge. In Fig. 9.10, we see the comparison between NLF(1)-0414F and NACA 67-314 airfoils in terms of geometry and pressure distribution — mainly in the aft portion of the airfoils in [429].

Results are shown for the same Mach number (0.4) and lift coefficient ($c_l = 0.46$) for slightly different angles of attack. Both the airfoils have the same maximum thickness.

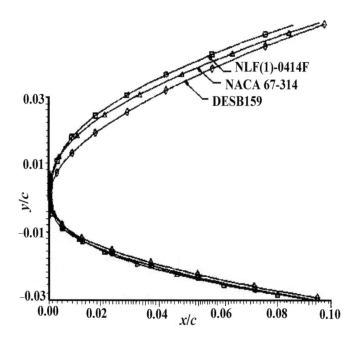

FIGURE 9.9
Comparison of indicated airfoil leading edges [429].

The favorable pressure gradient regions are similar, with the NLF(1)-0414F airfoil having slightly higher acceleration on both the surfaces. The upper surface acceleration of the NLF (1)-0414F airfoil was optimized by the use of the flat spot [429] in the pressure distribution at $x/c = 0.10$. For this airfoil, a small chord trailing edge (cruise) flap is crucially utilized for enhanced low drag performance. The major difference between the two airfoils relates to the reduced thickness for the NLF airfoil in the aft part of it, which determines different pressure recovery for these two airfoils. While concave type pressure recoveries are used in the NLF airfoil, for the NACA 67-314 airfoil a linear pressure recovery is obtained. Such different aft geometry of the two airfoils results in significantly different behavior in the turbulent pressure recovery region. This is related to the problem of turbulent separation in the pressure recovery region.

For the purpose of attaining low drag, the pressure gradient of an NLF airfoil should be favorable up to the desired transition location. After the point of minimum pressure, the boundary layer becomes turbulent due to an unfavorable pressure gradient. The unfavorable pressure gradient increases the rate of growth of the turbulent boundary layer, causing enhanced drag. Inspection of the von Karman momentum integral equation [55] can reveal the differences among various pressure recoveries used for NLF airfoil designs. This equation is given by

$$\frac{d\theta}{dx} + \frac{dU_e}{dx}\frac{\theta}{U_e}(H + 2 - M^2) = \frac{\tau_0}{\rho_e U_e^2} \tag{9.1}$$

Here the momentum thickness (θ) growth rate is given by the first term on the left-

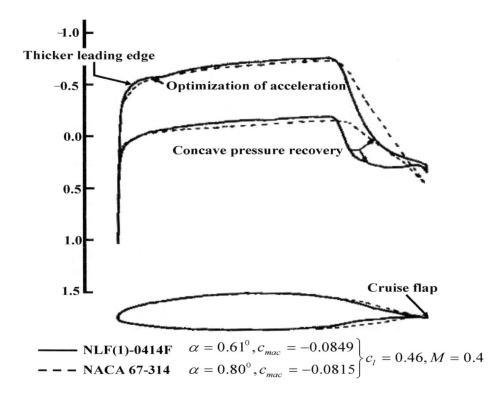

FIGURE 9.10

Comparison of pressure distribution between an NLF and six-series airfoil [429].

hand side, while the second term accounts for the effect of the pressure gradient that can be explained in terms of the boundary layer edge velocity (U_e) gradient in the streamwise direction. The skin friction term, as given by the term on the right-hand side, plays a secondary role in the region experiencing a steep pressure rise, as would be the case in the turbulent pressure recovery region in the aft portion of the airfoil. However, to obtain the benefit of low drag, [391] has shown that the boundary layer should be kept on the verge of separation throughout the pressure recovery (with the shape factor $H = 2.0$). A genuine pressure recovery of the Stratford type is, however, not suitable for NLF applications, because the turbulent boundary layer with Stratford recovery is likely to separate completely at off-design conditions. However, a milder concave pressure recovery resembling a Stratford recovery is preferred which provides lower drag without turbulent separation at off-design conditions.

In Fig. 9.11, a concave pressure recovery (that is typical of NLF designs) is compared with a linear recovery (that is typical of NACA six-series airfoils) from boundary layer calculations for a flow at a Reynolds number of one million. This figure is from [150] (also

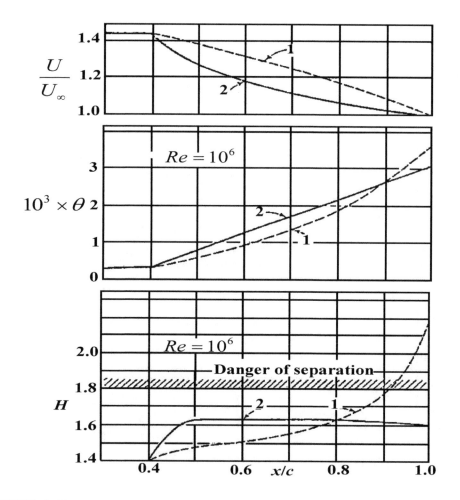

FIGURE 9.11
Effects of different pressure recoveries near the trailing edge on a turbulent boundary layer [150].

available in [427]). It is interesting to note that up to a certain distance the linear pressure recovery shows lesser drag as compared to the concave pressure recovery — somewhat closer to $x/c = 0.88$. The shape factor (H) also shows very spectacular growth for the linear pressure recovery case above $x/c = 0.8$. In the last frame, separation is indicated for $H = 1.8$ and it is clearly evident that the concave pressure recovery case is in no danger of flow separation, while the linear pressure recovery case indicates separation around $x/c = 0.92$. Schubauer & Spangenberg [317], however, have noted that incompressible turbulent boundary layers suffer separation when the shape factor grows to a value of 2.

In Fig. 9.12, the pressure distribution and shape factor are compared between the NLF(1)-0414F and NACA 67-314 airfoils. The shear layer properties were evaluated using a finite difference code [142]. In Fig. 9.10, it is shown that the six-series airfoil has a linear pressure recovery after the flow has become turbulent, while the NLF airfoil displays a concave pressure recovery after the minimum pressure point on the upper surface. In Fig. 9.12, both the airfoils show the flow is accelerated up to $x/c = 0.7$, as was also shown in Fig. 9.10. However, shape factor distribution for the NLF(1)-0414F airfoil shows that H grows to a maximum value of 1.9 at $x/c = 0.875$ and thereafter it reduces to 1.825 at the trailing edge. For the same conditions of the NLF airfoil ($M = 0.4, Re = 10^7$ and $c_l = 0.461$), the NACA 67- 314 airfoil suffers flow separation at $x/c = 0.90$ due to linear pressure recovery. The last correct calculated H was 1.83 for the NACA airfoil [429]. The results shown in Fig. 9.12 correspond to a flow that is turbulent right from the leading edge, as might happen due to leading edge contamination.

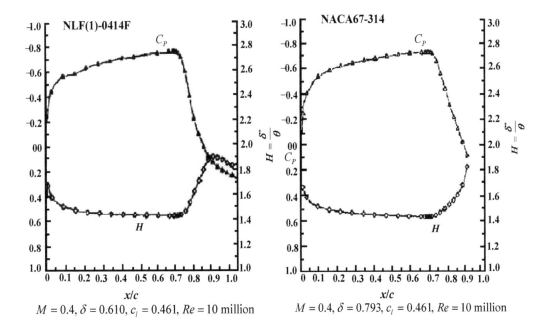

FIGURE 9.12
Comparison of pressure distribution and shape factor for NLF(1)-0414F and NACA 67-314 airfoils [429].

9.4   Flapping of Airfoils

There are two types of flaps which are used with NLF airfoils. In the first type, a small chord cruise flap deflection is used to alleviate the problem of turbulent separation in the recovery region at off-design lift coefficients. This cruise flap allows lift from the angle of attack to be traded with lift due to camber in [428]. Downward deflection of this flap adds aft camber, which achieves the same lift at a lower angle of attack, resulting in a milder pressure recovery region and thereby preventing turbulent separation. This effect can be seen from the inviscid pressure distribution shown in Fig. 9.13 for the DESB154 airfoil (from [428]) for almost the same angle of attack, with ($\delta_f = -2.6^0$) and without ($\delta f = 0$) the flap deflection. Notice the difference in lift coefficient values, but more importantly the sectional pitching moment is changed significantly by the deployment of the flap. This suggests that the flap can be used to alter the pitching moment and thereby reduce trim drag.

The other types of flaps can be used primarily to augment lift as high-lift devices. Thus, the flap affects airfoil performance by (i) augmenting the section lift coefficient and mainly to achieve $c_{l,max}$, (ii) altering the section pitching moment coefficient, including meeting the constraint on the pitching-moment coefficient ($c_{m,cruise} \geq -0.05$), (iii) shifting the low drag range for NLF airfoils to higher lift values. For higher speed applications, this can also be used to change the critical Mach number. In the above, the first and second requirements are conflicting in general and the flap can be used to alleviate this by deflecting the flap upwards. This allows an airfoil to be designed with significantly higher camber to achieve $c_{l,max}$ while retaining the ability to keep c_m smaller at cruise flight conditions.

The third reason for using a flap in an NLF airfoil is the most significant one; it is

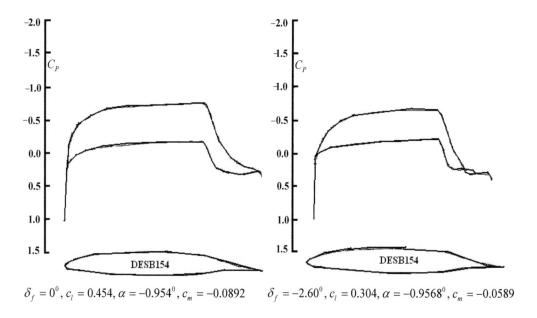

$$\delta_f = 0^0, c_l = 0.454, \alpha = -0.954^0, c_m = -0.0892 \qquad \delta_f = -2.60^0, c_l = 0.304, \alpha = -0.9568^0, c_m = -0.0589$$

FIGURE 9.13
Comparison of inviscid pressure distribution, with (right) and without (left) the flap deflection for DESB154 airfoil [428].

achieved by positive flap deflection to shift the low drag bucket to higher lift coefficients [379]. This happens as additional lift can be achieved via an increase in aft camber instead of an increase in angle of attack. This also allows achieving the same lift at lower angles of attack, resulting in the restoration of a less adverse pressure gradient for those c_l values which were outside the low drag bucket. Thus, the drag bucket will not only be shifted to higher lift coefficients, but also shifted to lower angles of attack due to induced pressure peaks on the upper surface.

For the NLF(1)-0414F airfoil, the drag polar is shown in Fig. 9.14, for very low speed operation (Mach number less than 0.12) at $Re = 10^7$ for flap deflection in the range $-10^0 \leq \delta_f \geq 12.5^0$. This figure shows that the use of a cruise flap helps to achieve a wider range of c_l for low drag operation, especially when viewed with respect to undeflected flap drag polar, for which one notes $c_{d,min} = 0.0027$ occurring at $c_l = 0.41$. For $\delta_f = -10^0$, minimum drag increases to 0.0030, occurring at $c_l = 0.01$. When the flap is deflected to 12.5^0, then the minimum drag coefficient increases to only 0.0033 at $c_l = 0.81$ yielding $L/D = 245$.

9.5 Effects of Roughness and Fixing Transition

The effects of fixing transition on section properties are shown in Fig. 9.15 for the NLF(1)-0414 airfoil with the flap undeflected for $M = 0.10$ at $Re = 10^7$ in [429]. It is noted that c_l and c_m do not change very much whether one is looking at the free transition case or when the transition is fixed near the leading edge.

However, the drag polar changes when one compares the free transition case with fixed transition cases. For the free transition case, the minimum profile drag coefficient is 0.0027 at $c_l = 0.41$. This profile drag is only 38% that of an unseparated fully turbulent airfoil ($c_d = 0.0083$). The maximum lift coefficient is marginally better at 1.83 for the free transi-

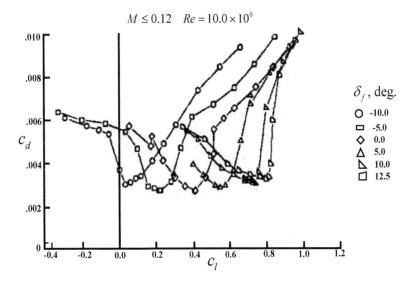

FIGURE 9.14
Effects of flap deflection on the NLF(1)-0414F airfoil as shown for the drag polar [429].

tion case at $\alpha = 18^0$. In contrast to lift and drag coefficients, the pitching moment coefficient is unaffected by fixing transition location on the airfoil. Transition is usually fixed in experiments by using a standard NACA roughness strip [1] or by fixing cylindrical trip wires fixed on identified locations.

In Fig. 9.16, the effects of roughness applied near the leading edge on section properties are shown for the NLF(1)-0414F airfoil [241] at a lower Reynolds number. The experiments were performed in the NASA Langley Low Turbulence Pressure Tunnel (LTPT) for $Re = 6 \times 10^6$.

Once again, lift and pitching moment coefficients do not change appreciably with a roughness element at $0.075c$. However, profile drag changes significantly for this lower Reynolds number, as compared to that shown in Fig. 9.15 for a Reynolds number of 10^7. Also, one notices that the drag bucket width reduces with an increase in Reynolds number, as can be seen from these two figures. We also point out another interesting airfoil characteristic in Fig. 9.16, which is independent of roughness effects. This is related to reduction of lift curve slope near 4^0 for the NLF(1)-0414F airfoil. This reduction in lift curve slope is due to a trailing edge separation in the pressure recovery region. Sometimes this is purposely done to control pitching moment − as reported in [104] for the design of the SHM1-airfoil. If this is considered undesirable, as stated in [241], then it can be eliminated by using boundary layer re-energizers or vortex generators placed on the upper surface of the airfoil.

The effects of fixing boundary layer transition near the leading edge on section aerodynamics and stability properties have been investigated in [241]. Fixing transition near the leading edge does not have serious effects on aerodynamic properties and additionally has no significant effects on lateral directional stability and control. It is understood that such transition can occur naturally due to insect or dirt contamination, etc., over the airfoil.

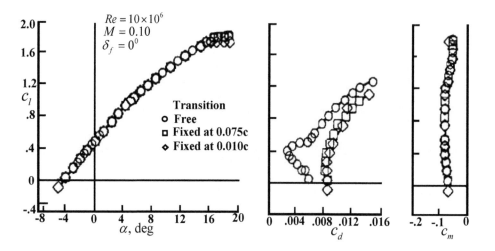

FIGURE 9.15
Effects of fixing transition for the NLF(1)-0414F airfoil. Perceptible effects are seen in the drag polar [429].

9.6 Effects of Vortex Generator or Boundary Layer Re-Energizer

In Fig. 9.16, we have noted the presence of an angle of attack above which the lift decreased relatively, which manifested itself via a change of lift curve slope for the NLF(1)-0414F airfoil after $\alpha = 4^0$. This is attributed to flow separation in the turbulent pressure recovery region. This can be avoided by turbulent mixing, as demonstrated in [317]. The authors used forced turbulent mixing by employing a simple plow and vortex generator whose function is to scour the boundary layer by creating an alternate strip of higher and lower velocity patterns. Such forced mixing causes the pressure recovery region to spread over twice the streamwise distance, causing the concave pressure recovery even milder, which precludes turbulent separation. Murri et al. [241] report results of using a vortex generator in an experimental investigation for the NLF(1)-0414F airfoil. Results are shown in Fig. 9.17, where the comparison is shown between the cases of with and without the vortex generator.

Here the vortex generators are small low aspect ratio wings, positioned on the upper surface of the airfoil at a high angle of attack with respect to the local flow direction. For the results shown in Fig. 9.17, vortex generators are positioned at $0.6c$, which are 0.2 inches high, spaced 1.6 inches apart, for the presented results at $Re = 3 \times 10^6$. While the trailing edge separation is indicated by the reduction of the lift curve slope for the case without the vortex generator for the case with the vortex generator the reduction of the lift curve slope disappears. But the maximum lift generated remains more or less the same. The drag performance for angles of attack greater than 4 degrees also improves, but at a lower lift coefficient range; a drag penalty is shown for the case with vortex generator. A similar drag reduction is obtained by applying tape turbulators for low Reynolds number operation by deploying them in the aft section of airfoils, as discussed in [241]. However, their purpose

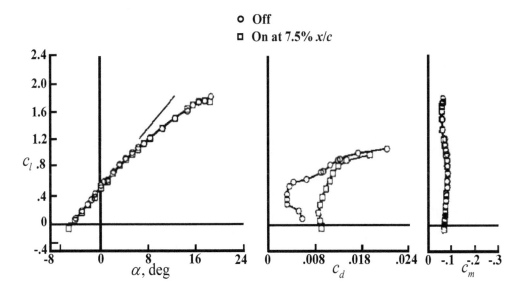

FIGURE 9.16
Effects of fixed roughness near the leading edge of the NLF(1)-0414F airfoil for $Re = 6 \times 10^6$ [241].

is to prevent laminar separation and is different from the vortex generator that is used to suppress turbulent flow separation.

9.7   Section Characteristics of Various Profiles

In the following, we compare properties of various other sections with NLF airfoils. In Fig. 9.18, we compare the maximum lift coefficient of the NLF(1)-0416 section with two NACA six-series airfoils of similar design lift coefficient and thickness, at various Reynolds numbers. There is a significant improvement of this NLF section with respect to the six-series section. The drag coefficient of the NLF(1)-0416 airfoil is compared with the NACA $63_2 - 415$ and $63_2 - 615$ airfoils in Figs. 9.19 and 9.20 for $c_l = 0.4$ and 1.0, respectively.

While for $c_l = 0.4$, six-series airfoils have a smaller drag coefficient for $c_l = 1.0$ the NLF airfoil has distinct advantages, specifically at higher Reynolds numbers, where the operational point is outside the drag bucket for the six-series airfoils. This is specifically achieved via multi-objective design considerations, as explained with the help of Fig. 9.5, and also discussed in [378].

The benefits of NLF airfoils do not persist if the airfoils are used for turbulent flows. This is shown in Figs. 9.21 to 9.23, where the data [378] compare two NLF airfoils with two classical NACA four- and five-digit airfoils.

The sections compared are the NLF(1)-0416, LS(1)-0417 (formerly GA (W)-1), NACA 23015 and NACA 4415 airfoils. Note that the NACA series airfoils have lower maximum thickness (15%) as compared to the NLF (16%) and LS (17%) airfoils. This difference will show up in the properties shown in these figures. The maximum lift coefficient compared in Fig. 9.21 indicates that $c_{l,max}$ for the NLF(1)-0416 airfoil, with transition free and fixed

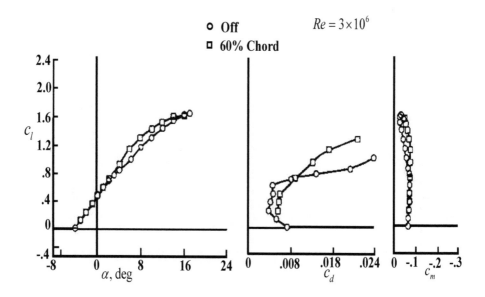

FIGURE 9.17
Effects of vortex generator used in the turbulence recovery zone for the NLF(1)-0414F airfoil for $Re = 3 \times 10^6$ [241].

cases, is comparable to the LS(1)-0417 airfoil. These $c_{l,max}$ values are significantly higher than those for the NACA 23015 and NACA 4415 airfoils. The drag coefficient for these four airfoils at lift coefficients of 0.4 and 1.0 are compared in Figs. 9.22 and 9.23, respectively. For $c_l = 0.4$, the NLF airfoil exhibits a substantially lower value of drag coefficient in Fig. 9.23, when transition is not fixed on the airfoil, for the cruise Reynolds number of four million. As the thickness ratio of the airfoils is different, corresponding drag coefficients are also different. However, the design of the NLF(1)-0414 airfoil is superior, as Somers [378] notes that the free transition drag value for this airfoil is the same as that for the LS(1)-0413 airfoil, which is only a 13% thick airfoil. This clearly shows that if the NLF(1)-0414 airfoil is used in an aircraft design and laminar flow is not achieved (e.g., due to premature transition by insect contamination near the leading edge), then the resultant penalty (even with respect to a thinner section like the NACA 23015 airfoil) is not too great. However, if laminar flow is achieved, then that would result in a significant reduction in profile drag.

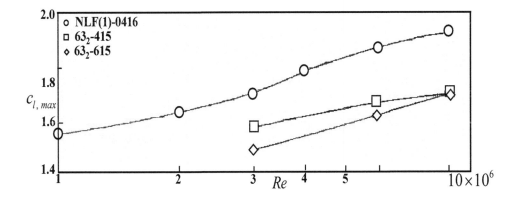

FIGURE 9.18
Comparison of maximum lift coefficients of the NLF(1)-0416 airfoil with two NACA six-series airfoils [378].

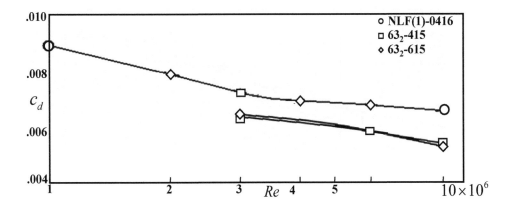

FIGURE 9.19
Comparison of the drag coefficient of the NLF(1)-0416 airfoil with two NACA six-series airfoils at $c_l = 0.4$ [378].

9.8   High Speed NLF Airfoils

It is known that laminar compressible boundary layers are more stable than incompressible boundary layers, so not much acceleration is needed to keep the flow laminar. Also, as lift increases, overall acceleration increases instead of negative pressure peaks forming at the leading edge [429]. However, with added acceleration comes the problem at the recovery region, with the transition point moving upstream. Also, with an increase in Mach number, the acceleration can lead to formation of shocks. Up to a Mach number of 0.4, section properties do not change appreciably, as has been shown experimentally for the NLF(1)-0416F airfoil [379] for moderate Reynolds numbers, without the flap deployed. However,

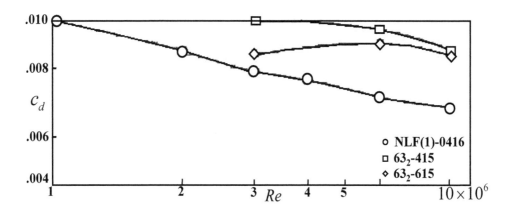

FIGURE 9.20
Comparison of the drag coefficient of the NLF(1)-0416 airfoil with two NACA six-series airfoils at $c_l = 1.0$ [378].

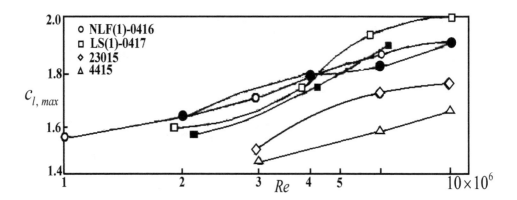

FIGURE 9.21
Comparison of maximum lift coefficients for NLF(1)-0416, LS(1)-0417, NACA 23015 and NACA 4415 airfoils. Open symbols are for data with free transition and solid symbols are data for fixed transition [378].

when the Mach number increases beyond 0.4, significant compressible effects come into play in determining aerodynamic section properties.

In Fig. 9.24, pressure distribution is shown over the NLF(1)-0416F airfoil at two Mach

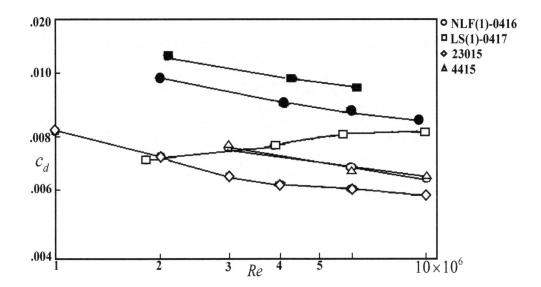

FIGURE 9.22
Comparison of drag coefficient at $c_l = 0.4$ for NLF(1)-0414, LS(1)-0416, LS(1)-0417, NACA 23015 and NACA 4415 airfoils. Open symbols are for free transition and solid symbols are for fixed transition data [378].

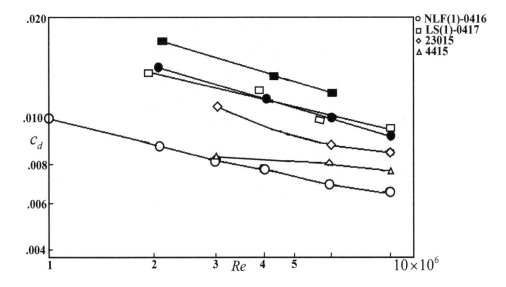

FIGURE 9.23
Comparison of drag coefficient at $c_l = 1.0$ for NLF(1)-0414, LS(1)-0416, LS(1)-0417, NACA 23015 and NACA 4415 airfoils. Open symbols are for free transition and solid symbols are for fixed transition data [378].

numbers, $M = 0.4$ and 0.7 [429]. For the higher Mach number case, flow accelerates strongly up to $x/c = 0.70$ which terminates in a shock with a very strong pressure recovery,. and the supersonic zone or the pocket is indicated over the airfoil profile by a dotted line. It is noted that lower surface pressure distribution does not change very much when the Mach number is increased from 0.4 to 0.7. The top surface pressure distribution causes unsteady flow separation and is responsible for very high drag. Camber and thickness cause enhanced acceleration at higher Mach numbers. One way to circumvent the problem is to deflect the cruise flap upwards to reduce effective camber. This is shown in Fig. 9.25, by the pressure distribution shown by the solid line when the flap of the $0.125c$ chord is deflected upward by 5.24^0. This de-cambering reduces flow acceleration seen from the reduced supersonic pocket in Fig. 9.25, once again shown by the solid line over the upper surface of the profile. If the flow is considered as fully turbulent, then calculations by the Harris–Blanchard code for $Re = 11 \times 10^6$, the boundary layer solution (not shown here), indicate flow separation for many variations of pressure recovery considered in [429].

The above failed exercise of possible de-cambering with a cruise flap leading to turbulent flow separation establishes the unsuitability of the NLF(1)-0414F airfoil for high Mach number applications. Thus, the upper surface of the NLF(1)-0414F airfoil was redesigned so that the pressure recovery starts from $x/c = 0.55$, without creating a supersonic pocket for $M = 0.7$. This redesigning led to a reduction of airfoil thickness from 14% to 13%. This was the rationale behind the design of HSNLF(1)-0213. Here the first two letters in the designation stand for high speed operation. Also, note the design cruise lift coefficient is

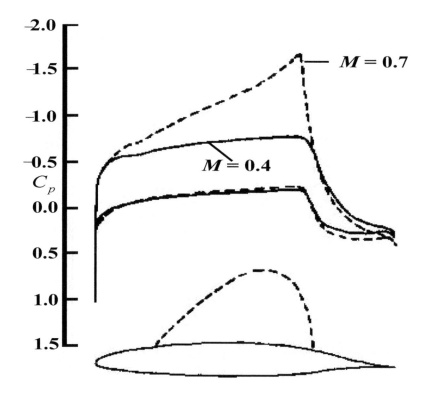

FIGURE 9.24
Effect of the Mach number on inviscid computed pressure distribution on the NLF(1)-0414F airfoil [429].

reduced from 0.4 to 0.26, for $M = 0.7$ and $Re = 11 \times 10^6$. This design allows a laminar boundary layer up to $0.55c$ on the upper surface and $0.65c$ on the lower surface. This airfoil was tested and the results reported in [364] for the basic airfoil; flap design considerations were reported in [429]. Experiments were performed in the 6×28 inch transonic tunnel at NASA Langley, a blowdown facility which usually results in overestimation of the drag value. In Fig. 9.26, variation of drag coefficient with Mach number is shown for HSNLF(1)-0213 for two different Reynolds numbers. For $Re = 4 \times 10^6$, a smooth model was tested, while for $Re = 10^7$ the transition was fixed at $0.05c$.

With the fixed transition case at a Reynolds number of 10^7, flow is essentially turbulent. In contrast, the smooth model data at a Reynolds number of 4×10^6 has only a limited amount of laminar flow due to tunnel flow quality. The drag divergence Mach number is seen to be 0.72 for the fixed transition case, as opposed to the design value of 0.70 for the higher Reynolds number. For the smooth model at the lower Reynolds number, the drag divergence Mach number is even higher, 0.74.

Apart from NASA series NLF airfoils, there is another NLF airfoil that has been actually used in the design of a business jet. This airfoil is designated SHM1 and reported in [105] with its additional features which makes it suitable for application in a general aviation jet class aircraft. Although the HSNLF(1)-0213F airfoil can be considered suitable for business jet application, it has a few drawbacks. For example, it has low $c_{l,max}$ at low Reynolds

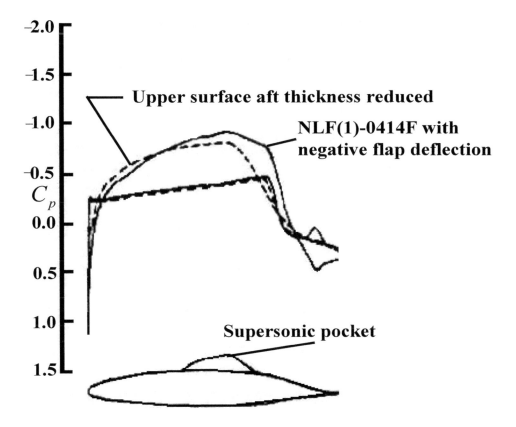

FIGURE 9.25
Redesign of the upper surface of the NLF(1)-0414F airfoil in terms of inviscid pressure distribution, which results in the HSNLF(1)-0213 airfoil, as shown by dotted lines [429].

numbers. More important, it has a thickness of 13%, which limits its ability to carry enough fuel in the wing. In contrast, the SHM1 airfoil has a thickness of 15% and at the same time it has a high value of drag divergence Mach number, so that its cruise Mach number is fixed at 0.69 for $Re = 11.7 \times 10^6$, with a cruise lift coefficient of 0.26. The larger thickness of the SHM1 airfoil allows a wing design with minimum planform area, thereby achieving low profile drag obtained through the NLF feature of the section. The low profile drag is desired over a range of $c_l = 0.18$ for $Re = 11.7 \times 10^6, M = 0.69$ to $c_l = 0.35$ for $Re = 13.6 \times 10^6, M = 0.31$. Note that to provide operational margin, the lower limit of the low drag range is reduced from the actual cruise lift coefficient. Higher values of lift coefficient and Reynolds number correspond to sea-level climb conditions. Additionally, the section has a low nose-down pitching moment tendency, docile stall characteristics and no significant maximum lift degradation due to an insect contamination problem. The sectional maximum lift coefficient without flap was set to a minimum of 1.6 for $Re = 4.8 \times 10^6, M = 0.134$ and this should not degrade by more than 7% due to leading edge contamination (LEC). The pitching moment constraint is given by $c_m \geq -0.04$ at $c_l = 0.38$ for $Re = 7.93 \times 10^6$ and $M = 0.70$ to minimize the trim-drag penalty at high altitude and high Mach number cruise condition. For this condition, the drag divergence Mach number should be higher than 0.7.

In Fig. 9.27, the profile and a typical potential flow pressure distribution is shown for this airfoil. According to results in [105], the flow accelerates on the upper surface up to $0.42c$, followed by a very mild concave pressure recovery and that, according to the authors, represents a compromise among the requirements of maximum lift, pitching moment and drag divergence. The pressure gradient on the lower surface is typical of all NLF airfoils and is favorable up to $x/c = 0.63$ for cruise conditions.

Also, the leading edge of the airfoil is so designed that there is flow transition near it, at high angles of attack. This alleviates the problem of loss of lift due to LEC. Additionally, the trailing edge portion of the airfoil was also redesigned, which confines the movement of

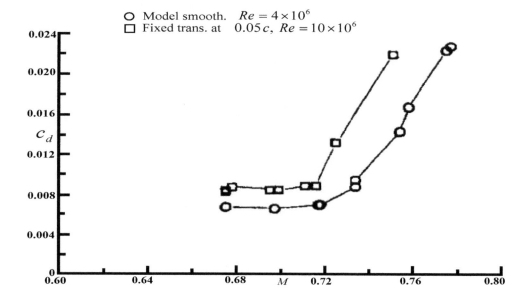

FIGURE 9.26
Variation of drag coefficient with Mach number for the HSNLF(1)-0213 airfoil for two Reynolds numbers with different surface conditions [364].

the separation point at high angles of attack. It is furthermore claimed that such a trailing edge design actually induces a small separation near the trailing edge, resulting in reduction of the pitching moment. An associated drag penalty caused by the separation is claimed to be negligible, as it induces a short and shallow bubble.

In Figs. 9.28 and 9.29, the transition location and lift curve are shown for the indicated Reynolds and Mach numbers. The transition point was located in the wind-tunnel experiment as well as by flight tests via an infrared flow-visualization technique, on the upper

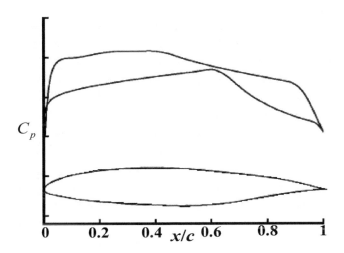

FIGURE 9.27
The SHM1 airfoil profile and typical pressure distribution at design conditions [105].

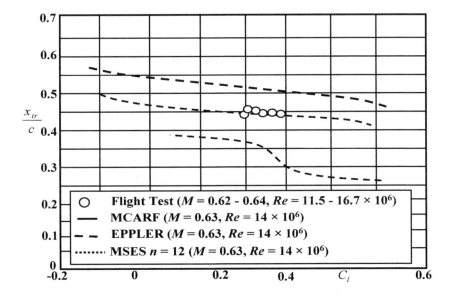

FIGURE 9.28
Experimental (flight test data) and theoretical prediction of transition location of the SHM1 airfoil [105].

surface. The lift curves also indicate that barring $Re = 2.8 \times 10^6$, at all higher Reynolds numbers the design constraint of a minimum value of 1.6 for $c_{l,max}$ was met from the wind tunnel test results.

When a roughness element of a 0.3 mm-high strip was installed on the upper surface at $0.05c$, the loss of maximum lift was about 5.6% compared to the clean case for $Re = 4.8 \times 10^6$. In Fig. 9.30, the drag polar of the SHM1 airfoil is shown for $Re = 10.3 \times 10^6, M = 0.27$, which is for a cruise flight condition.

A very interesting set of results given in [105] relates to the effects of steps and surface roughness on drag obtained from a flight test. A 0.2 mm high step located at $0.2c$ on the upper surface causes a drag increase of about 3 counts for $Re = 13 \times 10^6$ and $M = 0.62$. The drag increase is ten times greater if the same step is located at $0.1c$, demonstrating that the chordwise location of the step is critical with respect to drag. This can be taken as a guideline to position the upper skin parting line for wing structural design. It is also reported that the drag increases by 17 counts when the surface is roughened by sandpaper S.P. 600, as compared to matte paint surface for $Re \leq 14 \times 10^6$. When the roughened surface was polished with wax, the drag decreased to the same prior level before roughening.

Drag divergence characteristics of the airfoil are shown in Fig. 9.31 for the indicated lift coefficient values. Figures 9.27 to 9.31 are taken from [105].

In discussing the above, we have not talked about the various design and analysis tools that have been used for most of the results presented here. Inviscid flows have been calculated either by solving the Laplace equation using panel methods for incompressible flows and by solving the Euler equation for compressible flows. In solving the boundary layer equations, both the differential and integral equations have been solved. Many of the successful design codes actually depend upon solving the integral equation form of the boundary layer for quick calculations. However, the differential formulations would be preferred for higher accuracy. Transition prediction is solely based upon the linear stability theory for the evo-

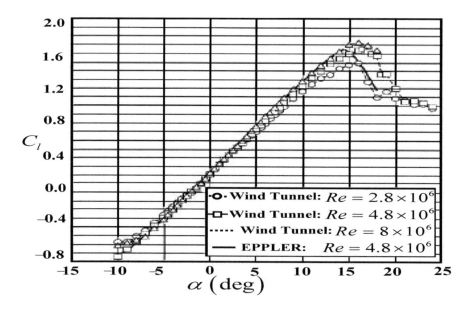

FIGURE 9.29
Wind tunnel data for the lift curve for the SHM1 airfoil for the indicated Reynolds number [105].

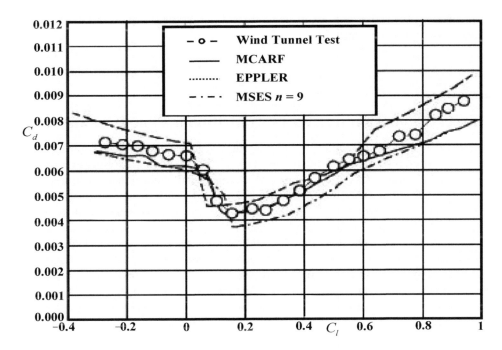

FIGURE 9.30
Experimental and theoretical drag polar for the SHM1 airfoil for near-climb flight conditions
– see text [105].

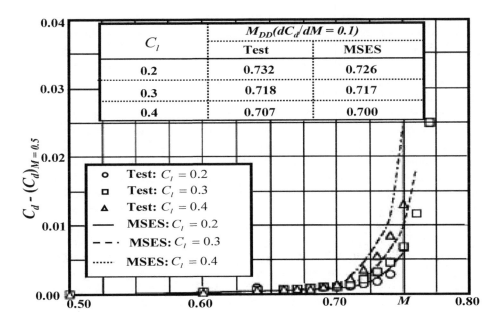

FIGURE 9.31
Experimental estimation of drag divergence Mach number for the SHM1 airfoil [105].

lution of TS waves and their exponential growth. A growth exponent is chosen for different codes and although they are backed by some experimental data, it is still very empirical in nature. In contrast, there are only a very few design tools based on the solution of the full Navier–Stokes equations currently in use.

9.9   Direct Simulation of Bypass Transitional Flow Past an Airfoil

Flow past an SHM1 airfoil is studied by solving the full Navier–Stokes equation for $Re = 10.3 \times 10^6$. This is designed as a natural laminar flow (NLF) airfoil and this Reynolds number corresponds to the cruise configuration, using the commercial codes indicated in Fig. 9.28. This is based on the classical view that transition is caused via the appearance and growth of Tollmien–Schlichting waves and the design process prevents transition in most part over the top and bottom surfaces of the airfoil. Thus, it is quite natural to compute the flow without any model(s) for transition and turbulence and the corresponding experimental results performed in a wind tunnel are specifically for two-dimensional flows. Simulated results shown in Fig. 9.32 for a zero angle of attack for the stream function do not indicate the presence of any wavy disturbances and vortical structures. One notes that at zero angle of attack the flow will accelerate on the top surface, almost up to the leading edge – as the maximum thickness is located aft of $0.30c$. Also, the boundary layer formed over the airfoil is very thin for this Reynolds number ($Re = 10.3 \times 10^6$), which also indicates flow separation over the airfoil at very small levels. This is noted when the figure is zoomed near the trailing edge portion; one notices formation and downstream convection of very small bubbles. These are physical in origin and capturing these satisfactorily is indicative of the DRP property of the numerical scheme used, as described in Chap. 2. The presence of unsteady separation bubble is characteristic of bypass transition, as described in Chap. 4. In Fig. 9.28, it is noted that such a transition occurs at around $0.45c$ for a range of C_l at elevated Reynolds numbers; also noted in Fig. 9.30 that the drag coefficient remains constant, with the drag bucket very narrow.

In the top frame of Fig. 9.33, we have shown the variation of displacement thickness of the flow for the conditions of Fig. 9.32 at $t = 4.50$. At this time, one notices that flow on the top surface shows instantaneous large fluctuations near the trailing edge, implying strong flow unsteadiness. This is not the case with the flow over the lower surface, although the boundary layer thickens near the trailing edge very rapidly. The behavior of the boundary layer can also be noted in terms of the Falkner–Skan pressure gradient parameter, $m = \frac{x}{U_e}\frac{\partial U_e}{\partial x}$, which can be used to compare an unsteady flow field with a steady flow. For the flow at $t = 4.50$, this parameter is plotted in the bottom frame. Steady flows indicate flow separation, when m decreases below -0.0904 and a dotted line is drawn in this frame. However, unsteady flows can sustain a higher adverse pressure gradient without showing separation. Note the truncated y-axis in this figure for the top surface data. Also, the unsteady separation noted here does not abide by the same criterion. These flows exhibit significantly larger swings for m, starting from $x = 0.60c$, while Fig. 9.32 show unsteady separation only after $x = 0.75c$. As one is solving the Navier–Stokes equation, one does not experience solution singularity. We note that the flow sequences depicted in Figs. 9.32 and 9.33 do not depend on explicit excitation. This may appear contradictory to the points made in Chaps. 3 to 7, where in discussing receptivity, we noted the central role of excitation. However, flows studied in Chaps. 3 to 5 corresponded to zero pressure gradient. Such flows require definite excitation for moderate Reynolds numbers. However, for higher Reynolds numbers for zero pressure gradient flow or flows experiencing an adverse pressure gradient, elements of instability are so strong that numerical disturbances related to round-off error can also trigger disturbance growth, as seen here near the trailing edge of the airfoil.

Thus, in triggering bypass transition on an NLF airfoil, one does not require explicit excitation. In Fig. 9.32, unsteady separation or bypass transition is seen to be caused by a strong adverse pressure gradient. The sequence by which the top surface experiences bypass transition is shown in Fig. 9.34. This airfoil has concavity on the top and bottom surfaces near the trailing edge. This concavity manifests itself in flow instability near the trailing edge on both surfaces. As a consequence, unsteady separation is set up locally and the region over which the disturbance is noted is seen to travel upstream. The upstream travel of disturbances appears as new packets created upstream, in exactly the same sequence, bypass transition is seen to occur for a zero pressure gradient boundary layer in Chap. 5, created by localized harmonic excitation. By $t = 4.5$, one notices the disturbed region on the top surface to extend from $x = 0.60c$ to the trailing edge. This also once again establishes that bypass transition is caused by upstream propagating disturbances, and the parabolized stability equation is going to be totally inadequate in analyzing such flows.

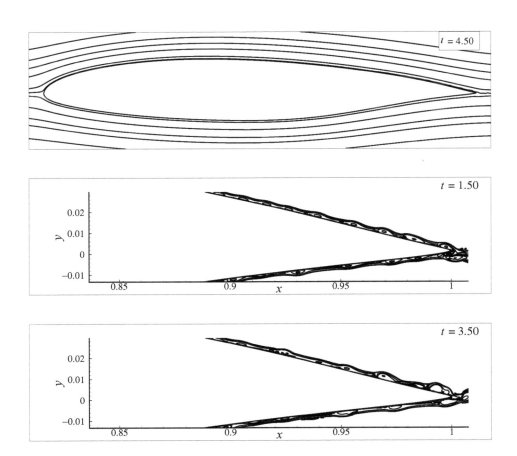

FIGURE 9.32

Stream function contours for flow past the SHM1 airfoil are shown at the indicated time instants for $Re = 10.3 \times 10^6$ when the airfoil is kept at zero AOA. The top frame shows the flow field around the full airfoil, while the bottom frames show the zoomed view of the flow field near the trailing edge, showing the presence of small bubbles indicative of bypass transition.

NLF airfoils are seen to suffer bypass transition without any excitation in Fig. 9.34. However, the computed bypass transition location is far aft of what has been noted in Fig. 9.28. What causes the difference? This is due to the fact that in computed flows, we consider a perfectly smooth geometry placed in a uniform flow and the transition is caused by the numerical disturbances acting as the seed for the adverse pressure gradient region over the airfoil surface. It is interesting to note the change of bypass transition caused by deterministic disturbances. In Fig. 9.35, SBS harmonic excitation is applied with a nondimensional frequency $F = 1.11441 \times 10^{-5}$ at the location in the figure indicated by an arrowhead. A little before $t = 1.75$, a spatio-temporal packet is noted to be created downstream of the harmonic exciter. With time this grows and convects downstream. A little after $t = 1.95$, this packet merges with main packet which existed without the SBS excitation. This enlarges the region over which bypass transition is seen to be active. The frame at $t = 6.50$ shows that the transitional location point moves to $x = 0.40c$, which is noted experimentally

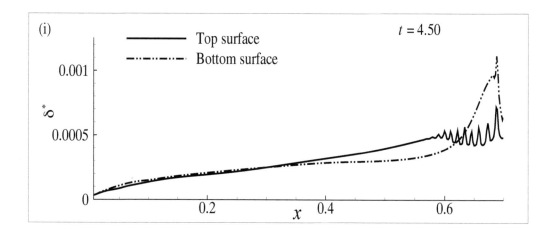

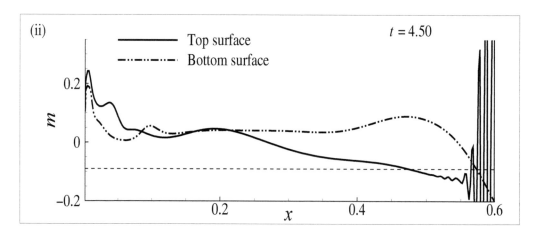

FIGURE 9.33

Variations of the displacement thickness (δ^*) on the top and bottom surfaces of the SHM1 airfoil are shown in the top frame at $t = 4.5$. The corresponding variation of steady flow separation parameter $m = \frac{x}{U_e} \frac{\partial U_e}{\partial x}$ on the top and bottom surfaces of the airfoil is shown in the bottom frame.

and shown in Fig. 9.28. This, once again, establishes the fact that if one is using a high accuracy method, then bypass transition is seen to occur computationally far downstream, as compared to what is observed in the experiment. To match the experimental value of drag, one must know the level of background disturbance environment and simulate flow accordingly. Such attempts at modeling free stream turbulence (FST) have been variously made in [329, 342] based on some data provided in [102] for wind tunnels. In [329], additional data have been obtained from flight tests to statistically characterize FST.

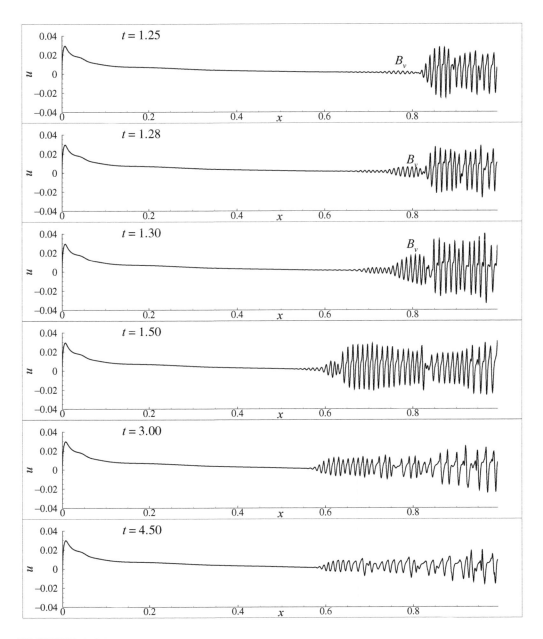

FIGURE 9.34

Variation of azimuthal component of velocity (u) at a height of 1.242×10^{-6} from the top surface of the SHM1 airfoil at the indicated time instants.

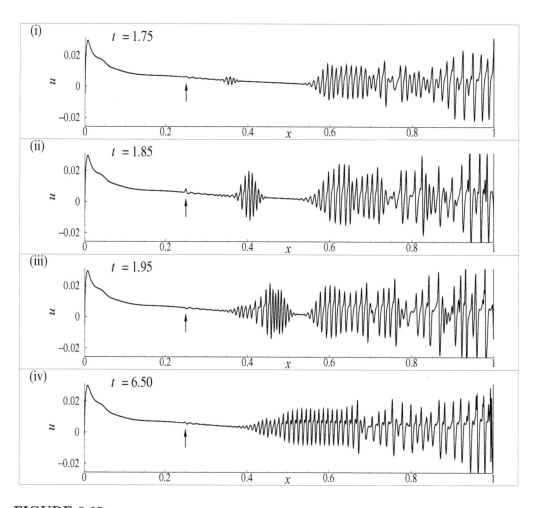

FIGURE 9.35
Variation of azimuthal component of velocity (u) at a height of 1.242×10^{-6} from the top surface of the SHM1 airfoil at the indicated time instants. Wall excitation corresponds to an SBS frequency of $F = 1.11441 \times 10^{-5}$ with an amplitude of 0.001.

10

Epilogue

10.1 Introduction

In this book we have focused upon the twin topics of flow instability and receptivity of flow to different types of disturbances acting inside and outside the shear layer, which represent the equilibrium flow. However, the distinctive feature of this book, apart from treating linear and nonlinear receptivity, is that transition to turbulence is followed from the onset of receptivity to fully developed turbulent two-dimensional flow (as exemplified by the energy spectrum). Why should it be considered special simulating two-dimensional turbulence from the first principle, when many researchers have surmised that turbulence is three dimensional?

Explaining and characterizing the turbulent state of fluid flow continue to eluded us as one of the unsolved problems of classical physics. The major issue in this regard is to understand the dynamical processes which transform a deterministic equilibrium laminar flow to the stochastic turbulent state. It is even uncertain whether one should study turbulence using stochastic or deterministic tools. While the turbulence appears as random and chaotic, the presence of coherent structures retains the possibility to view it as an admixture of deterministic and stochastic dynamics. This viewpoint is reinforced as laminar and turbulent states are both governed by the same nonlinear differential equation. Amidst all these, it is understood since the times of writing the Navier–Stokes equation [91, 288, 289, 297] that turbulence is a consequence of flow instability triggered by omnipresent background disturbances.

Thus, the subject has bifurcated in twin tracks — with flow instability studied using the deterministic route and fully developed turbulence studied theoretically and experimentally by utilizing stochastic tools. However, a study of the presented materials in the preceding chapters would convince readers that these twin tracks have been unified by taking a nominally two-dimensional flow from its early instability stage to the final turbulent stage following a strictly a deterministic approach. However, first and foremost, it is necessary to establish that studying two-dimensional turbulent flow is meaningful, as it is often stated that one of the defining characteristics of turbulence is its three dimensionality.

Any mechanism present in two-dimensional flows is more universal than any found in 3D. We have already noted that the Klebanoff mode (in Chap. 3) is 3D in nature and not supported as two-dimensional disturbance flow, although it can be supported by nominal two-dimensional equilibrium flows. We have also noted in Chaps. 4 and 5 that transient growth reported in Trefethen [411] is only relevant for 3D flows. For two-dimensional flows such transient growths are only marginally effective and do not cause flow transition.

10.2   Relevance of Two-Dimensional Turbulence

It has been shown in Chap. 2 that 3D flows have a vortex stretching term which is responsible for the energy cascade from large to small scales. This has been used in [180] to show the energy density spectrum is given by $E(k) \propto k^{-5/3}$ in an intermediate range and is considered to be a consequence of nonlinearity for 3D turbulent flow at those scales which are independent of the supplied energy at the largest scale and the dissipation at the Kolmogorov scale. This is known as the inertial subrange in turbulence studies. However, there is experimental evidence, including the results reported in Chap. 4, which point to the fact that two-dimensional turbulence in the main is possible with an energy spectrum given in an intermediate range by $E(k) \propto k^{-3}$. This has been explained in the theoretical works of [21, 184] for free and forced turbulence, respectively. Kraichnan considered creation of turbulence by random forcing centered around a fairly large wavenumber in Fourier space. In these theories, it has been suggested that fully developed two-dimensional turbulence should display (a) a direct cascade of enstrophy from large to small scales, (b) an inertial range with self-similar energy spectrum and (c) an inverse flow of energy from small to large scales. Special features of two-dimensional flows are the absence of a vortex stretching term and the presence of inverse migration of energy from small to large scales. This latter aspect of energy transfer has been identified as "Bose condensation" in [377]. A general description of two-dimensional turbulence can also be found in [79]. Although the migration of energy from small to large scales has also been termed "inverse cascade," this is not synonymous with the energy cascade of 3D flows, which is a step-by-step process by which energy migrates from large to small scales as a consequence of spatial equilibrium noted for fully developed turbulent flow.

The existence of attributes of two-dimensional turbulence is noted for large scale atmospheric flows in [244, 245] with data collected by aircraft. In Fig. 10.1, zonal and meridional wind spectra are shown again (as in Fig. 2.16) along with the spectrum of the potential temperature. It is readily noted that the k^{-3} variation precedes the $k^{-5/3}$ variation and one can estimate the energy content of these two spectra types; one notes the k^{-3} spectrum, which is associated with two-dimensional turbulence, accounts for 98.3% of the total energy and the rest shows variation following the $k^{-5/3}$ spectrum of the velocity field. According to an hypothesis due to Gage [107] the $k^{-5/3}$ variation is also an attribute of two-dimensional turbulence with a negative energy flux, which emerges in two-dimensional DNS of forced turbulence [219, 377]. It is well known [79] that strong magnetic field or rotation has the effects of suppressing one component of the velocity field. Additionally in atmosphere, stable density stratification inhibits vertical motions. One can show the effects of rotation in maintaining two-dimensionality, with the help of Taylor-Proudman theorem [265]. It has been correctly pointed out in [79], that two dimensionality is not perfect and residual non-planar motion is noted at small scales, as the case in Fig. 10.1 suggests for large k variation. Also, according to Kraichnan's theory, k^{-3} variation must occur for larger wavenumbers than those where one finds the $k^{-5/3}$ variation. Thus, the results in Fig. 10.1 have been noted in [102] as "paradoxical" (see also the sketch in [89] for the energy and enstrophy cascades for two-dimensional turbulence). We also note that according to Dewan [84] and VanZandt [422], the $k^{-5/3}$ range is due to long gravity waves breaking down to shorter waves, implying a positive energy flux, something akin to a 3D cascade of energy. Baroclinic instability [265] has also been identified as the source of enstrophy at large scales, which is responsible for the k^{-3} energy cascade. The k^{-3} spectrum has also been noted in [34, 67] from global weather data gathered. In [377], it has been conjectured that *there is an additional mechanism leading to the k^{-3} scaling at large scales that has previously been overlooked.* One of the main reasons for the reported investigation here is to underline such mechanisms for a flow computed by solving the Navier–Stokes equation without any artifice.

10.3 Role of Formulation in the Numerical Solution in Two-Dimensional DNS

At this stage, it is pertinent to point out the difference between formulation and methodology of the simulated Navier–Stokes equation here and typical formulations and methods used in the past. For example, in [377] the pseudo-spectral code of [157] has been used with

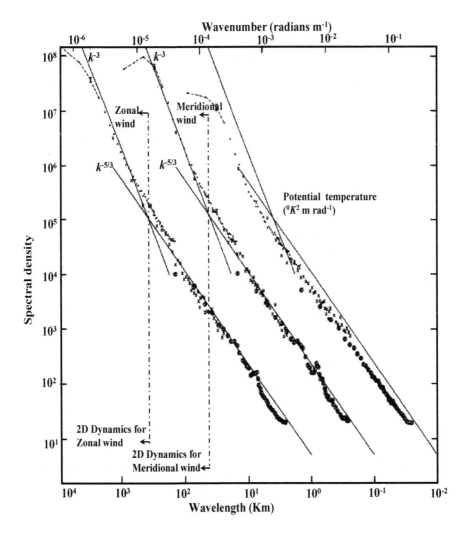

FIGURE 10.1
Kinetic energy and temperature spectrum near the tropopause from Nastrom and Gage [245].

512^2 and 2048^2 Fourier modes where the following equation has been solved.

$$\frac{\partial \omega}{\partial t} - \frac{\partial \psi}{\partial x}\frac{\partial \omega}{\partial y} + \frac{\partial \psi}{\partial y}\frac{\partial \omega}{\partial x} = \nabla \times f + \nu(-1)^{p+1}(\nabla^2)^p \omega - \alpha_e \omega$$

where ψ is the streamfunction, $\omega = -\nabla^2 \psi$ is the vorticity, f is an applied force, ν is the viscosity and α_e is a constant damping frequency representing Ekmann friction. If one uses $p = 1$ and $\alpha_e = 0$, then the actual vorticity transport equation is recovered. These simulations were performed in a periodic domain and thus any forcing through boundary condition(s) is not possible. To circumvent this, an explicit forcing is provided by $\nabla \times f$, which should be viewed as an artifice of modeling. The authors have also used a very high order hyperviscosity term with $p = 8$, which alters the dissipation spectrum of the transport equation drastically, thereby completely changing the dynamics of the turbulent flow. Fully developed turbulence is actually dictated by a fine balance between the energy and dissipation spectra and it is essential that these are represented accurately together. Unfortunately, in the literature, there is an overemphasis on discussing the energy spectrum, without detailed discussion of the dissipation spectrum. Such artifices used in periodic DNS of the modified Navier–Stokes equation allowed the authors to claim that they could *"restrict the parameter values of ... numerical simulations such that the forward cascade of enstrophy to small scales is unimportant."*

It is also well to remember that Kraichnan's [183] conjecture was that if energy is injected at some intermediate scale, the inverse cascade follows the $k^{-5/3}$ law following the Euler equation, without forcing and viscous action [102]. In contrast, in numerical simulations reported in [103, 146, 370, 377], the inverse cascade reached the size of the periodic box and the $k^{-5/3}$ range is disrupted, with vortices formed of a size comparable to that of forcing. With the constant Ekmann friction term $\alpha_e \omega$, a k^{-3} spectrum is observed at the largest scales as a consequence of balance between convection and friction. It has been specifically noted in [103, 203] that DNS do not strongly support the k^{-3} law for forced turbulence. According to Frisch [102], the simulations which obtain a clean k^{-3} spectrum do not *integrate the Navier–Stokes equation, but a modified equation with a high power of the Laplacian as the dissipation term*. In the presented two-dimensional DNS results here, no such artificial prescriptions are necessary in simulating an inhomogeneous flow with realistic boundary conditions, while the k^{-3} spectrum is cleanly obtained with (Chap. 7) and without (Chap. 5) heat transfer and in the presence of a variable pressure gradient over an airfoil (Chap. 9).

For the case of free stream excitation discussed in Chap. 4, we have seen that two-dimensional turbulence is created [199, 331], where flow transition was triggered by a convecting vortex in the free stream for a subcritical flow (with respect to TS waves) over a flat plate in the absence of a pressure gradient. The indicated flow visualization pictures show that the convecting vortex is created by a rotating and translating circular cylinder with a sufficiently high rotation rate, which destabilized the boundary layer forming underneath by a scouring mechanism created by circulatory effects of the convecting vortex. It is important to note that parallel dye filaments indicate the two dimensionality of the disturbance flow at the onset stage. Again, when the vortex has migrated past any station, the flow recovers the two-dimensional nature of the equilibrium flow. For subcritical flow, definitive forcing is necessary to strongly destabilize the flow without the appearance of TS waves, which is also known as bypass transition. However, in [323] and in Chaps. 5 and 7, it has been indicated that the mere presence of a TS wave does not imply that the actual transition is caused by it; instead it is dictated by the spatio-temporal wave front.

10.4 Dynamical System Representation of Turbulent Flows

In [102], an attempt has been made to formulate the Navier–Stokes equation as a stochastic dynamical system, with the statutory warning that the existence and uniqueness of the solution is not guaranteed for 3D flows. Establishment of an invariant measure to characterize the dynamical system is equally difficult. For finite dimensional chaotic systems, this is possible, but according to Frisch [102] such discrete maps are indeed *poor models* of the Navier–Stokes equation, as the invariant measure fills all of the available space, which cannot be true for realistic turbulence. Additionally, despite the presence of strange attractors for such low dimensional dynamical systems, different statistical properties are obtained even for the same initial conditions. While it is possible to represent stochastic dynamical systems by proper orthogonal decomposition (POD) [181], here we have shown transition to

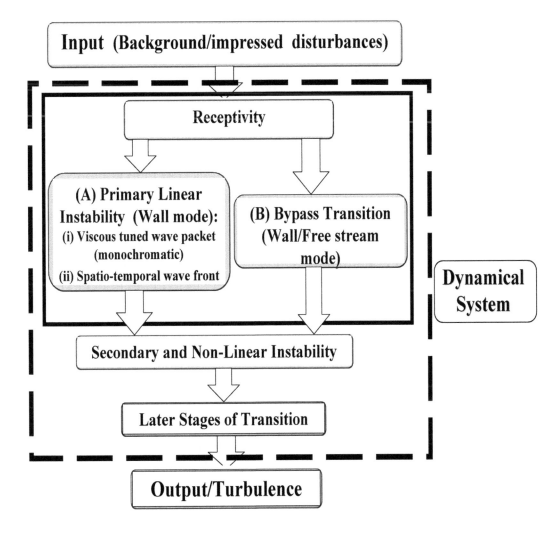

FIGURE 10.2
Block diagram showing different routes of turbulence generation.

fully developed turbulence by using a deterministic dynamical system for boundary layers with different pressure gradients excited harmonically at a spatial location where the flow is laminar. Thus, the input is also strictly deterministic and unique.

Different stochastic and deterministic dynamical system approaches to fluid turbulence are given in [34, 235, 151]. In Chap. 6, we have used a stochastic dynamical system approach using POD to characterize bluff body flow instability. For flow past a wall-bounded shear layer here, we look for a deterministic dynamical system model. A block diagram showing the way fluid flow transition occurs in a deterministic route due to the receptive nature of fluid flow is shown in Fig. 10.2. In this figure, the transfer function portrays both the classical TS waves and bypass routes. In a scenario projected in [235], the primary instability depends upon the amplitude of excitation, where it is expected that a low amplitude wall excitation gives rise to TS waves, while higher amplitude excitation need not show the presence of TS waves. However, in Chap. 5 and [349, 350], it has been shown that harmonic monochromatic excitation produces the spatio-temporal wave front, apart from the TS wave predicted by the linear spatial instability theory. This is indicated in block (A) of Fig. 10.2 corresponding to wall excitation. One of the features of the development of linear receptivity analysis in Chap. 3 based on the Orr–Sommerfeld equation was to show that the blocks representing receptivity and linear instability can be obtained by a unified approach and in essence are one and the same for both blocks. This figure also indicates the bypass route in block (B), which can be due to either wall or free stream excitation (as shown in [331]). The latter instability has been shown in Chap. 4 to be triggered by a convecting vortex which imposes an adverse pressure gradient causing unsteady small scale separations, depending upon the vortex strength, sign and distance from the shear layer. In [323] and Chaps. 5 and 7, a case of wall excitation has been studied and briefly shown that even when TS waves are present, the actual transition process is dominated by the spatio-temporal front. Although this spatio-temporal front has been shown in [349, 350] to originate via a linear mechanism, its continual spatio-temporal growth leads to nonlinear saturation which induces unsteady separation at the wall. Thus, during the later stages of the transition process, unsteady separation leads to intermittency and fully developed turbulent flow which are the common features for both routes indicated by blocks (A) and (B) in Fig. 10.2.

In the present book, details of this deterministic route of transition to fully developed turbulence have been described unambiguously and the dynamical system needs a definitive input excitation for flow transition, which can or cannot be quantified. The complete block shown by the dashed line in Fig. 10.2 defines the transfer function of the deterministic dynamical system. Also, the accuracy of the adopted numerical scheme ensures that the same procedure can be used for evaluating the unexcited equilibrium flow and the turbulent flow created by a deterministic input as indicated by the route(s) in Fig. 10.2. Thus, the computed transition route requires extreme accuracy of the numerical methods, which must obey the physical dispersion relation [347, 354]. Most accurate methods suffer from numerical instability at high wavenumbers and to prevent it the adaptive multi-dimensional filtering technique of [33] has been used for all results reported in Chaps. 5, 7 and 9 and in [323]. In Chap. 5 and in the study reported in [323], it has been been shown that two-dimensional flows have the expected k^{-3} spectrum as shown in geophysical flow in Fig. 10.1. Such a spectrum also emerges in the case of vortex-induced instability, discussed in Chap. 4, obtained numerically. However, in making such an assertion, one must also realize the role of the chosen computational domain for DNS, as discussed next.

10.5 Role of the Computational Domain

Use of a simultaneous blowing suction (SBS) strip to excite a zero pressure gradient was reported in [98], yet no local solution component or spatio-temporal front was reported in this reference, as compared to the result shown in Fig. 5.14, which identifies all three components clearly. The role of the computational domain is explained here with the linear spatial stability analysis results in Fig. 10.3. The linear instability studies use a parallel flow model and the connection between stability and receptivity approaches is established for the flow shown in Fig. 10.3(a). This is identical to Fig. 5.12, except that we identify the computational domain used in [98] and in [323]. The linear unstable region in the (Re_{δ^*}, β_0)-plane formed within the neutral curve is shown in Fig. 10.3(b). If the flow adjusts to local conditions and assumptions of linear spatial theories are valid, then the neutral curve is useful in tracking constant frequency disturbances in a growing boundary layer.

In Fig. 10.3(b), the computational domain used here is compared with the narrow domains used in [98] where the main intention was to exhibit the presence of TS waves. As shown in the schematic, for the chosen frequency identified by $F = $ constant line, the exciter is placed where the TS wave is damped. However, as the disturbance travels downstream, it enters the unstable region and exhibits a range over which the disturbance grows. Upon exiting the unstable region, the disturbance will decay again following the linear theory. When the Navier–Stokes equation is solved numerically, location of outflow is important as the fluid elements upon exiting the domain do not affect the flow inside the domain. To circumvent this, the outflow is placed where $Re_{\delta^*} = 6000$ and one tracks the disturbance evolution over an extended region, especially after the nonlinear stage of disturbance growth has started. Following the nonlinear action onset, upstream influence is noted and it is due to this reason that domain size is important. If the disturbance packet leaves the computational domain before the onset of nonlinearity, then such upstream influences will not be felt inside the computational domain. Also, the leading edge of the plate is important due to the nonparallel effects due to growth of the shear layer in the vicinity of the leading edge and flow singularity at the leading edge for disturbance evolution. In Chap. 5 and in [331, 323], the inflow of the domain is thus positioned always upstream of the leading edge. The physical problem considered here is a two-dimensional flow over a semi-infinite flat plate with the leading edge as the stagnation point. The computational domain is defined by, $-0.05 \leq x \leq 120$; $0 \leq y \leq 1.5$. In the wall-normal direction, a stretched tangent hyperbolic grid is used, given for the jth point as

$$y_j = y_{max} \left[1 - \frac{\tanh[b_y(1 - \eta_j)]}{\tanh b_y} \right]$$

Here $y_{max} = 1.5$ and $b_y = 2$ have been used for appropriate grid clustering and a total of 401 points are taken in the wall-normal direction. In the streamwise direction, the domain is divided in two segments: the first segment extends from $x_{in} = -0.05$ to $x_m = 10$, which contains points distributed by a tangent hyperbolic function, and for the ith point it is obtained as

$$x_i = x_{in} + (x_m - x_{in}) \left[1 - \frac{\tanh[b_y(1 - \xi_i)]}{\tanh b_y} \right]$$

In the second segment from $x_m = 10$ to $x_{out} = 120$, uniformly distributed points are taken. Altogether, 4501 points have been used in the streamwise direction.

In Fig. 3.40, a typical result is shown for the case $\alpha_1 = 0.1$; $F = 3 \times 10^{-4}$, i.e., the maximum wall-normal input disturbance $((v_d)_{max})$ is taken as $0.1U_\infty$, which is larger than the value considered in [98], which studied a case of $(v_d)_{max} \simeq 0.003$. For this higher amplitude of excitation, one notices three distinct components of solution at an early time of $t = 8$ in Fig. 3.40. Of specific interest is the presence of the spatio-temporal front, which was not possible to capture with such a narrow domain taken in [98]. When the (V, ω)-formulation was used as in [98], exactly the same results were obtained as were obtained using the (ψ, ω)-formulation. However for reasons unknown, in [98] the local component of the solution was also not shown.

To understand the transition process clearly, a lower amplitude case with $\alpha_1 = 0.002$; $F = 1 \times 10^{-4}$ was considered in Chap. 5, for which the amplitude of excitation is lower than that considered in [98]. In Fig. 5.17, we have noted the evolution of disturbance at $y = 0.0057$ (which is the tenth line from the wall) by plotting streamwise disturbance velocity as a function of x for the indicated time instants. The exciter is located at $x_{ex} = 1.5$, near the leading edge of the flat plate, where disturbance is created at a location which is stable

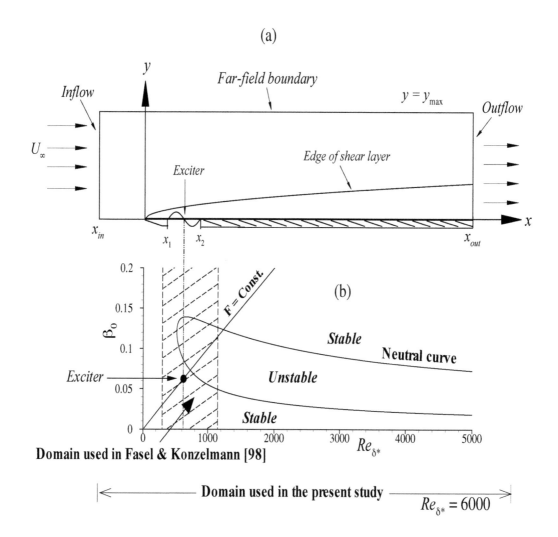

FIGURE 10.3
Theoretical neutral curve in the (Re_{δ^*}, β_0)-plane with stable and unstable parts identified.

according to the linear spatial stability theory. For this lower amplitude case, the spatio-temporal front is shown in the top frame at $t = 20$. In [98], the computational domain was so small that even the complete TS wave packet (as seen in the left of all frames in Fig. 5.17) was not captured and finding the spatio-temporal front is out of the question.

It is also noted in Fig. 5.17 that the TS wave packet changes slightly with time — an attribute of the nonlinear, nonparallel effect. The main conclusion in [98] was that there are no nonlinear and nonparallel effects when one solves the Navier–Stokes equation as the receptivity problem. This TS wave packet is essentially a progressive wave which attenuates due to stability properties (modified by nonlinear, nonparallel effects) shown in Fig. 10.3, and remains virtually localized in space, as the wave is continually created by excitation at the wall. In contrast, the spatio-temporal front propagates in space and time while growing significantly to a larger amplitude, as compared to the TS wave packet. In Chap. 5, it has been shown that the origin of the spatio-temporal front is due to a linear mechanism whose onset can be captured by the solution of the Orr–Sommerfeld equation by the Bromwich contour integral method. It is readily evident that by $t = 200$ the spatio-temporal front amplitude grows so large that nonlinear effects are noted. With the passage of time, this nonlinear effect also distorts the symmetry of the packet and induces unsteady separation on the wall below it, as noted in the bottom frame at $t = 260$. Thus, it is apparent that the disturbance evolution is dominated more by the spatio-temporal front, rather than the TS wave packet - as was commonly believed.

In Fig. 10.4, the maximum amplitude (u_{dm}) variation with time is shown for the case of Fig. 5.17. Based on the physical processes of this disturbance growth, five different stages have been identified. As noted in Chap. 5, this front can be traced to the linear mechanism governed by the Orr–Sommerfeld equation, which are Stages I and II in Fig. 10.4. First and foremost, we note that despite the excitation at the wall being at a single frequency, the spatio-temporal front is not monochromatic and/or multi-periodic. There are qualitative differences between Stages I and II. In Fig. 10.5, a corresponding explanatory bilateral Laplace transform of u_d for the spatio-temporal front is shown for Stage I. One readily observes the spectrum of the front to be wide band. At $t = 20$, there are few distinct lobes around central wavenumbers with the peak located at $k \simeq 15$; however, at the later time of $t = 30$, the peak shifts to a lower value and a larger number of narrower lobes appear due to dispersion which causes u_{dm} to reduce with time in Fig. 10.4 in Stage I, up to $t \simeq 60$.

In Stage II during $t = 60$ to 85, the spatio-temporal front remains localized almost at the same value of k, while the amplitude grows due to linear instability, as shown in Fig. 10.6. Also in Stage II, the amplitude at very low wavenumbers increases by almost two orders of magnitude, as compared to that in Stage I. In Stage III of disturbance evolution from $t \simeq 85$ to 200, the maximum amplitude grows nonlinearly with the appearance of higher wavenumbers with time. Also, the disturbance amplitude at low wavenumbers keeps increasing by a factor of nearly ten. This explains the nonlinear growth of u_{dm} during Stage III in Fig. 10.7. The following Stage IV corresponds to the rapid growth of disturbances which is characterized by the appearance of a distinct k^{-3} spectrum (at around $t = 233$) due to nonlinear growth and its saturation, as shown in Fig. 10.8. In the physical plane, this stage corresponds to the appearance of unsteady separations at the wall. As noted earlier, the bypass route of transition reported in [323, 331] for free stream and wall excitations exhibits this k^{-3} spectrum, as shown in Fig. 10.8.

10.5.1 Renewal Mechanism and Intermittent Nature of Turbulence

In discussing the results of Fig. 5.17, we have noted that the turbulence is a deterministic consequence of a single spatio-temporal front convecting downstream, i.e., once the spatio-temporal front is created, its growth and nonlinear saturation lead to small scale unsteady separations on the wall, which culminate in the intermittent nature of the wall bounded flow. In this figure, the evolution of a single front for the case of $\alpha_1 = 0.002$ was tracked for a moderate frequency excitation ($F = 10^{-4}$). This could lead one to conclude that this turbulence creation is a buffeting problem, i.e., sustenance and creation of turbulence would require subsequent creation of spatio-temporal fronts, as suggested in [229], by a continuous stream of input disturbances due to free stream turbulence. It is relevant to ask whether the renewal mechanism of spatio-temporal fronts is due to intrinsic or extrinsic dynamics. This question was answered in Chap. 5 by simulating the cases shown in Figs. 5.19a and 5.19b, for a couple of higher amplitude excitation cases.

In Fig. 5.19a, the streamwise disturbance velocity component for the case of $\alpha_1 = 0.01$ and $F = 10^{-4}$ is shown when the input excitation level of wall-normal velocity located at the same streamwise position is increased to 1% of the free stream velocity. In the top frame of the figure at $t = 94$, a single spatio-temporal front is marked **A**. In the following frame at $t = 164$, nonlinear saturation of this spatio-temporal front is noted. It is evident from the figure that unsteady separations are created on the wall, underneath this front, during

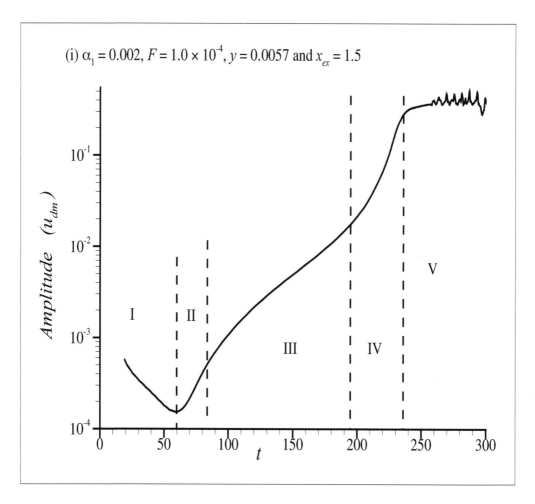

FIGURE 10.4
Maximum amplitude of streamwise disturbance velocity (u_{dm}) of the spatio-temporal front shown as a function of time. Indicated are five stages of disturbance evolution to turbulence.

this stage of disturbance evolution. The most important aspect of this propagating spatio-temporal front during this stage is the spawning of another spatio-temporal front, marked **B** in the frame. The fact that this front is created upstream of **A** is very significant, as this event negates the assumptions used in the parabolized stability equation (PSE) often used to study flow instability. As time progresses, the amplitude of front **B** also grows rapidly and its nonlinear distortion is more rapid due to the presence of **A**, while the gap between the two fronts reduces with time, as shown in the frames at $t = 194$ and 244. Also in the frame at $t = 244$ we noted spawning of another front, marked **C**. In the following frame at $t = 324$, one notices amalgamation of fronts **A**, **B** and **C**, while two new fronts **D** and **E** are formed upstream of the leading packet. These trailing packets grow with time, as the leading part of the first packet leaves the computational domain. By $t = 345$ (not shown), trailing fronts **D** and **E** have grown and become part of the leading packet. Thereafter, a weaker front is seen to form in the time sequence shown in Fig. 5.19a.

When the input wall-normal velocity at the wall is increased to 5% of the free stream

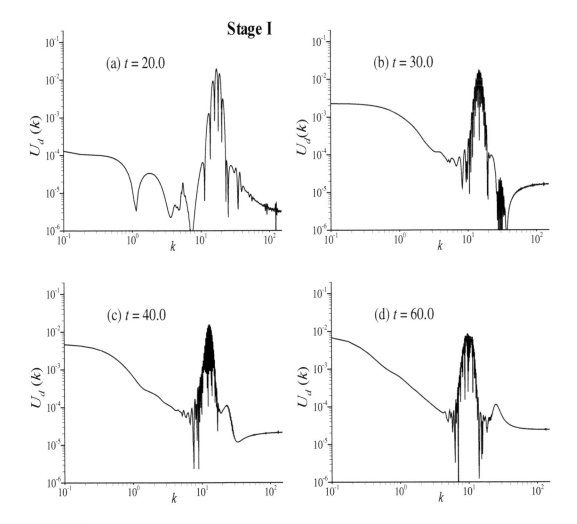

FIGURE 10.5
Bilateral Laplace transform of u_{dm} of the spatio-temporal front shown for Stage I.

velocity, results in Fig. 5.19b show the appearance and nonlinear growth of the first spatio-temporal front (**A**) earlier at $t = 34$. In the second frame at $t = 184$, four spatio-temporal fronts are noted at different levels of evolution. In the following frame at $t = 239$, one notes the amalgamation of **A**, **B**, **C** and **D**, while two new fronts are spawned upstream, shown as **E** and **F**. The nonlinear growth is so strong that at $t = 254$ another front is noted which also has suffered nonlinear growth and saturation. This process continues, as the calculations up to $t = 390$ confirm (not shown), with new spatio-temporal fronts appearing, establishing the renewal of spatio-temporal fronts, while the distinct identities of the TS wave packet and the spatio-temporal fronts are maintained. It has now been established that a time harmonic excitation at a fixed location gives rise to a self-regenerating sequence of spatio-temporal fronts; responsible for the k^{-3} spectrum noted in two-dimensional turbulence. As there appears to be some difference between *free* and *forced turbulence*, we have studied other wall bounded flows where no explicit forcing is needed for the ensuing transition process and this has been discussed in Chap. 9.

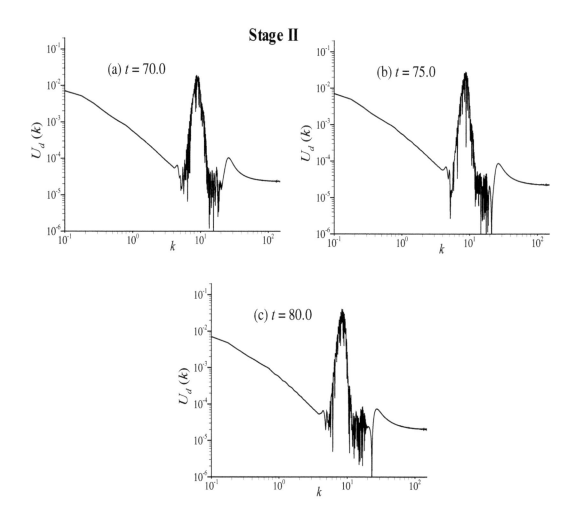

FIGURE 10.6
Bilateral Laplace transform of u_{dm} of the spatio-temporal front shown for Stage II.

10.6   Free and Forced Turbulence

In Chap. 9, flow past a natural laminar flow (NLF) SHM1 airfoil has been computed and compared with experimental results in [104]. Information about orthogonal grid generation around the airfoil, the solver for the vorticity transport equation and stream function equation and the multi-dimensional filters (including adaptive filters) are given in [32]. In computing a case for zero angle of attack for the airfoil, an O-grid with 5169×577 points for a Reynolds number (based on chord (c) and free stream speed) of 10.3 million is used along with the same formulation and numerical methods employed for the semi-infinite zero pressure gradient flat plate flow. As the flow over the airfoil experiences a variable pressure gradient, with a prominent adverse pressure gradient noted near the trailing edge of the airfoil, it is not necessary to excite the flow explicitly – as numerical errors are adequate to cause bypass transition over the top and bottom surfaces of the airfoil.

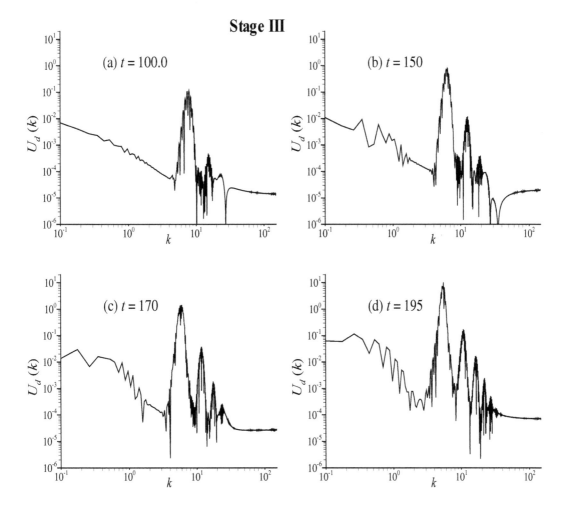

FIGURE 10.7
Bilateral Laplace transform of u_{dm} of the spatio-temporal front shown for Stage III.

In Fig. 10.9, the absolute magnitude of the bilateral Laplace transform of the azimuthal component of velocity for a grid line which is at a distance of $0.0000124c$ from the surface of the airfoil is shown at the indicated times. The top two frames are for the unexcited case and one notices clearly the presence of k^{-3} spectrum for $t = 3$ and 4. Bypass transition is noted after $x = 0.6c$; presence of the k^{-3} spectrum is indicative of it.

This flow has been also calculated when an SBS exciter is employed on the top surface of the airfoil at $x = 0.25c$ with a wall-normal velocity amplitude of 0.1% of the free stream speed at a very low nondimensional frequency of $F = 1.11441 \times 10^{-5}$, starting with the undisturbed flow calculated up to $t = 1.5$ and thereafter exciting the flow by the SBS strip. Computed solutions at $t = 5.5$ and 7 are shown in the bottom two frames of Fig. 10.9. Once

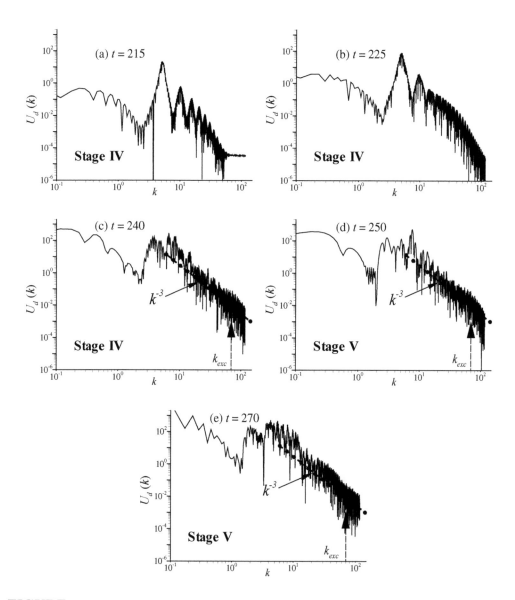

FIGURE 10.8
Bilateral Laplace transform of u_{dm} of the spatio-temporal front shown for Stages IV and V.

again one notes the k^{-3} spectrum, indicative of bypass transition. Also, the computed flow field indicates that this bypass transition occurs upstream at $x = 0.4c$, as compared to the case where no explicit excitation is applied. The size of the SBS strip dictates the excitation wavenumber (k_{exc}) and this is marked in the bottom frames for excited cases by a vertical arrowhead. It is noted that the energy input through excitation is at an intermediate length scale. The presence of explicit excitation causes some differences of the spectrum at lower

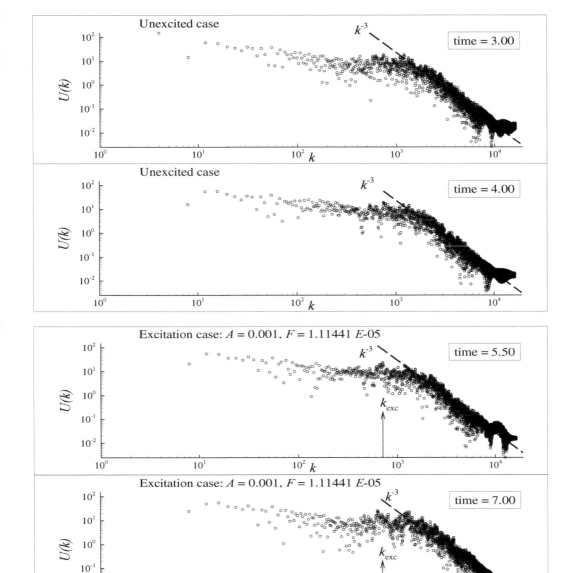

FIGURE 10.9
Bilateral Laplace transform of the azimuthal component of velocity for a grid line at a distance of $0.0000124c$ from the airfoil surface, shown at the indicated times.

wavenumbers, as compared to the unexcited case; still the k^{-3} spectrum is present in both cases. This is viewed along with the spectrum shown in Fig. 10.8 for the case of a zero pressure gradient boundary layer, where the excitation wavenumber (k_{exc}) is in Stages IV and V are at high end of the spectrum, although in absolute terms the airfoil case k_{exc} is an order of magnitude higher.

We have purposely included the airfoil case to show the similarity between the different two-dimensional flow configurations with respect to the finite streamwise length of the computational domain of the airfoil with the semi-infinite flat plate case. However, in both these cases we have been able to show the presence of the k^{-3} spectrum, as noted in the large scale atmospheric dynamics shown in Fig. 10.1. Thus, we can conclude that the presence of the k^{-3} spectrum for lab scale phenomenon is similar to what has been recorded in atmospheric data, which is indicative of the universal nature of the spectrum for both free and forced turbulence cases.

11

Selected Problems

Chapter 1:

1.1 For water flowing over a flat surface with velocity U_∞, a long water vapor bubble is formed. The equilibrium flow can be considered two-dimensional. Discuss the Kelvin–Helmholtz instability of the bubble interface (neglecting curvature and surface tension effects at the interface) for two- and three-dimensional disturbances. Which of the two is more conducive in breaking up the bubble by instability. Take the density ratio to be approximately 1:1000 and the velocity inside the bubble to be about $-0.005U_\infty$.

1.2 For the Kelvin–Helmholtz interfacial instability of the shear flow of the same fluid, i.e., $\rho_1 = \rho_2$, obtain the phase speed and the group velocity of the disturbance field.

1.3 a) For the Kelvin–Helmholtz instability of two-dimensional flows caused by shear of the same liquid across an interface, find the wavenumber range for instability.

b) What is the direction of propagation of the unstable wave obtained in (a)?

1.4 For the two velocity profiles shown in Fig. P1, which is more unstable? What would be the phase speed and group velocity of the disturbance field for each case? Use inviscid theory.

1.5 Show that Kelvin–Helmholtz (K-H) instability due to shear is most severe for two-dimensional disturbances. What is the role of K-H instability for the energy spectrum for general flow fields and especially for a two-dimensional mean flow field.

1.6 For air flowing with velocity U_1 over water (which is flowing with velocity U_2), show the lowest wavenumber (k_{min}) for disturbances arising out of 2D Kelvin–Helmholtz

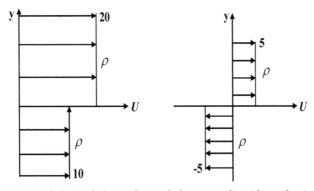

Fig. P1 Kelvin–Helmholtz instability of shear flow of the same liquid at the interface shown at $y = 0$.

instability which will be amplified at the interface is given by

$$k_{min} = \frac{g}{(U_1 - U_2)^2} \left(\frac{\rho_{\text{water}}}{\rho_{\text{air}}} - \frac{\rho_{\text{air}}}{\rho_{\text{water}}} \right)$$

1.7 Starting from the mathematical formulation and analysis of Kelvin–Helmholtz instability, describe the Rayleigh–Taylor instability by using $U_1 = U_2 = 0$.

1.8 In an estuary, salty water is over fresh water locally. The densities of fresh and salty water can be taken as 1020 kg/m^3 and 1040 kg/m^3, respectively. Calculate the cut-off wavenumber below which unstable waves grow at the interface, considering surface tension at the interface be 0.020 N/m. Find the wavelength at which the growth rate is maximum.

1.9 Is there an upper limit for critical Reynolds number for pipe flow?

1.10 Obtain the dispersion relation for Rayleigh–Taylor instability when the fluids on either side of the interface are of finite depths, H_1 and H_2.

1.11 If the vessel for problem 1.10 experiences an added acceleration f in the direction of gravity, what will be the dispersion relation?

Chapter 2:

2.1 What are the mechanisms for two- and three-dimensional flows by which energy is created at higher wavenumbers for turbulent flows?

2.2 Derive the energy and dissipation spectra for high Reynolds number turbulent flows in the inertial subrange.

2.3 Relate the time scales of mean and fluctuation fields of an equilibrium turbulent flow. Does this justify the usage of *unsteady RANS* in solving unsteady flows?

2.4 What is the closure problem in turbulence modeling for Reynolds-averaged Navier-Stokes (RANS) equation? What is the basis by which it is solved in engineering codes?

2.5 Explain what is meant by equilibrium turbulence. How can this be used in justifying "Unsteady RANS"? How is this used in constructing sub-grid scale (SGS) model in LES?

2.6 What are the justifications for relating turbulent stresses to the mean strain rates in turbulence models or subgrid scale stress models?

2.7 How is vortex stretching responsible for energy cascade in three-dimensional turbulence?

2.8 What is backscatter or inverse energy cascade? Discuss it in the context of enstrophy cascade in two-dimensional turbulence.

2.9 A viscous incompressible flow is established by a constant pressure gradient between two parallel planes, one of which is also imparted a constant velocity U_0. The two planes are at H distance apart. Find the velocity distribution and the limiting pressure gradient for the flow to separate.

2.10 How is the Kolmogorov length scale important in DNS of turbulent flows?

2.11 Derive the asymptotic suction required at the wall of a two-dimensional boundary layer to maintain the flow to be similar.

2.12 Is the dispersion relation always dependent on the governing differential equation? Obtain the dispersion relation for Rayleigh–Taylor instability.

2.13 Show that for an irrotational flow, nonlinearity of the convection term merely redefines the pressure and thus seldom gives rise to instability.

2.14 Solve the Blasius boundary layer equation $f''' + \frac{1}{2}ff'' = 0$ in $0 \leq \eta \leq \eta_{max}$ computationally as a boundary value problem, where η is the similarity profile introduced in the text. Choose a value of $\eta_{max} = 12$ for your calculation. Is there an inflection point for the Blasius profile?

2.15 Consider the following 1D convection equation,

$$\frac{\partial u}{\partial t} + c\frac{\partial u}{\partial x} = 0, \qquad c > 0 \tag{2.1}$$

Obtain the numerical amplification factor $|G|$ and normalized group velocity $|V_{gN}/c|$, for the following space-time discretization schemes used for this model equation: (i) OUCS3-RK_4, (ii) CD2-RK_4 and (iii) QUICK-RK_4 schemes.

2.16 Consider a wave packet with $k_0 = 50$ in a domain of length 10, with 4096 uniformly distributed points following the initial condition given by

$$u(x,t)|_{t=0} = e^{-\alpha(x-x_0)^2} \cos[k_0(x - x_0)] \tag{2.2}$$

where x_0 is the center of the wave packet at $t = 0$, whose central wavenumber is given by k_0. Solve Eq. (2.1) using the CD_2 and OUCS3 spatial discretization schemes with the RK_4 time integration method. Choose $\alpha = 30000$, $c = 0.1$ and CFL number $N_c = 0.1$.

Chapter 3:

3.1 For a periodic vortex train convecting at $Y = 20\delta^*$ over a zero pressure gradient (ZPG) boundary layer, compare the receptivity of the boundary layer when the train moves at (a) $c = U_\infty$ and (b) $c = U_\infty/3$. The distance between successive vortices is $a = 100\pi\delta^*$.

3.2 Why is the local solution upstream of a harmonic point source placed on the wall of a ZPG boundary layer independent of Reynolds number?

3.3 Why do ZPG boundary layers display Klebanoff or breathing mode in response to a very low frequency disturbance field? Is there a lower cut-off frequency for this?

3.4 A ribbon is placed outside a boundary layer over a ZPG flow and is vibrated electromagnetically at a constant physical frequency. Formulate the corresponding receptivity problem and state qualitatively the nature of the response field within the boundary layer.

3.5 Identify the eigenvalue(s) in the complex wavenumber plane for a ZPG boundary layer for $Re = 1000$, for a constant frequency excitation in the ranges (i) $0.003 < \bar{\omega}_0 < 0.03$ and (ii) $0.03 < \bar{\omega}_0 < 0.06$. Give justifications for your identification.

3.6 A pusher propeller driven aircraft wing is being affected due to tip vortices of the propeller blades. How will it affect the transition on the wing? Give a qualitative answer with justification.

3.7 From the neutral curves for boundary layers experiencing different pressure gradients (favourable and adverse), identify viscous and inviscid instabilities of these flows.

3.8 A boundary layer is excited at the wall and the free stream, simultaneously. For this flow, one uses the compound matrix method (developed for wall excitation problems) and obtains eigenvalues also on the left half of wavenumber plane. What is the significance of this result?

3.9 Show the coupling between wall modes and the free stream modes of the Orr–Sommerfeld equation.

3.10 Explain the role of essential singularity in creating a local solution. How does it help in explaining receptivity to any arbitrary excitation for boundary layers?

3.11 How do you obtain a differential equations governing the eigenfunctions of the Orr–Sommerfeld equation (ϕ) using the compound matrix method?

3.12 Suction is used effectively to control transition. A uniform suction is applied on a surface of negligible curvature for two-dimensional flow. Derive the flow field including displacement and momentum thickness for the case where the normal velocity induced by suction is the same everywhere. What can you say about the stability property of the flow field based on the shape parameter?

3.13 Write a computer program that solves for the eigenvalues of the Orr–Sommerfeld equation by grid search method using the compound matrix method for the Falkner-Skan similarity profile.

3.14 Compare the neutral stability curves for two-dimensional free shear layer and boundary layer flows.

3.15 By an analysis of the governing equation at the wall, discuss the effects of heat transfer and wall temperature at the wall for the instability of incompressible flows.

3.16 Write the non-dimensional viscous stability equations (Orr–Sommerfeld equation) for the combined Couette–Poisseuille flow of problem 2.9. Apply also Rayleigh's stability theorem for this velocity profile.

3.17 Using the Rayleigh stability equation, show that oscillations with a purely real c and complex α are excluded.

3.18 Would you solve for the eigenfunction by solving the following equation suggested in the compound matrix method

$$y_1\phi''' - y_3\phi' + y_5\phi = 0$$

3.19 It is often suggested in the compound matrix method that you can solve for the eigenfunction of the Orr–Sommerfeld equation using the compound matrix method by solving the equation given above in problem 3.18 or

$$y_1\phi'' - y_2\phi' + y_4\phi = 0$$

. Which one would you recommend for receptivity to the wall excitation problem?

3.20 Eigenvalues obtained using the compound matrix method for spatial analysis show wavenumber to have negative real and imaginary values. What can you say about the disturbance field?

Chapter 4:

4.1 For the Beltrami flow ($\vec{V} \times \vec{\omega} \equiv 0$), show that the non-linear convection term merely redefines pressure. For such a flow, write the governing equation for disturbance mechanical energy.

4.2 A single irrotational vortex of strength $\Gamma = 30U_\infty\delta^*$ is located instantaneously at $x_\nu = 100\delta^*$, $y_\nu = 20\delta^*$ from the leading edge of a flate plate over which a ZPG shear layer forms. Calculate the imposed pressure gradient parameter $m = \frac{x}{U_e}\frac{dU_e}{dx}$ as a function of x (measured with respect to the leading edge) numerically. Identify the location where the effects of the vortex will be most destabilizing.

4.3 How do you track disturbances in the $(Re, \bar{\omega}_0)$-plane for: i) a constant frequency wall excitation and ii) a steadily convecting disturbance in the free stream? Examples with appropriate diagrams for both cases are to be shown.

4.4 For a flat plate boundary layer with uniform suction, find (i) the velocity profile; (ii) the shape factor and (iii) find the condition for unsteady pressure generation.

Chapter 5:

5.1 Derive the governing equation for E_d given in Eq. (5.8).

5.2 What are the causes for the formation of spots in transitional and turbulent flows?

5.3 Compare the N factors in Fig. 5.25 with that used in the e^N-method in Sec. 3.5.3 and 3.5.4.

5.4 From the data provided in Fig. 5.25, develop an empirical correlation of amplifications factor with the amplitude control factor $(\alpha_1) : N = a + b\ln(\alpha_1)$ for moderate/high frequency cases. Describe quantitatively the effects of x_{ex} and F.

Chapter 6:

6.1 Using Landau's equation, explain vortex shedding behind a bluff body as one of supercritical stability following linear instability. Obtain the relation for the Strouhal number with frequency obtained by the linear instability theory.

6.2 a) What is the source of anomalous modes of the first kind (T_1) obtained by POD for universal instability of flows? Explain it with the help of the Navier–Stokes equation.
    b) Identify its nature as the linear and/or nonlinear mechanism.
    c) Can these modes be related to the Stuart–Landau equation for supercritical stability?
    d) Using the Stuart–Landau equation, explain subcritical instability of channel flow.

6.3 For a flow in a channel, for which $(Re)_{cr} = 5768$, subcritical instability may be triggered by large amplitude disturbance. This large disturbance is a fraction $(= \beta)$ of the threshold amplitude required for the given pressure gradient that is driving the flow. Let the corresponding Reynolds number another fraction $(= \theta)$ of the flow $(Re)_{cr}$. Find the pressure gradient at which this amplitude disturbance will trigger subcritical instability.

6.4 Simplify Eq. (6.22) for regular mode pairs.

6.5 Generalize Eqs. (6.23)–(6.25), when more than two domain modes are present.

Chapter 7:

7.1 Derive the boundary layer equations given in Eqs. (7.8)–(7.11) starting from the Navier–Stokes equation given in Eqs. (7.1)–(7.3).

7.2 Obtain the governing equation for Schneider's profile given in Eqs. (7.12)–(7.13).

7.3 Solve the system of Eqs. (7.14) and (7.15) numerically for different values of K.

7.4 Using the second components described in Eq. (7.49), obtain the dispersion relation for the wall excitation case given in Eq. (7.114). Comment on one possible variation of the dispersion relations depending on the receptivity to wall heat transfer processes.

7.5 Write a computer program for the grid search method to obtain the eigenspectrum shown in Fig. 7.8.

Chapter 8:

8.1 What are the major roles played by the pressure gradient on flow transition over an aircraft wing for two- and three-dimensional mean flows?

8.2 What are the reasons for the formation of stationary streaks found by surface flow visualization techniques on a swept wing?

8.3 Discuss various types of three-dimensional nonlinear breakdowns detected for wall bounded shear flows.

8.4 Discuss the role of the three dimensionality of disturbances in transitional flow.

Bibliography

[1] I.H. Abbot and A.E. von Doenhoff. *Theory of Wing Section.* Dover Publications, New York, 1959.

[2] N. Afzal and T. Hussain. Mixed convection over a horizontal plate. *ASME J. Heat Transfer,* 106:240-241, 1984.

[3] L.B. Aizin and N.F. Polyakov. Acoustic generation of Tollmien-Schlichting waves over local unevenness of surfaces immersed in stream. *Preprint 17, Akad. Nauk USSR, Siberian Div., Inst. Theor. Appl. Mech., Novosibirsk* (In Russian), 1979.

[4] L. Allen and T.J. Bridges. Numerical exterior algebra and the compound matrix method. *Numer. Math.,* 92:197-232, 2002.

[5] L. Allen and T.J. Bridges. Hydrodynamic stability of the Ekman boundary layer including interaction with a compliant surface. *Euro. J. Mechanics B/ Fluids,* 22:239-258, 2003.

[6] L.D. Allen and F.M. Burrows. Flight experiments on the boundary layer characteristics of a swept back wing. *Cranfield, College of Aero. Rept. No. 104,* 1956.

[7] J.D. Anderson Jr. *Fundamentals of Aerodynamics,* McGraw Hill, New York, 1991.

[8] A. Anscombe and L.N. Illingworth. Wind tunnel observations of boundary layer transition on a wing at various angles of sweep back. *ARC R and M 2968,* 1952.

[9] G. Arfken. *Mathematical Methods for Physicists,* 3rd edition. Academic Press, Orlando, 1985.

[10] D. Arnal. Description and prediction of transition in two-dimensional incompressible flow. *AGARD Report* No. 709, 2.1-2.71, 1984.

[11] D. Arnal. Three-dimensional boundary layer: Laminar-turbulent transition. *AGARD Report* No. 741:1-34, 1986.

[12] D. Arnal, G. Casalis and J.C. Jullien. Experimental and theoretical analysis of natural transition on infinite swept wing. In the IU-TAM Symp. Proc. on *Laminar-Turbulent Transition,* (Eds.: D. Arnal and R. Michel), Springer-Verlag, 311-326, 1989.

[13] D. Arnal, R. Coustols and J.C. Jullien. Experimental and theoretical study of transition phenomena on infinite swept wing. *La Recherche Aerospatiale,* No. 1984-4, 275-290, 1984.

[14] D. Arnal, M. Habiballah and E. Coustols. Laminar instability theory and transition criteria in two- and three-dimensional flow. *La Recherche Aerospatiale,* No. 1984-2, 1984.

[15] D. Arnal and J.C. Jullien. Three-dimensional transition studies at ONERA/CERT. *AIAA Paper* No. 87-1335, 1987.

[16] D.E. Ashpis and E. Reshotko. Vibrating problem revisited. *J. Fluid Mech.*, 213:531-547, 1990.

[17] B. Auld. *Acoustic Fields and Waves in Solids*. Wiley-Interscience, New York, 1973.

[18] D. Barkley. Linear analysis of the cylinder wake mean flow. *Europhys. Lett.*, 75(5):750-756, 2006.

[19] D.W. Bartlett. Application of a supercritical wing to an executive type jet transport model. *NASA*, TM-X 3251 (June 1975), 1975.

[20] G.K. Batchelor. Note on a class of solutions of the Navier-Stokes equations representing steady non-rotationally symmetric flow. *Quart. J. Mech. Appl. Math.*, 4:29-41, 1951.

[21] G.K. Batchelor. Computation of the energy spectrum in homogeneous two-dimensional decaying turbulence. *Phys. Fluids,* 12 (suppl. II):233-239, 1969.

[22] G.K. Batchelor. *An Introduction to Fluid Dynamics,* Cambridge Univ. Press, UK, 1988.

[23] F. Bauer, P. Garabedian and D. Korn. *A Theory of Supercritical Wing Sections, with Computer Programs and Examples,* Springer Verlag, New York, 1972.

[24] B.J. Bayly, S.A. Orszag and T. Herbert. Instability mechanisms in shear-flow transition. *Annu. Rev. Fluid Mech.,* 20:359-391, 1988.

[25] J.A. Beasley. Calculation of the laminar boundary layer and the prediction of transition on a sheared wing. *ARC R and M 3787,* 1973.

[26] M. Beckers and G.J.F. van Heijst. The observation of a triangular vortex in a rotating fluid. *Fluid Dyn. Res.,* 22:265-279, 1998.

[27] C.M. Bender and S.A. Orszag. *Advanced Mathematical Methods for Scientists and Engineers.* McGraw Hill Book Co., International Edition, Singapore, 1987.

[28] T.B. Benjamin and J.E. Feir. The disintegration of wave trains on deep water. Part 1. Theory. *J. Fluid Mech.*, 27(3):417-430, 1967.

[29] A. Bers. *Physique Des Plasmas.* (Eds.: C. Dewitt and J. Peyraud) Gordon and Breach, New York, 1975.

[30] R. Betchov and W.O. Criminale Jr. *Stability of Parallel Flows.* Academic Press, New York, 1967.

[31] T.R. Bewley, P. Moin and R. Temam. DNS-based predictive control of turbulence. *J. Fluid Mech.*, 477:179-225, 2001.

[32] Y.G. Bhumkar. High performance computing of bypass transition. Ph.D. Thesis, (submitted), 2011.

[33] Y.G. Bhumkar and T.K. Sengupta. Adaptive multi-dimensional filters. *Computers Fluids,* 49(1):128-140, 2011.

[34] T. Bohr, M. H. Jensen, G. Paladin and A. Vulpiani. *Dynamical Systems Approach to Turbulence.* Cambridge University Press, UK, 2005.

[35] F.W. Boltz, G.C. Kenyon and C.Q. Allen. Effect of sweep angle on the boundary layer stability characteristics of an untapered wing at low speeds. *NASA TN D-338,* 1960.

[36] M. Bouthier. Stabilite lineare des ecoulements presques paralleles, IILa couche limite de Blasius. *J. de Macaniqe*, 12:75-95, 1973.

[37] P. Bradshaw, G. Mizner and K. Unsworth. Calculation of compressible turbulent boundary layers on straight-tapered swept wing. *AIAA J.,* 14:399, 1976.

[38] K.S. Breuer and J.H. Haritonidis. The evolution of a localized disturbance in a laminar boundary layer. Part I. Weak disturbances. *J. Fluid Mech.*, 220:569-594, 1990.

[39] K.S. Breuer and M.T. Landahl. The evolution of a localized disturbance in a laminar boundary layer. Part II. Strong disturbances. *J. Fluid Mech.*, 220:595-621, 1990.

[40] K.S. Breuer and T. Kuraishi. Bypass transition in two and three dimensional boundary layers. *AIAA paper no. 93-3050*, 1993.

[41] R.A. Brewstar and B. Gebhart. Instability and disturbance amplification in a mixed-convection boundary layer. *J. Fluid Mech.,* 229:115-133, 1991.

[42] T.J. Bridges and P. Morris. Differential eigenvalue problems in which the parameters appear nonlinearly. *J. Comput. Phys.,* 55:437-460, 1984.

[43] L. Brillouin. *Wave Propagation and Group Velocity.* Academic Press, New York, 1960.

[44] K.W. Brinckman and J.D.A. Walker. Instability in a viscous flow driven by streamwise vortices. *J. Fluid Mech.,* 432:127-166, 2001.

[45] A.N. Brooks and T.J.R. Hughes. Streamline upwind/Petrov-Galerkin formulation for convection dominated flows with particular emphasis on the incompressible Navier-Stokes equations. *Comput. Methods Appl. Mech. and Eng.*, 32:199-259, 1982.

[46] C.-H. Bruneau and M. Saad. The 2D lid-driven cavity problem revisited. *Comput. Fluids*, 35:326-348, 2006.

[47] F.M. Burrows. A theoretical and experimental study of the boundary layer flow on a 45^0 swept back wing. *Cranfield College of Aero. Rept.* No. 109, 1956.

[48] T.A. Buter and H.L. Reed. Numerical investigation of receptivity to freestream vorticity. *AIAA-93-0073*, 1993.

[49] K.M. Butler and B.F. Farrell. Three-dimensional optimal perturbations in viscous shear flows. *Phys. Fluids A*, 4:1637-1650, 1992.

[50] R.L. Campbell and L.A. Smith. A hybrid algorithm for transonic airfoil and wing design. *AIAA paper 87-2552*, 1987.

[51] G.F. Carnevale and R.C. Kloosterziel. Emergence and evolution of triangular vortices. *J. Fluid Mech.,* 259:305-331, 1994.

[52] K.M. Case. Stability of inviscid plane Couette flow. *Phys. Fluids*, 3:143, 1961.

[53] W. Cazemier, R.W.C.P. Verstappen and A.E.P. Veldman. Proper orthogonal decomposition and low dimensional models for driven cavity flow. *Phys. Fluids,* 10(7):1685-1699, 1998.

[54] T. Cebeci. *Stability and Transition: Theory and Application.* Horizons Publishing Inc. Springer, Long Beach, CA, 2004.

[55] T. Cebeci and P. Bradshaw. *Momentum Transfer in Boundary Layers.* Hemisphere Publishing Corp., Washington DC, 1977.

[56] T. Cebeci, H.H. Chen and K. Kaups. Further consideration of the effect of curvature on the stability of three-dimensional flows. *Comput. Fluids,* 21(4):491-502, 1992.

[57] T. Cebeci and J. Cousteix. *Modelling and Computation of Boundary-Layer Flows.* Second Edn. Horizon Publishing Inc., Springer, Long Beach, CA, 2005.

[58] T. Cebeci and K. Stewartson. Stability and transition in three-dimensional flows. *AIAA J.,* 18:398-405, 1980.

[59] S. Chandrasekhar. *Radiative Transfer.* Dover Publications Inc., New York, 393, 1960.

[60] J.G. Charney, R. Fjørtoft and J. Von Neumann. Numerical integration of the barotropic vorticity equation. *Tellus,* 2(4):237-254, 1950.

[61] M. Chattopadhyay. Instability and Transition of Wall Bounded Flows by Free-Stream Excitation. M. Eng. Thesis, Dept of Mech. Eng. National University of Singapore, 2001.

[62] M. Chaudhari. Boundary-layer receptivity to three-dimensional unsteady vortical disturbances in free stream. *AIAA-96-0181,* 1996.

[63] M. Chaudhari and C.L. Streett. Boundary layer receptivity phenomena in three-dimensional and high speed boundary layer. *AIAA-90-5258,* 1990.

[64] T.S. Chen and Moutsoglu. Wave instability of mixed convection flow on inclined surfaces. *Num. Heat Transfer,* 2:497-509, 1979.

[65] T.S. Chen and A. Mucoglu. Wave instability of mixed convection flow over a horizontal flat plate. *Int. J. Heat Mass Transfer,* 22:185-196, 1979.

[66] T.S. Chen, E.M. Sparrow and A. Mucoglu. Mixed convection in boundary layer flow on a horizontal plate. *ASME J. Heat Transfer,* 99:66-71, 1977.

[67] T.C. Chen and A. Wiin Nielsen. Nonlinear cascades of atmospheric energy and enstrophy in a two-dimensional spectral index. *Tellus,* 30:313-322, 1978.

[68] W.G. Cochran. The flow due to a rotating disk. *Proc. Cambridge Phil. Soc.,* 30:365-375, 1934.

[69] F.S. Collier, D.W. Bartlett, R.D. Wagner, V.V. Tat and B.T. Anderson. Correlation of boundary-layer stability analysis with flight transition data. *IUTAM Symp. Proc.: Laminar-Turbulent Transition.* Toulouse, France, Sept. 11-15, 1985.

[70] S.S. Collis and S.K. Lele. A computational approach to swept leading-edge receptivity. *AIAA-96-0180,* 1996.

[71] L.F. Crabtree, D. Kuchemann and L. Sowerby. Three-dimensional boundary layers. In *Laminar Boundary Layers* (Ed.: L. Rosenhead). Clarendon Press, UK, 1963.

[72] J. Crank and P. Nicolson. A practical method for numerical evaluation of solutions of partial differential equations of the heat conduction type. *Proc. Cambridge Phil. Soc.,* 43(50):50-67, 1947.

[73] C.C. Croom and B.J. Holmes. Flight evaluation of an insect contamination protection system for laminar flow wings. *SAE Paper* No. 850860, 1985.

[74] J.D. Crouch. Non-localized receptivity of boundary layers. *J. Fluid Mech.*, 244:567-581, 1992.

[75] J.D. Crouch. Receptivity to three-dimensional boundary layers. *AIAA-93-0074*, 1993.

[76] M.J.P. Cullen. A finite-element method for a non-linear initial value problem. *J. Int. Math. Appl.*, 31:233-247, 1974.

[77] R.F. Dam, J.F. Egmond and J.W. Sloof. Optimization of target pressure distribution. In *Optimum Design Methods for Aerodynamics*. AGARD-R-803, 1994.

[78] A. Davey. A difficult numerical calculation concerning the stability of the Blasius boundary layer. *Proc. IUTAM Symp. on Stability in Mechanics of Continua* (Ed. F.H. Schroeder), Springer Verlag, 1982.

[79] P.A. Davidson. *Turbulence: An Introduction for Scientists and Engineers*. Oxford Univ. Press, UK, 2004.

[80] S.J. Davies and C.M. White. An experimental study of the flow of water in pipes of rectangular section. *Proc. Royal Soc. London Ser. A*, 119:92, 1928.

[81] A.E. Deane, I.G. Kevrekidis, G.E. Karniadakis and S.A. Orszag. Low-dimensional models for complex geometry flows: Application to grooved channels and circular cylinders. *Phys. Fluids A*, 3:2337-2354, 1991.

[82] A.T. Degani, J.D.A. Walker and F.T. Smith. Unsteady separation past moving surfaces. *J. Fluid Mech.*, 375:1-38, 1998.

[83] F.E. Dendy. Sediment trap efficiency of small reservoirs. *Trans. Amer. Soc. Agric. Eng.*, 17(5):898-908, 1974.

[84] E. M. Dewan. Stratospheric spectra resembling turbulence. *Science*, 204:832-835, 1979.

[85] F. Diaz, J. Gavalda, J.G. Kawall and F. Giralt. Vortex shedding from a spinning cylinder. *Phys. Fluids*, 26(12):3454-3460, 1983.

[86] A.J. Dietz. Local boundary layer receptivity to a convected free stream disturbance. *J. Fluid Mech.*, 378:291-317, 1999.

[87] A. Dipankar, T.K. Sengupta and S.B. Talla. Suppression of vortex shedding behind a circular cylinder by another control cylinder at low Reynolds numbers. *J. Fluid Mech.*, 573:171-190, 2007.

[88] S.S. Dodbele, C.P. Van Dam, P.M.H.W. Vijgen and B.J. Holmes. Shaping of airplane fuselages for minimum drag. *J. Aircraft*, 24(5):298-304, 1987.

[89] C.R. Doering and J.D. Gibbon. *Applied Analysis of the Navier-Stokes Equations*. Cambridge Univ. Press, UK, 1995.

[90] T.L. Doligalski, C.R. Smith and J.D.A. Walker. Vortex interaction with wall. *Ann. Rev. Fluid Mech.*, 26:573-616, 1994.

[91] P.G. Drazin and W.H. Reid. *Hydrodynamic Stability*. Cambridge Univ. Press, UK, 1981.

[92] H.L. Dryden. Transition from laminar to turbulent flow. In *Turbulent Flows and Heat Transfer, High Speed Aerodynamics and Jet Propulsion*. Edited by C.C. Lin, 1959.

[93] E.R.G. Eckert and E. Soehngen. Interferometric studies on the stability and transition to turbulence in a free-convection boundary-layer. *Proc. Gen. Disc. Heat Transfer, ASME and IME,* London, 321, 1951.

[94] W. Eckhaus. *Studies in Nonlinear Stability Theory.* Springer, New York, 1965.

[95] E.A. Elsaadawy and C.P. Britcher. The extent of a laminar boundary layer under the influence of a propeller slipstream. AIAA Applied Aerodynamics Conf. paper No. *AIAA 2002-2930,* 2002.

[96] R. Eppler and D. Somers. A computer program for the design and analysis of low-speed airfoils. *NASA TM-80210,* 1980.

[97] V.M. Falkner and S.W. Skan. Some approximate solutions of the boundary layer equations. *Phil. Mag.,* 12(7):865-869, 1931.

[98] H. Fasel and U. Konzelmann. Non-parallel stability of a flat plate boundary layer using the complete Navier-Stokes equation. *J. Fluid Mech.,* 221:331-347, 1990.

[99] R. Fjørtoft. Application of integral theorems in deriving criteria of stability for laminar flows and for baroclinic circular vortex. *Geofys. Publ. Oslo,* 17(6):1-51, 1950.

[100] A. Fortin, M. Jardak, J.J. Gervais and R. Pierre. Localization of Hopf bifurcations in fluid flow problems. *Int. J. Num. Methods Fluids,* 24(11):1185-1210, 1997.

[101] J.H.M. Fransson. Turbulent spot evolution in spatially invariant boundary layers. *Phys. Rev. E,* 81:035301-1-4, 2010.

[102] U. Frisch. *Turbulence.* Cambridge Univ. Press, Cambridge, UK, 1995.

[103] U. Frisch and P. L. Sulem. Numerical simulation of the inverse cascade in two-dimensional turbulence. *Phys. Fluids,* 27:1921-1923, 1984.

[104] M. Fujino and Y. Kawamura. Wave drag characteristics of an over-the-wing nacelle business-jet configuration. *J. Aircraft,* 40(6):1177-1184, 2003.

[105] M. Fujino, Y. Yoshizaki and Y. Kawamura. Natural-laminar-flow airfoil development for a lightweight business jet. *J. Aircraft,* 40(4):609-615, 2003.

[106] W.A. Fuller. *Introduction to Statistical Time Series.* Wiley, New York, 1978.

[107] K.S. Gage. Evidence for a $k^{-5/3}$ law inertial range in mesoscale two-dimensional turbulence. *J. Atmos. Sci.,* 36:1950-1954, 1979.

[108] M. Gaster. A note on the relation between temporally increasing and spatially increasing disturbances in hydrodynamic stability. *J. Fluid Mech.,* 14:222-224, 1962.

[109] M. Gaster. On the generation of spatially growing waves in a boundary layer. *J. Fluid Mech.,* 22:433, 1965.

[110] M. Gaster. A simple device for preventing turbulent contamination on swept leading edges. *J. Roy. Aero. Soc.,* 69:788-789, 1965.

[111] M. Gaster. On the effect of boundary-layer growth on flow stability. *J. Fluid Mech.,* 66(3):465-480, 1974.

[112] M. Gaster. Is the dolphin a red herring? Proc. IUTAM Symp. *Turbulence Management and Relaminarization* (Eds. Liepmann & Narasimha), Springer Verlag, Berlin, 1987.

[113] M. Gaster, C.E. Grosch and T.L. Jackson. The velocity field created by a shallow bump in a boundary layer. ICASE Rept. 94-21, NASA Langley RC (also published in *Physics of Fluids*) 6(9):3079-3085, 1994.

[114] M. Gaster and T.K. Sengupta. The generation of disturbance in a boundary layer by wall perturbation: The vibrating ribbon revisited once more. In *Instabilities and Turbulence in Engineering Flows* (Eds.: D.E. Ashpis, T.B. Gatski and R. Hirsch), 31-49, Kluwer, Dordrecht, 1993.

[115] T.B. Gatski. Review of incompressible fluid flow computations using the vorticity-velocity formulation. *Appl. Numer. Math.*, 7:227-239, 1991.

[116] T.B. Gatski, C.E. Grosch and M.E. Rose. A numerical solution of the Navier-Stokes equations for three-dimensional, unsteady, incompressible flows by compact schemes. *J. Comput. Phys.*, 82:298-329, 1989.

[117] T.B. Gatski, C.E. Grosch and M.E. Rose. A numerical study of the two-dimensional Navier-Stokes equations in vorticity-velocity variables. *J. Comput. Phys.*, 48:1-22, 1982.

[118] B. Gebhart, Y. Jalurai, R. L. Mahajan and B. Sammakia. *Buoyancy-Induced Flows and Transport.* Hemisphere Publications, Washington, DC, 1988.

[119] W.K. George. Some thoughts on similarity, the POD, and finite boundaries. In *Fundamental Problematic Issues in Turbulence* (Eds.: A. Gyr, W. Kinzelbach and A. Tsinober), Birkhauser Verlag, Basel, 1999.

[120] F. Giannetti and P. Luchini. Structural sensitivity of the first instability of the cylinder wake. *J. Fluid Mech.*, 581:167-197, 2007.

[121] M. Giles and M. Drela. A two-dimensional transonic aerodynamic design method. *AIAA paper no. 86-1793*, 1986.

[122] R.R. Gilpin, H. Imura and K. C. Cheng. Experiments on the onset of longitudinal vortices in horizontal Blasius flow heated from below. *ASME J. Heat Transfer*, 100:71-77, 1978.

[123] O.A. Godina. Restless rays, steady wave fronts. *J. Acoust. Soc. Am.*, 122(6):3353-3363, 2007.

[124] S. Goldstein. *Modern Development in Fluid Mechanics*, 1 and 2, Clarendon Press, Oxford, UK, 1938.

[125] M.E. Goldstein. The evolution of Tollmien-Schlichting waves near a leading edge. *J. Fluid Mech.*, 127:59-81, 1983.

[126] M.E. Goldstein and L.S. Hultgren. Boundary-layer respectivity to long-wave free-stream disturbances. *Ann. Rev. Fluid Mech.*, 21:137-166, 1989.

[127] M.E. Goldstein. Scattering of acoustic waves into Tollmien-Schlichting waves by small streamwise variation in surface geometry. *J. Fluid Mech.*, 154:509-529, 1985.

[128] M.E. Goldstein, P.M. Sockol and J. Sanz. The evolution of Tollmien-Schlichting waves near a leading edge. Part 2. Numerical determination of amplitudes. *J. Fluid Mech.*, 129:443-453, 1983.

[129] M. Golubitsky and D.G. Schaeffer. *Singularities and Groups in Bifurcation Theory.* Springer, New York, 1984.

[130] W.E. Gray. Transition in flight on a laminar-flow wing of low waviness (King Cobra). *RAE Report* No. 2364, (March 1950), 1950.

[131] W.E. Gray. The effect of wing sweep on laminar flow. *RAE TM Aero*, 255, 1952.

[132] W.E. Gray and P.W.J. Fullam. Comparison of flight and wind tunnel measurements of transition on a highly finished wing (King Cobra). *RAE Report* No. 2383, (June 1950), 1950.

[133] N. Gregory, J.T. Stuart and J.S. Walker. On the stability of three dimensional boundary layer with application to the flow due to a rotating disc. *Phil. Trans. Roy. Soc. London*, Series A, 248:155-199, 1955.

[134] P.M. Grescho and R.L. Sani. *Incompressible Flow and the Finite Element Method.* John Wiley and Sons, Chichester, UK, 1998.

[135] S.E. Haaland and E.M. Sparrow. Vortex instability of natural convection flows on inclined surfaces. *Int. J. Heat Mass Transfer*, 16:2355-2367, 1973.

[136] G. Hagen. Über die bewegung des wassers in engen zylinderschen Rohren. *Pogendorff's Ann. Phys. Chem.*, 46:423-442, 1839.

[137] P. Hall and M.R. Malik. On the instability of three-dimensional attachment-line boundary layer: weakly non-linear theory and numerical approach. *J. Fluid Mech.*, 395:229-245, 1986.

[138] P. Hall, M.R. Malik and D.I.A. Poll. On the stability of an infinite swept attachment-line boundary layer. *Proc. R. Soc. Lond. A*, 395:229-245, 1984.

[139] P. Hall and H. Morris. On the instability of boundary layers on heated flat plates. *J. Fluid Mech.*, 245:367-400, 1992.

[140] W.R. Hamilton. *The Collected Mathematical Papers.* 4, Cambridge Univ. Press, 1839.

[141] A.G. Hansen and P.L. Yohner. Some numerical solutions of similarity equations for three-dimensional, laminar incompressible boundary-layer flows. *NACA TN*, 4370, 1958.

[142] J.E. Harris and D.K. Blanchard. Computer program for solving laminar, transitional, or turbulent compressible boundary-layer equations for two-dimensional and axisymmetric flow. *NASA Tech. Memo.*, 83207, 1982.

[143] W. Heisenberg. Über stabilität und turbulenz von flüssigkeitsströmen. *Ann. Phys. Lpz.*, 379:577-627, 1924. (Translated as 'On stability and turbulence of fluid flows'. NACA Tech. Memo. Wash. No 1291, 1951).

[144] H. von Helmholtz. On discontinuous movements of fluids. *Phil. Mag.*, (4)36:337-346, 1868.

[145] D.S. Henningson, A. Lundbladh and A.V. Johansson. A mechanism for bypass transition from localized disturbances in wall-bounded shear flows. *J. Fluid Mech.*, 250:169-207, 1993.

[146] J.R. Herring and J. C. McWilliams. Comparison of direct-numerical simulation of two-dimensional turbulence with two-point closure: the effect of intermittency. *J. Fluid Mech.*, 153:229-242, 1985.

[147] R.M. Hicks, E.M. Murman and G.N. Vanderplaats. An assessment of airfoil design by numerical optimization. *NASA TMX 3092*, 1976.

[148] C. Hirsch. *Numerical Computation of Internal and External Flows. Vols. I and II: Computational Methods for Inviscid and Viscous Flows.* Wiley, Chichester, UK, 1990.

[149] F.R. De-Hoog, B. Laminger and R. Weiss. A numerical study of similarity solutions for combined forced and free convection. *Acta Mech.*, 51:139-149, 1984.

[150] S.F. Hoerner. Base drag and thick trailing edges. *J. Aerodyn. Sci.*, 17(10), 1950.

[151] P. Holmes, J.L. Lumley and G. Berkooz. *Coherent Structures, Dynamical Systems and Symmetry*, Cambridge Univ. Press, Cambridge, UK, 1996.

[152] F. Homann. Einfluss grosser zähigkeit bei strömung um zylinder. *Forsch. auf dem Gebiete des Ingenieurwesens*, 7(1):1-10, 1936.

[153] P. Huerre and P.A. Monkewitz. Absolute and convective instabilities in free shear layers. *J. Fluid Mech.*, 159:151, 1985.

[154] T.H. Hughes and W.H. Reid. On the stability of the asymptotic suction boundary layer profile. *J. Fluid Mech.*, 23:715-735, 1965.

[155] P.A. Iyer and R.E. Kelly. The instability of the laminar free convection flow induced by a heated, inclined plate. *Int. J. Hear Mass Transfer*, 17:517-525, 1974.

[156] C.P. Jackson. A finite-element study of the onset of vortex shedding in flow past variously shaped bodies. *J. Fluid Mech.*, 182:23-45, 1987.

[157] P.E. Jackson, Z. S. She and S. A. Orszag. A case study in parallel computing: homogeneous turbulence on a hypercube. *J. Sci. Comput.*, 6:27-45, 1991.

[158] E.N. Jacobs. Preliminary report on laminar-flow airfoils and new methods adopted for airfoil and boundary layer investigations. *NACA WR L-345*, 1939.

[159] R.G. Jacobs and P.A. Durbin. Simulation of bypass transition. *J. Fluid Mech.*, 428:185-212, 2001.

[160] N.A. Jaffe, T.T. Okamura and A.M.O. Smith. Determination of spatial amplification factors and their application to predicting transition. *AIAA*, 8(2):301-308, 1970.

[161] A. Jameson. Aerodynamic design via control theory. *J. Scientific Computing*, 3:233-260, 1988.

[162] F. Jenkins and H. White. *Fundamentals of Physical Optics.* McGraw-Hill, New York, 1973.

[163] J. Jimenez. Transition to turbulence in two-dimensional Poiseuille flow. *J. Fluid Mech.*, 218:265-297, 1990.

[164] B.H. Jorgensen, J.N. Sorenson and M. Brons. Low-dimensional modeling of a driven cavity flow with two free parameters. *Theoret. Comput. Fluid Mech.*, 16(4):299-317, 2003.

[165] R.D. Joslin. Direct simulation of evolution and control of three-dimensional instabilities in attachment-line boundary layers. *J. Fluid Mech.*, 291:369-392, 1995.

[166] R.D. Joslin. Simulation of non-linear instabilities in an attachment-line boundary layer. *Fluid Dyn. Res.*, 18:81-97, 1996.

[167] R.D. Joslin. Aircraft laminar flow control. *Ann. Rev. Fluid Mech.*, 30:1-29, 1998.

[168] A. Juel, A.G. Darbyshire and T. Mullin. The effect of noise on pitchfork and Hopf bifurcations. *Proc. R. Soc. Lond. A*, 453:2627-2647, 1997.

[169] Th. von Karman. Über laminare und turbulente reibung. *Z. Angew. Math. Mech.*, 1:233-252, 1921.

[170] K. Kaups and T. Cebeci. Compressible laminar boundary layers with suction on swept and tapered wings. *J. Aircraft*, 14(7), 1977.

[171] Lord Kelvin. Hydrokinetic solutions and observations. *Phil Mag.(4)*, 42:362-377, 1871.

[172] J.M. Kendall. Experimental study of laminar boundary layer receptivity to a traveling pressure field. *AIAA paper no.*, 87-1257, 1987.

[173] J.M. Kendall. Boundary layer receptivity to free stream turbulence. *AIAA paper no. 90-1504*, 1990.

[174] E.J. Kerschen. Linear and non-linear receptivity to vortical free-stream disturbances. In *Boundary Layer Stability and Transition to Turbulence*. (Eds.: D.C. Reda, H.L. Reed and R.K. Kobayashi). *ASME FED*, 114:43-48, 1991.

[175] M. Kiya, Y. Suzuki, A. Mikio and M. Hagino. A contribution to the free stream turbulence effect on the flow past a circular cylinder. *J. Fluid Mech.*, 115:151-164, 1982.

[176] P.S. Klebanoff and K.D. Tidstrom. Evolution of amplified waves leading to transition in a boundary layer with zero pressure gradient. *NASA TN D- 195*, 1959.

[177] P.S. Klebanoff, K.D. Tidstrom and Sargent L.M. The three-dimensional nature of boundary-layer instability. *J. Fluid Mech.*, 12:1-34, 1962.

[178] P.S. Klebanoff. Effect of free stream turbulence on the laminar boundary layer. *Bull. Am. Phys. Soc.*, 10:1323, 1971.

[179] W. Koch. Local instability characteristics and frequency determination of self excited wake flow. *J. Sound Vib.*, 99:53-83, 1985.

[180] A.N. Kolmogorov. Dissipation of energy in locally isotropic turbulence. *Dokl. Akad. Nauk SSSR*, 32:16-18, 1941. (Reprinted in *Proc. R. Soc. Lond. A*, 434:15-17, 1991).

[181] D.D. Kosambi. Statistics in function space. *J. Indian Math. Soc.*, 7:76-88, 1943.

[182] L.S.G. Kovasznay. Hot-wire investigation of the wake behind cylinders at low Reynolds numbers. *Proc. Roy. Soc. London A.*, 198:174-190, 1949.

[183] R.H. Kraichnan. Inertial ranges in two-dimensional turbulence. *Phys. Fluids*, 67:1417-1423, 1967.

[184] R. Kraichnan and D. Montgomery. Two-dimensional turbulence. *Reports in Progress in Physics*, 43:547-619, 1980.

[185] H. Kreiss and J. Oliger. Comparison of accurate methods for the integration of hyperbolic equations. *Tellus*, 24:199-215, 1972.

[186] I.M. Kroo. *Applied Aerodynamics: A Digital Textbook.* Desktop Aeronautics Inc. Stanford, CA, 1995.

[187] M.T. Landahl. A note on an algebraic instability of inviscid parallel shear flows. *J. Fluid Mech.,* 98:243-251, 1980.

[188] M.T. Landahl. Wave mechanics of boundary layer turbulence and noise. *Journal of the Acoustical Society of America* 57:824-831 1975.

[189] M.T. Landahl and E. Mollo-Christensen. *Turbulence and Random Processes in Fluid Mech.,* Cambridge Univ. Press, New York, 1992.

[190] L.D. Landau. On the problem of turbulence. *C.R. Acad. Sci. USSR,* 44:311-315, (1944)

[191] L.D. Landau. *Collected Papers,* Pergamon Press, Oxford, UK, 1965.

[192] L.D. Landau and E.M. Lifshitz. *Fluid Mechanics,* 6. Addision - Wesley. Pergamon Press, London, 1959.

[193] L.G. Leal. Steady separated flow in a linearly decelerated free stream. *J. Fluid Mech.,* 59:513-535, 1973.

[194] P. Leehey and P. Shapiro. Leading edge effect in laminar boundary layer excitation by sound. *Laminar Turbulent Transition* (Eds.: R. Eppler and H. Fasel). 321-331, Springer Verlag, Berlin, 1979.

[195] S.J. Leib, D.W Wundrow and M.E. Goldstein. Generation and growth of boundary-layer disturbances due to free-stream turbulence. *AIAA-99-0408,* 1999.

[196] B.P. Leonard, M.A. Leschziner and J. McGuirk. Third order finite-difference method for steady two-dimensional convection. *Numerical Methods in Laminar and Turbulent Flows* (Eds. C. Taylor, K. Morgan and C.A. Brebbia). 807-819, Pentech Press, London, 1978.

[197] B.P. Leonard. A stable and accurate convective modeling procedure based on quadratic upstream interpolation. *Comp. Meth. App. Mech. Eng.,* 19:59-98, 1979.

[198] M.J. Lighthill. *Fourier Analysis and Generalized Functions.* Cambridge Univ. Press, Cambridge, UK, 1978.

[199] T.T. Lim, T.K. Sengupta and M. Chattopadhyay. A visual study of vortex-induced subcritical instability on a flat plate laminar boundary layer. *Expts. Fluids,* 37:47-55, 2004.

[200] C.C. Lin. *Theory of Hydrodynamic Stability.* Cambridge Univ. Press, Cambridge, UK, 1955.

[201] N. Lin, H.L. Reed and W.S. Saric. Effect of leading edge geometry on boundary-layer receptivity to freestream sound. In *Instability, Transition and Turbulence* (Eds.: M.Y. Hussaini, A. Kumar and C.L. Streett). Springer, New York, 1992.

[202] N. Lin, G.K. Stuckert and Th. Herbert. Boundary layer receptivity to freestream vortical disturbances. *AIAA-95-0772,* 1995.

[203] E. Lindborg. Can the atmospheric kinetic energy spectrum be explained by two-dimensional turbulence? *J. Fluid Mech.,* 388:259-288, 1999.

[204] Z. Liu and C. Liu. Fourth order finite difference and multigrid methods for modeling instabilities in flat plate boundary layer – two-dimensional and three-dimensional approaches. *Comput. Fluids,* 23:955-982, 1994.

[205] X. Liu and W. Rodi. Experiments on transitional boundary layers with wake induced unsteadiness. *J. Fluid Mech.,* 231:229-256, 1991.

[206] J.R. Lloyd and E.M. Sparrow. On the instability of natural convection flow on inclined plates. *J. Fluid Mech.,* 42:465-470, 1970.

[207] D. Lucor and G. Em. Karniadakis. Noisy inflow cause a shedding-mode switching in flow past an oscillating cylinder. *Phys. Rev. Letters,* 92(15):154501-1-154501-4, 2004.

[208] A. Lundbladh and A.V. Johansson. Direct simulation of turbulent spots in plane Couette flow. *J. Fluid Mech.,* 229:499-576, 1991.

[209] X. Ma and G.E. Karniadakis. A low-dimensional model for simulating three-dimensional cylinder flow. *J. Fluid Mech.,* 458:181-190, 2002.

[210] L.M. Mack. Boundary layer stability theory. *Doc. 900-277, Rev.* A, Jet Prop. Lab, Pasadena, CA, 1969.

[211] L.M. Mack. A numerical study of the temporal eigenvalue spectrum of the Blasius boundary layer. *J. Fluid Mech.,* 73(3):497, 1976.

[212] L.M. Mack. Transition and laminar instability. *JPL Publication,* 77-15, 1977.

[213] L.M. Mack. Boundary layer stability theory. Special course on. *Stability and Transition of Laminar Flow.* AGARD Report No. 709, 1984.

[214] L.M. Mack. Stability of three-dimensional boundary layers on swept wings at transonic speeds. *IUTAM Symp., Transonic III,* Contingent, Germany, 1988.

[215] M.D. Maddalon and A.L. Braslow. Simulated-airline-service flight tests of laminar-flow control with perforated-surface suction system. *NASA PT 2966,* 1990.

[216] M.R. Malik. COSAL – a black box compressible stability analysis code for transition prediction in three-dimensional boundary layer. *NASA CR 165 925,* 1982.

[217] M.R. Malik and D.I.A. Poll. Effect of curvature on three-dimensional boundary layer stability. *AIAA Pap. No. 84-0490,* 1984.

[218] M.R. Malik and D.I.A. Poll. Effect of curvature on three-dimensional boundary-layer stability. *AIAA J.,* 23:1362-1369, 1985.

[219] M.E. Maltrud and G. K. Vallis. Energy spectra and coherent structures in forced two-dimensional and beta-plane turbulence. *J. Fluid Mech.,* 228:321-342, 1991.

[220] J.A. Masad. Effect of surface waviness on transition in three-dimensional boundary layer flow. *NASA CR 201641,* 1996.

[221] R.J. McGhee and W.D. Beasley. Low-speed aerodynamic characteristics of a 17-percent thick airfoil section designed for general aviation applications. *NASA TN D-7428,* 1973.

[222] R.J. McGhee, W.D. Beasley and R.T. Whitcomb. NASA low- and medium-speed airfoil development. In *Advanced Technology Airfoil Research,* II, NASA CP 2046, 1980.

[223] J.H. Merkin and D. Ingham. Mixed convection similarity solutions on a horizontal surface. *J. Appl. Math. Phys (ZAMP)*, 38:102-116, 1987.

[224] R. Michel. Etude de la transition sur les profils d'aile − establishment d'un point de transition et calcul de la trainee de profil en incompressible. *ONERA report no. 1/1578A*, 1952.

[225] R. Michel, D. Arnal, E. Coustols and J.C. Jullien. Experimental and theoretical studies of boundary layer transition on a swept infinite wing. In Proc. of IUTAM Symp. *On Laminar-Turbulent Transition*, Novosibirsk, USSR, Springer Verlag, 1984.

[226] S.J. Miley and E. von Lavante. Propeller propulsion integration − state of technology surveys. *NASA CR-3882*, 1985.

[227] S.J. Miley, R.M. Howard and B.J. Holmes. Wing laminar boundary layer in the presence of a propeller slipstream. *J. Aircraft*, 25(7):606-611, 1988.

[228] F. Moens, J. Perraud, A. Krumbein, T. Toulorge, P. Lannelli, P. Eliasson and A. Hanifi. Transition prediction and impact on 3D high-lift wing configuration. *AIAA Paper* No. 2007-4302, 2007.

[229] A.S. Monin and A.M. Yaglom. *Statistical Fluid Mechanics: Mechanics of Turbulence*. The MIT Press, Cambridge, MA, 1971.

[230] P. Moresco and J.J. Healey. Spatio-temporal instability in mixed convection boundary layers. *J. Fluid Mech.*, 402:89-107, 2000.

[231] M.V. Morkovin. Transition from laminar to turbulent flow − a review of the recent advances in its understanding. *ASME Trans.*, 80:1121-1128, 1958.

[232] M.V. Morkovin. Critical evaluation of transition from laminar to turbulent shear layer with emphasis on hypersonically traveling bodies. *AFFDL-TR-68-149*, 1969.

[233] M.V. Morkovin. Instability, transition to turbulence and predictability. *AGARDO-GRAPH* No. 236, NATO Doc., 1978.

[234] M.V. Morkovin. On receptivity to environmental disturbances. *Instability and Transition* (Eds.: M.Y. Hussaini and R.G. Voigt). 272-280, Springer Verlag, New York, 1990.

[235] M.V. Morkovin. Panoramic view of changes in vorticity distribution in transition instabilities and turbulence. In *Transition to Turbulence* (Eds.: D.C. Reda, H.L. Reed and R. Kobayashi). ASME FED Publication, 114:1-12, 1991.

[236] M. Morzynski, K. Afanasiev and F. Thiele. Solution of the eigenvalue problems resulting from global nonparallel flow stability analysis. *Comput. Meth. Appl. Mech. Eng.*, 169:161-176, 1999.

[237] A. Moutsoglu, T.S. Chen and K.C. Cheng. Vortex instability of mixed convection flow over a horizontal flat plate. *ASME J. Heat Transfer*, 103:257-261, 1981.

[238] A. Mucoglu and T.S. Chen. Wave instability of mixed convection flow along a vertical flat plate. *Num. Heat Transfer*, 1:267-283, 1978.

[239] J.W. Murdock. The generation of Tollmien-Schlichting wave by a sound wave. *Proc. R. Soc. London A*, 372:517-534, 1980.

[240] E.W. Mureithi and J. P. Denier Absolute-convective instability of mixed forced-free convection boundary layers. *Fluid Dyn. Res.*, 372:517-534, 2010.

[241] D.G. Murri, R.J. McGhee, F.L. Jordan, P.J. Davis and J.K. Viken. Wind tunnel results of the low speed NLF(1)-0414F airfoil. *NASA CP 2487, Part 3*, 673-695, 1987.

[242] M.T. Nair and T.K. Sengupta. Unsteady flow past elliptic cylinders. *J. Fluids Struct.*, 11:555-595, 1997.

[243] M.T. Nair, T.K.Sengupta and U.S. Chauhan. Flow past rotating cylinders at high Reynolds numbers using higher order upwind scheme. *Comput. Fluids*, 27(1):47-70, 1998.

[244] G.D. Nastrom and K.S. Gage. A climatology of atmospheric wave number spectra of wind and temperature observed by commercial aircraft. *J. Atmos. Sci.*, 42:950-960, 1985.

[245] G.D. Nastrom, K. S. Gage and W. H. Jasperson. Kinetic energy spectrum of large-scale and mesoscale atmospheric processes. *Nature.* 310:36-38, 1984.

[246] A.H. Nayfeh. Stability of three-dimensional boundary layers. *AIAA J.*, 18:406-416, 1980.

[247] M. Nemec and D.W. Zingg. Newton-Kryolv algorithm for aerodynamic design using the Navier-Stokes equations. *AIAA J.*, 40(6):1146-1154, 2002.

[248] B.S. Ng and W.H. Reid. On the numerical solution of the Orr-Sommerfeld problem: Asymptotic initial conditions for shooting method. *J. Comput. Phys.*, 38:275-293, 1980.

[249] B.S. Ng and W.H. Reid. The compound matrix method for ordinary differential systems. *J. Comput. Phys.*, 58:209-228 1985.

[250] M. Nishioka and H. Sato. Measurements of velocity distributions in the wake of a circular cylinder at low Reynold numbers. *J. Fluid Mech.*, 65:97-112, 1973.

[251] M. Nishioka and H. Sato. Mechanism of determination of the shedding frequency of vortices behind a cylinder at low Reynolds numbers. *J. Fluid Mech.*, 89:49-60, 1978.

[252] M. Nishioka and M.V. Morkovin. Boundary-layer receptivity to unsteady pressure gradients: Experiments and overview. *J. Fluid Mech.*, 171:219-261, 1986.

[253] B.R. Noack, K. Afanasiev, M. Morzynski, G. Tadmor and F. Thiele. A hierarchy of low-dimensional models for the transient and post-transient cylinder wake. *J. Fluid Mech.*, 497:335-363, 2003.

[254] C. Norberg. Fluctuating lift on a circular cylinder: review and new measurement. *J. Fluids Struct.*, 17:57-96, 2003.

[255] V. Noshadi and W. Schneider. A numerical investigation of mixed convection on a horizontal plate. *Advances in Fluid Mechanics and Turbomachinery* (Eds.: H.J. Rath and C. Egbers), Springer, Berlin, 87-97, 1998.

[256] A.V. Obabko and K.W. Cassel. Navier-Stokes solutions of unsteady separation induced by a vortex. *J. Fluid Mech.*, 465:99-130, 2002.

[257] C.J. Obara, E.C. Hastings, J.A. Schoenster, T.L. Parrott and B.J. Holmes. Natural laminar flow flight experiments on a turbine engine nacelle fairing. *AIAA Paper*, 86-9756, 1986.

[258] H.G. Obremski, M.V. Morkovin and M. Landahl. A portfolio of stability characteristics of incompressible boundary layer. *AGARDOGRAPH*, No. 134, 1969.

[259] D. Obrist and P.J. Schmid. On the linear stability of swept attachment-line boundary layer flow. Part 2. Non-modal effect and receptivity. *J. Fluid Mech.*, 493:31-58, 2003.

[260] W. McF. Orr. The stability or instability of the steady motions of a perfect liquid and of a viscous liquid. Part I: A perfect liquid. Part II: A viscous liquid. *Proc. Roy. Irish Acad.*, A27:9-138, 1907.

[261] G.A. Oswald, K.N. Ghia and U. Ghia. Direct solution methodologies for the unsteady dynamics of an incompressible fluid. S.N. Atluri and G. Yagawa Eds., *Int. Conf. Comput. Eng. Sci.*, 2, Atlanta, Springer Verlag, Berlin, 1988.

[262] P.R. Owen and D.G. Randall. Boundary layer transition on a sweptback wing. *RAE TM*, 277, 1952.

[263] P.R. Owen and D.G. Randall. Boundary layer transition on a sweptback wing: a further investigation. *RAE TM*, 330, 1953.

[264] A. Papoulis. *Fourier Integral and Its Applications*. McGraw Hill, New York, 1962.

[265] J. Pedlosky. *Geophysical Fluid Dynamics*. Springer Verlag, New York, 1979.

[266] Y.-F. Peng, Y.-H. Shiau and R.R. Hwang. Transition in a 2-D lid driven cavity flow. *Comput. Fluid*, 32:337-352, 2003.

[267] V.J. Peridier, F.T. Smith and J.D.A. Walker. Vortex-induced boundary-layer separation. Part 1. The unsteady limit problem. $Re \to \infty$. *J. Fluid Mech.*, 232:99-131, 1991.

[268] V.J. Peridier, F.T. Smith and J.D.A. Walker. Vortex-induced boundary-layer separation. Part 2. Unsteady interacting boundary-layer theory. *J. Fluid Mech.*, 232:133-165, 1991.

[269] W. Pfenninger. Investigations on reductions of friction on wings, in particular by means of boundary layer suction. *NACA TM 1181*, 1947.

[270] W. Pfenninger, L. Gross and J.W. Bacon Jr. Experiments on a 30^0 swept 12% thick symmetrical laminar section in the 5ft by 7ft Michigan tunnel. *Northrop Aircraft Inc. Rep. NAI-57-317* (BLC-93), 1957.

[271] W. Pfenninger. Some results from the X-21A program. Part I. Flow phenomenon at the leading edge of swept wings. *AGARDOGRAPH 97*, 1965.

[272] W. Pfenninger and J.W. Bacon. Amplified laminar boundary-layer oscillation and transition at the front attachment-line of a 45^0 swept flat-nosed wing with and without suction. *Viscous Drag Reduction* (Ed.: C.S. Wells). 85-105, Plenum Press, New York, 1969.

[273] O.M. Phillips. *The Dynamics of Upper Ocean*. Cambridge Univ. Press, New York, 1977.

[274] T. Poinsot and D. Veynante. *Theoretical and Numerical Combustion*. Edwards, PA, 2nd Edition, 2005.

[275] J.L.M. Poiseuille. Recherches expèrimentelles surle mouvment des liquids dans les tubes detrès petits diamètres. *Computes Rendus,* 11:961-967, 1840.

[276] D.I.A. Poll. Transition in the infinite swept attachment-line boundary layer. *Aero. Quart,* 30:607-629, 1979.

[277] D.I.A. Poll. Transition description and prediction in three-dimensional flows. *AGARD Rep. No. 709.* (Special Course on Stability and Transition of Laminar Flows, VKI, Belgium), 1984.

[278] D.I.A. Poll. Some observations of the transition process on the windward face of a long yawed cylinder. *J. Fluid Mech.,* 150:329-356, 1985.

[279] D.I.A. Poll, M. Danks and M. Yardley. The effects of suction and blowing on stability and transition at a swept attachment-line. *Transitional Boundary Layers in Aeronautics* (Eds.: R. Henkes and J. Van Ingen). Elsevier, Amsterdam, Holland, 1996.

[280] M. Poliashenko and C.K. Aidun. A direct method for computation of simple bifurcations. *J. Comput. Phys.,* 121(2):246-260, 1995.

[281] L. Prandtl. Uber flussigkeitsbewegung bei sehr kleiner reibung. *Proc. of 3rd Internat. Math. Cong. Heidelberg.* (Also available in English as NACA Tech. Memo. 452,) 1904.

[282] L. Prandtl. In *Aerodynamic Theory* (Ed.: W.F. Durand). 3:178-190. Springer, Berlin, 1935.

[283] L. Prandtl. On boundary layers in three-dimensions. *Rept. Aero. Res. Council, London No. 9828,* 1946.

[284] M. Provansal, C. Mathis and L. Boyer. Benàrd-von Kàrmàn instability: transient and forced regimes. *J. Fluid Mech.,* 182:1-22, 1987.

[285] M.K. Rajpoot, T.K. Sengupta and P.K. Dutt. Optimal time advancing dispersion relation preserving schemes. *J. Comput. Phys.,* 229(10):3623-3651, 2010.

[286] Lord Rayleigh. On the stability or instability of certain fluid motions. In *Scientific Papers,* 1:361-371, 1880.

[287] Lord Rayleigh. On the stability or instability of certain fluid motions. In *Scientific Papers,* 3:17-23, 1887.

[288] Lord Rayleigh. *Scientific Papers 1,* Cambridge Univ. Press, Cambridge, 1889.

[289] Lord Rayleigh. *Scientific Papers 2,* Cambridge Univ. Press, Cambridge, 1890.

[290] W.H. Raymond and A. Garder. Selective damping in a Galerkin method for solving wave problems with variable grids. *Mon. Weather Rev.,* 104:1583-1590, 1976.

[291] S.C. Reddy and D.S. Henningson. Energy growth in viscous channel flows. *J. Fluid Mech.,* 252:209-238, 1993.

[292] S.C. Reddy, P.J. Schmid, J.S. Baggett and D.S. Henningson. On the stability of streamwise streaks and transition thresholds in plane channel flows. *J. Fluid Mech.,* 365:269-303, 1998.

[293] H.L. Reed. Disturbance-wave interactions in flows with cross-flow. *AIAA,* 85-0494, 1985.

[294] H.L. Reed. Wave interactions in swept-wing flow. *Phys. Fluids,* 30(11):3419-3426, 1988.

[295] H.L. Reed and W.S. Saric. Stability of three-dimensional boundary layers. *Ann. Rev. Fluid Mech.,* 21:235-285, 1989.

[296] E. Reshotko. Boundary layer stability and transition. *Ann. Rev. Fluid Mech.,* 8:311-349 [425, 465, 472] 1976.

[297] O. Reynolds. An experimental investigation of the circumstances which determine whether the motion of water shall be direct or sinuous and of the law of resistance in parallel channels. *Phil. Trans. Roy. Soc.,* 174:935-982, 1883.

[298] O. Reynolds. On the dynamical theory of incompressible viscous fluids and the determination of the criterion. *Phill. Trans. Roy. Soc. London Ser. A,* 186:123-164, 1895.

[299] J.M. Robertson. *Hydrodynamics in Theory and Application.* Dover Publications, New York, 1969.

[300] G.E. Robertson, J.H. Seinfeld and L.G. Leal. Combined forced and free convection flow past a horizontal flat plate. *AICHE,* 19(5):998-1008, 1973.

[301] S.K. Robinson. Coherent motions in the turbulent boundary layer. *Ann. Rev. Fluid Mech.,* 23:601-639, 1991.

[302] H.L. Rogler and E. Reshotko. Disturbances in a boundary layer introduced by a low intensity array of vortices. *SIAM J. Appl. Maths.,* 28:431-462, 1975.

[303] A. Roshko. On the drag and shedding frequency of two-dimensional bluff bodies. *NACA TN 3169,* 1954.

[304] A.I. Ruban. On the generation of Tollmien-Schlichting waves by sound. *Fluid Dyn.,* 19:709-716, 1985.

[305] M. Sahin and R.G. Owen. A novel fully-implicit finite volume method applied to the lid-driven cavity problem: Parts I and II. *Int J. Num. Methods Fluids,* 42:57-88, 2003.

[306] W.S. Saric and L.G. Yeates. Generation of crossflow vortices in a three-dimensional flat-plate flow. *Laminar-Turbulent Transition. Vol. II* (Ed. V. Kozlov). Springer Verlag, Berlin, 1984.

[307] W.S. Saric and L.G. Yeates. Experiments on the stability of crossflow vortices in swept-wing flows. *AIAA Paper* No. 85-0493, 1985.

[308] W.S. Saric, E.B. White and H.L. Reed. Boundary-layer receptivity to freestream disturbances and its role in transition. *AIAA-99-3788,* 1999.

[309] T. Sarpakaya. Nonimpulsively started steady flow about a circular cylinder. *AIAA J.* 29:1283-1289, 1991.

[310] H. Schlichting. Zur entstehung der turbulenz bei der plattenströmung. *Nach. Gesell. d. Wiss. z. Gött., MPK,* 181-208, 1933.

[311] H. Schlichting. Amplitudenverteilung und energiebilanz der kleinen storungen bei der plattenströmung. *Nach. Gesell. d. Wiss. z. Gött., MPK,* 1:47-78, 1935.

[312] H. Schlichting. *Boundary Layer Theory,* Seventh Ed. McGraw Hill, New York, 1979.

[313] P.J. Schmid. Linear stability theory and bypass transition in shear flows. *Phys. Plasmas*, 7:1788, 2000.

[314] P.J. Schmid and D.S. Henningson. *Stability and Transition in Shear Flow*. Springer Verlag, New York, 2001.

[315] W. Schneider. A similarity solution for combined forced and free convection flow over a horizontal plate. *Int. J. Heat Mass Transfer*, 22:1401-1406, 1979.

[316] G.B. Schubauer and H.K. Skramstad. Laminar boundary layer oscillations and the stability of laminar flow. *J. Aero. Sci.*, 14(2):69-78, 1947.

[317] G.B. Schubauer and W.G. Spangenberg. Forced mixing in boundary layers. *National Bureau of Standards Rept.*, 6107 (Aug. 1958), 1958.

[318] M.R. Scott and H.A. Watts. Computational solution of linear two point boundary value problems via orthonormalization. *SIAM J. Num. Analysis*, 14, 40, 1977.

[319] T.K. Sengupta. Impulse response of laminar boundary layer and receptivity. In *Proc. 7th Int. Conf. Numerical Meth. for Laminar and Turbulent Layers*, (Ed.: C. Taylor), 1991.

[320] T.K. Sengupta. Solution of the Orr-Sommerfeld equation for high wave numbers. *Comput. Fluids*, 21(2):302-304, 1992.

[321] T.K. Sengupta. *Fundamentals of Computational Fluid Dynamics*. Universities Press, Hyderabad, India, 2004.

[322] T.K. Sengupta, M. Ballav and S. Nijhawan. Generation of Tollmien-Schlichting waves by harmonic excitation. *Phys. Fluids*, 6(3):1213-1222, 1994.

[323] T.K. Sengupta and S. Bhaumik. Onset of turbulence from the receptivity stage of fluid flows. *Phys. Rev. Lett.*, 154501:1-5, 2011.

[324] T.K. Sengupta, S. Bhaumik and V. Lakshmanan. Design and analysis of a new filter for LES and DES. *Comput. Struct.*, 87:735-750, 2009.

[325] T.K. Sengupta, S. Bhaumik, V. Singh and S. Shukl. Nonlinear and nonparallel receptivity of zero-pressure gradient boundary layer. *Int. J. Emerging Multidisciplinary Fluid Sci.*, 1(1):19-35, 2009.

[326] T.K. Sengupta, S. Bhumkar and V. Lakshmanan. Design and analysis of a new filter for LES and DES. *Comput. Struct.*, 87:735-750, 2009.

[327] T.K. Sengupta, M. Chattopadhyay, Z-Y. Wang and K.S. Yeo. By-pass mechanism of transition to turbulence. *J. Fluid Struct.*, 16:15-29, 2002.

[328] T.K. Sengupta, V. Chaturvedi, P. Kumar and S. De. Computation of leading edge contamination. *Comput. Fluids*, 33:927-951, 2004.

[329] T.K. Sengupta, D. Das, P. Mohanamuraly, V.K. Suman and A. Biswas. Modelling free stream turbulence based on wind tunnel and flight data for instability studies. *Int. J. Emerging Multidiscip. Fluid Sci.*, 1(3):181-201, 2009.

[330] T.K. Sengupta, S. De and K. Gupta. Effect of free-stream turbulence on flow over airfoil at high incidences. *J. Fluids Struct.*, 15(5):671-690 2001.

[331] T.K. Sengupta, S. De and S. Sarkar. Vortex-induced instability of an incompressible wall-bounded shear layer. *J. Fluid Mech.*, 493:277-286, 2003.

[332] T.K. Sengupta and S. Dey. Proper orthogonal decomposition of direct numerical simulation data of by-pass transition. *Comput. Struct.*, 82:2693-2703, 2004.

[333] T.K. Sengupta and A. Dipankar. Comparative study of time advancement methods for solving Navier-Stokes equations. *J. Sci. Comp.*, 21(2):225-250, 2004.

[334] T.K. Sengupta and A. Dipankar. Subcritical instability on the attachment-line of an infinite swept wing. *J. Fluid Mech.*, 529:147-171, 2005.

[335] T.K. Sengupta, A. Dipankar and P. Sagaut. Error dynamics: Beyond von Neumann analysis. *J. Comput. Phys.*, 226:1211-1218, 2007.

[336] T.K. Sengupta, G. Ganeriwal and S. De. Analysis of central and upwind compact schemes. *J. Comput. Phys.*, 192:677-694, 2003.

[337] T.K. Sengupta, A. Guntaka and S. Dey. Navier-Stokes solution by new compact scheme for incompressible flows. *J. Sci. Comput.*, 21:269-282, 2004.

[338] T.K. Sengupta, R. Jain and A. Dipankar. A new flux-vector splitting compact finite volume scheme. *J. Comput. Phys.*, 207:261-281, 2005.

[339] T.K. Sengupta, A. Kasliwal, S. De and M. Nair. Temporal flow instability for Robins-Magnus effect at high rotation rates. *J. Fluids Struct.*, 17, 941-953, 2003.

[340] T.K. Sengupta, V. Lakshmanan and V.V.S.N. Vijay. A new combined stable and dispersion relation preserving compact scheme for non-periodic problems. *J. Comput. Phys.*, 228(8):3048-3071, 2009.

[341] T.K. Sengupta and S.G. Lekoudis. Calculation of turbulent boundary layer over moving wavy surface. *AIAA J.*, 23(4), 1985.

[342] T.K. Sengupta, T.T. Lim and M. Chattopadhyay. An experimental and theoretical investigation of a by-pass transition mechanism. *IIT Kanpur Rept. No. IITK/Aero/AD/2001/02*, 2001.

[343] T.K. Sengupta, T.T. Lim, S.V. Sajjan, S. Ganesh and J. Soria. Accelerated flow past a symmetric aerofoil: experiments and computations. *J. Fluid Mech.*, 591:255-288, 2007.

[344] T.K. Sengupta and M.T. Nair. A new class of waves for Blasius boundary layer. In Proc. 7th *Asian Cong. Fluid Mech.*, 785-788, Allied Publishers, Chennai, India, 1997.

[345] T.K. Sengupta, M.T. Nair and V. Rana. Boundary layer excited by low frequency disturbances – Klebanoff mode. *J. Fluids Struct.*, 11:845-853, 1997.

[346] T.K. Sengupta and T. Poinsot. *Instabilities of Flows: With and Without Heat Transfer and Chemical Reaction*. Springer Wien, New York, 2010.

[347] T.K. Sengupta, M.K. Rajpoot and Y.G. Bhumkar Space-time discretizing optimal DRP schemes for flow and wave propagation problems. *Comput. Fluids*, 47(1):144-154, 2011.

[348] T.K. Sengupta and A.K. Rao. Spatio-temporal receptivity of boundary-layers by Bromwich contour integral method. In Proc. Boundary And Interior Layers Conf. (BAIL 2006) held at Goettingen, Germany (Eds.: G. Lube and G. Rapin), 2006.

[349] T.K. Sengupta, A.K. Rao and K. Venkatasubbaiah. Spatio-temporal growing wave fronts in spatially stable boundary layers. *Phys. Rev. Letters*, 9;96(22):224504, 2006.

[350] T.K. Sengupta, A.K. Rao and K. Venkatasubbaiah. Spatio-temporal growth of disturbances in a boundary layer and energy based receptivity analysis. *Physics of Fluids*, 18:094101, 2006.

[351] T.K. Sengupta and R. Sengupta. Flow past an impulsively started circular cylinder at high Reynolds number. *Comput. Mech.*, 14(4):298-310, 1994.

[352] T.K. Sengupta and A.P. Sinha. Surface mass transfer: A receptivity mechanism for boundary-layers. In *Proc. 6th Asian Cong. Fluid Mech.* (Eds.: Y.T. Chew and C.P. Tso), 1242-1245, 1995.

[353] T.K. Sengupta, N. Singh and V.K. Suman. Dynamical system approach to instability of flow past a circular cylinder. *J. Fluid Mech.*, 656:82-115, 2010.

[354] T.K. Sengupta, S.K. Sircar and A. Dipankar. High accuracy compact schemes for DNS and acoustics. *J. Sci. Comput.*, 26(2):151-193, 2006.

[355] T.K. Sengupta, V.K. Suman and N. Singh. Solving Navier-Stokes equation for flow past cylinder using single-block structured and overset grids. *J. Comput. Phys.*, 229(1):178-199, 2010.

[356] T.K. Sengupta and S.B. Talla. Robins-Magnus effect: A continuing saga. *Current Science*, 86, No. 7, April 2004.

[357] T.K. Sengupta, S.B. Talla and S.C. Pradhan. Galerkin finite element methods for wave propagation problems. *Sadhana; Indian Academy of Sciences Proc.*, 30(5):611-624, 2005.

[358] T.K. Sengupta and K. Venkatasubbaiah. Spatial stability for mixed convection boundary layer over a heated horizontal plate. *Stud. Appl. Math.*, 117:265-298, 2006.

[359] T.K. Sengupta, K. Venkatasubbaiah and S.S. Pawar. Nonlinear instability of mixed convection flow over a horizontal cylinder. *Acta Mechanica*, 201:197-210, 2008.

[360] T.K. Sengupta, V.V.S.N. Vijay and N. Singh. Universal instability modes in internal and external flows. *Comput. Fluids*, 40:221-235, 2011.

[361] T.K. Sengupta, V.V.S.N. Vijay and S. Bhaumilk. Further improvement and analysis of CCD scheme: Dissipation discretization and de-aliasing properties. *J. Comput. Phys.*, 228(17):6150-6168, 2009.

[362] T.K. Sengupta, Z.-Y. Wang, L.Y. Pin and M. Chattopadhyay. Receptivity to stationary pulsating vortex. *Proc. 8th Asian Cong. Fluid Mech.* (Ed.: E. Cui), 959-963, 1999.

[363] T.K. Sengupta, Z.-Y. Wang, K.S. Yeo and M. Chattopadhyay. Receptivity to convected vortices – bypass route. *Proc. 8th Asian Cong. Fluid Mech.* (Ed.: E. Cui), 964-968, 1999.

[364] W.G. Sewall, R.J. McGhee, D.E. Hahne and F.L. Jordan. Wind tunnel results of the high-speed NLF(1)-0213 airfoil. *NASA CP 2487,* Part 3:697-726, 1987.

[365] S. Saurabh. Study of instability of flow past two side-by-side circular cylinders at low Reynolds number. M. Tech. thesis, IIT Kanpur, 2011.

[366] H. Shaukatullah and B. Gebhart. An experimental investigation of natural convection flow on an inclined surface. *Int. J. Heat Mass Transfer,* 21:1481-1490, 1978.

[367] Y.I. Shokin. *The Method of Differential Approximation.* Springer Verlag, Berlin, 1983.

[368] C. Shu and Y.T. Chew. On the equivalence of generalized differential quadrature and highest order finite difference scheme. *Comp. Meth. Appl. Mech. Eng.,* 155:249-260, 1998.

[369] S.G. Siegel, J. Seidel, C. Fagley, D.M. Luchtenburg, K. Cohen and T. Mclaughlin. Low-dimensional modelling of a transient cylinder wake using double proper orthogonal decomposition. *J. Fluid Mech.,* 610:1-42, 2008.

[370] E.D. Siggia and H. Aref. Point-vortex simulation of the inverse cascade in two-dimensional turbulence. *Phys. Fluids,* 24:171-173, 1981.

[371] L. Sirovich. Turbulence and the dynamics of coherent structures. Part I-III. *Quart. J. Appl. Math.,* 45(3):561-590, 1987.

[372] J. Smagorinsky. Some historical remarks on the use of nonlinear viscosities. In *Large Eddy Simulation of Complex Engineering and Geophysical Flows* (Eds. Galperin B., Orszag, S. A.), Cambridge Univ. Press, New York, 1993.

[373] C.R. Smith. Use of 'Kernel' experiments for modeling of near-wall turbulence. In *Near Wall Turbulent Flows* (Eds.: R.M.C. So, C.G. Speziale and B.E. Launders), 33-42. Elsevier, Amsterdam, Holland, 1993.

[374] A.M.O. Smith and N. Gamberoni. Transition, pressure gradient and stability theory. *Douglas Aircraft Rept. No. ES-26338,* 1956.

[375] F. Smith and D.J. Higton. Flight test on a King Cobra FZ-440 to investigate the practical requirements for the achievement of low profile drag coefficients on a low drag airfoil. British *ARC R and M 2375.* Also *RAE* Report No. 2043, A.R.C. 9043, August 1945.

[376] C.R. Smith, J.D.A. Walker, A.H. Haidari and U. Soburn. On the dynamics of near-wall turbulence. *Phil. Trans. R. Soc. Lond. A,* 336:131-175, 1991.

[377] L. M. Smith and V. Yakhot. Finite-size effects in forced two-dimensional turbulence. *J. Fluid Mech.,* 274:115-138, 1994.

[378] D. Somers. Design and experimental results for a natural-laminar-flow airfoil for general aviation applications. *NASA Tech. Mem.,* No. 1861, 1981.

[379] D. Somers. Design and experimental results for a flapped natural-laminar-flow airfoil for general aviation applications. *NASA Tech. Mem.,* No. 1865, 1981.

[380] A. Sommerfeld. Ein Beitrag zur hydrodynamiscen Erklarung der turbulenten Flussigkeitsbewegung. In *Proc. 4th Int. Cong. Mathematicians,* Rome, 116-124, 1908.

[381] A. Sommerfeld. *Partial Differential Equation in Physics.* Academic Press, New York, 1949.

[382] P. Spalart. Direct numerical study of leading edge contamination. *AGARD CP,* 438:5.1-5.13, 1988.

[383] J.G. Spangler and C.S. Wells Jr. Effect of free stream disturbances on boundary layer transition. *AIAA J.,* 6:534-545, 1968.

[384] S.P. Sparks and S.J. Miley. Development of a propeller afterbody analysis with contracting slipstream. *SAE TP 830743,* 1983.

[385] E.M. Sparrow and R.B. Husar. Longitudinal vortices in natural convection flow on inclined plates. *J. Fluid Mech.,* 37:251-255, 1969.

[386] E.M. Sparrow and W.J. Minkowycz. Buoyancy effects on horizontal boundary-layer flow and heat transfer. *Int. J. Heat Mass Transfer,* 5:505-511, 1962.

[387] K.R. Sreenivasan, P.J. Strykowski and D.J. Olinger. Hopf bifurcation, Landau equation and vortex shedding behind circular cylinders. *Forum on Unsteady Flow Separation* (Ed.: K.N. Ghia), 1-13. ASME, 1987.

[388] A.J. Srokowski and S.A. Orszag. Mass flow requirements for LFC wing design. *AIAA Paper* No. 77-1222, 1977.

[389] H. Steinrueck. Mixed convection over a cooled horizontal plate: Non-uniqueness and numerical instabilities of the boundary layer equations. *J. Fluid Mech.,* 278:251-265, 1994.

[390] G.G. Stokes. On the theories of inertial friction of fluids in motion. *Trans. Camb. Phil. Soc.,* 8:287-305, 1845.

[391] B.S. Stratford. The prediction of separation of the turbulent boundary. *J. Fluid Mech.,* 5(1):1-16, 1959.

[392] P.J. Strykowski and K.R. Sreenivasan. On the formation and suppression of vortex shedding at low Reynolds number. *J. Fluid Mech.,* 218:74-104, 1990.

[393] P.J. Strykowski. The control of absolutely and convectively unstable shear flows. PhD thesis, Yale University, 1986.

[394] J.T. Stuart. The basic theory of the stability of three-dimensional boundary layer. *ARC Report* No. 15904, 1953.

[395] J.T. Stuart. On the nonlinear mechanics of wave disturbances in stable and unstable parallel flows. Part 1. The basic behaviour in plane Poiseuille flow. *J. Fluid Mech.,* 9:353-370, 1960.

[396] J.T. Stuart. *Laminar Boundary Layer* (Ed.: L. Rosenhead). Clarendon Press, Oxford, UK, 1963.

[397] B. Swartz and B. Wendroff. The relation between the Galerkin and collocation methods using smooth splines. *SIAM J. Num. Analysis* 11(5):994-996, 1974.

[398] C.K.W. Tam. Directional acoustic radiation from a supersonic jet generated by shear layer instability. *J. Fluid Mech.,* 46:757-768, 1971.

[399] I. Tani. On the design of airfoils in which the transition of the boundary layer is delayed. *NACA TM 1351,* 1952.

[400] G.I. Taylor. Statistical theory of turbulence. V. Effects of turbulence on boundary layer. *Proc. Roy. Soc. A,* 156, No. 888:307-317, 1936.

[401] G.I. Taylor. Some recent developments in the study of turbulence. *Proc. V^{th} Int. Conf. Appl. Mech.* (Eds.: J.P. Den Hartog and H. Peters), 1939.

[402] D.P. Telionis. *Unsteady Viscous Flows.* Springer Verlag, New York, 1981.

[403] H. Tennekes and J.L. Lumley. *First Course in Turbulence.* MIT Press, Cambridge, MA, 1971.

[404] V. Theofilis. On linear and non-linear instability of the incompressible swept attachment-line boundary layer. *J. Fluid Mech.*, 355:193-227, 1998.

[405] V. Theofilis, A. Fedorov, D. Obrist and U. Dallmann. The extended Goertler-Haemmerlin model for linear instability of three-dimensional incompressible swept attachment-line boundary layer flow. *J. Fluid Mech.*, 487:271-313, 2003.

[406] P.A. Thompson *Compressible-Fluid Dynamics.* McGraw Hill, New York, 1972.

[407] P.T. Tokumaru and P.E. Dimotakis. The lift of a cylinder rotary motion in a uniform flow. *J. Fluid Mech.*, 255:1-10, 1993.

[408] W. Tollmien. Über die enstehung der turbulenz. I, English translation. *NACA TM 609*, 1931.

[409] D. Tordella and C. Cancelli. First instabilities in the wake past a circular cylinder: Comparison of transient regimes with Landau's model. *Meccanica*, 26:75-83, 1991.

[410] L.N. Trefethen. Group velocity in finite difference schemes. *SIAM Review*, 24(2):113-136, 1982.

[411] L.N. Trefethen, A.E. Trefethen, S.C. Reddy and T.A. Driscoll. Hydrodynamic stability without eigenvalues. *Science*, 261(5121):578-584, 1993.

[412] D.J. Tritton. A note on vortex streets behind circular cylinders at low Reynolds numbers. *J. Fluid Mech.*, 45(1):203-208, 1970.

[413] D.J. Tritton. *Physical Fluid Dynamics.* Van Nostrand Reinhold Co., New York, 1977.

[414] C. Tropea, A.L. Yarin and J.F. Foss. *Springer Handbook of Experimental Fluid Mechanics.* Springer Verlag, Berlin, 2007.

[415] A. Tumin. The spatial stability of natural convection flow on inclined plates. *ASME J. Fluid Eng.*, 125:428-437, 2003.

[416] S. Unnikrishnan. Linear stability analysis and nonLinear receptivity study of mixed convection boundary layer developing over a heated flat plate. M. Tech. thesis, IIT Kanpur, 2011.

[417] E.R. Van Driest. Turbulent boundary layer in compressible fluids. *J. Aero. Sci.*, 18:145-160, 1951.

[418] B. Van der Pol and H. Bremmer. *Operational Calculus Based on Two-Sided Laplace Integral,* Cambridge Univ. Press, Cambridge, UK, 1959.

[419] H.A. Van der Vorst. Bi-CGSTAB: A fast and smoothly converging variant of Bi-CG for the solution of non-symmetric linear systems. *SIAM J. Sci. Stat. Comput.*, 12:631-644, 1992.

[420] M. Van Dyke. *Perturbation Methods in Fluid Mechanics.* Academic Press, New York, 1964.

[421] J.L. Van Ingen. A suggested semi-empirical method for the calculation of the boundary layer transition region. *Inst. Of Technology, Dept. of Aeronautics and Eng. Rept. VTH-74*, 1956.

[422] T.E. VanZandt. A universal spectrum of buoyancy waves in atmosphere. *Geophys. Res. Lett.*, 9:575-578, 1982.

[423] K. Venkatasubbaiah and T.K. Sengupta. Mixed convection flow past a vertical plate: Stability analysis and its direct simulation. *Int. J. Thermal Sci.*, 48(3):461-474, 2008.

[424] A. Vicini and D. Quagliarella. Inverse and direct airfoil design using a multiobjective genetic algorithm. *AIAA J.*, 35(9):1499-1505, 1997.

[425] R. Vichnevetsky and J.B. Bowles. *Fourier Analysis of Numerical Approximations of Hyperbolic Equations.* Society for Industrial and Applied Mathematics, Philadelphia, 1982.

[426] P.M.H.W. Vijgen and B.J. Holmes. Experimental and numerical analysis of laminar boundary-layer flow stability over an aircraft fuselage forebody. In *Research in Natural Laminar Flow and Laminar-Flow Control.* NASA CP 2487, Part 3:861-886, 1987.

[427] P. Vijgen, K. Williams, B. Williams and J. Roskam. Design considerations of natural laminar flow airfoils for medium-speed regional aircraft. Report No. *KU-FRL-6131-1*, Flight Research Laboratory, Univ. of Kansas Center for Research, 1984.

[428] J.K. Viken. Aerodynamic design considerations and theoritical results for a high Reynolds number natural laminar flow airfoil. M.S. Thesis. The School of Engineering and Applied Science, George Washington Univ., 1983.

[429] J.K. Viken, S.A. Viken, W. Pfenninger, H.L. Morgan and R.L. Campbell. Design of the low-speed NLF(1)-0414F and the high-speed HSNLF(1)-0213 airfoils with high lift systems. Symp. Proc. *Research in Natural Laminar Flow and Laminar-Flow Control,* NASA Langley Res. Center, NASA CP 2487, Part 3, 1987.

[430] G. Volpe and R.E. Melnik. A method for designing closed airfoils for arbitrary super-critical speed distribution. *AIAA* Paper No. 85-5023, 1985.

[431] L.B. Wahlbin. A dissipative numerical method for the numerical solution of first order hyperbolic equations. *Mathematical Aspects of Finite Elements in Partial Deferential Equations* (Ed. C. de Boor). Academic Press, New York, 147-170, 1974.

[432] L.B. Wahlbin. A modified Galerkin procedure with cubics for hyperbolic problems. *Math. Comp.*, 29:978-984, 1975.

[433] X.A. Wang. An experimental study of mixed, forced, and free convection heat transfer from a horizontal flat plate to air. *ASME J. Heat Transfer,* 104:139-144, 1982.

[434] J. Watson. On the nonlinear mechanics of wave disturbances in stable and unstable parallel flows. Part 2. The development of a solution for plane Poiseuille flow and for plane Couette flow. *J. Fluid Mech.*, 9:371-389, 1960.

[435] A.R. Wazzan, C. Gazley Jr. and A.M.O. Smith. H-R, method for predicting transition. *AIAA J.,* 19(6):109-114, 1981.

[436] C.S. Wells Jr. Effects of freestream turbulence on boundary layer transition. *AIAA J.*, 5(1):172-174, 1967.

[437] A.J. Wheeler and J.P. Johnston. Three-dimensional turbulent boundary layers − data sets for two-space-coordinate flows. *Stanford Univ. Rept.*, MD 32, 1972.

[438] F.M. White. *Viscous Fluid Flow.* McGraw Hill Int. Edn., New York, 1991.

[439] F.M. White. *Fluid Mechanics.* Sixth Edn. McGraw Hill, New York, 2008.

[440] G.B. Whitham. *Linear and Nonlinear Waves.* Wiley-Intescience, New York, 1974.

[441] E.T. Whittaker and G.N. Watson. *Modern Analysis.* Oxford Univ. Press, London, 1946.

[442] K. Wiesenfeld and B. McNamara. Small-signal amplification in bifurcating dynamical systems. *Physical Review A*, 33(1):629-642, 1986.

[443] C.H.K. Williamson. Oblique and parallel modes of vortex shedding in the wake of a circular cylinder at low Reynold numbers. *J. Fluid Mech.*, 206:579-627, 1989.

[444] C.H.K. Williamson. Vortex dynamics in cylinder wake. *Ann. Rev. Fluid Mech.*, 28:477-539, 1996.

[445] R.S. Wu and K.C. Cheng. Thermal instability of Blasius flow along horizontal plates. *Int. J. Heat Mass Transfer*, 19:907-913, 1976.

[446] X. Wu, R.G. Jacobs, J.C.R. Hunt and P.A. Durbin. Simulation of boundary layer transition induced by periodically passing wakes. *J. Fluid Mech.*, 399:109-153, 1999.

[447] A. Zebib. Stability of viscous flow past a circular cylinder. *J. Eng. Math.*, 21:155-165, 1987.

[448] D.W. Zingg. Comparison of high-accuracy finite-difference schemes for linear wave propagation. *SIAM J. Sci. Comp.*, 22(2):476-502, 2000.

[449] S. Zuccher, A. Bottaro and P. Luchini. Algebraic growth in a Blasius boundary layer. *Eur. J. Mech. B/Fluids*, 25, 1, 2006.

[450] E.J. Zuercher, J.W. Jacobs and C.F. Chen. Experimental study of the stability of boundary-layer flow along a heated inclined plate. *J. Fluid Mech.*, 367:1-25, 1998.

Index

T - #0877 - 101024 - C16 - 254/178/24 - PB - 9781138076211 - Gloss Lamination